AF443452

Small Cell Networks

Deployment, PHY Techniques, and Resource Management

This comprehensive resource explores state-of-the-art advances in the successful deployment and operation of small cell networks.

A broad range of technical challenges, and possible solutions, are addressed, including practical deployment considerations and interference management techniques, all set within the context of the most recent cutting-edge advances. Key aspects covered include 3GPP standardization, applications of stochastic geometry, PHY techniques, MIMO techniques, handover, and radio resource management, including techniques designed to make the best possible use of the available spectrum.

Detailed technical information is provided throughout, with a consistent emphasis on real-world applications. Bringing together world-renowned experts from industry and academia, this is an indispensible volume for researchers, engineers, and systems designers in the wireless communication industry.

Tony Q. S. Quek is an Assistant Professor at the Singapore University of Technology and Design, and a Scientist at the Institute for Infocomm Research. He has chaired numerous IEEE workshops on heterogeneous, femtocell, and small cell networks, and is an Editor for *IEEE Transactions on Communications* and *IEEE Wireless Communications Letters*.

Guillaume de la Roche is a wireless systems engineer at Mindspeed Technologies. He has been involved in organising numerous IEEE small cell workshops, and is the author of four books on wireless networking.

İsmail Güvenç is an Assistant Professor at Florida International University. He has contributed to several IEEE standards, is a co-author of *Ultra-wideband Positioning Systems* (2008), a co-editor of *Reliable Communications for Short-range Wireless Systems* (2011), and editor for *IEEE Communications Letters* and *IEEE Wireless Communications Letters*.

Marios Kountouris is an Assistant Professor at the Department of Telecommunications, Supélec (École Supérieure d'Électricité), France, and has chaired several IEEE workshops on femtocell networks, heterogeneous networks, and small cell networks.

Small Cell Networks

Deployment, PHY Techniques, and Resource Management

Edited by

TONY Q. S. QUEK
Singapore University of Technology and Design

GUILLAUME DE LA ROCHE
Mindspeed Technologies Inc.

İSMAIL GÜVENÇ
Florida International University

MARIOS KOUNTOURIS
Supélec (École Supérieure d'Électricité)

CAMBRIDGE
UNIVERSITY PRESS

CAMBRIDGE UNIVERSITY PRESS
Cambridge, New York, Melbourne, Madrid, Cape Town,
Singapore, São Paulo, Delhi, Mexico City

Cambridge University Press
The Edinburgh Building, Cambridge CB2 8RU, UK

Published in the United States of America by Cambridge University Press, New York

www.cambridge.org
Information on this title: www.cambridge.org/9781107016781

© Cambridge University Press 2013

This publication is in copyright. Subject to statutory exception
and to the provisions of relevant collective licensing agreements,
no reproduction of any part may take place without the written
permission of Cambridge University Press.

First published 2013

Printed and bound in the United Kingdom by the MPG Books Group

A catalog record for this publication is available from the British Library

ISBN 978-1-107-01678-1 Hardback

Cambridge University Press has no responsibility for the persistence or
accuracy of URLs for external or third-party internet websites referred to
in this publication, and does not guarantee that any content on such
websites is, or will remain, accurate or appropriate.

Contents

Contributors

Ansuman Adikhary
University of Southern California

Salam Akoum
University of Texas at Austin

Tansu Alpcan
University of Melbourne

Eitan Altman
INRIA

Jeffrey G. Andrews
University of Texas at Austin

Huseyin Arslan
University of South Florida

Gunther Auer
Docomo Europe

Mehdi Bennis
University of Oulu

Zubin Bharucha
Docomo Europe

Giuseppe Caire
University of Southern California

Kwang-Cheng Chen
National Taiwan University

Holger Claussen
Bell Labs Ireland, Alcatel-Lucent

Mérouane Debbah
Supélec (École Supérieure d'Électricité)

Mustafa Cenk Erturk, Jr.
University of South Florida

İsmail Güvenç
Florida International University

Robert W. Heath, Jr.
University of Texas at Austin

Lester Ho
Bell Labs Ireland, Alcatel-Lucent

Jakob Hoydis
Supélec (École Supérieure d'Électricité)

Han Shin Jo
University of Texas at Austin

Veeraruna Kavitha
INRIA

Marios Kountouris
Supélec (École Supérieure d'Électricité)

Shao-Yu Lien
National Taiwan University

David Lopéz-Pérez
Kings College, London

Sayandev Mukherjee
NTT Docomo USA

Dusit Niyato
Nanyang Technological University

Vasileios Ntranos
University of Southern California

Florian Pivit
Bell Labs Ireland, Alcatel-Lucent

Tony Q. S. Quek
Singapore University of Technology and Design

Alberto Rabbachin
European Commission

Sreenath Ramanath
INRIA

Sundeep Rangan
Polytechnic Institute of New York University

Guillaume de la Roche
Mindspeed Technologies Inc.

Sam Samuel
Bell Labs Ireland, Alcatel-Lucent

Hyundong Shin
Kyung Hee University

Chee Wei Tan
City University of Hong Kong

Moe Z. Win
Massachusetts Institute of Technology

Ping Xia
University of Texas at Austin

Foreword

Mobile networks have undergone a major transformation in recent years, shifting from primarily delivering voice and text services to transporting data and connecting to the Internet. In many mobile networks today, data traffic already constitutes more than 97% of the total bits transmitted. Demand for data continues to be strong, driven by high take up and ease of use of smartphone, tablets, and related devices.

Long term evolution (LTE) promises one answer to the impending data capacity crunch, with lightening fast data rates and impressive spectral efficiency. However, this new technology – even with associated new spectrum – won't come close to satisfying forecast demand.

The industry now recognizes that the greatest capacity gains will be achieved by spectrum reuse through deployment of large numbers of small cells. A variety of radio technologies including 3rd Generation (3G), (LTE), and wireless fidelity (WiFi) will be relevant depending on device capabilities, spectrum availability, and price point.

No longer is femtocell technology considered purely as a coverage solution for the home. Today, it is difficult to find a network operator that does not have small cells somewhere on their roadmap.

Over the next five years, we can expect to see a major shift in network equipment investment. Analyst research already points to the majority of new cell-site investment moving across from macrocells into small cells in that timeframe. With some $50 billion of CapEx spent annually on mobile access networks, that will create both significant industry re-alignment and tremendous opportunity.

While the foundations of small cell technology have been laid, and the solution proven with several million femtocells already in commercial use, there are many aspects ripe for technical improvement and refinement.

Technical research in this field, as exemplified in this book, will make the difference between a good and excellent customer experience. It will help drive the cost down by increasing efficiency and network performance.

I commend those who through their work help the industry take one more step toward small cell solutions, which will satisfy the thirst for data communications on the move.

David Chambers
Editor, *ThinkSmallCell*

Acknowledgments

First, we would like to express our sincere gratitude to all our contributors, without whom this book would never have been accomplished. Indeed, it was for us a great pleasure to have such high-quality contributions from major researchers in the field of small cell networks. Our contributors, as they appear in the book, are as follows: Tony Q. S. Quek, Guillaume de la Roche, İsmail Güvenç, Marios Kountouris, Ping Xia, Han-Shin Jo, Jeffrey G. Andrews, Sayandev Mukherjee, Alberto Rabbachin, Hyundong Shin, Moe Z. Win, Salam Akoum, Robert W. Heath, Jr., Giuseppe Caire, Ansuman Adikhary, Vasileios Ntranos, Holger Claussen, Lester Ho, Sam Samuel, Florian Pivit, Jakob Hoydis, Merouane Debbah, Sreenath Ramanath, Veeraruna Kavitha, Eitan Altman, Kwang-Cheng Chen, Shao-Yu Lien, Mehdi Bennis, Dusit Niyato, Tansu Alpcan, Chee Wei Tan, Sundeep Rangan, Zubin Bharucha, Gunther Auer, Mustafa Cenk Erturk, Jr., Huseyin Arslan, and David Lopéz-Pérez.

We would like to thank Cambridge University Press staff and in particular Elizabeth Horne and Phil Meyler for their continuous encouragement and support during the course of this project.

Our sincere thanks go to Dr. Xiaoli Chu, Dr. Lorenza Giupponi, Dr. David Lopéz-Pérez, and Dr. Alvaro Valcarce, who agreed to review parts of the book.

Last, but not least, we also wish to thank David Chambers for writing the foreword of the book.

Tony Q. S. Quek would like to thank his family and colleagues at I2R, A*STAR and at the Singapore University of Technology and Design (SUTD) for their encouragement and support. Guillaume de la Roche is very grateful to his family and friends for their encouragement during the time devoted to compiling this book. He also wishes to say thank you to his colleagues at Mindspeed for their support. İsmail Güvenç would like to thank his family and colleagues at DOCOMO Innovations, Inc., for their encouragement and support. Marios Kountouris would like to thank his family and all those who were a source of encouragement, support, and enthusiasm.

Acronyms

3G	3rd Generation
3GPP	3rd Generation Partnership Project
3GPP2	3rd Generation Partnership Project 2
4G	4th Generation
ABS	almost blank subframe
AMC	adaptive modulation and coding
ANR	automatic neighbor relation
AOPC	adaptive outage-based power control
ARQ	automatic repeat request
ASE	area spectral efficiency
AWGN	additive white Gaussian noise
BLER	block error rate
BS	base station
BSS	base station switching
CA	carrier aggregation
CapEx	capital expenditure
CC	component carrier
CCDF	complementary cumulative distribution function
CCE	control channel element
CDF	cumulative distribution function
CDMA	code division multiple access
CEM	certainty equivalent margin
CF	characteristic function
CFI	control format indicator
CoMP	coordinated multi-point transmission
CoV	coefficient of variation
C-RNTI	cell radio network temporary identifier
CQI	channel quality indicator
CRRM	cognitive radio resource management
CRS	common reference symbol
CSG	closed subscriber group
CSI	channel state information
CSIT	channel state information at the transmitter

DE	deterministic equivalent
DECT	digital enhanced cordless telecommunications
DIA	downlink interference alignment
DL	downlink
DPC	distributed power control
DSL	digital subscriber line
EBF	eigenbeamforming
EGT	evolutionary game theory
eNB	enhanced NodeB
EVDO	evolution data only
ES	elastic
FAP	femtocell access point
FBS	femtocell base station
FDD	frequency division duplex
FDMA	frequency division multiple access
FDTD	finite-difference time-domain
FIFO	first in first out
FIR	finite impulse response
FMS	femtocell mobile station
FFR	fractional frequency reuse
FU	femtocell user
FUE	femtocell user equipment
FP	fictitious play
GP	geometric program
GPS	global positioning system
GSM	global system for mobile communication
GT	game theory
HARQ	hybrid automatic repeat request
HeNB	Home eNodeB
HetNet	heterogeneous network
HII	high interference indicator
HK	Han–Kobayashi
HNB	Home NodeB
HO	handover
HPPP	homogeneous Poisson point process
HSDPA	high speed downlink packet access
HSPA	high speed packet access
HSUPA	high speed uplink packet access
HUE	home user equipment
IA	interference alignment
IC	interference cancellation
ICI	inter-cell interference
ICIC	inter-cell interference coordination
IFA	inverted-F antenna

i.i.d.	independent and identically distributed
ILP	integer linear programming
IM	interference minimization
InH	indoor hotspot
IoT	interference to thermal
IR	indoor ratio
ISP	Internet service provider
JCCDF	joint complementary cumulative distribution function
JFI	Jain's fairness index
LHS	left-hand side
LMMSE	linear minimum mean square error
LTE	long term evolution
LTE-A	LTE-advanced
LUT	look up table
MAC	medium access control
MBMS	multimedia broadcast multicast services
MBS	macrocell base station
MBSFN	multimedia broadcast single frequency network
MCB	main circuit board
MC-CDMA	multicarrier CDMA
MCS	modulation and coding scheme
MF	matched filter
MIMO	multiple input multiple output
MISO	multiple input single output
MMS	macrocell mobile station
MMSE	minimum mean squared error
MR	measurement report
MRC	maximum ratio combining
MS	mobile station
MUE	mobile user equipment
multi-RAT	multiple radio access technology
MU-MIMO	multi-user multiple input multiple output
NB	NodeB
NE	Nash equilibrium
NES	non-elastic
NLM	network listening mode
OA	open access
OFDM	orthogonal frequency division multiplexing
OFDMA	orthogonal frequency division multiple access
OpEx	operating expenditure
PBCH	physical broadcast channel
PCB	printed circuit board
PCFICH	physical control format indicator channel
PCI	physical cell identity

PDCCH	physical downlink control channel
PDF	probability density function
PDSCH	physical downlink shared channel
PFS	proportional fair scheduling
PHICH	physical hybrid-ARQ indicator channel
PHY	physical
PoA	price of anarchy
PPP	Poisson point process
PRT	priority region threshold
PSS	primary synchronization signal
PUS	path-loss-aware user selection
QAM	quadrature amplitude modulation
QCA	quantization cell approximation
QoS	quality of service
QPSK	quadrature phase shift keying
QTFI	QoS-oriented tiered fairness index
RACH	random access channel
RAP	resource allocation problem
RB	resource block
RE	resource element
REG	resource element group
RF	radio frequency
RHS	right-hand side
RL	reinforcement learning
RMT	random matrix theory
RNTP	relative narrowband transmit power
RRAA	radio resource allocation architecture
RRH	remote radio heads
RSRP	reference signal received power
RSRQ	reference signal received quality
RSS	received signal strength
RSSI	receive strength signal indicator
RVQ	random vector quantization
RZF	regularized zero-forcing
SAE	system architecture evolution
SC	small cell
SCN	small cell network
SIC	successive interference cancellation
SIMO	single input multiple output
SINR	signal to interference plus noise ratio
SIR	signal to interference ratio
SISO	single input single output
SMS	short message service
SNR	signal to noise ratio

SON	self-organizing network
SPAP	subchannel and power allocation subproblem
SSR	spectrum splitting ratio
SSS	secondary synchronization signal
SU-MIMO	single user multiple input multiple output
TDD	time division duplex
TDMA	time division multiple access
TFI	tiered fairness index
TTI	transmission time interval
UE	user equipment
UL	uplink
UMi	urban microcell
UMTS	universal mobile telecommunication system
UT	user terminal
VoIP	voice over IP
WiFi	wireless fidelity
WiMAX	wireless interoperability for microwave access
WJFI	weighted sum JFI
WLAN	wireless local area network
ZF	zero-forcing

1 Small cell networks overview

Tony Q. S. Quek, Guillaume de la Roche, İsmail Güvenç,
and Marios Kountouris

Driven by a new generation of wireless user equipment and the proliferation of bandwidth-intensive applications, user data traffic and the corresponding network load are increasing in an exponential manner. Most of this new data traffic is being generated indoors, which requires increased link budget and coverage extension to provide satisfactory user experience. As a result, current cellular networks are reaching their breaking point and conventional cellular architectures, which are devised to cater to large coverage areas and optimized for homogeneous traffic, are facing unprecedented challenges to meet these user demands.

In this context, there has been an increasing interest to deploy small cellular access points in residential homes, subways, and offices. These network architectures, which may be either operator deployed and/or consumer installed and are comprised of a mix of low-power cells underlying the macrocell network, are commonly referred to as small cell networks. By deploying additional network nodes within the local area range and bringing the network closer to end users, spatial reuse and coverage can be significantly improved, thus allowing future cellular systems to achieve higher data rates, while retaining the seamless connectivity and mobility of cellular networks.

Inspired by the attractive features and potential advantages of small cell networks, their development and deployment are gaining momentum in the wireless industry and research communities during the last few years. It has also attracted the attention of standardization bodies such as the Third Generation Partnership Project (3GPP) Long Term Evolution (LTE)-Advanced (see Chapter 14) and the IEEE 802.16 Wireless Metropolitan Area Networks. However, these networks also come with their own challenges, and there are significant technical issues that still need to be addressed for successful rollout and operation. Our goal in editing this book is to fill an important and emerging need for a comprehensive book that addresses deployment, physical-layer (PHY) techniques, and resource management aspects of small cell networks, by soliciting contributions from key researchers in this area.

Small Cell Networks: Deployment, PHY Techniques, and Resource Management, ed. Tony Q. S. Quek,
Guillaume de la Roche, İsmail Güvenç, and Marios Kountouris. Published by Cambridge University Press.
© Cambridge University Press 2013.

Table 1.1 Different types of small cells and comparison with macrocell.
©2011 IEEE. Reprinted, with permission, from [2].

Types of nodes	Transmit power	Coverage	Backhaul
Macrocell	46 dBm	Few km	S1 interface
Picocell	23–30 dBm	<300 m	X2 interface
Femtocell	<23 dBm	<50 m	Internet IP
Relay	30 dBm	300 m	Wireless
RRH	46 dBm	Few km	Fiber

1.1 Overview of small cell networks

Deploying small cells in a wireless network aims at offloading the macrocells, improving indoor coverage and cell-edge user performance, and boosting spectral efficiency per unit area via spatial reuse. They can be deployed with relatively low network overhead, and have high potential for reducing the energy consumption of future wireless networks. Also, this new palette of low-power "miniature" base stations (BSs) requires little or no upfront planning and lease costs, therefore drastically reducing the operational (OpEx) and capital (CapEx) expenditures of networks [1].

As shown in Table 1.1, details of the different types of small cells and their comparison with macrocells are as follows:

- Macrocellular networks consist of conventional operator-installed BSs, providing open public access and wide area coverage typically on the order of few kilometers. In LTE, they are also called enhanced NodeBs (eNBs). Usually destined to provide a guaranteed minimum data rate under maximum tolerable delay and outage constraints, macrocells typically emit up to 46 dBm, serving thousands of customers and using a dedicated backhaul.
- Picocells are low-power, operator-installed cell towers with the same backhaul and access features as macrocells. They are usually deployed in a centralized way, serving a few tens of users within a radio range of 300 m or less, and have a typical transmit power range from 23 dBm to 30 dBm. Picocells are mainly utilized for capacity and outdoor or indoor coverage infill, i.e., in environments with insufficient macro penetration (e.g., office buildings).
- Femtocells, also known as home BSs or home eNBs, are low-cost, low-power, user-deployed access points, offloading data traffic using consumers' broadband connections (digital subscriber line (DSL), cable, or fiber), and serving a dozen of active users in homes or enterprises. Typically, the femtocell range is less than 50 m and its transmit power less than 23 dBm. They operate in open or restricted (closed subscriber group (CSG)) access.
- Relays are usually operator-deployed access points that route data from the macro BS to end users and vice versa. They are positioned so as to increase signal strength and to improve reception in poor coverage areas and dead spots in the existing networks

(e.g., cell edges, tunnels). They can operate in transparent or non-transparent modes (e.g., IEEE 802.16m), with little or no incremental backhaul expense and with similar transmit power as picocells.

- Remote radio heads (RRH) are compact-size, high-power, low-weight units, which are mounted outside the conventional macro BS, thus creating a distributed BS. They are linked to a central unit that performs control and baseband signal processing, therefore adding flexibility to network deployments with either site acquisition or physical limitation constraints.

Small cells entail a significant paradigm shift transitioning from "traditional", centralized macrocell/microcell approaches to more autonomous, uncoordinated, and intelligent rollouts. However, this paradigm shift, which can be seen as an excellent opportunity for enhancements, also introduces challenges. For instance, although the advantage of deploying femtocells is supported by recent studies suggesting that 50 % of all voice calls and more than 70 % of data traffic originate indoors, cross-tier interference and traffic load variability may become a barrier to a successful deployment of this type of network. Hence, small cells bring into play significant technical issues and raise substantial challenges that are presented in the following sections.

1.2 Technical and deployment challenges in small cell networks

Self-organization, backhauling, handover, and interference are identified here as the key technical challenges facing small cell networks.

1.2.1 Self-organization

Since some cells such as picocells and femtocells will be user deployed without operator supervision, their proper operation mainly depends on their self-organizing features [3] (see Chapter 16). The self-organizing capability of small cell networks can be generally classified into three processes:

- *Self-configuration*, where newly deployed cells are automatically configured by down-loaded software before entering into the operational state.
- *Self-healing*, where cells can automatically perform failure recovery or execute compensation mechanisms whenever failures occur.
- *Self-optimization*, where cells constantly monitor the network status and optimize their settings to improve coverage and reduce interference.

The deployment of self-organizing small cell networks is an intricate task due to the various types of coexisting cells and the increasing number of network parameters that need to be considered. The random, uneven, and time-varying nature of user arrivals and their resulting traffic load also exacerbate the difficulties associated with deploying a completely self-organized small cell network.

1.2.2   Backhauling

Backhaul network design will be a major issue because of the complex topology of the various type of coexisting cells. For instance, the deployment of picocells will require access to utility infrastructure with power supply and wired network backhauling, which may be potentially expensive. Femtocells, which in contrast have relatively lower backhauling costs, may face difficulties in maintaining quality of service (QoS) since backhauls rely on consumers' broadband connections. Hence, operators need to plan small cell network backhaul carefully to identify the most cost effective and QoS guaranteed solution. Such a solution is likely to be a mixture of both wireless and wired backhaul technologies, in which some cells may have dedicated interfaces to the core network, some other cells may form a cluster to aggregate and forward the traffic to the core, and other cells may rely on relays as an alternative interface.

1.2.3   Handover

Handovers and mobility management (see Chapter 9) are essential in order to provide a seamless uniform service when users move in or out of the cell coverage. Furthermore, handovers are efficient for traffic load balancing, by shifting users at the border of adjacent/overlapping cells from the more congested cells to the less congested ones. Nevertheless, this comes at the expense of system overhead, which is likely to be significant in small cell networks due to the large number of base station cells and the different types of backhaul links available for each type of cell. In addition, the probability of handover failure increases the probability of user outage [3].

1.2.4   Interference

The deployment of small cells overlying the macrocells will originate new cell boundaries, in which end users may suffer from strong inter-cell interference, degrading the performance of the overall cellular network (see Chapter 4). Unlike traditional single-tier cellular networks, in small cell networks, the cross-tier and intra-tier interference problems are significantly challenging. The backhaul network supporting different types of cells may have different bandwidth and delay constraints. For instance, femtocells are unlikely to be connected directly to the core network and thus only limited backhaul signaling for interference coordination is possible. Moreover, restricted access control associated with picocells and femtocells may lead to strong interference scenarios in both uplink and downlink since users may not handover to the nearest cells. The self-organizing capability of cells also requires continuous sensing and monitoring of the radio environment around them in order to dynamically and adaptively mitigate or avoid interference.

The main sources of interference in small cell networks can be classified as follows.

- *Unplanned deployment:* Low-power nodes such as femtocells are typically deployed in an ad hoc manner by users. They can even be moved or switched on/off at any time. Hence, traditional network planning and optimization become inefficient because

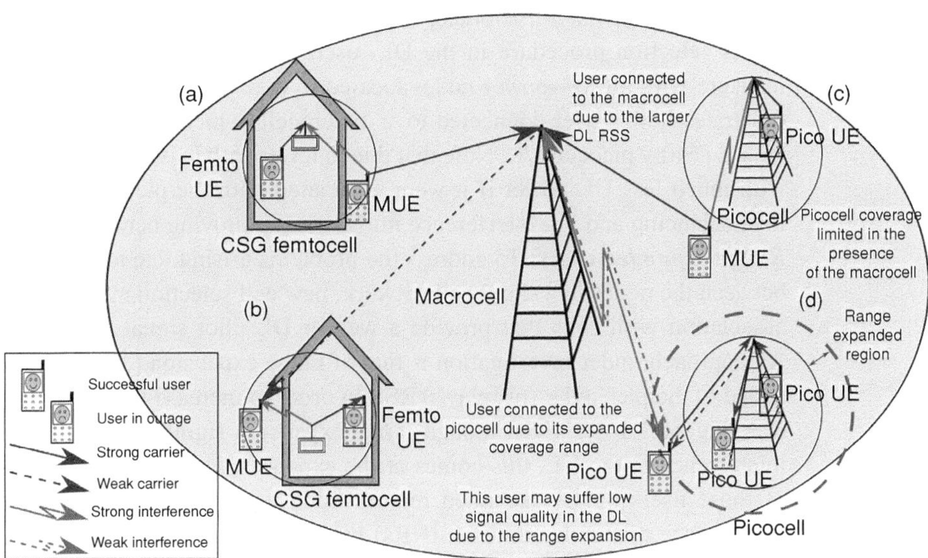

Figure 1.1 Some of the dominant DL and UL cross-tier interference scenarios in small cell networks: (a) macrocell user jamming the UL of a femtocell; (b) femtocell jamming the DL of a macro user; (c) macrocell user jamming the UL of a nearby picocell; (d) range-expanded picocell (mitigates UL interference from macrocell to picocells). ©2011 IEEE. Reprinted, with permission, from [2].

operators do not control either the number or the location of these cells. This motivates the need for new decentralized interference avoidance schemes that operate independently in each cell, utilizing only local information, while achieving an efficient solution for the entire network (see Chapter 11).

- *CSG access:* The fact that some cells may operate in CSG mode (see Chapter 2), in which cell access is restricted to certain user equipment (UE) and non-subscribers are thus not always connected to the nearest BS, originates significant cross-tier interference components [4]. Figure 1.1 depicts a challenging scenario for inter-cell interference coordination (ICIC), in which different non-subscribers walk nearby houses hosting a CSG femtocell. In the uplink (UL), non-subscriber (a) transmits at high power to compensate for the path losses to its far serving macrocell, jamming the UL of the nearby CSG femtocell(s). In the downlink (DL), a CSG femtocell interferes with the DL reception of non-subscriber (b) connected to the far macrocell. Hence, this DL macrocell user equipment (MUE) becomes a victim user.

- *Power difference between nodes:* Picocells and relays usually operate in open access mode, meaning that all users of a given operator can access to them. Open access helps to minimize DL interference as end users always connect to the strongest cell, thus avoiding the CSG interference issue. However, in small cell networks, being attached to the cell that provides the strongest DL received signal strength (RSS) may not be the best strategy since users will tend to connect to macrocells, and not to those cells being at the shortest path loss distance. This is due to the large difference in transmission power between macrocells and low-power nodes. In that way, traffic load will be unevenly distributed, thus overloading macrocells. Moreover, due to this server selection procedure in the DL, users connected to macrocells will severely interfere with all low-power nodes located in their vicinity in the UL. Figure 1.1(c) illustrates how a user connected to a macrocell, which provides the best DL RSS, jams a nearby picocell UL. Note that due to lower path loss, this MUE would transmit with much less UL power if it were associated with the picocell. This would allow load balancing and UL interference mitigation, improving network performance.

- *Range expanded users:* To address the problems arising due to the power difference between the nodes in a small cell network, new cell selection methods that allow user association with cells that provide a weaker DL pilot signal quality are necessary. An approach under investigation is that of range expansion [5], in which an offset is added to the picocell's (or relay's) RSS in order to increase its DL coverage footprint (see Figure 1.1(d)). Even though range expansion significantly mitigates cross-tier interference in the UL, this comes at the expense of reducing the DL signal quality of those users in the expanded region. Such users may suffer from DL signal to interference plus noise ratios (SINRs) below $0\,\mathrm{dB}$ since they are connected to cells that do not provide the best DL RSS (see Figure 1.1(d)).

1.3   Overview of contributions in the book

The rest of the book is composed of 15 chapters covering the issues discussed in the previous section. Note that while certain chapters may be more tailored toward femtocells, the general concepts, in most cases, can also be applied to other types of small cells in Table 1.1.

Chapter 2 titled "Fundamentals of access control in femtocells" by P. Xia, H.-S. Jo, and J. G. Andrews compares *open access* and *closed access* configurations of femtocells. Open access allows an arbitrary nearby cellular user to use the femtocell, whereas closed access restricts the use of the femtocell to users explicitly approved by the owner. The choice on femtocell access control is fundamental in femtocell deployments and involves many important issues such as legal/privacy concerns, cross-tier interference, and handoffs. In this chapter, the authors investigate the best approaches in *co-spectrum uncoordinated* femtocell networks for both the femtocell owners (interested in higher home users' rates) and the network operators (interested in higher cellular users' rates). In the downlink, the preferences of the two parties depend strongly on the distance between the macrocell base station and the femtocell. By characterizing signal to interference ratio (SIR) distributions for various zones within the macrocell as a function of this macro–femto distance, the chapter quantifies the sum throughput of home users and cellular users under open and closed access. The chapter also shows that the interests of home and cellular users are in conflict, and then investigates *shared access* (open access), but with limits on femtocell resources allocated to cellular users. Then, the authors derive the optimal resource allocation that maximizes the network throughput and also satisfies rate requirements of both parties. In the uplink, the best approach depends heavily on whether the multiple access scheme is orthogonal (time division multiple access (TDMA) or orthogonal frequency division multiple access (OFDMA), per subband) or non-orthogonal (code division multiple access (CDMA)). In a TDMA/OFDMA network, the best approaches for both the femtocell owner and the network operator depend on cellular user density, whereas in CDMA, open access is conclusively preferred by both parties.

Chapter 3 by S. Mukherjee is titled "Coverage analysis using the Poisson point process model" and it focuses on the spatial distribution of coverage and capacity in cellular wireless networks. The traditional approach to determining, say, the probability of coverage at an arbitrary location in the deployment region relies on exhaustive and detailed simulations for a large number of deployment scenarios. This is time consuming, slow, and expensive. While feasible for the traditional macrocellular networks, it is no longer practical for macro–femto overlay networks to investigate every deployment scenario (including every combination of densities of the tiers, their relative transmit powers, and the fraction of open access femtocells) via simulation. The chapter shows how base station location modeling and analytical methods from stochastic geometry yield closed-form analytical results on SINR distribution. This in turn enables us to obtain insights into the dependence of system-wide metrics such as coverage probability on the deployment parameters and, specifically, identify different deployment scenarios that are equivalent with regard to these metrics. For a chosen operating value of the system metric, the service provider can choose one set of deployment parameters from the corresponding equivalence class based on business or economic criteria. This chapter illustrates the applicability of these analytical techniques to the case of a macro–femto heterogeneous network.

Chapter 4 "Interference modeling for cognitive femtocells" is authored by A. Rabbachin, T. Q. S. Quek, H. Shin, and M. Z. Win. Destructive interference between femtocells and the macrocell occurs due to the universal frequency reuse in heterogeneous networks. Since conventional interference avoidance procedures through centralized radio resource allocation are infeasible due to unacceptably high complexity, cognitive radio resource management (CRRM) can provide a solution to autonomously mitigate the interference. This chapter considers a CRRM procedure based on spectrum sensing and presents an interference model, which accounts for the CRRM procedure, secondary spatial reuse protocol, and environment-dependent conditions such as path loss, shadowing, and channel fading. The authors then show how this model can be used to have a deep understanding of the effect of femtocell deployment on the macrocell uplink channel and therefore accurately design the femtocell access to the radio channel.

In **Chapter 5** "Multiple antenna techniques in small cell networks" by S. Akoum, M. Kountouris, and R. W. Heath, Jr., the potential role that multiple antenna communication and linear precoding are bound to play in femtocell networks is evaluated. The coverage performance of linear beamforming, both single user and multi user (zero-forcing (ZF) precoding), is studied under perfect and imperfect channel state information at the transmitter (CSIT). This chapter also presents the state of the art in femtocell and spatial network literature in terms of multiple input multiple output (MIMO) precoding and coordination techniques.

Chapter 6 by G. Caire, A. Adikhary, and V. Ntranos is titled "Physical layer techniques for cognitive femtocells". The authors consider "cognitive" femtocells, where the *cognition* attribute consists of the ability of decoding the macrocell downlink control channel broadcasted by the base station and making decisions on their transmission opportunities based on the scheduling, power, and rate allocation of the overlaying macrocell system. The operation of such femtocells is somehow reminiscent of "Type II" relays, as specified in the IEEE 802.16j standard. The chapter considers the tradeoff region of the macrocell versus the aggregate femtocell throughput. For a baseline system architecture, including only single-antenna terminals, the authors evaluate the impact of open versus closed access and interference cancellation at the femtocells. Then, two possible extensions of the basic architecture are considered. The first extension is based on the "downlink interference alignment" approach, in order to eliminate completely the cross-tier interference from the macrocell base station to the femtocell mobile stations. As an alternative, the authors also consider the use of multiple antennas at the femtocell base stations (FBSs) and simple linear minimum mean squared error (MMSE) receivers/beamformers. In this case, by aligning the macrocell downlink with the femtocells uplink and, vice versa, the macrocell uplink with the femtocells downlink, authors obtain a single input multiple output (SIMO)/multiple input single output (MISO) interference channel that can be optimized by exploiting power control and uplink–downlink duality. A careful combination of these techniques achieves a remarkably large aggregate femtocell throughput, and incurs a very low penalty in terms of the precious and scarce macrocell throughput. The good performance achieved by the proposed schemes, with relatively elementary linear/adaptive signal processing techniques, advocates for a system design that places femtocells at the *core* of future wireless networks and provisions for facilitating the required cognitive features through the design of an appropriate macrocell tier frame structure and control channel, rather than treating femtocells as an add-on "afterthought," to fill in the coverage bad spots of an already existing and independently designed macrocell network.

"Femtocell coverage optimization" is investigated in **Chapter 7** by the authors H. Claussen, L. Ho, S. Samuel, and F. Pivit. In femtocell deployments where cells are deployed by the end users in an uncontrolled way, the pilot signal leakage to the outside of its prescribed indoor radiating environment results in patchy outdoor coverage. This can lead to a highly increased signaling load to the core network as a result of the higher number of mobility events caused by passing users. In this chapter, the impact of such capture effects on the core network signaling load is investigated. It is shown that without further optimization, the resulting increase in signaling is unacceptably high and as a result prevents the large-scale deployment of femtocells. Different self-optimization methods for residential femtocell coverage are considered that can significantly reduce the resulting mobility events. First, mobility event based pilot power self-optimization is examined for femtocells with a single antenna that matches the coverage to the size of the indoor environment and thereby can provide both good indoor coverage and a reduced level of mobility events. It is shown that the proposed coverage optimization can significantly outperform simpler methods that aim to achieve a constant cell radius. Then, this concept is further refined by using a low cost switched pattern multi-element antenna solution. Antenna gain pattern measurements of a prototype with two patches and two inverted F antennas are presented and a corresponding feeder network is discussed. Self-optimization methods that jointly select an appropriate antenna pattern and optimize the pilot power are proposed. Finally, a second multi-element antenna solution is investigated where lobes in different directions, generated by several patch antennas, can be individually attenuated to shape the coverage. Both multi-element antenna solutions allow for a better match of femtocell coverage to the shape of each individual house, and result in a further improvement of both indoor coverage and core network signaling resulting from mobility events.

Chapter 8 is on "Random matrix methods for cooperation in small cell networks," and the authors are J. Hoydis and M. Debbah. Satisfying the exponentially increasing demand for mobile data services requires a significant network densification. However, with decreasing cell sizes, the inter-cell interference grows and must be mitigated. The cooperation of several BSs, which jointly process the user data from multiple cells, is one possible way to counter interference, essentially turning harmful into useful signals. Clusters of cooperating BSs, which can be seen as large virtual antenna arrays, also enable user mobility in dense networks where frequent handovers between BSs would cause a prohibitive amount of signaling overhead. The aim of this chapter is to provide a methodology for the performance analysis of BS cooperation in dense networks based on large random matrix theory (RMT). After a short introduction to several fundamental lemmas and recent results of RMT, the authors derive deterministic approximations of the mutual information and achievable rates with linear receivers and precoders for arbitrary clusters of cooperating BSs. These results are then extended to account for more realistic conditions, such as imperfect channel state information (CSI), Rician fading channels with deterministic line-of-sight components, and random user locations.

"Mobility in small cell networks" is investigated by V. Kavitha, S. Ramanath, and E. Altman in **Chapter 9**. The authors consider a large macrocell divided into a number of small cells, and study the impact of mobility in such systems, especially the effect of frequent handovers. One of the goals of the chapter is to model the small cell system so as to compute its mobility performance measures. It develops tools that take into account not only the instantaneous geometry, but also the way it varies in time. The first important quantity computed is the moments of the *service time*. It is the time taken by the system to serve the users, i.e., the time the BS spends on a call. Then, the chapter derives expressions for these service times, during which the information is exchanged between the moving user and the set of appropriate BSs (which it encounters during its journey), using variable rate of transmission.

Chapter 10 is titled "Cognitive radio resource management in autonomous femtocell networks," and is authored by K.-C. Chen and S.-Y. Lien. Destructive interference between femtocells and the macrocell occurs due to the universal frequency reuse in small cell networks. Conventionally, effective interference avoidance through the centralized radio resources allocation is infeasible due to an unacceptably high complexity. The characteristic of haphazard deployment of femtocells also burdens the interference control due to the lack of interfaces between the macrocells and femtocells. To tackle this challenge, in this chapter, the authors consequently introduce a CRRM technique for femtocells to autonomously mitigate interference. The ultimate goals of the CRRM for femtocells are to autonomously and effectively mitigate interference to/from the macrocell, provide guaranteed QoS, and achieve a fully effective radio resources utilization in each femtocell, thus enabling a successful femtocell deployment in 3GPP.

The title of **Chapter 11** is "Decentralized reinforcement learning techniques for interference management in heterogeneous networks" and it is authored by M. Bennis, D. Niyato, and T. Alpcan. In this chapter, the *strategic* coexistence between macrocell and femtocell tiers is studied using tools from evolutionary game theory (EGT) and reinforcement learning (RL), pertaining to different deployment scenarios (networked vs. stand-alone femtocells). In the first case, FBSs exchange information through a central controller, and adapt their strategies based on their instantaneous and average payoffs of the femtocell population. A fictitious play (FP) formulation and its low-complexity variant are also examined where FBSs maximize their performance taking into account the empirical frequency of other femtocells' transmission strategies. In the second case, when information exchange among femtocells is no longer possible, each femtocell gradually learns by interacting with its local environment through trial and error, and adapts its strategies until reaching a steady state. The overall performance of the network in terms of spectral efficiency and convergence behavior is shown to be adamantly driven by the type of information available at femtocells. Finally, investigations show that thanks to a variety of reinforcement learning tools, femtocells are able to self-organize in dense networks relying only on local information while mitigating their interference toward the legacy macrocell network.

"Resource allocation optimization in heterogeneous wireless networks" is tackled in **Chapter 12** by C. W. Tan. One first important resource to allocate is the transmission power, which has an impact on the size of each cell and therefore on the overlaps between cells. Hence, Chapter 12 covers power control as a key degree of freedom for the resource allocation problem in small cell networks, where interference management and performance are important characteristics. In this chapter, a class of practical and important power control problems for resource allocation is studied. This approach aims at enforcing fairness among femtocells to minimize the worst outage probability under Rayleigh fading and to reduce the total power consumption under outage specification constraints for all the users. The proposed method, which combines tools from non-linear Perron–Frobenius theory and optimization theory, will be highlighted to solve these seemingly non-convex resource allocation problems. This approach exploits the problem structure to unveil the hidden convexity that leads to the analytical characterization of the optimal value and solution through the spectral property of matrices induced by non-linear positive mappings. It provides a systematic way to derive distributed algorithms that are computationally fast and often free of parameter tuning. Useful bounds and connections between related resource allocation problems can also be obtained to evaluate the optimality and fairness of resource allocation. This theory-driven approach can also give useful insights into designing adaptive protocols for performance optimization in heterogeneous wireless networks.

Chapter 13 by S. Rangan introduces "New strategies for femto–macro cellular interference control." Since power control is not sufficient by itself, this chapter presents sophisticated interference control mechanisms that depart from conventional macrocellular interference control. In particular, the chapter revisits the basic paradigm for interference control in CDMA macrocellular systems based on universal frequency reuse, treating interference as noise, and using power and rate control to manage and adapt to the interference. The idea in this chapter is to take into account the fundamental issues that femtocell networks introduce, i.e., (i) due to access restrictions and the unplanned nature of deployments, interference can be much stronger and varied than traditional macrocellular networks; and (ii) since femtocells are typically deployed in an ad hoc and decentralized manner, interference control must be significantly more autonomous and distributed. In this chapter, alternative methods based on subband partitioning and interference cancelation are presented. It is shown that these techniques can offer significant improvements in throughput and can be implemented in a decentralized manner with minimal coordination with the macrocellular network.

Following on from the contribution on interference management is **Chapter 14** by Z. Bharucha and G. Auer, and titled "Femtocell interference control in standardization." The chapter is dedicated to solving the problem of resource allocation in the context of LTE networks. User-deployed femtocells, each exclusively serving a set of registered users and sharing the same frequency spectrum as the overlay, are considered. Such a co-channel and random deployment can cause heavy DL control and data channel interference especially to MUE in the vicinity of one or more femtocells and not belonging to their CSG. This chapter is dedicated to addressing this issue, termed ICIC, paying particular attention to the control channel. In systems that employ full frequency reuse, such as long term evolution (LTE), the issue of interference is a very serious one and can severely compromise cell-edge performance. While scheduling strategies do not come under the purview of LTE standardization, LTE does provide standardized signaling methods so that an appropriate signaling strategy may be employed to avoid excessive inter-cell interference for the data channels. However, these signaling methods are developed to be exchanged between macro NodeBs (NBs) over the X2 interface. It is expected that LTE will not have access to such an interface. Furthermore, unlike the data channel, the various control channels cannot be conveniently relocated in order to avoid interference. In this chapter ICIC techniques for LTE are presented. Since this chapter deals with the LTE control channel, the various control channels are introduced and the methods by which they are mapped to resources in the frequency domain are described. Methods for femto-to-macro ICIC are detailed and simulations are used to evaluate the performance. It is also shown that the performance of one of the control channels is particularly poor and hence a new solution is proposed.

If most of the previous techniques aim at reducing interference and thus increasing capacity, there is also a need to ensure that resource allocation for each cell will optimize the QoS for the users. That is why **Chapter 15** "Spectrum assignment and fairness in femtocell networks," by M. C. Erturk, İ. Güvenç, S. Mukherjee, and H. Arslan focuses on varying levels of resource allocation to different tiers in a heterogeneous network, which should meet the QoS requirements of different tiers, and at the same time maximize the total system capacity. In this chapter, a modified QoS-oriented fairness metric is proposed, which captures important characteristics of tiered network architectures that are not captured by the Jain's fairness index (JFI). A heterogeneous network composed of femtocells deployed within a macrocell network is considered, and optimization of the resource splitting ratio is investigated by using the proposed metric for co-channel, dedicated-channel and hybrid scenarios. First, using homogeneous Poisson processes, sum-capacities in such a network are expressed in closed form for co-channel, dedicated-channel, and hybrid resource allocation methods. Then, a resource-splitting strategy that simultaneously considers capacity maximization, fairness constraints, and QoS constraints is proposed. Detailed computer simulations utilizing 3GPP simulation assumptions show that a hybrid allocation strategy with a well-designed resource split ratio enjoys the best cell-edge user performance, with minimal degradation in the sum-throughput of macrocell users when compared with that of co-channel operation.

Finally, the last chapter in the book authored by D. López-Pérez and G. de la Roche is **Chapter 16**, and is titled "Self-organization and interference avoidance for LTE femtocells." One important capability femtocells should include is the concept of the self-organization, which is also partially discussed in other chapters. Chapter 16 focuses on the current state of the art on self-organizing network (SON) capability types and standardization. When implementing SON in a network, there are three ways to do it: (i) with a global manager which is in charge of a given number of femtocells and their optimization (a.k.a. centralized optimization); (ii) with distributed algorithms included in each femtocell access point (FAP); or (iii) with a hybrid approach where both centralized and distributed techniques are used. This chapter discusses the main SON techniques and then focuses more on the interference management aspect, which is probably the hardest challenge in large femtocell deployments. For this purpose, a novel and efficient distributed radio resource management technique is presented and its performance is evaluated in enterprise femtocell scenarios. In this approach, each cell independently optimizes its modulation and coding scheme and subchannel and transmit power allocations, while achieving a good degree of coordination at the network level.

References

[1] V. Chandrasekhar, J. G. Andrews, and A. Gatherer, "Femtocell networks: a survey," *IEEE Commun. Mag.*, vol. 46, no. 9, pp. 59–67, Sep. 2008.

[2] D. López-Pérez, I. Guvenc, G. de la Roche, M. Kountouris, T. Q. S. Quek, and J. Zhang, "Enhanced inter-cell interference coordination challenges in heterogeneous networks," *IEEE Wireless Commun. Mag.*, vol. 18, no. 3, pp. 22–30, June 2011.

[3] H. Claussen, L. T. W. Ho, and L. G. Samuel, "An overview of the femtocell concept," *Bell Labs Technical J.*, vol. 13, no. 1, pp. 221–46, Mar. 2008.

[4] D. López-Pérez, A. Valcarce, G. de la Roche, and J. Zhang, "OFDMA femtocells: a roadmap on interference avoidance," *IEEE Commun. Mag.*, vol. 47, no. 9, pp. 41–8, Sep. 2009.

[5] NTT DOCOMO, *R1-103264, Performance of eICIC with Control Channel Coverage Limitation*, 3GPP Std., Montreal, Canada, May 2010.

2 Fundamentals of access control in femtocells

Ping Xia, Han-Shin Jo, and Jeffrey G. Andrews

2.1 Access control in femtocell deployments

One unique trait of femtocells is that they are paid, installed, and managed by the end users. Compared with macrocells, which are installed by the network operators and thus can be accessed by any user, femtocells can choose the set of users that is allowed for access. In the simplest scenario, the femtocell can be configured for either: (i) *closed access*, where only registered home users can use the femtocell; or (ii) open access where any nearby users are allowed to use the femtocell. The choice of femtocell access control involves many important issues in two-tier femtocell networks [1–6].

The first important issue is cross-tier interference. Unlike wireless fidelity (WiFi) access points, femtocells serve users in licensed spectrum to guarantee quality of service (QoS) and because the devices they communicate with are developed for those frequencies. Compared to allocating separate channels inside the licensed spectrum exclusively to femtocells, sharing spectrum would be preferred from an operator perspective [1, 5, 7]. However, the co-channel spectrum sharing between femtocells and macrocells potentially gives rise to serious cross-tier interference in closed access. As shown in Figures 2.1 and 2.2, in closed access a cellular user, even when it is geographically close to a femtocell, is forced to communicate with the distant macro base station (BS). Therefore, this cellular user suffers from strong downlink interference from the nearby femtocell (see Figure 2.1), and likewise causes strong uplink interference to that femtocell (see Figure 2.2) [8–10]. These effects are akin to the well known near–far problem, but exacerbated by the decentralization and the lack of coordinated power control inherent in a two-tier network [4, 7, 9–13]. On the other hand, open access can avoid such strong cross-tier interference by allowing nearby cellular users simply to use the femtocell and coordinate with the existing users through it (see Figures 2.1 and 2.2) [7, 8, 11].

Another argument in favor of open access is that it allows traffic offloading or user balancing in the network. As mobile data traffic has undergone a striking increase recently, network operators strongly prefer open access, to offload users from heavily loaded macrocells to other infrastructure including femtocells [14, 15].

Small Cell Networks: Deployment, PHY Techniques, and Resource Management, ed. Tony Q. S. Quek, Guillaume de la Roche, İsmail Güvenç, and Marios Kountouris. Published by Cambridge University Press.
© Cambridge University Press 2013.

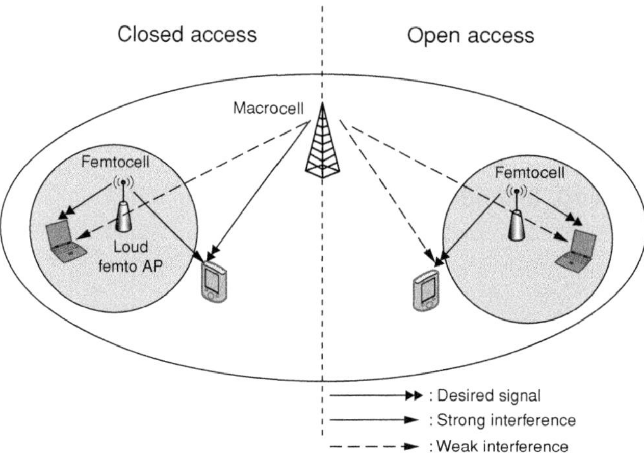

Figure 2.1 Open vs. closed access in downlink two-tier femtocell networks.

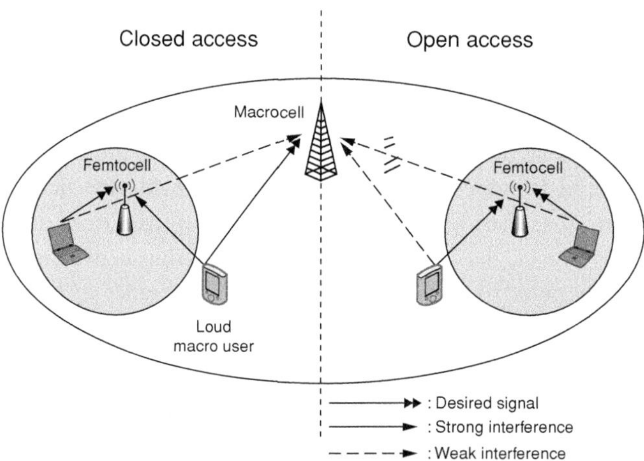

Figure 2.2 Open vs. closed access in uplink two-tier femtocell networks.

On the other hand, there are also many arguments in favor of closed access. Home users can monopolize their own femtocell and its backhaul in closed access. In particular, the monopoly of the femtocell backhaul can be a significant benefit, because femtocells use IP based backhaul (e.g., digital subscriber line (DSL) and cable modem) with very limited capacity (e.g., 1–10 Mbps). The loss of femtocell resource (which is already quite limited) in open access can be a serious concern of femtocell owners. Besides, closed access is preferred for concerns such as privacy, security, and handoffs [4, 16].

Besides the major issues mentioned above, the femtocell access control also affects many other aspects of two-tier femtocell networks [6, 17–20]. In short, the choice of femtocell open and closed access is fundamental in femtocell deployments.

Table 2.1 Concerns related to open vs. closed access.

Concerns	Preferred access control
Cross-tier interference	Open access
Traffic offloading	Open access (for network operators)
Femtocells' limited resource	Closed access (for femtocell owners)
Privacy and security concerns	Closed access
Handoffs	Closed access

As seen from Table 2.1, each access mode has its merits and disadvantages for both network operators and femtocell owners. Previous studies have investigated the best approach by simulating the cellular user rate (the interest of network operators) and the home user rate (the interest of femtocell owners) in each access mode [4, 7, 9–12, 21–24]. They suggested a hybrid access approach with an upper limit to the femtocell resources used by unregistered users to access the femtocell. We term this approach *shared access* since the femtocell is shared with cellular users, but within limits and hence not fully open.

In this chapter, we investigate the rates of home and cellular users analytically and through numerical simulations in each access mode. We also quantify their dependence on key system parameters such as the macrocell size, the densities and distributions of femtocells and users, and the distance between the macro BS and the target femtocell. We first derive the distribution of the signal to interference ratio (SIR)[1] as a function of these key system parameters under closed, shared, and open access. We then use this SIR distribution to evaluate the rates of the home and cellular users, and therefore quantify the best approaches for both femtocell owners and network operators.

The rest of this chapter is organized as follows. In Section 2.2, we describe general models and assumptions applicable to both the downlink and the uplink analysis. In Section 2.3 and 2.4, after describing additional assumptions specific for the downlink and uplink analysis, we derive the users' rates in femtocell open, shared, and closed access. Further discussions and conclusions are shown in Section 2.5.

2.2 System model

We consider a macrocell area of radius R_m with one macro BS (located at the center) and multiple co-spectrum femtocells. Suppose there are N cellular users, denoted as $U_1, U_2, \ldots, U_N$, roaming inside the macrocell. Their positions are independent and identically distributed (i.i.d.) random variables, uniformly distributed in the macrocell area. These users may use the macro or femto BSs in shared (with restrictions) and open access, but are forced to use the macro BS in closed access.

---

[1] Conventional cellular networks are generally interference limited while thermal noise is negligible. Interference is even more significant in two-tier femtocell networks due to the overlaid femtocells, generally of high density [25–27]. Therefore, in this analysis we neglect thermal noise.

To show the impact of macro–femto distance on femtocell access control, we analyze a target femtocell, denoted as fBS_0, which is located at a distance D away from the macro BS. By investigating this target femtocell throughput, we quantify the impact of access control strategies on a femtocell with an arbitrary distance from the macro BS. Since femtocells are installed by the end users, the locations of other femtocells are distributed randomly rather than in a regular pattern. Suppose they are distributed according to a Poisson point process[2] (PPP) $\Phi = \{X_j, j = 1, 2, \ldots\}$ with intensity λ, where each X_j is the location of the j-th femtocell denoted as fBS_j. The mean number of femtocells per macrocell is given as $N_f = \pi \lambda R_m^2$. The femtocell owners, or alternatively the home users, are uniformly deployed indoors (in the home), a disk $\mathcal{H} \subset \mathbb{R}^2$ of radius R_i centered at their femtocells. In any type of femtocell access control, the home users can freely choose either the macro BS or their own femtocells as serving base stations.

2.2.1   Channel model and interference characterization

We consider path loss and short-term fading in the channel model. The path-loss exponent of outdoor (indoor) transmission is denoted by α (β). In particular, the channel model is given by

$$H(|x|) = \begin{cases} h|x|^{-\alpha} & \text{outdoor \& cross-wall transmission} \\ h|x|^{-\beta} & \text{indoor transmission.} \end{cases} \tag{2.1}$$

Here, h is assumed as Rayleigh fading with unit average power, $|x|$ is the distance from the transmitter to the respective receiver. Setting $\alpha > \beta$ incorporates wall penetration loss in the channel model.

Assumption 2.1 *We assume that there is no coordination between the macro and any femto BSs, neither between different femtocells, in terms of power control or resource scheduling.*

In the *uncoordinated* femtocell network, the macro and femto BS use fixed transmit powers P_m^t and P_f^t respectively in the downlink, and require received powers P_m^r and P_f^r (through uplink power control) in the uplink. Thus the interference in the femtocell network can then be calculated. For example, a cellular user served by the macro BS experiences downlink interference as

$$\text{Interf}^{\text{UL}} = \sum_{x_j \in \Phi \setminus \{x_0\}} P_f^t H(|x_j|) + P_f^t H(|x_0|), \tag{2.2}$$

where $|x_0|$ and $\{|x_j|, j = 1, 2 \ldots\}$ are its distances to the target femtocell and other femtocells, respectively. In the uplink, this user causes interference, for example, to the

---

[2] See Chapter 3 for more details about the Poisson point process model.

target femtocell as

$$\text{Interf}^{\text{UL}} = P_m^r H(|x_0|)/H(|x_m|), \tag{2.3}$$

where $|x_m|$ is its distance to the macro BS.

Compared with the downlink, the uplink interference is harder to characterize, because it is proportional to the ratio of a user channel to different BSs. We thus have the following definition on *uplink interference factor*.

Definition 2.1 *A mobile user uplink interference factor I is defined as $\frac{H(|x_0|)}{H(|x_m|)}$, where $H(|x_0|)$ and $H(|x_m|)$ are its channels to the target femtocell and the macro BS, respectively. For the N cellular users $\{U_1, \ldots, U_N\}$, their uplink interference factors $\{I_1, I_2, \ldots, I_N\}$ are i.i.d. random variables, and we define their ordered statistics as*

$$I_{(1)} = \min(I_1, \ldots, I_N), \quad I_{(N)} = \max(I_1, I_2, \ldots, I_N)$$

and for $1 < k < N$,

$$I_{(k)} = \min(\{I_1, \ldots, I_N\}\setminus\{I_{(1)}, \ldots, I_{(k-1)}\}).$$

Accordingly, cellular users are reordered as $\{U_{(1)}, U_{(2)}, \ldots, U_{(N)}\}$.

It is clear that the uplink interference is fully characterized by the uplink interference factor, whose cumulative distribution function (CDF) will be derived in Lemma 2.4 in Section 2.4.

2.2.2   Femtocell coverage and cell association

Assumption 2.2 *In two-tier femtocell networks, a mobile user prefers to associate with: (i) in the downlink, the BS from which it receives the strongest long term average power (i.e. fading is averaged out); and (ii) in the uplink, the BS with the lowest path loss.*

The cell association preference above is reasonable in that it maximizes the mobile user SIR in both downlink and uplink, compared with other association policy. We therefore define the *femtocell coverage* as the area inside which a mobile user prefers (but may or may not be allowed by the femtocell) to connect with the femtocell. In the uplink, the femtocell coverage is easy to quantify, while in the downlink it is derived in the following lemma.

Lemma 2.1 *(Femtocell downlink coverage) For a femtocell located at distance D from the macro BS, the downlink femtocell coverage is approximated to within a circle centered at that femtocell, and its radius R_f is*

$$R_f = \kappa^{-1/\alpha} D, \tag{2.4}$$

where $\kappa = P_m^t / P_f^t \gg 1$.

Table 2.2 Notation and simulation values used for both the downlink and the uplink.

Symbol	Description	Sim. value
R_m	Macrocell radius	500 m
R_i	Indoor (home) area radius	20 m
R_f	Downlink femtocell coverage area radius	Not fixed
D	Distance between the target femtocell and the macro BS	Not fixed
P_m^t	Macro BS downlink transmit power	43 dBm
P_f^t	Femto BS downlink transmit power	13 dBm
P_m^r, P_f^r	Macro and femto BS uplink received powers	$P_m^r = P_f^r$
α	Outdoor path-loss exponent	4
β	Indoor path-loss exponent	3
η_L	Home users portion of femtocell resource with L handed over cellular users	Not fixed
$\mathcal{R}_h^{CA}, \mathcal{R}_h^{OA}$	Home users' sum rate in closed and open access	Not fixed
$\mathcal{R}_c^{CA}, \mathcal{R}_c^{OA}$	Cellular users' sum rate in closed and open access	Not fixed

Proof. Denote the location of the macro BS as $(0, 0)$ and, without loss of generality, the location of the femtocell as $(D, 0)$. For a mobile user at a position (x, y), its received average power from the macrocell is $P_m^t(\sqrt{x^2 + y^2})^{-\alpha}$ and that from the femtocell is $P_f^t(\sqrt{(D - x)^2 + y^2})^{-\alpha}$. The area where the mobile user prefers the femtocell over the macrocell is governed by $\frac{x^2+y^2}{(D-x)^2+y^2} > \kappa^{\alpha/2}$, where $\kappa = P_m^t/P_f^t \gg 1$. The inequality can be rewritten as

$$\left(x - \frac{\kappa^{2/\alpha} D}{\kappa^{2/\alpha} - 1} \right)^2 + y^2 < \frac{\kappa^{2/\alpha} D^2}{(\kappa^{2/\alpha} - 1)^2}. \tag{2.5}$$

Note that $\frac{\kappa^{2/\alpha}}{\kappa^{2/\alpha}-1}$ is very close to 1 because $\kappa \gg 1$. For example, $\frac{\kappa^{2/\alpha}}{\kappa^{2/\alpha}-1} = 1.02$ for the values in Table 2.2. Thus (2.5) can be simplified to $(x - D)^2 + y^2 < k^{-2/\alpha} D^2$, which is the area within a circle and completes the proof. $\qquad\square$

Remark 2.1 *As shown in Lemma 2.1, a femtocell downlink coverage area is increased as the macro–femto distance becomes larger.*

In all femtocell access strategies, the home users can freely choose between the macro BS and their own femtocell according to their preferences. Thus they will be served by their own femtocell if they are in its coverage and by the macro BS otherwise. On the other hand, whether a cellular user will be served by a femtocell is more complicated, dependent on femtocell access control: (i) in closed access, a femtocell will not serve any cellular users, even if they are in its coverage; (ii) in the open access, a femtocell will serve any cellular user as long as it is in its coverage; (iii) as a compromise between these two extremes, a femtocell with shared access will serve cellular users in its coverage, but with restrictions on how many and which cellular users will be served. Note that the motivations of letting nearby cellular users use femtocells are quite different between the uplink and the downlink: reducing co-channel interference for the home users in the

Table 2.3 Femtocell resource allocation in closed, shared, and open access.

	Closed	Shared	Open
Value of η_L when L guest users	1	Any value in $[0, 1]$	$\eta_L = M/(L + M)$ when M home users
Maximum number of guest users	0	A finite value specified by the femtocell	∞ (i.e., no restrictions)

uplink and for these cellular users themselves in the downlink. As a result, the models on such restrictions are also not the same in the downlink and the uplink, and they will be described in detail in Section 2.3 and 2.4.

2.2.3 Resource allocation

We here describe how macro and femto BSs allocate their over-the-air resources (time/frequency) among their users. In time division multiple access (TDMA), the end user time-shares the entire spectrum. In 4th Generation (4G) long term evolution (LTE) and wireless interoperability for microwave access (WiMAX), the end user is assigned a portion of the spectrum for a (sub)frame, which is identical to being allocated the entire spectrum for certain time slots (i.e. TDMA) from an analysis perspective. Besides, each subband in orthogonal frequency division multiple access (OFDMA) is orthogonal and allocated in a TDMA fashion along the time axis. Therefore in this chapter, the BSs are thus assumed to perform only time-slot allocation among users, which can also be viewed as TDMA or OFDMA on a per subband basis. From Assumption 2.1, we consider distributed resource allocation in two-tier femtocell networks.

In the macrocell network, since all the users are i.i.d. located inside the macrocell and thus have i.i.d. channels, the time resources will be fairly allocated among them. Therefore, when the macrocell BS is serving M cellular users, each user enjoys an average time fraction[3] of $1/M$. On the other hand, when the femtocell serves additional L guest cellular users, the time fraction allocated to the home users and the L cellular users should be η_L and $1 - \eta_L$ respectively. The home users fairly share the η_L fraction of time slots allocated to them, and so do the L guest femtocellular users. As shown in Table 2.3, different femtocell access strategies have different specifications on the values of η_L and the maximum number of L.

In summary, in this section we describe the general system model for our following analysis. Because of the key differences between the downlink and the uplink as listed in Table 2.4, we also have additional models specific for the uplink and downlink analysis respectively, which will be elaborated on in Section 2.3 and 2.4 below.

---

[3] Although in practice the time slot assigned to a mobile user can only be a group of discrete values, the average time fraction of the mobile user can be any value between 0 and 1. The same argument is applicable to the value of η_L.

Table 2.4 Key differences between downlink and uplink.

	Downlink	Uplink
Strong cross-tier interference	Femtocells to nearby cellular users	Nearby cellular users to femtocells
Motivation for open and shared access	Helping nearby cellular users	Helping home users
Cell association preference	Cell with strongest received power	Cell with lowest path loss

Table 2.5 Additional notations used in downlink analysis.

Notation	Description
G	Shannon gap in AMC
M	Number of different rates in AMC
D_{th}	Threshold macro–femto distance (radius of inner region)
$\mathcal{F}_{\text{a}}, \mathcal{F}_{\text{b}}$	Two zones inside the femtocell coverage when the femtocell is in the inner region
$\mathcal{F}_{\text{i}}, \mathcal{F}_{\text{o}}$	Two zones inside the femtocell coverage when the femtocell is in the outer region
$T_x, x \in \{\text{a, b, i, o}\}$	Throughput of zone $\mathcal{F}_x$ (if allocated with all time slots)

2.3   Femtocell access control in the downlink

In this section, we evaluate the open, closed, and shared access of the target femtocell in downlink, by quantifying the rate of the home users (the interest of femtocell owners) and cellular users (the interests of the network operator). We first provide expressions for the SIR distributions for various zones inside the macrocell as a function of the macro–femto distance. The rate of home users and cellular users in each access mode can be readily determined from these expressions. Refer to Table 2.5 for additional notations used in the downlink analysis.

2.3.1   Additional models for downlink analysis

We consider adaptive modulation and coding (AMC) in the downlink, which is standard practice. By measuring its channel and interference, the end user can estimate its instantaneous SIR γ and then provide feedback to its transmitter. The transmitter can then use M-level variable-rate transmission by dividing the SIR range into M regions, $[\Gamma_m, \Gamma_{m+1}), m = 1, \ldots, M$, where Γ_1 is the minimum SIR providing the lowest discrete rate and $\Gamma_{M+1} = \infty$. Then, the instantaneous transmission rate (in bps/Hz) is

$$r_m = \log_2 \left(1 + \frac{\Gamma_m}{G} \right) \text{ for } \gamma \in [\Gamma_m, \Gamma_{m+1}), 1 \leq m \leq M, \tag{2.6}$$

where G is the Shannon gap for M-level variable-rate transmission. Therefore if allocated with all the time slots, the long-term rate of a user is given as

$$\mathcal{R} = \sum_{m=1}^{M} r_m \cdot \mathbb{P}[\Gamma_m \leq \gamma < \Gamma_{m+1}]. \tag{2.7}$$

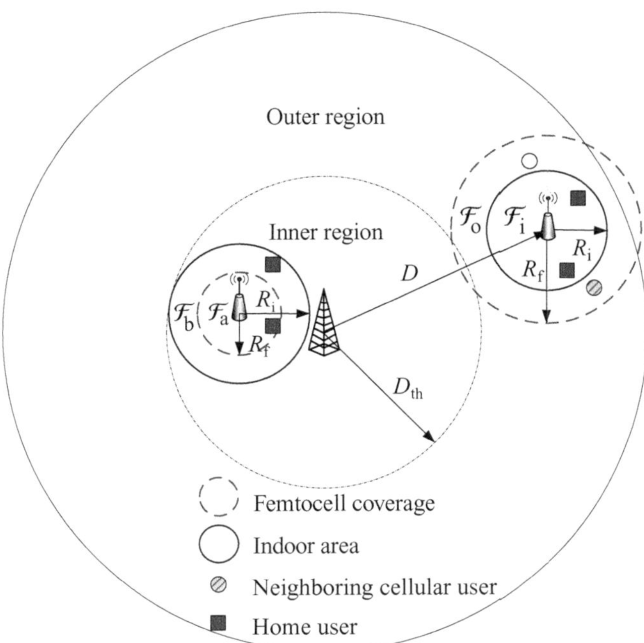

Figure 2.3 Femtocell coverage variation for the macro–femto BS distance D; in the outer region $(D > D_\mathrm{th})$, femtocell coverage is larger than the indoor area $(R_f > R_i)$, while $R_f < R_i$ in the inner region. The femtocell coverage is classified into four geometrical zones $\mathcal{F}_\mathrm{a}$, $\mathcal{F}_\mathrm{b}$, $\mathcal{F}_\mathrm{i}$, and $\mathcal{F}_\mathrm{o}$.

2.3.2 Coverage geographic zones

In Lemma 2.1, (2.4) states that the target femtocell coverage area $\mathcal{F}$ is a function of its distance D to the macro BS: for a femtocell very close to the macro BS, femtocell coverage can be smaller than the indoor area, whereas for a femtocell far from the macro BS, the coverage is larger than the indoor area and leaks into the outdoor area. Thus there is a threshold distance D_th such that femtocell coverage area is exactly equal to the indoor (home) area (see Figure 2.3). We accordingly partition the macrocell into two regions, *inner region* where $D \le D_\mathrm{th}$ and *outer region* where $D \ge D_\mathrm{th}$.

In the inner region, some indoor "home users" actually communicate with the macro BS, while in the outer region, outdoor neighboring cellular users would like to connect to the femtocell. To characterize an arbitrary end user SIR, we classify femtocell coverage into four geographic zones, whereby users in the same zone are of the same type (home vs. cellular users) associate with the same BS (femto vs. macro BS), and thus have the same SIR expression. Refer to Figure 2.3.

When the target femtocell is in the inner region, its coverage is divided into $\mathcal{F}_\mathrm{a}$ and $\mathcal{F}_\mathrm{b}$ as follows.

- $\mathcal{F}_\mathrm{a}$: indoor area where home users are served by the femtocell – a disk with radius R_f.
- $\mathcal{F}_\mathrm{b}$: indoor area where home users are served by the macro BS – a circular annulus with inner radius R_f and outer radius R_i.

When the target femtocell is in the outer region, its coverage is divided into $\mathcal{F}_i$ and $\mathcal{F}_o$ as follows.

- $\mathcal{F}_i$: indoor area where home users are served by the femtocell – a disk with radius R_i.
- $\mathcal{F}_o$: nearby outdoor area where neighboring cellular users are served by the femtocell (in open access) or the macrocell (in closed access) – a circular annulus with inner radius R_i and outer radius R_f.

The cell coverage model based on multiple geographic zones is essential for a comparative study of open, closed, and shared access, although it is more intricate than conventional models [25, 26]. Note that the femtocell access control does not affect the rates of cellular users outside the four zones. Thus we only need to compare the rates of nearby cellular users to quantify the network operator's preference on femtocell access in the downlink.

2.3.3   Per-zone average SIR and throughput

As described above, the femtocell downlink coverage area can be classified into four geographic zones, according to the macro–femto distance. We here characterize the spatially averaged SIR and throughput of each zone, which will be used to derive the rates of home and cellular users in Section 2.3.4. In the femtocell coverage, we denote R as the distance between the femtocell and an arbitrary home user, and D_m as the distance between the femtocell and an arbitrary nearby cellular user. We assume small sized femtocell $R \ll D$ resulting $D_m \approx D$.

Cellular user SIR in zone $\mathcal{F}_o$

Since the neighboring cellular users prefer to associate with the femtocell, they communicate with the femtocell (open access) or the macro BS (closed access), which results in different SIR according to the access scheme as follows:

$$\gamma(R) = \begin{cases} \dfrac{P_m^t h_m D^{-\alpha}}{P_f^t h_{f,0} R^{-\alpha} + \sum_{j \in \Phi} P_f^t h_{f,j} |X_j|^{-\alpha}} & \text{Closed access} & (2.8) \\[2em] \dfrac{P_f^t h_{f,0} R^{-\alpha}}{P_m^t h_m D^{-\alpha} + \sum_{j \in \Phi} P_f^t h_{f,j} |X_j|^{-\alpha}} & \text{Open access,} & (2.9) \end{cases}$$

where h_m and $h_{f,j}$ are Rayleigh fading of the macrocell and the j-th femtocell, and $|X_j|$ denotes the distance between the user and the j-th femtocell.

Lemma 2.2   *The CDF of spatially averaged SIR over $\mathcal{F}_o$ is given as follows: (i) Closed access:*

$$S_o^{\mathrm{CA}}(\Gamma) = \mathbb{E}_R \left[\mathbb{P}[\gamma(R) \leq \Gamma] \right], \quad R \in [R_i, R_f]$$

$$= 1 - \frac{e^{-\lambda C_\alpha (K\Gamma)^{2/\alpha}}}{R_f^2 - R_i^2} \left(A\left(R_f, \alpha\right) - A\left(R_i, \alpha\right) \right) \qquad (2.10)$$

$$A(x, y) = x^2 \left(1 - {}_2F_1 \left[2/y, 1; 1 + 2/y; -x^y/(K\Gamma) \right] \right)$$

where $C_\alpha = \frac{2\pi^2}{\alpha} \csc(\frac{2\pi}{\alpha})$, $K = \frac{P_f^t D^\alpha}{P_m^t}$, and ${}_2F_1[\cdot]$ is the Gauss hypergeometric function.

(ii) Open access:

$$S_{\mathrm{o}}^{\mathrm{OA}}(\Gamma) = 1 - \frac{1}{R_f^2 - R_i^2} \int_{R_i^2}^{R_f^2} \frac{e^{-r\lambda C_\alpha \Gamma^{2/\alpha}}}{K^{-1}\Gamma r^{\alpha/2} + 1} \, \mathrm{d}r. \tag{2.11}$$

Proof. In closed access, (2.8) is rewritten as $\gamma(R) = \frac{h_m}{K(I_1 + I_2)}$, where $K = \frac{P_f^t D^\alpha}{P_m^t}$ and $I_1 = h_{f,0} R^{-\alpha}$ and $I_2 = \sum_{j \in \Phi} h_{f,j} |X_j|^{-\alpha}$. The complementary cumulative distribution function (CCDF) of the user SIR at distance R from the femtocell is given as [28]

$$\mathbb{P}[\gamma(R) \geq \Gamma] = \mathcal{L}_{I_1}(\Gamma K)\mathcal{L}_{I_2}(\Gamma K),$$

where $\mathcal{L}_{I_1}(\Gamma K)$ is the Laplace transform of I_1 (exponential random variable scaled by $R^{-\alpha}$), which is given as

$$\mathcal{L}_{I_1}(\Gamma K) = \int_0^\infty e^{-\Gamma K s} f_{I_1}(s)\mathrm{d}s = \frac{1}{\Gamma K R^{-\alpha} + 1}.$$

Moreover, $\mathcal{L}_{I_2}(\Gamma K)$ is the Laplace transform of the Poisson shot-noise process I_2. For exponential $h_{f,j}$ with unit mean, $\mathcal{L}_{I_2}(\Gamma K)$ is given in [28]. Thus we get

$$\mathbb{P}[\gamma(R) \geq \Gamma] = \frac{e^{-\lambda C_\alpha (K\Gamma)^{2/\alpha}}}{\Gamma K R^{-\alpha} + 1}, \tag{2.12}$$

where $C_\alpha = \frac{2\pi^2}{\alpha} \csc(\frac{2\pi}{\alpha})$. Assuming the cellular users are uniformly located at the zone $\mathcal{F}_{\mathrm{o}}$, the probability density function (PDF) of the distance R is $f_R(r|R_i \leq R \leq R_f) = \frac{2r}{R_f^2 - R_i^2}$. The spatially averaged SIR distribution over $\mathcal{F}_{\mathrm{o}}$ is given as

$$S_{\mathrm{o}}^{\mathrm{CA}}(\Gamma) = \mathbb{E}_R\left[\mathbb{P}[\gamma(R) \leq \Gamma]\right] = 1 - \frac{2e^{-\lambda C_\alpha (K\Gamma)^{2/\alpha}}}{R_f^2 - R_i^2} \int_{R_i}^{R_f} \frac{R}{\Gamma K R^{-\alpha} + 1}\mathrm{d}R. \tag{2.13}$$

Desired result (2.10) is given by calculating (2.13) with an integration formula given by

$$\int \frac{t}{at^{-\alpha} + 1}\mathrm{d}t = \frac{t^2}{2}\left(1 - \,_2F_1\left[\frac{2}{\alpha}, 1; 1 + \frac{2}{\alpha}; -\frac{t^\alpha}{a}\right]\right). \tag{2.14}$$

Next, in open access, (2.9) is rewritten as $\gamma(R) = \frac{h_{f,0}}{R^\alpha(I_3 + I_4)}$, where $I_3 = K^{-1}h_m$ and $I_4 = \sum_{j \in \Phi} h_{f,j}|X_j|^{-\alpha}$. Following the way used to obtain (2.12), CCDF of the user SIR at distance R from the femtocell is given as

$$\mathbb{P}[\gamma(R) \geq \Gamma] = \frac{e^{-\lambda C_\alpha \Gamma^{2/\alpha} R^2}}{\Gamma K^{-1} R^\alpha + 1}.$$

Using the same way to obtain (2.13), we obtain the desired result (2.10). $\qquad\square$

Home user SIRs in zones $\mathcal{F}_{\mathrm{i}}$, $\mathcal{F}_{\mathrm{a}}$, and $\mathcal{F}_{\mathrm{b}}$

The received SIR at a home user in $\mathcal{F}_{\mathrm{i}}$ or $\mathcal{F}_{\mathrm{a}}$ is given as

$$\gamma(R) = \frac{P_f^t h_{f,0} R^{-\beta}}{P_m^t h_m D^{-\alpha} + \sum_{j \in \Phi} P_f^t h_{f,j}|X_j|^{-\alpha}}.$$

The SIR of a home user in $\mathcal{F}_b$ is given as

$$\gamma(R) = \frac{P_m^t h_m D^{-\alpha}}{P_f^t h_{f,0} R^{-\beta} + \sum_{j \in \Phi} P_f^t h_{f,j} |X_j|^{-\alpha}}.$$

Lemma 2.3 *The CDF of spatially averaged SIR over the zone $\mathcal{F}_i$, $\mathcal{F}_a$, and $\mathcal{F}_b$ is given as follows.*

(1) Zone $\mathcal{F}_i$: Using the similar way to that in proof of (2.11), we obtain

$$S_i(\Gamma) = 1 - \frac{2}{R_i^2} \int_0^{R_i} \frac{r K \cdot e^{-\lambda C_\alpha (L^2 \Gamma)^{2/\alpha} r^{2\beta/\alpha}}}{\Gamma r^\beta + 1} dr. \tag{2.15}$$

(2) Zone $\mathcal{F}_a$: The user SIR in the zone $\mathcal{F}_a$ is given above and $\mathcal{F}_a$ is a disk with radius R_f. Thus the CDF of spatially averaged SIR over the zone $\mathcal{F}_a$ is given by (2.15) with R_f replacing R_i, i.e.,

$$S_a(\Gamma) = S_i(\Gamma)|_{R_i = R_f}. \tag{2.16}$$

(3) Zone $\mathcal{F}_b$: Using the similar way to that in proof of (2.10), we obtain

$$S_b(\Gamma) = 1 - \frac{e^{-\lambda C_\alpha (L^2 K \Gamma)^{2/\alpha}}}{R_i^2 - R_f^2} \left(A\left(R_i, \beta\right) - A\left(R_f, \beta\right) \right). \tag{2.17}$$

In (2.15),(2.16) and (2.17), $K = \frac{P_f^t D^\alpha}{P_m^t}$ as in Lemma 2.2.

Proof. The proof is very similar to Lemma 2.2, and is thus omitted. $\qquad\square$

Per-zone throughput

From the above SIR characterizations, we can readily derive the throughput of each zone – the spatially averaged user rate if all time slots are allocated to users in this zone. For zone $\mathcal{F}_x$, $x \in \{a, b, i, o\}$, we have

$$\begin{aligned}
T_x &\stackrel{(a)}{=} \sum_{i=1}^{U_x} \frac{1}{U_x} \mathbb{E}_R \left[\sum_{m=1}^{M} r_m \mathbb{P}(\Gamma_m \le \gamma(R) < \Gamma_{m+1}) \right] \\
&\stackrel{(b)}{=} \sum_{m=1}^{M} r_m \mathbb{E}_R \left[\mathbb{P}(\Gamma_m \le \gamma(R) < \Gamma_{m+1}) \right] \\
&= \sum_{m=1}^{M} r_m \left[S_x(\Gamma_{m+1}) - S_x(\Gamma_m) \right],
\end{aligned} \tag{2.18}$$

where U_x is the number of users in that zone. (a) holds because the time slots are equally shared among U_x users in the zone $\mathcal{F}_x$. (b) holds because the locations of all users are i.i.d., so are their SIRs.

2.3.4   Per-tier throughput

In this section, we analyze the per-tier rate of closed, open, and shared access based on the results derived previously.

Closed or open access?

Denote U_a, U_b, U_i, and U_o as numbers of users in zones $\mathcal{F}_a$, $\mathcal{F}_b$, $\mathcal{F}_i$, and $\mathcal{F}_o$, respectively. Let U_c and U_h denote the number of outdoor cellular users and the number of home users per femtocell, respectively, and λ_c and λ_h are the corresponding user densities. The average number of users in a macrocell is then given by

$$U = U_c + N_f U_h = U_c + (N_{f1} + N_{f2})U_h \overset{(a)}{=} U_c + N_{f1}(U_a + U_b) + N_{f2}U_i,$$

where $N_{f1} = N_f(D_{th}/R_c)^2$ and $N_{f2} = N_f(1 - (D_{th}/R_c)^2)$ is respectively the average number of femtocells in the inner region ($D \le D_{th}$) and the outer region ($D_{th} < D \le R_c$). Furthermore, (a) follows from $U_h = U_i = U_a + U_b$.

Closed access. For the inner region ($D \le D_{th}$), home users in $\mathcal{F}_i$ connect to the femtocell, while the neighboring cellular users of the femtocell (i.e., users in $\mathcal{F}_o$) are served by the macro BS. For the outer region ($D > D_{th}$), regardless of femtocell access scheme, the home users in $\mathcal{F}_a$ connect to the femtocell, but the remaining home users in $\mathcal{F}_b$ communicate to the macro BS. The home users in $\mathcal{F}_b$ share the same frequency channel with cellular users by using different time slots. Based on the femtocell/macrocell access scenario of the users, the following theorem quantifies the per-tier user throughput in closed access.

Theorem 2.1 *In closed access, the average sum throughput of home users $\mathcal{R}_h^{CA}$ and neighboring cellular users $\mathcal{R}_c^{CA}$ with respect to the target femtocell is given as*

$$\mathcal{R}_h^{CA} = \begin{cases} T_a + \rho_b^{CA} T_b & \text{inner region} \\ T_i & \text{outer region} \end{cases} \tag{2.19}$$

$$\mathcal{R}_c^{CA} = \rho_o T_o^{CA}, \qquad \text{outer region} \tag{2.20}$$

where the per-zone throughput T_a, T_b, T_i, and T_o^{CA} is given from (2.18). ρ_b^{CA} and ρ_o are the fractions of time slots allocated to users in $\mathcal{F}_b$ and $\mathcal{F}_o$, respectively, from their serving macro BS, which are given as

$$\rho_b^{CA} = \frac{U_b}{U_c + N_{f1}\overline{U}_b} = \frac{R_i^2 - \kappa^{-2/\alpha} D^2}{\frac{\lambda_c}{\lambda_h}(R_c^2 - N_f R_i^2) + \frac{1}{2}N_{f1} R_i^2}, \tag{2.21}$$

$$\rho_o = \frac{U_o}{U_c + N_{f1}\overline{U}_b} = \frac{\kappa^{-2/\alpha} D^2 - R_i^2}{(R_c^2 - N_f R_i^2) + \frac{\lambda_h}{2\lambda_c}N_{f1} R_i^2}. \tag{2.22}$$

Open access. In the outer region, the target femtocell in open access provides service to neighboring cellular users in $\mathcal{F}_o$ as well as home users in $\mathcal{F}_i$. Thus the home users share the downlink radio resource of the femtocell with the cellular users in time division manner. On the other hand, in the inner region, the femtocell/macrocell access scenario of the home users in open access is the same as that in closed access. The following theorem quantifies the per-tier user throughput in open access.

Theorem 2.2 *In open access, the average sum throughput of home users, $\mathcal{R}_{\mathrm{h}}^{\mathrm{OA}}$, and neighboring cellular users, $\mathcal{R}_{\mathrm{c}}^{\mathrm{OA}}$, with respect to the target femtocell is given as*

$$\mathcal{R}_{\mathrm{h}}^{\mathrm{OA}} = \begin{cases} T_{\mathrm{a}} + \rho_{\mathrm{b}}^{\mathrm{OA}} T_{\mathrm{b}} & \textit{inner region} \\ \rho_{\mathrm{i}} T_{\mathrm{i}} & \textit{outer region} \end{cases} \tag{2.23}$$

$$\mathcal{R}_{\mathrm{c}}^{\mathrm{OA}} = (1 - \rho_{\mathrm{i}}) T_{\mathrm{o}}^{\mathrm{OA}} \qquad \textit{outer region}, \tag{2.24}$$

where $T_{\mathrm{o}}^{\mathrm{OA}}$ is given from (2.18). $\rho_{\mathrm{b}}^{\mathrm{OA}}$ is the fraction of time slots allocated to the home users in $\mathcal{F}_{\mathrm{b}}$ from their serving macro BS, and ρ_{i} is the fraction of time slots allocated to the home users in $\mathcal{F}_{\mathrm{i}}$ from their serving femto BS. They are given as

$$\rho_{\mathrm{b}}^{\mathrm{OA}} = \frac{U_{\mathrm{b}}}{U_{\mathrm{c}} - N_{\mathrm{f2}}\overline{U}_{\mathrm{o}} + N_{\mathrm{f1}}\overline{U}_{\mathrm{b}}}$$

$$= \frac{R_{i}^{2} - \kappa^{-2/\alpha} D^{2}}{\frac{\lambda_{\mathrm{c}}}{\lambda_{\mathrm{h}}}(R_{\mathrm{c}}^{2} - N_{\mathrm{f1}} R_{i}^{2} - \frac{\kappa^{-2/\alpha}}{2} N_{\mathrm{f2}}(R_{\mathrm{c}}^{2} + D_{\mathrm{th}})) + \frac{1}{2} N_{\mathrm{f1}} R_{i}^{2}} \tag{2.25}$$

$$\rho_{\mathrm{i}} = \frac{U_{\mathrm{i}}}{U_{\mathrm{i}} + U_{\mathrm{o}}} = \left(1 + \frac{\lambda_{\mathrm{c}}}{\lambda_{\mathrm{h}}}\left(\frac{\kappa^{-2/\alpha} D^{2}}{R_{i}^{2}} - 1\right)\right)^{-1}. \tag{2.26}$$

Comparing the home users rate in Theorem 2.1 and 2.2, we have the following observations:

1. When the femtocell is in the inner region, it serves no cellular users in either open or closed access, because cellular users in the inner region prefer the macro BS for a better signal. So open access will not cause any cost to such femtocells. In fact, some of their home users also connect to the macro BS (i.e., users in zone $\mathcal{F}_{\mathrm{b}}$). These home users have more time resource in open access than closed access, i.e., $\rho_{\mathrm{b}}^{\mathrm{OA}} \geq \rho_{\mathrm{b}}^{\mathrm{CA}}$, because in open access some cellular users are offloaded from the macrocell to femtocells in the outer region. Therefore, we have $\mathcal{R}_{\mathrm{h}}^{\mathrm{CA}} \leq \mathcal{R}_{\mathrm{h}}^{\mathrm{OA}}$ for femtocells in the inner region.
2. When the femtocell is in the outer region, it serves all its home users in each access mode. Besides, in open access, it has to serve additional cellular users, which only means loss of femtocell resources (i.e., $\rho_{\mathrm{i}} \leq 1$ in Theorem 2.2) without any benefits to the home users (e.g., reduced macro–femto interference in the uplink). Therefore we have $\mathcal{R}_{\mathrm{h}}^{\mathrm{CA}} \geq \mathcal{R}_{\mathrm{h}}^{\mathrm{OA}}$ for femtocells in the outer region.

Based on the observations above and from the numerical simulation in Figure 2.4, we have the following remark.

Remark 2.2 *In the downlink, femtocells in the inner region prefer open access, while femtocells in the outer region prefer closed access.*

We now compare the nearby cellular users' rate in open and closed access from Theorem 2.1 and 2.2. In closed access, these users are allocated with time fraction of $\rho_{\mathrm{o}} = \frac{U_{\mathrm{o}}}{U_{\mathrm{c}} + N_{\mathrm{f1}}\overline{U}_{\mathrm{b}}}$ from their serving macro BS, and $1 - \rho_{\mathrm{i}} = \frac{U_{\mathrm{o}}}{U_{\mathrm{i}} + U_{\mathrm{o}}}$ from their serving femtocell BS in open access. Since the femtocell user load $U_{\mathrm{i}} + U_{\mathrm{o}}$ is much smaller than the macrocell user load $U_{\mathrm{c}} + N_{\mathrm{f1}}\overline{U}_{\mathrm{b}}$, we have that $1 - \rho_{\mathrm{i}} > \rho_{\mathrm{o}}$. Besides, these users

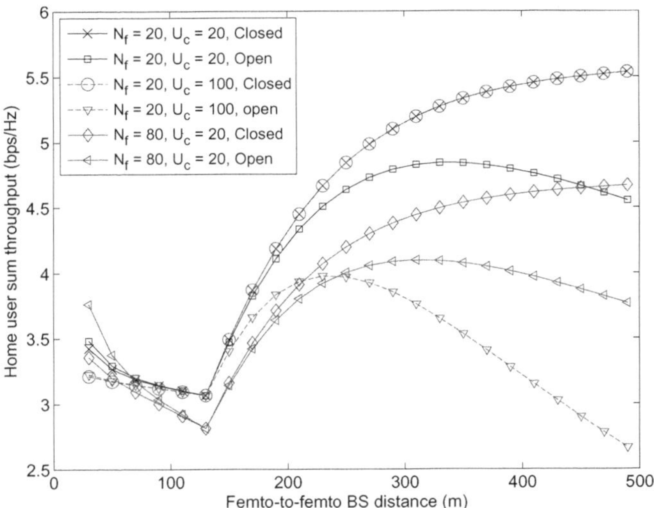

Figure 2.4 Average sum throughput of the home users for different number of femtocells and cellular users ($D_{\mathrm{th}} = 130\,\mathrm{m}$).

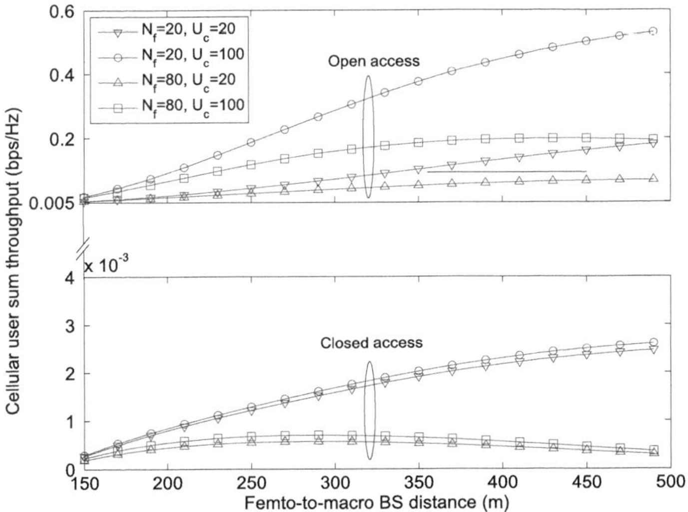

Figure 2.5 Average sum throughput of the neighboring cellular users for different numbers of femtocells and cellular users ($D_{\mathrm{th}} = 130\,\mathrm{m}$).

cannot associate with the femtocell with better signal in closed access, leading to lower SIRs and rates than open access (i.e., $T_0^{\mathrm{OA}} \leq T_0^{\mathrm{OA}}$). So overall we have $\mathcal{R}_c^{\mathrm{CA}} < \mathcal{R}_c^{\mathrm{OA}}$.

Remark 2.3 *Open access improves the cellular users' rates by allowing them to associate with the nearby femtocells with better signals and fewer users.*

Figure 2.5: Numerical simulations show that the benefits from open access can be significant to the cellular users. The throughput ($\mathcal{R}_c^{\mathrm{CA}} < 0.003$ bps/Hz) in closed access is too low to offer typical services (0.003 bps/Hz is equivalent to 15 kbps for 5 MHz bandwidth), which can be boosted up to 0.5 bps/Hz in open access.

Alternative: shared access

Since neither open nor closed access can completely satisfy the need of both user groups, we now turn our attention to the shared access approach. As described in Section 2.2, the key advantage of shared access is the flexible choice of η_L – the portion of femtocell resources allocated to the home users when the femtocell serves L cellular users. In the following, we derive the optimal value of η_L, to maximize network throughput under users' QoS requirements. For analytical convenience, here we consider the case that all $\{\eta_L, L = 1, \ldots\}$ are equal, i.e., the resources allocation is independent of the number of cellular users supported at that femtocell. We thus simply denote η as the portion of femtocell resources allocated to the home users, and $1 - \eta$ as the rest allocated to admitted cellular users.

The network throughput is given as

$$\mathcal{R}^{\mathrm{SA}} = \eta T_{\mathrm{i}} + (1 - \eta)T_{\mathrm{o}}^{\mathrm{OA}}, \quad \eta \in [0, 1]. \tag{2.27}$$

We define the QoS requirement as the average user throughput $\overline{\mathcal{R}}_{\mathrm{h}}$ (home user) and $\overline{\mathcal{R}}_{\mathrm{c}}$ (cellular user) being larger than the required minimum throughput Ω_{h} (home user) and $\Omega_{\mathrm{c}} = \varepsilon\Omega_{\mathrm{h}}$ with $\varepsilon \in (0, 1]$ (cellular user), respectively. Satisfying the QoS, the time slot allocation problem to maximize the network throughput $\mathcal{R}^{\mathrm{SA}}$ is formulated as

$$\max_{0 \leq \eta \leq 1} \eta T_{\mathrm{i}} + (1 - \eta)T_{\mathrm{o}}^{\mathrm{OA}} \quad \texttt{subject to}: \overline{\mathcal{R}}_{\mathrm{h}} \geq \Omega_{\mathrm{h}}, \ \overline{\mathcal{R}}_{\mathrm{c}} \geq \Omega_{\mathrm{c}} = \varepsilon\Omega_{\mathrm{h}} \tag{2.28}$$

where $\overline{\mathcal{R}}_{\mathrm{h}} = \frac{\eta T_{\mathrm{i}}}{U_{\mathrm{i}}}$ and $\overline{\mathcal{R}}_{\mathrm{c}} = \frac{(1-\eta)T_{\mathrm{o}}^{\mathrm{OA}}}{U_{\mathrm{o}}}$.

Theorem 2.3   *The optimal value η^* of the time slot allocation in (2.28) is given as*

$$\eta^* = 1 - \varepsilon\frac{\Omega_{\mathrm{h}}U_{\mathrm{o}}}{T_{\mathrm{o}}^{\mathrm{OA}}}. \tag{2.29}$$

The solution η^* is feasible when it is equal or larger than $\frac{\Omega_{\mathrm{h}}U_{\mathrm{i}}}{T_{\mathrm{i}}}$.

Proof.   Denote $\mathcal{Q}_1$ and $\mathcal{Q}_2$ as a set of η satisfying the QoS requirement (2.28) in the order of description, respectively. We then obtain the intersection of the three sets as $\mathcal{Q} = \mathcal{Q}_1 \cap \mathcal{Q}_2 = \{\eta \mid \frac{\Omega_{\mathrm{h}}U_{\mathrm{i}}}{T_{\mathrm{i}}} \leq \eta \leq 1 - \varepsilon\frac{\Omega_{\mathrm{h}}U_{\mathrm{o}}}{T_{\mathrm{o}}^{\mathrm{OA}}}\}$ for $\frac{\Omega_{\mathrm{h}}U_{\mathrm{i}}}{T_{\mathrm{i}}} \leq 1 - \varepsilon\frac{\Omega_{\mathrm{h}}U_{\mathrm{o}}}{T_{\mathrm{o}}^{\mathrm{OA}}}$. Define a function of η as $f(\eta) = \eta T_{\mathrm{i}} + (1 - \eta)T_{\mathrm{o}}^{\mathrm{OA}} = (T_{\mathrm{i}} - T_{\mathrm{o}}^{\mathrm{OA}})\eta + T_{\mathrm{o}}^{\mathrm{OA}}$. Since $T_{\mathrm{i}} > T_{\mathrm{o}}^{\mathrm{OA}}$, $f(\eta)$ monotonically increases with increasing η. Thus η^* is the maximum $\eta \in \mathcal{Q}$, which yields (2.29). Moreover, since $\mathcal{Q} = \emptyset$ for $\frac{\Omega_{\mathrm{h}}U_{\mathrm{i}}}{T_{\mathrm{i}}} > 1 - \varepsilon\frac{\Omega_{\mathrm{h}}U_{\mathrm{o}}}{T_{\mathrm{o}}^{\mathrm{OA}}}$, η^* is feasible for $\eta^* \geq \frac{\Omega_{\mathrm{h}}U_{\mathrm{i}}}{T_{\mathrm{i}}}$. $\qquad\square$

Figure 2.6 compares the network throughput for different femtocell access schemes. In the outer region, closed access always provides higher throughput than shared access because shared access with $\eta = 1$, which does not satisfy the QoS requirement, is the same as closed access. For $D \leq D_{\mathrm{th}}$, shared access obtains the same throughput as open access regardless of ε. The reason is that, like open access, the shared access with time slot allocation allows access from all neighboring cellular users located in the

Table 2.6 Preferred access and throughput comparison of open, closed, and shared access.

	Femtocells in the inner region	Femtocells in the outer region
Home users	Open access	Closed access
Cellular users	Indifferent	Open access
Both users	Shared access	Shared access

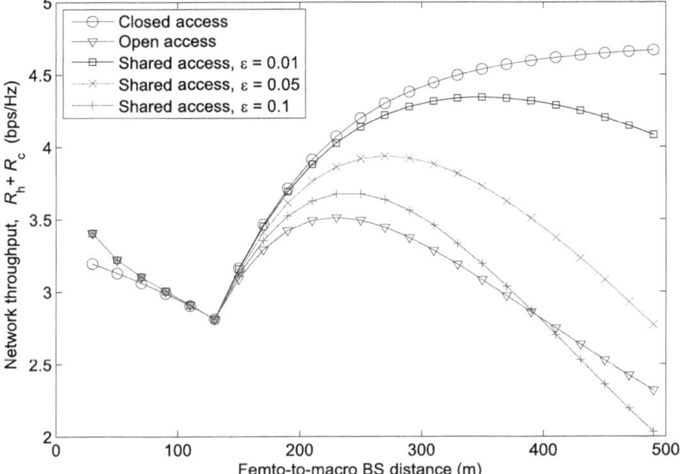

Figure 2.6 Network throughput for different femtocell access, ($N_f = 80$, $U_c = 100$, $D_{th} = 130$ m, $\Omega_h = 0.01$, and $\Omega_c = 0.01\varepsilon$).

zone $\mathcal{F}_o$. Note that the shared access with appropriate value of ε achieves higher (at $D > D_{th}$) or equal (at $D \leq D_{th}$) network throughput than open access. We summarize these observations in Table 2.6. These results motivate shared access – i.e., open access, but with limits – in femtocell-enhanced cellular networks with universal frequency reuse.

2.3.5 Conclusion for the downlink analysis

Unlike the uplink results in Section 2.4, our results show that in the downlink the preferred access scheme for home and cellular users is incompatible. In particular, closed access provides higher throughput for home users and lower throughput for neighboring cellular users; vice versa with open access. We derive the optimal femtocell resource allocations in shared access so as to maximize the network performance while satisfying a network-wide QoS requirement. Under such allocation policy, shared access achieves higher network throughput than open access for femtocells within the outer area, and meets the QoS requirements of both home and cellular users. These results motivate shared access with appropriate femtocell resource allocation in downlink two-tier femtocell networks.

Table 2.7 Additional notations used in uplink analysis.

Notation	Description
$\mathcal{R}_t$	User target rate
$\Gamma_{h,L}$	SIR target of home users
$\Gamma_{g,L}$	SIR target of guest femtocellular users
$\Gamma_{m,L}$	SIR target of macrocellular users
K	Maximum number of guest users at the femtocell

2.4 Femtocell access control in the uplink

In this section we investigate the best access approach in the uplink, from the viewpoints of both the femtocell owner and the network operator. We first describe the simplifications and additional models for uplink analysis, and then quantify the preferences of both parties in orthogonal (TDMA or OFDMA, per subband) and non-orthogonal (code division multiple access (CDMA)) uplink multiple access schemes. Refer to Table 2.7 for additional notations used in the uplink analysis.

2.4.1 Simplifications of the general system model

For uplink analysis, the general system model described in Section 2.2 can be simplified as follows:

1. Because uplink femtocell transmissions typically originate and terminate indoors and are of low power, their contribution to the overall interference is expected to be negligible compared to the more numerous and high power outdoor (macrocellular) users. Thus in the uplink, the femto-to-femto interference can be neglected and the system is simplified as comprising only the macro BS and the target femtocell located at a distance D away.

2. Because the BSs usually have advanced receivers (e.g., RAKE receiver in CDMA and multi-antenna receive diversity) and fading does not have a large effect in a wideband system with sufficient diversity (e.g., distributed subcarrier allocation in OFDMA), the short-term fading can be ignored in the uplink channel model. Therefore, the channel model in the uplink is simplified as

$$
H(|x|) = \begin{cases} |x|^{-\alpha} & \text{outdoor \& cross-wall transmission} \\ |x|^{-\beta} & \text{indoor transmission.} \end{cases}
\tag{2.30}
$$

3. For the purpose of analytical tractability, we suppose that the target femtocell serves only one home user. In the uplink, this home user associates with the femtocell, because it is geographically closer (in the same house) and the indoor channel has a smaller path-loss exponent (i.e., $\alpha > \beta$ as in Table 2.2). We denote this home user as U_0 and its interference factor as I_0.

2.4.2   Additional models for uplink analysis

The rate model: Compared with the downlink, the uplink is generally of lower rate. Thus we assume that users in the uplink are conservative: each of them only requires a *fixed* target rate $\mathcal{R}_t$. Accordingly, each user has a target SIR. The user achieves the required rate $\mathcal{R}_t$ when the received SIR is above the respective target SIR. Otherwise it is in outage and the rate is zero. Thus the user long-term rate $\mathcal{R}$ is simply its target rate multiplied by its probability of success.

It is important to note that a user target SIR is dependent on its time/frequency resource: with less time/frequency resource, the target SIR must be higher to achieve the fixed target rate $\mathcal{R}_t$, and vice versa. As the femtocell serves different numbers of guest cellular users, the resource allocated to each user will be different. Therefore, when the femtocell serves L cellular users, we denote the target SIRs of the home users, L guest cellular users, and the remaining $N - L$ macrocellular users as $\Gamma_{h,L}$, $\Gamma_{g,L}$, and $\Gamma_{m,L}$, respectively.

Handover model for open and shared access: As mentioned in Table 2.4, the motivation of uplink open and shared access is to reduce the strong interference from nearby cellular users. In other words, the femtocell has the incentive to serve a cellular user, when it can potentially reduce the home user outage by reducing co-channel interference.

Assumption 2.3   *In the uplink when nearby cellular users cause outage to the home user (i.e., the home user cannot meet its rate requirement), the femtocell picks the noisiest interferer from the macrocell to serve. This procedure continues as long as the home user still experiences outage and the number of handed over cellular users does not exceed the upper limit K.*

In the rest of the uplink analysis, we simply call the shared access model open access, since it is indeed open access with a finite value of K and flexible choices of η_L, where arbitrarily large K and specific value of η_L (see Table 2.3) conform to fully open access.

2.4.3   Throughput in orthogonal multiple access

Interference characterization

In an arbitrary time slot in TDMA or OFDMA per subband, suppose users U_i and U_j are active at the femtocell and the macrocell, respectively, and the received SIRs at two BSs are

$$\begin{cases} \dfrac{P^r_f}{P^r_m I_j} & \text{SIR at the target femtocell} \\[2mm] \dfrac{P^r_m}{P^r_f / I_i} & \text{SIR at the macro BS.} \end{cases} \qquad (2.31)$$

Note that the locations of cellular users are i.i.d, so are their interference factors. The following lemma derives the CDF of an arbitrary cellular user interference factor.

Lemma 2.4 *The CDF of an arbitrary cellular user interference factor is*

$$F_I(i) = \begin{cases} (r/R_m)^2 & 0 \le i < (\frac{R_m}{R_m+D})^\alpha \\ L(i) & (\frac{R}{R_m+D})^\alpha \le i < 1 \\ \frac{\pi - \varphi + 0.5\sin(2\varphi)}{\pi} & i = 1 \\ 1 - L(i) & 1 < i \le (\frac{R_m}{R_m-D})^\alpha \\ 1 - (r/R_m)^2 & (\frac{R_m}{R_m-D})^\alpha < i \end{cases} \tag{2.32}$$

where r, L(i), φ are given by:

$$r = \frac{i^{1/\alpha} D}{|1 - i^{2/\alpha}|}, \quad \varphi = \arccos\left(\frac{D}{2R_m}\right),$$

$$L(i) = \frac{(\theta - 0.5\sin 2\theta)r^2 + (\phi - 0.5\sin 2\phi)R_m^2}{\pi R_m^2},$$

where $x_c = \frac{Di^{2/\alpha}}{|1-i^{2/\alpha}|}$, $\theta = \arccos\left(\frac{r^2+x_c^2-R_m^2}{2rx_c}\right)$ *and* $\phi = \arccos\left(\frac{x_c^2+R_m^2-r^2}{2x_c R_m}\right)$.

Proof. Denote (x, y) as the location of a cellular user. Thus the CDF of its uplink interference factor I is

$$F_I(i) \triangleq \mathbb{P}(I \le i)$$

$$= \mathbb{P}\left((1 - i^{\frac{2}{\alpha}})x^2 + (1 - i^{\frac{2}{\alpha}})y^2 + 2Di^{\frac{2}{\alpha}}x - D^2 i^{\frac{2}{\alpha}} \le 0\right)$$

$$= S/(\pi R_m^2). \tag{2.33}$$

where S is the area inside the macrocell and is governed by

$$(1 - i^{2/\alpha})x^2 + (1 - i^{2/\alpha})y^2 + 2Di^{2/\alpha}x - D^2 i^{2/\alpha} \le 0.$$

When $i \neq 1$, the above equation defines a circle area with center of $x_c = \frac{Di^{2/\alpha}}{|1-i^{2/\alpha}|}$ and radius of $r = \frac{i^{1/\alpha} D}{|1-i^{2/\alpha}|}$. Moreover, when $i < 1$, S is the area inside the circle, while when $i > 1$, S is the area outside the circle. Using basic geometry to calculate S under different values of i, the expression of $F_I(i)$ follows. $\qquad\square$

Using the CDF $F_I(\cdot)$ derived above, we are able to characterize the uplink interference and quantify the home user rate (Theorem 2.4) and cellular users' sum rate (Theorem 2.5).

The home user rate in open and closed access
Theorem 2.4 *In TDMA or OFDMA uplink, the home user rate in femtocell closed access is*

$$\mathcal{R}_h^{CA} = \mathcal{R}_t F_I\left(\frac{P_f^r}{P_m^r \Gamma_{h,0}}\right). \tag{2.34}$$

In open access with $K = 1$ (i.e., the femtocell can serve at most one guest cellular user), the home user rate is

$$\mathcal{R}_{\mathrm{h}}^{\mathrm{OA}} = \mathcal{R}_t F_I^N \left(\frac{P_f^r}{P_m^r \Gamma_{h,0}} \right) + \mathcal{R}_t p_{h,1}, \tag{2.35}$$

where $p_{h,1}$ is given by

$$p_{h,1} = \frac{N}{N-1} F_I \left(\frac{P_f^r}{P_m^r \Gamma_{h,1}} \right) \left[1 - F_I^{N-1} \left(\frac{P_f^r}{P_m^r \Gamma_{h,0}} \right) \right]. \tag{2.36}$$

Proof. In closed access, the macro BS serves all the N cellular users and fairly shares the time slots among them. In other words, each macrocellular user causes interference to the femtocell for $1/N$ of the time. The home user rate is then

$$\mathcal{R}_{\mathrm{h}}^{\mathrm{CA}} = \mathcal{R}_t \sum_{j=1}^{N} \frac{1}{N} \mathbb{P} \left(\frac{P_f^r}{P_m^r I_j} \geq \Gamma_{h,0} \right) = \mathcal{R}_t F_I \left(\frac{P_f^r}{P_m^r \Gamma_{h,0}} \right). \tag{2.37}$$

In open access, the home user can achieve the target rate $\mathcal{R}_t$ in two circumstances: (i) (scenario I) all the N cellular users are served by the macro BS, none of which can cause outage to the home user during its active time slots; and (ii) (scenario II) the strongest interferer $U_{(N)}$ (which causes the strongest interference $P_m^r I_{(N)}$) causes outage to the home user. After the femtocell hands over $U_{(N)}$, the remaining $N-1$ macrocellular users do not cause outage to the home user. The probabilities of these two scenarios are

$$\mathbb{P}(\text{scenario I}) = \mathbb{P} \left(\frac{P_f^r}{P_m^r I_1} > \Gamma_{h,0}, \ldots, \frac{P_f^r}{P_m^r I_N} > \Gamma_{h,0} \right) = F_I^N \left(\frac{P_f^r}{P_m^r \Gamma_{h,0}} \right)$$

$$\mathbb{P}(\text{scenario II}) = \frac{1}{N-1} \sum_{j=1}^{N-1} \mathbb{P} \left(I_{(j)} \leq \frac{P_f^r}{P_m^r \Gamma_{h,1}}, I_{(N)} > \frac{P_f^r}{P_m^r \Gamma_{h,0}} \right)$$

$$= \frac{N}{N-1} F_I \left(\frac{P_f^r}{P_m^r \Gamma_{h,1}} \right) \left[1 - F_I^{N-1} \left(\frac{P_f^r}{P_m^r \Gamma_{h,0}} \right) \right]. \qquad \square$$

Two interesting observations from Theorem 2.4 are in order.

1. In closed access, the home user uplink long-term rate $\mathcal{R}_{\mathrm{h}}^{\mathrm{CA}}$ is independent of the number of cellular users N. This is because in TDMA the macro-to-femto interference is time shared, i.e., only one cellular user is scheduled in one time slot. Therefore, the home user long-term rate, which is averaged over time, is not scaled by N.
2. In open access, the home user rate $\mathcal{R}_{\mathrm{h}}^{\mathrm{OA}}$ contains two terms, meaning the rates before and after the femtocell hands over a cellular user. Because $F_I(\cdot)$ is a CDF smaller than 1 and consequently $F_I^N(x) < F_I(x)$, the first item in $\mathcal{R}_{\mathrm{h}}^{\mathrm{OA}}$ is strictly smaller than $\mathcal{R}_{\mathrm{h}}^{\mathrm{CA}}$. Therefore, whether open access improves the home user rate is dependent on the value of $p_{h,1}$, the home user probability of success after the femtocell hands over guest cellular users. As seen from (2.36), $p_{h,1}$ increases when the home user target SIR $\Gamma_{h,1}$ gets lower, which requires allocating more femtocell resource to the home user after the handover.

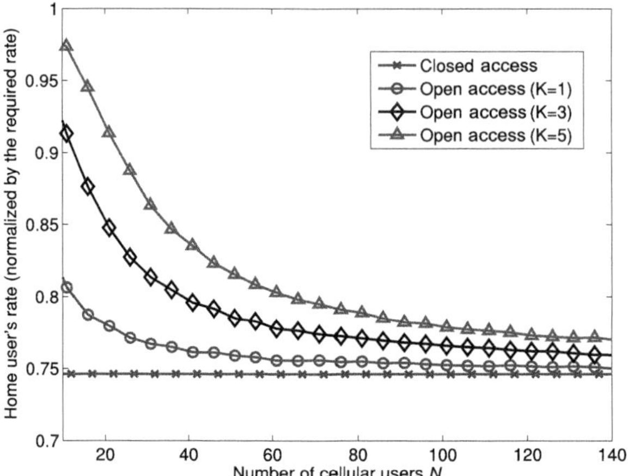

Figure 2.7 The home user rate (normalized by the target rate $\mathcal{R}_t$) in TDMA or OFDMA uplink. Since in closed access each cellular user is allocated with a time fraction of $\frac{1}{N}$ from the macrocell, we have $\eta_L = 1 - \frac{L}{N}$ in open access for fair comparison (i.e., the guest cellular user still has a time fraction of $\frac{1}{N}$ after the handover), $0 \leq L \leq K$.

Remark 2.4 *In TDMA or OFDMA uplink, whether open access improves the home user rate is dependent on the femtocell resource allocation after the handover. To the home user, the loss of femtocell resource in open access can be more dominant than the benefit of reduced cross-tier interference.*

The following corollary shows that when the number of cellular users N goes to infinity, even a tiny loss of femtocell resource leads to a lower home user rate in open access.

Corollary 2.1 *As the number of cellular users becomes arbitrarily large, that is $N \to \infty$, the home user rate in (2.34) and (2.35) becomes*

$$\mathcal{R}_h^{\text{CA}} = \mathcal{R}_t F_I \left(\frac{P_f^r}{P_m^r \Gamma_{h,0}} \right), \tag{2.38}$$

$$\mathcal{R}_h^{\text{OA}} = \mathcal{R}_t F_I \left(\frac{P_f^r}{P_m^r \Gamma_{h,1}} \right). \tag{2.39}$$

Note that after the femtocell serves one guest cellular user, the home user cannot monopolize all the femtocell resource as before. Thus we have $\eta_1 < \eta_0 = 1$. With less time/frequency resource, the home user has to increase the SIR target to achieve the same target rate, i.e., $\Gamma_{h,1} > \Gamma_{h,0}$. As a result, $\mathcal{R}_h^{\text{OA}}$ in (2.39) is smaller than $\mathcal{R}_h^{\text{CA}}$ in (2.38).

The conclusion from Corollary 2.1 is confirmed by Figure 2.7 from numerical simulations, which shows that open access provides an appreciable rate gain to the home user when N is small, but a very small rate gain when N is large (e.g., on the order of hundreds of users per macrocell).

Remark 2.5 *In TDMA or OFDMA uplink, open access provides a diminishing rate gain to home users as the number of cellular users N becomes large. In the asymptotic case of $N \to \infty$, home user rate is smaller in open access than closed access.*

The reason why open access is not a suitable choice in densely populated scenarios is explained as follows. When the number of cellular users increases, the amount of time occupied by each interferer decreases. Thus, in high cellular user density, handing over a small group of interferers (Corollary 2.1 is derived in the case of $K = 1$, but the argument can be extended) lowers the macro-to-femto interference merely for a small portion of time, i.e., the very short time length originally occupied by them. The home user signal quality is still inhibited by the residual interference from the remaining cellular users. Such an observation implies that the femtocells should consider the active time lengths of the interferers for handover in open access.

Remark 2.6 *In TDMA or OFDMA, femtocells in open access should admit the interferers that cause strong interference for a long time period.*

Cellular users' rate in open and closed access

In the following theorem, we derive the sum rate of cellular users (including cellular users in the macrocell and the femtocell) in open and closed access.

Theorem 2.5 *In TDMA or OFDMA uplink, the cellular users' sum throughput in closed access is*

$$\mathcal{R}_c^{\mathsf{CA}} = N\mathcal{R}_t \mathbb{P}\left(\frac{P_m^r}{P_f^r/I_0} \geq \Gamma_{m,0}\right) \tag{2.40}$$

In open access with $K = 1$ (i.e., the femtocell can serve at most one guest cellular user), the cellular users' sum throughput is

$$\mathcal{R}_c^{\mathsf{CA}} = N\mathcal{R}_t \mathbb{P}\left(\frac{P_m^r}{P_f^r/I_0} \geq \Gamma_{m,0}\right) F_I^N\left(\frac{P_f^r}{P_m^r \Gamma_{h,0}}\right) + \mathcal{R}_t[(N-1)p_{m,1} + p_{g,1}], \tag{2.41}$$

where $p_{m,1}$ and $p_{g,1}$ are

$$p_{m,1} = \eta_1 \mathbb{P}\left(\frac{P_m^r}{P_f^r/I_0} \geq \Gamma_{m,1}\right)$$

$$+ (1 - \eta_1)\left[1 - \max\left(F_I^N\left(\frac{P_f^r}{P_m^r \Gamma_{h,0}}\right), F_I^N\left(\frac{P_f^r \Gamma_{m,1}}{P_m^r}\right)\right)\right]$$

$$p_{g,1} = \frac{N}{N-1} F_I\left(\frac{P_f^r}{P_m^r \Gamma_{g,1}}\right)\left[1 - F_I^{N-1}\left(\frac{P_f^r}{P_m^r \Gamma_{h,0}}\right)\right].$$

Proof. The proof is similar to that of Theorem 2.4 and so is omitted. $\square$

Similar to Theorem 2.4, the cellular users' sum rate $\mathcal{R}_c^{\mathsf{OA}}$ contains two terms, meaning the sum rates before and after the femtocell hands over to a cellular user. Again the first term in $\mathcal{R}_c^{\mathsf{OA}}$ is strictly smaller than $\mathcal{R}_h^{\mathsf{CA}}$. Therefore, in TDMA or OFDMA uplink, our conjecture is that open access may lower the cellular users' rate as well.

As shown in Figure 2.8 from numerical simulations, open access does reduce the cellular users' sum rate, especially when the number of cellular users N is small. The reason is explained as follows. In closed access, the femto-to-macro interference is

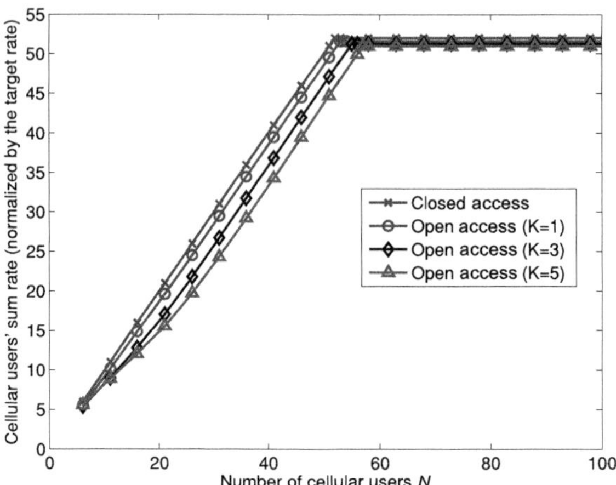

Figure 2.8 Cellular users' sum rate (normalized by the target rate $\mathcal{R}_t$) in TDMA or OFDMA uplink. We choose $\eta_L = 1 - \frac{L}{N}$ for fair comparison, $0 \leq L \leq K$.

caused by the home user during all time slots. In open access after handover, some femtocell time slots are allocated to the guest cellular users, which cause interference to the macrocell as well. Note that because the home user is closer to the femtocell (both located in the same house) and the indoor channel has a smaller path-loss exponent, the home user transmitting power and thus interference to the macrocell are smaller than guest cellular users. As a result, the femto-to-macro interference is increased in open access.

Remark 2.7 *In TDMA or OFDMA uplink, open access reduces the cellular users' sum rates, because of the increased femto-to-macro interference in open access. The rate reduction is appreciable when the number of cellular users N is small.*

There is a turning point in each curve in Figure 2.8, which means that the network serves the maximum number of cellular users. As more cellular users add in, the network has to turn off these additional users and the sum rate curves become flat. Comparing the turn points of all the curves in Figure 2.8, we have the following remark.

Remark 2.8 *By offloading cellular users from the macrocell to the femtocell, open access can increase the maximum number of cellular users in the network.*

Note that the remark above is also true for CDMA uplink, as shown in Figure 2.10 in Section 2.4.4.

The choices of femtocell owners and network operators

Summarizing our analytical and numerical results in Table 2.8, below we describe the choices of open vs. closed access in TDMA or OFDMA uplink.

1. When the number of cellular users N is small, open access provides an appreciable rate gain to the home user, but reduces the cellular users' rates. Therefore, the choices

Table 2.8 Choices of two parties in orthogonal multiple access.

Number of cellular users	Femtocell owners	Network operators
Small	Open access	Closed access
Large	Indifferent	Indifferent

of the two parties are in disagreement: the femtocell owners prefer open access, while network operators prefer closed access.

2. When the number of cellular users N is large, open access provides a marginal rate gain to the home users. For the network operators, it slightly reduces the cellular user rate but also increases the maximum number of cellular users in the network. In short, neither femtocell owners nor network operators have a strong incentive for either open or closed access.

2.4.4 Throughput in non-orthogonal multiple access

3rd Generation (3G) CDMA networks have been launched worldwide in recent years and will be in wide service for at least a decade. This necessitates research and standardization for incorporating femtocells in CDMA cellular networks [29–32]. Even if both TDMA and CDMA are part of the medium access (e.g., high speed packet access (HSPA) in 3rd Generation Partnership Project (3GPP) and evolution data only (EVDO) in 3GPP2), we restrict our attention to the CDMA aspect here, and this analysis would thus be valid per time or frequency slot. We show that in CDMA the interests of the femtocell owner and the network operator are compatible: open access is the appropriate approach for both parties.

Interference characterization

In CDMA, suppose L cellular users are served by the femtocell and the received SIRs at the two BSs are

$$
\begin{cases}
\dfrac{P_f^r}{L P_f^r + P_m^r \sum_{j=1}^{N-L} I_{(j)} + \sigma^2} & \text{SINR at the femtocell} \\[2ex]
\dfrac{P_m^r}{P_f^r/I_0 + \sum_{j=N-L+1}^{N} P_f^r/I_{(j)} + (N-L-1)P_m^r + \sigma^2} & \text{SINR at the macro BS.}
\end{cases}
\tag{2.42}
$$

It is seen that in CDMA the interference from cellular users is additive. For a set of arbitrary k cellular users $\{U_{n_1}, \ldots, U_{n_k}\}$, denote function $G_I(k, \cdot)$ as the CDF of $\sum_{j=n_1}^{n_k} I_j$, which is given by

$$
G_I(k, i) = \mathbb{P}\left(\sum_{j=n_1}^{n_k} I_j \leq i \right).
$$

$G_I(k, \cdot)$ is the same for whatever k cellular users, since their uplink interference factors are i.i.d.

Lemma 2.5 An upper bound on the CDF function $G_I(k, \cdot)$ is given by

$$G_I(k, i) \le G_I^{ub}(k, i) = (F_I(i))^k \triangleq F_I^k(i). \tag{2.43}$$

Proof. Without loss of generality, we assume I_{n_1} is the maximum *interference factor* of $\{I_{n_1}, \ldots, I_{n_k}\}$. Therefore $\mathbb{P}(I_{n_1} \le i) = F_I^k(i)$, which provides an upper bound on $G_I(k, \cdot)$.

$$G_I(k, i) = \mathbb{P}\left(\sum_{j=n_1}^{n_k} I_j \le i\right) = 1 - \mathbb{P}\left(\sum_{j=n_1}^{n_k} I_j > i\right) \le 1 - \mathbb{P}\left(I_{n_1} > i\right) = F_I^k(i).$$

$\square$

Using the CDF $G_I(\cdot)$ derived above, we can quantify the home user rate (Theorem 2.6) and cellular users' sum rate (Theorem 2.7) in open and closed access.

Home user rate in open and closed access

Theorem 2.6 *In CDMA, the home user rate in closed access is*

$$\mathcal{R}_h^{CA} = \mathcal{R}_t G_I\left(N, \frac{P_f^r}{P_m^r \Gamma_{h,0}}\right), \tag{2.44}$$

and that in open access is lower bounded as

$$\mathcal{R}_h^{OA} \ge \mathcal{R}_t G_I\left(N, \frac{P_f^r}{P_m^r \Gamma_{h,0}}\right). \tag{2.45}$$

Proof. In closed access, no cellular user is served by the femtocell, meaning that the value of L in equation (2.42) is zero. Thus the success probability of the home user is

$$\mathbb{P}\left(\sum_{j=1}^{N} I_j \le \frac{P_f^r}{P_m^r \Gamma_{h,0}}\right) = G_I\left(N, \frac{P_f^r}{P_m^r \Gamma_{h,0}}\right). \tag{2.46}$$

In open access, L in equation (2.42) can be any integer between 0 and K. When $L = 0$, the home user probability of success is the same as (2.46) and its rate is $\mathcal{R}_t G_I\left(N, \frac{P_f^r}{P_m^r \Gamma_{h,0}}\right)$. For other values of L, the home user probability of success is non-negative. Thus in open access, the home user rate $\mathcal{R}_h^{OA}$ is larger than $\mathcal{R}_t G_I\left(N, \frac{P_f^r}{P_m^r \Gamma_{h,0}}\right)$. $\square$

Theorem 2.6 shows that in CDMA, open access has better performance than closed access in terms of the home user rate, irrespective of the femtocell resource allocation after handover. Figure 2.9 shows from numerical simulation that such a rate gain is very significant.

Remark 2.9 *In CDMA, open access always provides a positive rate gain to the home users, independent of femtocell resource allocation after the handover. Besides, such a rate gain is usually significant.*

It is interesting to compare the results in CDMA (Section 2.4.4) and in TDMA/OFDMA (Section 2.4.3). In CDMA the interference is additive. So handing over a small group of strongest interferers always means a significant reduction in

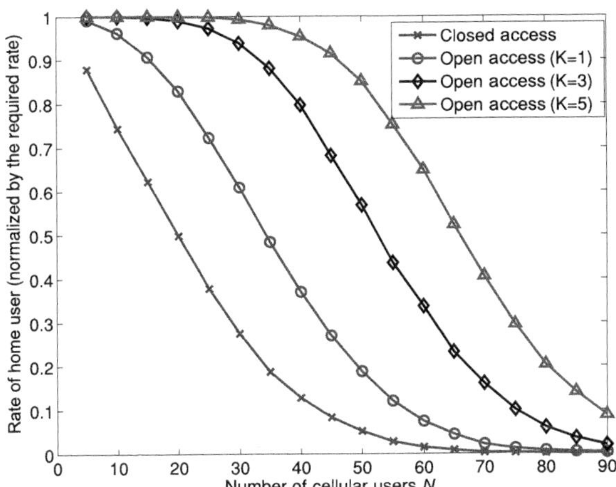

Figure 2.9 The home user rate (normalized by the target rate $\mathcal{R}_t$) in CDMA.

macro-to-femto interference and consequently an improvement in the home user rate. On the contrary, interference is time shared in TDMA. Thus handing over a small group of interferers for just part of the time does not guarantee an appreciable reduction of cross-tier interference. In other words, reducing cross-tier interference is more dominantly important in CDMA, compared with TDMA or OFDMA.

Cellular users' rate in open and closed access

Theorem 2.7 *In CDMA, the cellular users' sum rate in closed access is*

$$\mathcal{R}_c^{\mathrm{CA}} = N\mathcal{R}_t \mathbb{P}\left(\frac{P_m^r}{P_f^r/I_0 + (N-1)P_m^r} \geq \Gamma_{m,0} \right). \tag{2.47}$$

In open access with $K = 1$ (i.e., the femtocell can serve at most one guest cellular user), the cellular users' sum throughput is

$$\mathcal{R}_c^{\mathrm{CA}} = N\mathcal{R}_t \mathbb{P}(\frac{P_m^r}{P_f^r/I_0 + (N-1)P_m^r} \geq \Gamma_{m,0})G_I\left(N, \frac{P_f^r}{P_m^r \Gamma_{h,0}} \right)$$
$$+ \mathcal{R}_t[(N-1)p_{m,1} + p_{g,1}], \tag{2.48}$$

where $p_{m,1}$ and $p_{g,1}$ are the success probabilities of the macrocellular users and the guest femtocellular user, respectively, after handover in open access.

Similar to TDMA/OFDMA, Theorem 2.7 shows that in CDMA the cellular users' sum rate in open access can be lower or higher than closed access. Figure 2.10 shows from numerical simulation that the rate loss and gain in open access are both very marginal. Therefore, the network operators are quite neutral about open vs. closed access, in terms of cellular users' sum rate. They may slightly prefer open access because it can increase the maximum number of cellular users in the network, as seen from Figure 2.10.

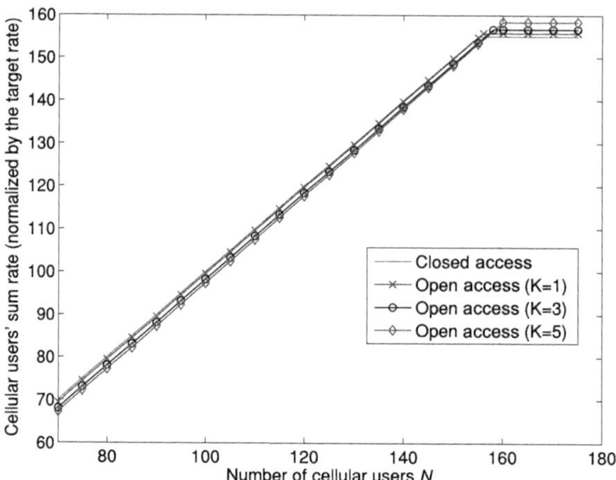

Figure 2.10 Cellular users' sum throughput (normalized by the target rate $\mathcal{R}_t$) in CDMA.

The choices of femtocell owners and network operators

In summary, the choices of femtocell owners and network operators in CDMA uplink
are compatible: open access is preferred by both parties. In particular, the femtocell
owners are highly incentivized for open access. Therefore, CDMA femtocells should be
configured for open access in the uplink.

2.5   Summary and conclusions

This chapter provides an analytical framework for evaluating femtocell access schemes
in co-spectrum uncoordinated two-tier femtocell networks. The framework quantifies
femtocell-site-specific "loud neighbor" effects and can be used to compare other techniques such as power control and spectrum allocation. The key conclusions are listed
below.

1. In the downlink, the network operators prefer open access, which allows cellular users
   using nearby femtocells with better signals and lighter user load. On the other hand,
   the femtocell owners' choice depends on the locations of their femtocells: cell-center
   femtocells prefer open access (because macro users will not choose these femtocells
   in the first place, no matter whether open or closed access) while cell-edge femtocells
   prefer closed access.
2. In non-orthogonal uplink multiple access (CDMA), open access is the preferred
   choice of both parties. In particular, the femtocell owners have a strong incentive for
   open access, because inference reduction from open access is dominantly important.
3. In orthogonal uplink multiple access (TDMA or OFDMA), the choices of the two
   parties are dependent on the number of cellular users in the network and, more
   importantly, can be incompatible. Our conjecture is that the performance of open

access will be improved in TDMA or OFDMA, if there is coordination between femtocells and macrocells.

As seen above, the network operator generally prefers open access in most scenarios, for cross-tier interference reduction and traffic offloading. This aligns with our intuitive conjectures listed in Table 2.1. However, the choices of femtocell owners are more complicated. They prefer open access only if cross-tier interference is dominantly strong in closed access (as in CDMA uplink). In other scenarios, their choices are influenced by many other factors, such as femtocell locations, user density, and resource allocations in femtocell open access.

Note that these conclusions are possibly contingent on our modeling assumptions, among which the following two may be the most critical: (i) the assumption of no inter-BS coordination both among and across the two tiers. Possible coordination schemes include time/spectrum splitting between femtocells and macrocells [26]; (ii) the assumption on handoff policy. More sophisticated handoff policy may further improve the performance of open access [23]. Investigations on the impact of these assumptions are important related topics for future research.

References

[1] V. Chandrasekhar, J. G. Andrews, and A. Gatherer, "Femtocell networks: a survey," *IEEE Commun. Mag.*, vol. 46, no. 9, pp. 59–67, Sep. 2008.

[2] A. Golaup, M. Mustapha, and L. Patanapongpibul, "Femtocell access control strategy in UMTS and LTE," *IEEE Commun. Mag.*, vol. 47, no. 9, pp. 117–23, Sep. 2009.

[3] R. Y. Kim, J. S. Kwak, and K. Etemad, "WiMAX femtocell: requirements, challenges, and solutions," *IEEE Commun. Mag.*, vol. 47, no. 9, pp. 84–91, Sep. 2009.

[4] G. de la Roche, A. Valcarce, D. López-Pérez, and J. Zhang, "Access control mechanisms for femtocells," *IEEE Commun. Mag.*, vol. 48, no. 1, pp. 33–9, Jan. 2010.

[5] J. G. Andrews, H. Claussen, M. Dohler, S. Rangan, and M. C. Reed, "Femtocells: past, present, and future," *IEEE J. Sel. Areas Commun. (JSAC)*, vol. 30, no. 3, pp. 497–508, Apr. 2012.

[6] S. Yun, Y. Yi, D.-H. Cho, and J. Mo, "Open or close: on the sharing of femtocells," in *Proc. IEEE Int. Conf. on Computer Commun. (INFOCOM)*, Apr. 2011, pp. 116–20.

[7] D. Choi, P. Monajemi, S. Kang, and J. Villasenor, "Dealing with loud neighbors: the benefits and tradeoffs of adaptive femtocell access," in *Proc. IEEE Global Telecommun. Conf. (GLOBECOM)*, Nov.–Dec. 2008.

[8] S. Joshi, R. Cheung, P. Monajemi, and J. Villasenor, "Traffic-based study of femtocell access policy impacts on HSPA service quality," in *Proc. IEEE Global Telecommun. Conf. (GLOBECOM)*, Nov. 2009, pp. 1–6.

[9] "Initial Home NodeB coexistence simulation results," Nokia Siemens Networks, 3GPP Document Reference R4-070902, 3GPP TSG-RAN WG4 Meeting 43bis, June 2007.

[10] "3rd Generation Partnership Project; Technical Specification Group Radio Access Networks; 3G Home NodeB Study Item Technical Report (Release 8)," 3GPP, 3GPP TR 25.820, Mar. 2008.

[11] "Open and closed access for Home NodeBs," Nortel, Vodafone, 3GPP document Reference R4-071231, 3GPP TSG-RAN WG4 Meeting 44, Aug. 2007.

[12] D. López-Pérez, A. Valcarce, G. D. L. Roche, E. Liu, and J. Zhang, "Access methods to WiMAX femtocells: a downlink system-level case study," in *IEEE Singapore Int. Conf. on Commun. Systems (ICCS)*, Nov. 2008, pp. 1657–62.

[13] V. Chandrasekhar, J. G. Andrews, T. Muharemovic, Z. Shen, and A. Gatherer, "Power control in two-tier femtocell networks," *IEEE Trans. Wireless Commun.*, vol. 8, no. 8, pp. 4316–28, Aug. 2009.

[14] "Cisco visual networking index: global mobile data traffic forecast update, 2010–2015," White Paper, Cisco, Feb. 2011.

[15] A. Khandekar, N. Bhushan, J. Tingfang, and V. Vanghi, "LTE advanced: heterogeneous networks," in *European Wireless Conference*, June 2010, pp. 978–82.

[16] D. López-Pérez, A. Valcarce, A. Ladanyi, G. de la Roche, and J. Zhang, "Intracell handover for interference and handover mitigation in OFDMA two-tier macrocell-femtocell networks," *EURASIP J. Wireless Commun. Networking*, vol. 2010 (2010), pp. 1–15, Feb. 2010.

[17] P. Tarasak, T. Q. S. Quek, and F. P. S. Chin, "Uplink timing misalignment in open and closed access OFDMA femtocell networks," *IEEE Commun. Lett.*, vol. 15, no. 9, pp. 926–8, Sep. 2011.

[18] T. Elkourdi and O. Simeone, "Outage and diversity-multiplexing trade-off analysis of closed and open-access femtocells," in *Proc. IEEE Global Telecommun. Conf. (GLOBECOM)*, Dec. 2010.

[19] O. Simeone, E. Erkip, and S. Shamai, "Robust communication against femtocell access failures," in *Proc. IEEE Inform. Theory Workshop (ITW)*, Oct. 2009, pp. 263–7.

[20] S. Barbarossa, A. Carfagna, S. Sardellitti, M. Omilipo, and L. Pescosolido, "Optimal radio access in femtocell networks based on Markov modeling of interferers' activity," in *Proc. IEEE Int. Conf. on Acoustics, Speech, and Sig. Proc. (ICASSP)*, May 2011, pp. 3212–15.

[21] S. Mukherjee, "UE coverage in LTE macro network with mixed CSG and open access femto overlay," in *Proc. IEEE Int. Conf. on Commun. (ICC)*, June 2011.

[22] S. Mukherjee, "Analysis of UE outage probability and macrocellular traffic offloading for WCDMA macro network with femto overlay under closed and open access," in *Proc. IEEE Int. Conf. on Commun. (ICC)*, June 2011.

[23] Y. Chen, J. Zhang, and Q. Zhang, "Utility-aware refunding framework for hybrid access femtocell network," *IEEE Trans. Wireless Commun.*, vol. 11, no. 5, pp. 1688–97, May 2012.

[24] Y. Y. Li, L. Yen, and E. S. Sousa, "Hybrid user access control in HSDPA femtocells," in *Proc. IEEE Global Telecommun. Conf. (GLOBECOM)*, Dec. 2010, pp. 679–83.

[25] V. Chandrasekhar, M. Kountouris, and J. G. Andrews, "Coverage in multi-antenna two-tier networks," *IEEE Trans. Wireless Commun.*, vol. 8, no. 10, pp. 5314–27, Oct. 2009.

[26] V. Chandrasekhar and J. G. Andrews, "Spectrum allocation in two-tier networks," *IEEE Trans. on Communications*, vol. 57, no. 10, pp. 3059–68, Oct. 2009.

[27] M.-S. Alouini and A. Goldsmith, "Area spectral efficiency of cellular mobile radio systems," *IEEE Trans. Veh. Technol.*, vol. 48, no. 4, pp. 1047–66, July 1999.

[28] F. Baccelli, B. Blaszczyszyn, and P. Muhlethaler, "An ALOHA protocol for multihop mobile wireless networks," *IEEE Trans. Inform. Theory*, vol. 8, no. 6, pp. 569–86, Feb. 2006.

[29] D. N. Knisely, T. Yoshizawa, and F. Favichia, "Standardization of femtocells in 3GPP," *IEEE Commun. Mag.*, vol. 47, no. 9, pp. 68–75, Sep. 2009.

[30] D. N. Knisely and F. Favichia, "Standardization of femtocells in 3GPP2," *IEEE Commun. Mag.*, vol. 47, no. 9, pp. 76–82, Sep. 2009.

[31] P. Humblet, B. Raghothaman, A. Srinivas, S. Balasubramanian, C. Patel, and M. Yavuz, "System design of cdma2000 femtocells," *IEEE Commun. Mag.*, vol. 47, no. 9, pp. 92–100, Sep. 2009.

[32] M. Yavuz, F. Meshkati, S. Nanda, A. Pokhariyal, N. Johnson, B. Raghothaman, and A. Richardson, "Interference management and performance analysis of UMTS/HSPA+ femtocells," *IEEE Commun. Mag.*, vol. 47, no. 9, pp. 102–9, Sep. 2009.

3 Coverage analysis using the Poisson point process model

Sayandev Mukherjee

3.1 Introduction

The rapid rate of increase in data traffic means that future wireless networks will have to support a large number of users with high data rates. A promising way to achieve this is by spectrum reuse through the deployment of cells with small range, such that the same time-frequency resources may be reused simultaneously in multiple cells. At the same time, the traditional *coverage* requirement for wireless users (supporting a modest rate at cell-edge users) is most economically met with cells having large range, i.e., the traditional *macrocellular* architecture. Thus the wireless cellular networks of the future are likely to be *heterogeneous*, i.e., have one or more *tiers* of small cells *overlaid* on the macrocellular tier.

We consider the problem of network design from the point of view of a service provider considering the deployment of a network in a certain region. The primary metric we shall focus on is the *coverage* on the *downlink*: namely, the probability that a user at an arbitrary location in the deployed network has coverage. Later, we shall briefly discuss the distribution of the maximum rate that such a user could support from one of the base stations (BSs).

When we talk of the service provider *designing* a network to satisfy a certain coverage criterion, we mean the choice of deployment parameters for such a network, including:

1. Number of tiers of the network
2. Densities of the BSs in the tiers
3. Transmit powers of the BSs in the tiers.

Note that the service provider seeks to satisfy the coverage criterion with a deployment that is compatible with the service provider's targets for expected revenue, capital expenditure (CapEx) and operating expenditure (OpEx). In other words, there is a complicated *utility function* that depends on economic variables such as the pricing model for service, CapEx, OpEx, and expected revenue, in addition to the coverage criterion. In our analysis, we focus only on the relationship between the deployment parameters of the network and the coverage criterion.

Small Cell Networks: Deployment, PHY Techniques, and Resource Management, ed. Tony Q. S. Quek, Guillaume de la Roche, İsmail Güvenç, and Marios Kountouris. Published by Cambridge University Press.
© Cambridge University Press 2013.

Since coverage and capacity are both related to the signal to interference plus noise ratio (SINR) at the user's receiver, we shall focus on the statistical properties of the SINR in the sequel. Our analysis will show that there are many sets of deployment parameters that are equivalent from the view of the SINR distribution. Thus, once an operating SINR distribution has been chosen, the service provider can choose one set of deployment parameters from the corresponding equivalence class based on its utility function, which takes into account the economic variables.

3.2 Distribution of SINR

Owing to fading and interference, the SINR at an arbitrary user (user equipment (UE), in 3GPP terminology, which we adopt here) is a random variable. The cumulative distribution function (CDF) of the wideband SINR at an arbitrary user is often referred to as the *geometry* in the literature. It is a representation of the "environment" that an arbitrary UE in the network will encounter.

3.2.1 Determining the CDF of SINR via simulation

The traditional approach to determining the CDF of the SINR over all locations in the deployment region is through simulation. In fact, for a single-tier macrocellular network with "regular" placement of BSs on the familiar hexagonal lattice, the most common way to compute the interference at any location is via brute-force numerical calculation, which is implemented in the simulation. Further, for the usual values of path-loss exponents for macrocellular wireless links, it has been observed that the total downlink interference power is essentially that due to the two "rings" of BSs (6 BSs in the first ring, 12 BSs in the second ring) around the cell containing the UE. This permits the simulation to be efficient with regard to memory requirements and run time by restricting the deployment region to the "19-cell wraparound region" [1], where *wraparound* means that the upper and lower, and the right and left, boundaries of the region are assumed contiguous, so there are no edge effects in the simulation, and statistics on the SINR may be drawn from all locations in the region without introducing bias.

It is important, however, to recall that the 19-cell wraparound model for simulation was proposed for the study of macrocellular networks. Let us examine the assumptions behind this model:

1. *Regular placement of BSs*: While this assumption was never applicable to any real-world deployment, it could at least be claimed to be more-or-less accurate for macrocells, because of careful planning of cell sites by operators. However, it is increasingly inapplicable as the number of cells increases, and their size decreases, as is the case for future networks. The reason for this is that careful planning of cell sites is often impossible if the number of sites is large, and the deployment map then takes on a more "random" appearance.

2. *Interference limited to two rings of BSs*: With the placement of BSs no longer regular, the interference is effectively that due to all BSs within some distance from the UE, instead of those BSs located in certain "rings." However, the value of this effective range depends upon the parameters of the path-loss model and the transmit powers of the BSs, which are both different for the smaller cell sizes of future networks. In particular, for a macro–femto overlay network, these ranges are different for the two tiers, and a suitably sized region would need to be considered such that edge effects due to both tiers can be eliminated. This is likely to enlarge the deployment region to be simulated, increase memory requirements, and reduce the speed of the simulator.

To summarize, in future heterogeneous networks with smaller cell sizes, where the BS locations are more irregular, the simulations need to be done over larger deployment regions and will get more time consuming.

A single-tier network has only two deployment parameters: the transmit powers of the BSs, and their density (number of BSs per unit area). Thus the total number of scenarios required to be simulated for a single-tier network is not large, especially since the choices for transmit power and density of the BSs required for system operation are tightly constrained. However, each additional tier in the network multiplies the number of possibilities for the overall combination of transmit powers and densities of the various tiers. In other words, if our goal is exhaustive simulation of all feasible operating scenarios for a multi-tier heterogeneous network, then the number of scenarios to be simulated rises exponentially with the number of tiers. If one of the tiers is that of user-owned (not operator-owned) femtocells, then the number of scenarios is further augmented by the fact that some femtocells may be switched on and off by the user depending on the need for coverage at that user's premises. In other words, the layout of BSs may even become dynamic. Clearly, an exhaustive simulation-based investigation of all possible operating scenarios for a multi-tier heterogeneous network is challenging at best and infeasible at worst.

3.2.2   The role of analytic modeling

The analytic modeling-based investigation of deployment scenarios has two phases. In the first phase, we use probabilistic models for the locations of the BSs to determine analytic expressions for the CDF of the SINR in the deployment region. As we shall see, this has the benefit of providing insights into the *combinations* of deployment parameters that affect the CDF of the SINR, and therefore the different sets of deployment parameters that are *equivalent* in that they yield the same CDF of the SINR. This analytic phase allows us to sift through the large space of combinations of deployment parameters to quickly settle on certain equivalence classes of deployment parameters, each class corresponding to some desired CDF of the SINR. The service provider may then choose a set of deployment parameters from one of these equivalence classes based on its economic utility function.

In the second phase of the investigation, the shortlist of deployment scenarios (as defined by the deployment parameters) chosen in the first phase is then investigated in depth via simulation. This allows the power of detailed simulation, incorporating all relevant aspects whose behavior and impact on performance are to be investigated, to be harnessed for a few selected deployment scenarios.

We now describe in detail the analytic models that we shall employ to derive the CDF of the SINR.

3.3 The Poisson point process model for BS locations

Consider a single tier in a heterogeneous network (HetNet). Recall from the discussion above that as cell sizes get smaller and the number of cells required to be deployed in a given region gets larger, the overall map of the locations of the BSs in the deployment region begins to look more "random" in a real-world deployment. This observation can serve as the starting point for modeling the locations of the BSs as points of a *stochastic point process*. In this work, we shall focus only on BS locations in the two-dimensional plane. Given a certain number of BSs in (a subset of) the deployment region, what is the probability distribution of their locations within that region? The simplest possible probabilistic model would model the locations of these BSs as being *independent and identically distributed* (i.i.d.) and *uniform* over the region of interest. We are usually interested in subsets of the deployment region, e.g., the region served by a single BS. If the deployment region itself is finite, then the number of interfering BSs depends upon the location of the serving BS, and edge effects make the analysis of SINR distribution different for the region served by each BS. To make the analysis tractable, we propose to assume that the deployment region is infinite, i.e., the entire two-dimensional plane, while still retaining the property that conditioned on the number of BSs in any finite region, the locations of these BSs are i.i.d. uniform over that region. It remains to specify the distribution of the number of BSs in any region. Clearly, this could be modeled by any non-negative integer-valued random variable, but as will be seen below, the choice of the Poisson random variable yields great gains in tractability. In short, we shall model the BS locations for this tier in the plane $\mathbb{R}^2$ by a Poisson point process (PPP) Φ, which has the following properties:

1. The number of BSs in any finite region $\mathcal{A} \subset \mathbb{R}^2$, denoted $N_\mathcal{A}$, is a Poisson random variable with mean $\mu_\mathcal{A} \equiv \int_\mathcal{A} \lambda(x)\,dx$, where $\lambda(\cdot)$ is a non-negative-valued function of two variables, called the *intensity function* of the PPP Φ:

$$\mathbb{P}\{N_\mathcal{A} = n\} = \exp(-\mu_\mathcal{A})\frac{\mu_\mathcal{A}^n}{n!}, \quad n = 0, 1, 2, \ldots.$$

2. For two disjoint finite regions $\mathcal{A} \subset \mathbb{R}^2$ and $\mathcal{B} \subset \mathbb{R}^2$, the corresponding number of BSs, $N_\mathcal{A}$ and $N_\mathcal{B}$, are independent.
3. Given the number of BSs in any finite region $\mathcal{A} \subset \mathbb{R}^2$, i.e., conditioned on $N_\mathcal{A} = n$, say, the locations of these n BSs are i.i.d. uniform over $\mathcal{A}$.

Finally, we assume that the PPPs corresponding to BS locations in the different tiers of a HetNet are themselves independent.

In this work, we further restrict ourselves to PPP models for the locations of BSs in the tiers where the intensity function $\lambda(x)$ is a constant over the entire plane: $\lambda(x) \equiv \lambda > 0$ for all $x \in \mathbb{R}^2$. Such PPP models are called *homogeneous*. Note that for any finite region $\mathcal{A} \subset \mathbb{R}^2$, it then follows that $\mathbb{E} N_{\mathcal{A}} = \lambda \times \mathrm{area}(\mathcal{A})$, so λ (now just called the *intensity* of the homogeneous PPP) is just the mean number of points of the process per unit area, i.e., the *density* of the BSs in that tier. In the following analysis, we shall use the terms "intensity" and "density" interchangeably.

3.4 Wireless channel model

The specification of the model for BS locations in the tiers is only one part of the overall specification of the network. In order to analyze the distribution of SINR in the network, we also need to specify a model for the wireless links from the BSs in the tiers to the UE of interest. There are two components to the so-called *link loss*, i.e., the difference between transmitted and received powers on a given wireless link: the distance-dependent *path loss* (which has no random component, and is completely known if the distance between transmitter and receiver is known), and the *fading* (which is random). We now discuss the models for these two components.

3.4.1 Path-loss model

We adopt the popular slope–intercept model, which says that the path loss on the link between a BS in a given tier and an arbitrarily located UE is

$$\text{path loss in dB} = 10\alpha \log_{10}(\text{distance in m}) - 10 \log_{10} K, \qquad (3.1)$$

where $\alpha > 2$ is the *path-loss exponent* and K is a constant accounting for the differences in height of transmit and receive antennas, etc. Note that α and K both depend upon the frequency band of operation.

Note that the path-loss model only applies to the attenuation over the air, i.e., after the transmission has left the transmit antenna, and just before entering the receive antenna. The final expression for received power can account for factors such as antenna gains as decrements to the above path loss.

3.4.2 Fading model

We shall show in the subsequent sections that the calculation of both the joint and marginal complementary cumulative distribution functions (CCDFs) of the downlink SINR at an arbitrarily located UE in the network are examples of a class of problems that are made tractable by pairing the PPP model for BS locations with a particular class of i.i.d. fading processes on all links, namely a mixture of Nakagami-m [2] fading processes with positive integer-valued m. This causes the received power on all links to be a mixture of *Erlang* random variables, i.e., a mixture of gamma random variables with positive integer-valued shape parameters [3, p. 79]. Moreover, the probability density function (PDF) of an arbitrary positive-valued continuous random variable can be approximated uniformly to arbitrary accuracy by a mixture of Erlang PDFs [4, Problem 5.6, p. 14]. The distributions of received power in many wireless fading channels including lognormal shadowing are approximated in [5] using mixtures of Gamma PDFs (with non-integer order), and [6, eqns. (16)–(17)] shows how to approximate such mixtures of general Gamma PDFs by mixtures of Erlang PDFs.

Note that though it is customary in the literature to assume lognormal shadow fading on wireless links, we do not work with the lognormal distribution in our analysis.[1] Though it would be desirable to account for lognormal shadow fading as well, it greatly complicates the analysis. While the characteristic function (the Fourier transform of the PDF) of the total interference from all BSs in a tier (with no restriction on minimum distance) may be computed exactly for links with shadow fading [8, eqn. (5)] [9, eqn. (16)], it can only be approximately calculated [9, eqn. (19)] if we want the interference from all BSs more than a minimum distance away, as we shall require for our model. Further, the CCDF requires the PDF corresponding to this characteristic function, which cannot be determined in closed form, thus requiring further approximations [9, eqn. (72)] or numerical inversion of the Fourier transform. Alternative approaches include approximations for the distribution of the sum of lognormal random variables [10] [11, eqn. (12)], applying extreme value theory to estimate tail probabilities [12], or deriving bounds on the CCDF by considering only the single strongest interferer [13, 14], but none of them yield exact results.

Finally, we add that if the lognormal shadow fading is interpreted as the residual error of the linear regression model (in the logarithmic scale) yielding the slope–intercept path-loss model, then all qualitative results derived assuming the standard deviation of the lognormal shadowing is zero (i.e., no shadow fading) will continue to hold for positive standard deviation. Further, it has been observed, at least in a single-tier network, that lognormal shadowing has almost no effect on the mean interference factor [15] when the serving BS is that received strongest at the UE.

The simplest example of a mixture of Erlang PDFs is the exponential PDF, corresponding to the Rayleigh fading process. We shall see that the joint or marginal SINR CCDF for an arbitrary mixture of Erlang PDFs can be derived from the result for Rayleigh fading. For this reason, we begin by assuming i.i.d. Rayleigh fading on all links from all BSs (macro and femto) to any UE location. As we shall show, we can derive an *exact, closed-form* for the joint complementary cumulative distribution function (JCCDF) at an arbitrary UE location conditioned on the distances from the candidate serving BSs in the accessible tiers.

---

[1] Strictly speaking, the analysis for a single-tier network in [7] can handle i.i.d. lognormal fading on all links between the UE and interfering BSs, but its extension to a multi-tier HetNet requires the somewhat artificial assumption that there be no lognormal fading between the UE and all *candidate* serving BSs (one candidate serving BS per tier of the HetNet, as defined in the following section). Since any candidate serving BS that is ultimately not the BS serving the UE will then become interfering BS, this would introduce an arbitrary discrepancy in the fading model for interfering BSs, with lognormal fading on the links to non-candidate serving BSs and not on the links to candidate serving BSs.

3.5   Statement of the SINR calculation problem

With the above location model (independent homogeneous PPPs) for BSs in the tiers, and the above wireless channel model, we are ready to derive the expressions for the distribution of SINR at an arbitrarily located UE in the network. First, however, we need to define the candidate serving BSs, the criteria for their selection, the criterion for choosing the BS that will serve the UE, and some notation to represent received power from the candidate serving BSs and from all interferers.

3.5.1   Candidate serving BSs and the serving BS

We begin by considering a snapshot of the wireless network at a particular moment in time, with the locations of *co-channel* tiers of BSs in the HetNet modeled by independent homogeneous PPPs. This corresponds to the model for a single resource element (a single subcarrier over one transmission interval) in the long term evolution (LTE) standard, or to a single transmission interval for a frequency non-selective channel in the high speed packet access (HSPA) standard. Label the tiers of the network $1, \ldots, n_{\text{tier}}$, and assume that the UE is only allowed to access the BSs in tiers $1, \ldots, n_{\text{open}}$. For example, a macro–femto HetNet with a mixture of open access (OA) and closed subscriber group (CSG) femtocells would be represented as a HetNet with $n_{\text{tier}} = 3$ and $n_{\text{open}} = 2$, with tier 1 representing the macrocells, tier 2 the OA femtocells, and tier 3 the CSG femtocells.

Next, we assume that each BS in a tier transmits with the maximum power allowed for BSs in that tier. This immediately models the case of reference symbols (LTE) or pilot channels (HSPA), but also covers the case of data channels if we assume that the cells are all fully loaded. For any given UE, a single *candidate serving BS* is chosen from each tier according to some criterion, and the BS that actually serves the UE is chosen from among these candidate serving BSs according to some criterion.

3.5.2   Definition of SINR

Let Φ_i denote the homogeneous PPP representing the locations of the BSs in tier i with intensity λ_i, $i = 1, \ldots, n_{\text{tier}}$. For simplicity, we shall use Φ_i to denote both the point process itself and the set of all the points. A BS in tier i is identified by its location $b \in \Phi_i$. A BS $b \in \Phi_i$ at distance R_b from the UE transmits with power P_i^{tx} and is received at the UE with power

$$Y_b = \frac{H_b}{R_b^{\alpha_i}}, \quad H_b \sim \text{Exp}(P_i), \quad P_i = K_i P_i^{\text{tx}},$$

where the exponential distribution for H follows from the i.i.d. Rayleigh fading assumption, and K_i and α_i are the intercept and slope respectively of the slope–intercept path-loss model (3.1) describing links from BSs in tier i and the UE.

Finally, we also assume that thermal noise at the UE receiver (measured over the same bandwidth as the received powers) is N_0.

For any tier $i = 1, \ldots, n_{\text{tier}}$, the total received power at the UE from all BSs in tier i is given by

$$W_i \equiv \sum_{b \in \Phi_i} Y_b = \sum_{b \in \Phi_i} \frac{H_b}{R_b^{\alpha_i}}, \quad i = 1, \ldots, n_{\text{tier}}.$$

For each accessible tier $i = 1, \ldots, n_{\text{open}}$, let us denote the candidate serving BS from tier i by B_i. The received power at the UE from the candidate serving BS in accessible tier i is therefore $U_i \equiv Y_{B_i}$, while the interference from all other BSs in the accessible tier i is

$$V_i \equiv \sum_{b \in \Phi_i \setminus \{B_i\}} Y_b = \sum_{b \in \Phi_i \setminus \{B_i\}} \frac{H_b}{R_b^{\alpha_i}} = W_i - U_i, \quad U_i \equiv Y_{B_i}, \quad i = 1, \ldots, n_{\text{open}}.$$

One of the *candidate serving* BSs is chosen (based on any of several criteria to be defined later) as the *serving* BS. Thus the serving BS is B_I, where I is the *serving tier*. If $I = i$, then the SINR at the UE is

$$\Gamma_i \equiv \frac{U_i}{\displaystyle\sum_{\substack{j=1 \\ j \neq i}}^{n_{\text{open}}} U_j + \left[\displaystyle\sum_{k=1}^{n_{\text{open}}} V_k + \displaystyle\sum_{k=n_{\text{open}}+1}^{n_{\text{tier}}} W_k + N_0 \right]}, \quad i = 1, \ldots, n_{\text{open}}. \tag{3.2}$$

3.5.3 Marginal and joint complementary CDF (CCDF) of SINR

We focus on deriving the marginal CCDF of the SINR,

$$\mathbb{P}\{\Gamma_I > \gamma\}, \tag{3.3}$$

for all $\gamma \geq 0$ and the joint CCDF of the SINR,

$$\mathbb{P}\{\Gamma_{j_1} > \gamma_{j_1}, \ldots, \Gamma_{j_k} > \gamma_{j_k}\}, \tag{3.4}$$

for all subsets of $k \leq n_{\text{open}}$ distinct indices $(j_1, \ldots, j_k)$ such that $1 \leq j_1 < j_2 < \cdots < j_k \leq n_{\text{open}}$, and all positive-valued k-tuples $(\gamma_{j_1}, \ldots, \gamma_{j_k}) \in \mathbb{R}_+^k$. Note that (3.4) is equivalent to knowing the joint PDF of the SINR,

$$f_{\Gamma_{j_1}, \ldots, \Gamma_{j_k}}(\gamma_{j_1}, \ldots, \gamma_{j_k}) = \frac{(-1)^k \partial^k}{\partial \gamma_{j_1} \cdots \partial \gamma_{j_k}} \mathbb{P}\{\Gamma_{j_1} > \gamma_{j_1}, \ldots, \Gamma_{j_k} > \gamma_{j_k}\}. \tag{3.5}$$

3.5.4 Canonical form of joint CCDF

Observe that the joint CCDF (3.4) may be written as

$$\mathbb{P}\{\mathbf{AX} > \mathbf{Z}\boldsymbol{\rho}\}, \quad \mathbf{A} = \mathbf{I}_k - \boldsymbol{\rho}\mathbf{1}_k^{\mathsf{T}}, \quad \boldsymbol{\rho} = \left[\frac{\gamma_{j_1}}{1 + \gamma_{j_1}}, \ldots, \frac{\gamma_{j_k}}{1 + \gamma_{j_k}} \right]^{\mathsf{T}}, \quad \mathbf{1}_k = \underbrace{[1, \ldots, 1]}_{k}^{\mathsf{T}},$$

$$\tag{3.6}$$

where $\mathbf{I}_k$ is the $k \times k$ identity matrix, and

$$X = [U_{j_1}, \ldots, U_{j_k}]^{\top}, \quad Z = \sum_{l=1}^{k} V_{j_l} + \sum_{\substack{l=1 \\ l \notin \{j_1, \ldots, j_k\}}}^{n_{\text{tier}}} W_l + N_0. \tag{3.7}$$

Note that all off-diagonal entries of $\mathbf{A}$ are no greater than zero. The independence of fading on all links and the independence of BS locations across tiers mean that the received powers $\{U_{j_l}\}_{l=1}^{k}$ from the candidate serving BSs in tiers $j_1, \ldots, j_k$ are independent. Whether X and Z are independent will depend upon the criterion for choosing the candidate serving BSs in the tiers. For example, this is true if the candidate serving BS in each tier is the one that is nearest to the UE location, but not true if it is the one that is received strongest at the UE location. However, we shall see that even in the latter case, it is possible to rewrite the joint CCDF (3.4) as a sum of a finite number of terms, each of which contains a probability of the form (3.6) wherein X, Z are all independent (though they will no longer be defined as in (3.7)). Thus we will focus on computing in closed form the *canonical probability* $\mathbb{P}\{\mathbf{A}X > Z\boldsymbol{b}\}$, where all entries in X, Z are independent.

3.5.5 Specifying the location of the UE

Note that the JCCDF in (3.4) is at a single arbitrary UE location. However, we can specify the UE location in more detail while still leaving it arbitrary. For example, practical scenarios of interest usually involve UE locations that are a certain minimum distance away from the nearest BS in each tier. Further, we may only be interested in computing (3.4) at UE locations that are nearer than some maximum distance from their candidate serving BS in each tier. We thus assume that we are only interested in the distribution (3.4) at an arbitrary UE location that is at least at a distance of $d_{\min,i} \geq 0$ from the *nearest* BS of tier i, $i = 1, \ldots, n_{\text{tier}}$, and that is at most at a distance of $d_{\max,j} < d_{\min,j}$ from the candidate serving BS of tier j, $j = 1, \ldots, n_{\text{open}}$. For future reference, we therefore define

$$m_i = \pi \lambda_i d^2_{\min,i}, \quad i = 1, \ldots, n_{\text{tier}}, \tag{3.8}$$

$$M_i = \pi \lambda_i d^2_{\max,i}, \quad i = 1, \ldots, n_{\text{open}}. \tag{3.9}$$

3.6 Effectiveness of the PPP model for analysis

3.6.1 A basic result

In this section, we state and prove a key mathematical result that shows why the combination of the PPP model, together with the assumption of an Erlang mixture PDF for the received power on the wireless links, enables us to calculate in closed form the probability of a general class of events.

For any two vectors $v = [v_1, \ldots, v_n]^\top$ and $u = [u_1, \ldots, u_n]^\top$, define $v > u \Leftrightarrow v_l > u_l, l = 1, \ldots, n$, with similar definitions for $v \geq u$, $v < u$, and $v \leq u$. We also define

$$\mathbf{0}_n = [\underbrace{0, \ldots, 0}_{n}]^\top, \qquad e_k^{(n)} = [\underbrace{0, \ldots, 0}_{k-1}, 1, \underbrace{0, \ldots, 0}_{n-k}]^\top, \qquad k = 1, \ldots, n.$$

Given $u \in \mathbb{R}^n$ and $\mathcal{V} \subset \mathbb{R}^n$, we also define $u + \mathcal{V} = \{u + v : v \in \mathcal{V}\}$. With this notation, we begin with the following.

Lemma 3.1 *Given $\tilde{b} \in \mathbb{R}_+^n \setminus \{\mathbf{0}_n\}$ and an $n \times n$ Z-matrix $\mathbf{A}$ (i.e., every off-diagonal entry in $\mathbf{A}$ is ≤ 0), the set $\{x \in \mathbb{R}^n : \mathbf{A}x > \tilde{b}\} \cap (\mathbb{R}_+^n \setminus \{\mathbf{0}_n\})$ is non-empty if and only if $\mathbf{A}$ is an M-matrix (i.e., $\mathbf{A}^{-1}$ exists, and every entry in $\mathbf{A}^{-1}$ is ≥ 0).*

Proof: Suppose $\mathbf{A}$ is not an M-matrix. From (I_{28}) in [16, Ch. 6, Thm. 2.3, p. 136], it follows that if $\mathbf{A}x > \mathbf{0}_n$ for any $x \in \mathbb{R}^n$, then $x \notin \mathbb{R}_+^n \setminus \{\mathbf{0}_n\}$. Then

$$\{x \in \mathbb{R}^n : \mathbf{A}x > \tilde{b}\} \cap (\mathbb{R}_+^n \setminus \{\mathbf{0}_n\}) \subseteq \{x \in \mathbb{R}^n : \mathbf{A}x > \mathbf{0}_n\} \cap (\mathbb{R}_+^n \setminus \{\mathbf{0}_n\}) = \emptyset.$$

On the other hand, if $\mathbf{A}$ is an M-matrix, then $\mathbf{A}^{-1}$ exists and every entry of $\mathbf{A}^{-1}$ is non-negative, so

$$\{x \in \mathbb{R}^n : \mathbf{A}x > \tilde{b}\} = \mathbf{A}^{-1}\tilde{b} + \{x \in \mathbb{R}^n : \mathbf{A}x > \mathbf{0}_n\}$$
$$= \mathbf{A}^{-1}\tilde{b} + \{x \in \mathbb{R}^n : (\exists \beta \in \mathbb{R}_{++}^n)\, \mathbf{A}x = \beta\}$$
$$= \mathbf{A}^{-1}\tilde{b} + \{\mathbf{A}^{-1}\beta : \beta \in \mathbb{R}_{++}^n\} \subseteq \mathbb{R}_+^n \setminus \{\mathbf{0}_n\},$$

which is clearly non-empty. This concludes the proof. $\qquad\square$

If $\mathbf{A}$ is an M-matrix, we write $\mathbf{A}^{-1} = [a_1^{(-1)}, \ldots, a_n^{(-1)}]^\top$, where $a_k^{(-1)} = \mathbf{A}^{-1}e_k^{(n)} \in \mathbb{R}_+^n \setminus \{\mathbf{0}_n\}$ is the kth column of $\mathbf{A}^{-1}$, $k = 1, \ldots, n$. Note that $\mathbf{A}^{-1}\beta = \sum_{k=1}^n \beta_k a_k^{(-1)}$ for any $\beta = [\beta_1, \ldots, \beta_n]^\top \in \mathbb{R}^n$. Then, for any $c = [c_1, \ldots, c_n]^\top \in \mathbb{R}_{++}^n$, we have

$$\int_{\{x \in \mathbb{R}^n : \mathbf{A}x > \tilde{b}\}} \left(\prod_{k=1}^n c_k\right) \exp(-c^\top x)\, dx$$

$$= \exp\left(-c^\top \mathbf{A}^{-1}\tilde{b}\right) \int_{\{\mathbf{A}^{-1}\beta : \beta \in \mathbb{R}_{++}^n\}} \left(\prod_{k=1}^n c_k\right) \exp(-c^\top x)\, dx$$

$$= \exp\left(-c^\top \mathbf{A}^{-1}\tilde{b}\right) \det \mathbf{A}^{-1} \int_{\mathbb{R}_{++}^n} \left(\prod_{k=1}^n c_k\right) \exp\left\{-\sum_{k=1}^n \left[c^\top a_k^{(-1)}\right]\beta_k\right\} d\beta$$

$$= \left(\prod_{k=1}^n c_k\right) \exp\left(-c^\top \mathbf{A}^{-1}\tilde{b}\right) \det \mathbf{A}^{-1} \prod_{k=1}^n \int_0^\infty \exp\left\{-\left[c^\top a_k^{(-1)}\right]\beta_k\right\} d\beta_k$$

$$= \left(\prod_{k=1}^n c_k\right) \exp\left(-c^\top \mathbf{A}^{-1}\tilde{b}\right) h_{\mathbf{A}^{-1}}(c), \tag{3.10}$$

where for any arbitrary matrix $\mathbf{B} \in \mathbb{R}^{n \times n}$ with all non-negative entries,

$$h_{\mathbf{B}}(c) = \det\mathbf{B} \prod_{k=1}^{n} \frac{1}{c^{\top}\mathbf{B}e_k^{(n)}}, \quad c \in \mathbb{R}_{++}^{n}. \tag{3.11}$$

Then we have:

Lemma 3.2 *Suppose $X_1, \ldots, X_n$ are independent random variables with $X_k \sim Exp(1/c_k)$, $k = 1, \ldots, n$. Then for any $\tilde{b} \in \mathbb{R}_+^n \setminus \{\mathbf{0}_n\}$ and any $n \times n$ Z-matrix $\mathbf{A}$, we have*

$$\mathbb{P}\{\mathbf{A}X > \tilde{b}\} = \begin{cases} 0, & \text{if } \mathbf{A} \text{ is not an M-matrix,} \\ \left(\prod_{k=1}^{n} c_k\right) \exp\left(-c^{\top}\mathbf{A}^{-1}\tilde{b}\right) h_{\mathbf{A}^{-1}}(c), & \text{if } \mathbf{A} \text{ is an M-matrix.} \end{cases} \tag{3.12}$$

Lemma 3.3 *Now suppose $\tilde{b} = Zb$, where Z is a positive-valued random variable, and $b \in \mathbb{R}_+^n \setminus \{\mathbf{0}_n\}$ is a constant. Further, suppose $Z, X_1, \ldots, X_n$ are all independent. Then with the assumptions of Lemma 3.2, we have*

$$\mathbb{P}\{\mathbf{A}X > Zb\} = \mathbb{E}\left[\mathbb{P}\{\mathbf{A}X > Zb \mid Z\}\right]$$

$$= \begin{cases} 0, & \text{if } \mathbf{A} \text{ is not an M-matrix,} \\ \left(\prod_{k=1}^{n} c_k\right) h_{\mathbf{A}^{-1}}(c)\mathcal{L}_Z(c^{\top}\mathbf{A}^{-1}b), & \text{if } \mathbf{A} \text{ is an M-matrix,} \end{cases} \tag{3.13}$$

where

$$\mathcal{L}_Z(s) = \mathbb{E}[\exp(-sZ)], \quad s > 0$$

is the Laplace transform of (the PDF of) the random variable Z.

Finally, we get the following result, which is the generalization of [17, Thm. 1] to a multi-tier heterogeneous cellular network:

Theorem 3.1 *Suppose $Z, X_1, \ldots, X_n$ are independent non-negative-valued random variables and for each $k = 1, \ldots, n$, the PDF $f_{X_k}(\cdot)$ of X_k is a mixture of Erlang PDFs:*

$$f_{X_k}(x) = \sum_{l=1}^{p_k} a_{k,l} \frac{\left[c_k^{(l)}\right]^{m_k^{(l)}+1} x^{m_k^{(l)}}}{m_k^{(l)}!} \exp\left[-c_k^{(l)}x\right], \quad x \geq 0,$$

where $p_k \in \{1, 2, \ldots\}$, for each $l = 1, \ldots, p_k$, $a_{k,l} \geq 0$, $c_k^{(l)} \geq 0$, $m_k^{(l)} \in \{0, 1, \ldots\}$, and $\sum_{l=1}^{p_k} a_{k,l} = 1$. For any $b \in \mathbb{R}_+^n \setminus \{\mathbf{0}_n\}$ and any $n \times n$ Z-matrix $\mathbf{A}$,

$$\mathbb{P}\{AX > Zb\}$$

$$
= \begin{cases}
0, & \text{if } A \text{ is not an M-matrix,} \\[2ex]
\displaystyle\sum_{\substack{l=[l_1,\ldots,l_n]^\top \\ \in\{1,\ldots,p_1\}\times\cdots\times\{1,\ldots,p_n\}}} \left\{ \prod_{k=1}^{n} \frac{(-1)^{m_k^{(l_k)}} a_{k,l_k} \left[c_k^{(l_k)}\right]^{m_k^{(l_k)}+1}}{m_k^{(l_k)}!} \right\} \\[2ex]
\quad \times \dfrac{\partial^{m_1^{(l_1)}+\cdots+m_n^{(l_n)}}}{\partial(c_1^{(l_1)})^{m_1^{(l_1)}}\cdots\partial(c_n^{(l_n)})^{m_n^{(l_n)}}} h_{A^{-1}}\left(c^{(l)}\right) \mathcal{L}_Z\left(\left[c^{(l)}\right]^\top A^{-1} b\right), & \text{if } A \text{ is an M-matrix.}
\end{cases}
$$

$$(3.14)$$

Proof: From the independence of $X_1, \ldots, X_n$, we have their joint PDF given by

$$
f_X(x) = \prod_{k=1}^{n} f_{X_k}(x_k) = \prod_{k=1}^{n} \sum_{l_k=1}^{p_k} a_{k,l_k} \frac{\left[c_k^{(l_k)}\right]^{m_k^{(l_k)}+1} x_k^{m_k^{(l_k)}}}{m_k^{(l_k)}!} \exp\left[-c_k^{(l_k)} x_k\right]
$$

$$
= \sum_{\substack{l=[l_1,\ldots,l_n]^\top \\ \in\{1,\ldots,p_1\}\times\cdots\times\{1,\ldots,p_n\}}} \left\{ \prod_{k=1}^{n} a_{k,l_k} \frac{\left[c_k^{(l_k)}\right]^{m_k^{(l_k)}+1} x_k^{m_k^{(l_k)}}}{m_k^{(l_k)}!} \right\} \exp\left\{ -\left[c^{(l)}\right]^\top x \right\}
$$

$$
= \sum_{\substack{l=[l_1,\ldots,l_n]^\top \\ \in\{1,\ldots,p_1\}\times\cdots\times\{1,\ldots,p_n\}}} \left\{ \prod_{k=1}^{n} \frac{(-1)^{m_k^{(l_k)}} a_{k,l_k} \left[c_k^{(l_k)}\right]^{m_k^{(l_k)}+1}}{m_k^{(l_k)}!} \right\}
$$

$$
\quad \times \frac{\partial^{m_1^{(l_1)}+\cdots+m_n^{(l_n)}}}{\partial(c_1^{(l_1)})^{m_1^{(l_1)}}\cdots\partial(c_n^{(l_n)})^{m_n^{(l_n)}}} \exp\left\{ -\left[c^{(l)}\right]^\top x \right\}, \tag{3.15}
$$

where for any $l \in \{1, \ldots, p_1\} \times \cdots \times \{1, \ldots, p_n\}$,

$$
c^{(l)} = \left[c_1^{(l_1)}, c_2^{(l_2)}, \ldots, c_n^{(l_n)}\right]^\top.
$$

We can now redo the derivation of (3.10), with the quantity in curly braces in the first line of (3.15) taking the place of $\prod_{k=1}^{n} c_k$, and applying the mixed partial derivative in (3.15) to the integral over x, to obtain (3.14). $\qquad\square$

3.6.2   Key advantage of the PPP model: calculating $\mathcal{L}_Z(s)$

We shall see in the next section that the joint and marginal CCDFs of SINR at an arbitrary UE location are examples of the class of probabilities $\mathbb{P}\{AX > Zb\}$ discussed in this section, with the entries of X corresponding to the received power at the UE from certain (candidate serving) BSs, and Z being the total (interference) power at the UE location from all (other) BSs. Thus, these CCDFs can be computed in closed form provided the Laplace transform $\mathcal{L}_Z(s)$ is available in closed form. We now show that *$\mathcal{L}_Z(s)$ can indeed be computed in closed form when the locations of interferers in each* *tier are modeled by points of a PPP.* Further, it is not available in closed form (and is usually computed numerically from simulation) if the interferers are located at points of a hexagonal lattice. This is the tractability advantage of the PPP location model.

To illustrate this, consider any tier i and all BSs in this tier that are at least at distance d from the UE location (assumed without loss of generality to be at the origin). The Laplace transform of the total received power at the UE from all such BSs in this tier is

$$\mathbb{E}\exp\left[-s\sum_{b\in\Phi_i:R_b>d}Y_b\right] = \mathbb{E}\left[\prod_{b\in\Phi_i:R_b>d}\exp\left(-s\frac{H_b}{R_b^{\alpha_i}}\right)\right]$$

$$= \mathbb{E}_{\Phi_i}\left\{\prod_{b\in\Phi_i:R_b>d}\mathbb{E}_{H_b}\left[\exp\left(-s\frac{H_b}{R_b^{\alpha_i}}\right)\right]\right\} = \mathbb{E}_{\Phi_i}\left[\prod_{b\in\Phi_i}g_s(b)\right],$$

$$(3.16)$$

where we have used the independence of fading on the links, and also the independence of $\{H_b\}$ and $\{R_b\}$. Further,

$$g_s(b) = \mathbb{E}_{H_b}\left[\exp\left(-s\frac{H_b}{R_b^{\alpha_i}}1\{R_b>d\}\right)\right] = \mathcal{L}_H\left(\frac{s}{R_b^{\alpha_i}}1\{R_b>d\}\right), \quad b\in\Phi_i,$$

and H is a random variable with the same distribution as H_b for any $b\in\Phi_i$. For example, if H is an Erlang random variable with PDF

$$f_H(x) = \sum_{l=1}^{p}a_l\frac{c_l^{m_l+1}x^{m_l}}{m_l!}\exp(-c_lx), \; x\geq 0, \quad a_1,\dots,a_p\geq 0, \sum_{l=1}^{p}a_l=1,$$

then its Laplace transform is

$$\mathcal{L}_H(s) = \mathbb{E}\,e^{-sH} = \int_0^{\infty}e^{-sx}f_H(x)\,\mathrm{d}x = \sum_{l=1}^{p}a_l\left(\frac{c_l}{s+c_l}\right)^{m_l+1} \quad s\geq 0.$$

Finally, we evaluate (3.16) from Campbell's theorem [18, eqn. (3.35), p. 33]:

$$\mathbb{E}_{\Phi_i}\left[\prod_{b\in\Phi_i}g_s(b)\right] = \exp\left\{-2\pi\lambda_i\int_0^{\infty}r\left[1-\mathcal{L}_H\left(\frac{s}{r^{\alpha_i}}1\{r>d\}\right)\right]\mathrm{d}r\right\}$$

$$= \exp\left\{-2\pi\lambda_i\int_0^{\infty}r\left[1-\mathcal{L}_H\left(\frac{s}{r^{\alpha_i}}\right)\right]1\{r>d\}\,\mathrm{d}r\right\}$$

$$= \exp\left\{-2\pi\lambda_i\int_d^{\infty}r\left[1-\mathcal{L}_H\left(\frac{s}{r^{\alpha_i}}\right)\right]\mathrm{d}r\right\}. \qquad (3.17)$$

The Laplace transform of the total received power from all BSs in multiple tiers is the product of terms of the form (3.17) over all the tiers being considered.

3.6.3   Determining when a Z-matrix is an M-matrix

Note that Theorem 3.1 relies on the condition that the $n\times n$ Z-matrix $\mathbf{A}$ be an M-matrix. From (E_{17}) in [16, Ch. 6, Thm. (2.3), p. 135], this is equivalent to the condition that for every $k = 1, \ldots, n$, $\det\mathbf{A}_{[k]} > 0$, where $\mathbf{A}_{[k]}$ is the $k \times k$ submatrix of $\mathbf{A}$ comprising only the entries in the first k rows and first k columns. Note that $\mathbf{A}_{[n]} = \mathbf{A}$. For the CCDF problems we are interested in, we shall see that the Z-matrix $\mathbf{A}$ satisfies

$$\det\mathbf{A}_{[1]} > \det\mathbf{A}_{[2]} > \cdots > \det\mathbf{A}_{[n]} = \det\mathbf{A}, \tag{3.18}$$

hence such a matrix $\mathbf{A}$ is an M-matrix if and only if $\det\mathbf{A} > 0$.

3.7　Expressions for joint and marginal CCDF of SINR

For brevity, all derivations in this section assume i.i.d. Rayleigh fading between any BS and the UE location of interest. From Theorem 3.1, the CCDFs when the i.i.d. fading processes are mixtures of Nakagami-m processes with positive integer-valued m (e.g., when multi-user multiple input multiple output (MU-MIMO) with zero-forcing (ZF) beamforming is employed [19]) can be obtained from the expressions for the CCDFs for Rayleigh fading.

First, let us define some notation that will simplify the expressions that will be presented later.

Given any $(\gamma_{j_1}, \ldots, \gamma_{j_k}) \in \mathbb{R}^k_{++}$, we define

$$\rho_{j_l} = \frac{\gamma_{j_l}}{1 + \gamma_{j_l}}, \quad l = 1, \ldots, k.$$

Define $\mathbb{N} = \{1, 2, \ldots\}$. For any $\boldsymbol{n} = [n_1, \ldots, n_k]^\top \in \mathbb{N}^k$, define

$$\xi(\boldsymbol{n}) = \frac{1}{1 - \sum_{l=1}^k n_l \rho_{j_l}}.$$

Given the path-loss exponents, we also define

$$\delta_i = \frac{2}{\alpha_i}, \quad i = 1, \ldots, n_{\text{tier}},$$

and the quantities

$$\kappa = \lambda_1 P_1^{\delta_1} + \cdots + \lambda_{n_{\text{tier}}} P_{n_{\text{tier}}}^{\delta_{n_{\text{tier}}}} \tag{3.19}$$

and

$$\beta_i = \frac{\lambda_i P_i^{\delta_i}}{\kappa}, \quad i = 1, \ldots, n_{\text{tier}}. \tag{3.20}$$

For future reference, we also define the function

$$G_\delta(x) = \int_x^\infty \frac{1}{1 + t^{1/\delta}}\, dt, \qquad\qquad x \geq 0$$

$$= \begin{cases} 1/\text{sinc}(\delta), & x = 0, \\ \left(\dfrac{\delta}{1-\delta}\right)\left(\dfrac{x}{1+x^{1/\delta}}\right) {}_2F_1\left(1, 1; 2 - \delta; \dfrac{1}{1+x^{1/\delta}}\right), & x > 0, \end{cases} \tag{3.21}$$

where $\mathrm{sinc}(\cdot)$ and $_2F_1(\cdot,\cdot;\cdot;\cdot)$ are the sinc and hypergeometric functions, respectively:

$$\mathrm{sinc}(z) = \frac{\sin(\pi z)}{\pi z}, \qquad _2F_1(a,b;c;z) = 1 + \sum_{n=1}^{\infty} \frac{z^n}{n!} \prod_{m=0}^{n-1} \frac{(a+m)(b+m)}{c+m}.$$

From direct computation or as a special case of the above, we also have $G_{1/2}(x) = \cot^{-1} x, \, x \geq 0$.

3.7.1 Joint CCDF: candidate serving BS is "nearest"

We begin with the case where the candidate serving BS from each accessible tier is the BS in that tier that is nearest to the UE. In other words, for tier $i \in \{1, \ldots, n_{\mathrm{open}}\}$, the candidate serving BS is

$$B_i = \arg\min_{b \in \Phi_i} R_b, \quad i = 1, \ldots, n_{\mathrm{open}}.$$

Define $X_i = U_i$, the received power at the UE from the candidate serving BS in accessible tier i, $i = 1, \ldots, n_{\mathrm{open}}$. Further, given $1 \leq j_1 < \cdots < j_k \leq n_{\mathrm{open}}$, define Z to be the total interference plus thermal noise power at the UE from all BSs other than the candidate serving BSs from tiers $j_1, \ldots, j_k$:

$$Z = \sum_{l=1}^{k} V_{j_l} + \sum_{\substack{l=1 \\ l \notin \{j_1, \ldots, j_k\}}}^{n_{\mathrm{tier}}} W_l + N_0.$$

Then, conditioned on the distances to the candidate serving BSs in these tiers, the joint CCDF (3.4) is seen to be

$$\mathbb{P}\{\Gamma_{j_1} > \gamma_{j_1}, \ldots, \Gamma_{j_k} > \gamma_{j_k}\} = \mathbb{P}\{\mathbf{AX} > Z\boldsymbol{\rho}\},$$

where $X = [X_{j_1}, \ldots, X_{j_k}]^{\top}$, $\boldsymbol{\rho} = [\rho_{j_1}, \ldots, \rho_{j_k}]^{\top}$, and

$$\mathbf{A} = \mathbf{I}_k - \boldsymbol{\rho}\mathbf{1}_k^{\top}, \quad \mathbf{1}_k = [\underbrace{1, \ldots, 1}_{k}]^{\top}, \tag{3.22}$$

where $\mathbf{I}_k$ is the $k \times k$ identity matrix. It is easily seen that (3.18) holds for the Z-matrix $\mathbf{A}$. The joint CCDF of the SINR in this case was first determined for a single-tier network in [7] and subsequently generalized to two- and three-tier HetNets in [20] and then to HetNets with arbitrary numbers of tiers in [21].

Theorem 3.2 *Conditioned on the distances to the candidate serving BSs being* $R_{j_l} \equiv R_{B_{j_l}} = r_l (\geq d_{\mathrm{min},\,j_l})$, $l = 1, \ldots, k$, *we have*

$$\mathbb{P}\{\Gamma_{j_1} > \gamma_{j_1}, \ldots, \Gamma_{j_k} > \gamma_{j_k} \mid R_{j_1} = r_1, \ldots, R_{j_k} = r_k\}$$

$$= [\xi(\mathbf{1}_k)]_{+} \frac{\exp\left[-L\left(s, u_1, \ldots, u_k\right)\right]}{\prod_{l=1}^{k}\left[1 + s\left(\dfrac{\pi \kappa \beta_{j_l}}{u_l}\right)^{1/\delta_{j_l}}\right]}\Bigg|_{s=s_{\xi(\mathbf{1}_k),\boldsymbol{\rho}}(u_1,\ldots,u_k)},$$

$$\tag{3.23}$$

where $[x]_+ = \max\{x, 0\}$ *and*

$$u_l = \pi \lambda_{j_l} r_l^2, \quad l = 1, \dots, k, \tag{3.24}$$

$$L(s, u_1, \dots, u_k) = s N_0 + \sum_{l=1}^{k} \beta_{j_l} \pi \kappa s^{\delta_{j_l}} G_{\delta_{j_l}} \left(\frac{u_l}{\beta_{j_l} \pi \kappa s^{\delta_{j_l}}} \right)$$

$$+ \sum_{\substack{i=1 \\ i \neq j_1, \dots, j_k}}^{n_{\text{tier}}} \beta_i \pi \kappa s^{\delta_i} G_{\delta_i} \left(\frac{m_i}{\beta_i \pi \kappa s^{\delta_i}} \right), \tag{3.25}$$

$$\boldsymbol{\rho} = [\rho_{j_1}, \dots, \rho_{j_k}]^{\top}, \tag{3.26}$$

and for any $\xi > 0$ *and* $\boldsymbol{\mu} = [\mu_1, \dots, \mu_k]^{\top} \in \mathbb{R}_+^k,$

$$s_{\xi, \boldsymbol{\mu}}(u_1, \dots, u_k) = \xi \left[\mu_1 \left(\frac{u_1}{\pi \kappa \beta_{j_1}} \right)^{1/\delta_{j_1}} + \dots + \mu_k \left(\frac{u_k}{\pi \kappa \beta_{j_k}} \right)^{1/\delta_{j_k}} \right]. \tag{3.27}$$

Note that from [21, eqn. (25)], $\exp[-L(s, u_1, \dots, u_k)]$ is the Laplace transform (evaluated at s) of the total thermal noise plus interference power from all BSs other than the candidate serving BSs from tiers $j_1, \dots, j_k$. Recall that for any $\boldsymbol{u} = [u_1, \dots, u_k]^{\top}$, $\|\boldsymbol{u}\|_1 = \sum_{l=1}^{k} |u_l|$.

Since the candidate serving BS from each accessible tier is simply the BS in that tier that is nearest to the UE location, we have the PDF of R_i given by [18, eqn. (2.35), p. 21]

$$f_{R_i}(r) = 2\pi \lambda_i r \, \exp\left\{-\lambda_i \pi r^2\right\}, \, r \geq 0, \quad i = 1, \dots, n_{\text{open}}. \tag{3.28}$$

Since the PPPs modeling BS locations in the tiers are assumed independent, the joint PDF of $R_{j_1}, \dots, R_{j_k}$ is the product of the marginal PDFs given by (3.28), and thus

$$f_{R_{j_1}, \dots, R_{j_k}}(r_1, \dots, r_k) \, dr_1 \cdots dr_k = \prod_{l=1}^{k} f_{R_{j_l}}(r_l) \, dr_l = \prod_{l=1}^{k} \exp(-u_l) \, du_l.$$

Thus the unconditional joint complementary CDF is given by integrating the right hand side of (3.23) with respect to this joint PDF over the range of distances $[d_{\min, j_1}, d_{\max, j_1}] \times \cdots \times [d_{\min, j_k}, d_{\max, j_k}]$:

$$\mathbb{P}\{\Gamma_{j_1} > \gamma_{j_1}, \dots, \Gamma_{j_k} > \gamma_{j_k}\} = [\xi(\mathbf{1}_k)]_+$$

$$\times \int_{m_{j_1}}^{M_{j_1}} \cdots \int_{m_{j_k}}^{M_{j_k}} \frac{\exp[-L\left(s_{\xi(\mathbf{1}_k), \boldsymbol{\rho}}(u_1, \dots, u_k), u_1, \dots, u_k\right) - \sum_{l=1}^{k} u_l]}{\prod_{l=1}^{k} \left[1 + s_{\xi(\mathbf{1}_k), \boldsymbol{\rho}}(u_1, \dots, u_k) \left(\frac{\pi \kappa \beta_{j_l}}{u_l} \right)^{1/\delta_{j_l}} \right]} \, du_k \cdots du_1.$$

$$\tag{3.29}$$

This result can be generalized to the case where the measured interference power from all tiers other than $j_1, \dots, j_k$ is filtered at the UE receiver by a length-T finite impulse response (FIR) filter (still assuming i.i.d. Rayleigh fading across all links at any given time, but the fading between the user and a given base station may have arbitrary autocorrelation across the T samples). This extension is possible because we only need the Laplace transform of the total interference power from all such tiers, and this has been computed for the weighted average above in [22].

Note that although the joint CCDF conditioned on the distances to the candidate serving BSs, given by (3.23), is in closed form, the unconditional joint CCDF (3.29) is not. However, if $\gamma_{j_l} \geq 1$, $l = 1, \ldots, k$, then $[\xi(\mathbf{1}_k)]_+ = 0$ unless $k = 1$. Further, if we consider the slightly simplified case where thermal noise is assumed negligible ($N_0 = 0$), and $d_{\min,i} = 0$, $i = 1, \ldots, n_{\text{tier}}$ and $d_{\max,j} = \infty$, $j = 1, \ldots, n_{\text{open}}$, then for $k = 1$, the one-dimensional integral in (3.29) can be evaluated in closed form [21, eqn. (42)].

3.7.2 Joint CCDF: candidate serving BS is "strongest"

Now let us consider the case where the candidate serving BS B_i from any accessible tier $i \in \{1, \ldots, n_{\text{open}}\}$ is the one for which the received power at the UE is the strongest among all BSs from that tier:

$$Y_{B_i} = \max_{b \in \Phi_i} Y_b, \quad i = 1, \ldots, n_{\text{open}}.$$

In this case it is easy to see that the SINR at the UE when receiving from B_i is also the maximum SINR when receiving from any BS in tier i, $i = 1, \ldots, n_{\text{open}}$. In other words, the candidate serving BS in each accessible tier is the *max-SINR* BS in that tier. The SINR at the UE when served by this BS B_i is given by (3.2), as before.

As shown in [23], the joint CCDF of $\Gamma_{j_1}, \ldots, \Gamma_{j_k}$ can be written using the inclusion-exclusion formula in terms of integrals over the distances from the BSs of probabilities of the form $\mathbb{P}\{\mathbf{A}\mathbf{X} > Z\tilde{\rho}\}$, where if $\mathbf{X}$ is of dimension $n = n_1 + \cdots + n_k$, the first n_1 entries are the received powers at the UE from BSs in tier j_1, the next n_2 entries are the received powers at the UE from BSs in tier j_2, and so on. Further, $Z = \sum_{l=1}^{n_{\text{tier}}} W_l$ is the total power at the UE from all BSs in all tiers, $\tilde{\rho}$ is as defined in (3.33), and $\mathbf{A} = \mathbf{I}_n - \tilde{\rho}\mathbf{1}_n^\top$. The joint distribution of SINR in this case was derived in [23] for $d_{\min,i} = 0$ and $\alpha_i = \alpha$, $i = 1, \ldots, n_{\text{tier}}$, generalizing the result of [24] to apply to arbitrary positive values of $\gamma_{j_1}, \ldots, \gamma_{j_k}$. For the general case of arbitrary $d_{\min,i} \geq 0$, $i = 1, \ldots, n_{\text{tier}}$, the same steps in the derivation (see [23, eqn. (39)]) yield the following result:

Theorem 3.3 *Suppose the candidate serving BS from each accessible tier is selected as the BS from that tier that is received strongest at the UE. If we assume i.i.d. Rayleigh fading on the links from all BSs to the UE, then, with the same notation as for (3.29),*

$$\mathbb{P}\{\Gamma_{j_1} > \gamma_{j_1}, \ldots, \Gamma_{j_k} > \gamma_{j_k}\} = \sum_{\mathbf{n} = [n_1, \ldots, n_k]^\top \in \mathbb{N}^k} \frac{(-1)^{n-k}}{n_1! \cdots n_k!} [\xi(\mathbf{n})]_+$$

$$\times \int_{\tilde{m}_1}^{\tilde{M}_1} \cdots \int_{\tilde{m}_n}^{\tilde{M}_n} \frac{\exp\left[-\tilde{L}(\tilde{s}_{\xi(\mathbf{n}),\tilde{\rho}}(v_1, \ldots, v_n))\right]}{\prod_{l=1}^{n}\left[1 + \tilde{s}_{\xi(\mathbf{n}),\tilde{\rho}}(v_1, \ldots, v_n)\left(\dfrac{\pi\tilde{\kappa}\tilde{\beta}_l}{v_l}\right)^{1/\tilde{\delta}_l}\right]} \, dv_n \cdots dv_1, \quad (3.30)$$

where

$$n = \|\boldsymbol{n}\|_1 = n_1 + \cdots + n_k, \tag{3.31}$$

$$\tilde{L}(\tilde{s}) = \tilde{s} N_0 + \sum_{i=1}^{n_{\text{tier}}} \pi \kappa \beta_i \tilde{s}^{\delta_i} G_{\delta_i}\left(\frac{m_i}{\pi \kappa \beta_i \tilde{s}^{\delta_i}}\right), \tag{3.32}$$

$$\tilde{\boldsymbol{\rho}} = [\tilde{\rho}_1, \ldots, \tilde{\rho}_n]^\top = [\underbrace{\rho_{j_1}, \ldots, \rho_{j_1}}_{n_1}, \underbrace{\rho_{j_2}, \ldots, \rho_{j_2}}_{n_2}, \ldots, \underbrace{\rho_{j_k}, \ldots, \rho_{j_k}}_{n_k}]^\top, \tag{3.33}$$

with similar definitions for $\tilde{\lambda}_l, \tilde{P}_l, \tilde{\delta}_l, \tilde{m}_l, \tilde{M}_l, \, l = 1, \ldots, n,$ *and for any* $\xi > 0$ *and any* $\tilde{\boldsymbol{\mu}} = [\tilde{\mu}_1, \ldots, \tilde{\mu}_n]^\top \in \mathbb{R}_+^n,$

$$\tilde{s}_{\xi,\tilde{\mu}}(v_1, \ldots, v_n) = \xi \left[\tilde{\mu}_1 \left(\frac{v_1}{\pi \tilde{\kappa} \tilde{\beta}_1}\right)^{1/\tilde{\delta}_1} + \cdots + \tilde{\mu}_n \left(\frac{v_n}{\pi \tilde{\kappa} \tilde{\beta}_n}\right)^{1/\tilde{\delta}_n} \right], \tag{3.34}$$

$$\tilde{\kappa} = \tilde{\lambda}_1 \tilde{P}_1^{\tilde{\delta}_1} + \cdots + \tilde{\lambda}_n \tilde{P}_n^{\tilde{\delta}_n}, \tag{3.35}$$

$$\tilde{\beta}_l = \frac{\tilde{\lambda}_l \tilde{P}_l^{\tilde{\delta}_l}}{\tilde{\kappa}}, \quad l = 1, \ldots, n. \tag{3.36}$$

No thermal noise, no min/max distance constraints, and same path-loss exponent

For the slightly simplified but still practically relevant scenario where $N_0 = 0$ (i.e., thermal noise is assumed negligible compared to the total interference power at the UE receiver), $\delta_i = \delta$, $i = 1, \ldots, n_{\text{tier}}$, and $d_{\min,i} = 0$, $i = 1, \ldots, n_{\text{tier}}$ and $d_{\max,i} = \infty$, $i = 1, \ldots, n_{\text{open}}$, we can simplify the above expressions to obtain

$$\mathbb{P}\{\Gamma_{j_1} > \gamma_{j_1}, \ldots, \Gamma_{j_k} > \gamma_{j_k}\}$$

$$= \sum_{\boldsymbol{n}=[n_1,\ldots,n_k]^\top \in \mathbb{N}^k} \frac{(-1)^{n-k} \operatorname{sinc}(\delta)^n}{n \, \xi(\boldsymbol{n})^{\delta n}} [\xi(\boldsymbol{n})]_+ \, T_k(\boldsymbol{n}) \binom{n}{n_1, \ldots, n_k} \prod_{l=1}^{k} \left(\frac{\beta_{j_l}}{\rho_{j_l}^\delta}\right)^{n_l}, \tag{3.37}$$

In (3.37), if $n > 1$, $T_k(\boldsymbol{n})$ is given by

$$T_k(\boldsymbol{n}) = \delta^{n-1} \int_0^1 dv_1 \int_0^{1-v_1} dv_2 \cdots \int_0^{1-v_1-\cdots-v_{n-2}} dv_{n-1}$$

$$\times \frac{v_1^\delta}{[v_1 + \xi(\boldsymbol{n})\tilde{\rho}_1]} \frac{v_2^\delta}{[v_2 + \xi(\boldsymbol{n})\tilde{\rho}_2]} \cdots \frac{v_{n-1}^\delta}{[v_{n-1} + \xi(\boldsymbol{n})\tilde{\rho}_{n-1}]} \frac{(1 - v_1 - \cdots - v_{n-1})^\delta}{[(1 - v_1 - \cdots - v_{n-1}) + \xi(\boldsymbol{n})\tilde{\rho}_n]}, \tag{3.38}$$

and if $n = 1$ (i.e., if $k = 1$ and $n_1 = 1$),

$$T_1(1) = \frac{1}{1 + \xi(1)\rho_{j_1}} = 1 - \rho_{j_1} = \frac{1}{\xi(1)} = \frac{1}{1 + \gamma_{j_1}}. \tag{3.39}$$

Closed form for the case where $\gamma_{j_l} \geq 1, l = 1, \ldots, k$

Note that neither (3.30) nor (3.38) is in closed form. However, when $\gamma_{j_l} \geq 1, l = 1, \ldots, k$, we have $\rho_{j_l} \geq 1/2, l = 1, \ldots, k$. Then for any $\boldsymbol{n} = [n_1, \ldots, n_k] \in \mathbb{N}^k$,

$$1 - \sum_{l=1}^{k} n_k \rho_{j_l} \leq 1 - n \min_{1 \leq l \leq k} \rho_{j_l} \leq 1 - n/2 \leq 0 \text{ if } n \geq 2,$$

so $\xi(\boldsymbol{n}) > 0$ if and only if $n \equiv \|\boldsymbol{n}\|_1 = 1$, i.e. if and only if $k = 1, n_1 = 1$. In other words, $\mathbb{P}\{\Gamma_{j_1} > \gamma_{j_1}, \ldots, \Gamma_{j_k} > \gamma_{j_k}\} = 0$ unless $k = 1$, and in that case, with $n_1 = 1$, we have from (3.37) and (3.39)

$$\mathbb{P}\{\Gamma_j > \gamma\} = \mathrm{sinc}(\delta)\frac{\beta_j}{\gamma^\delta}, \quad \gamma \geq 1.$$

Thus if the SINR thresholds for coverage in the accessible tiers are $\gamma_1, \ldots, \gamma_{n_{\mathrm{open}}}$, and $\gamma_i \geq 1, i = 1, \ldots, n_{\mathrm{open}}$, the probability of coverage is available in closed form as

$$\mathbb{P}\left(\bigcup_{i=1}^{n_{\mathrm{open}}} \{\Gamma_i > \gamma_i\}\right) = \sum_{i=1}^{n_{\mathrm{open}}} \mathbb{P}\{\Gamma_i > \gamma_i\} = \mathrm{sinc}(\delta) \sum_{i=1}^{n_{\mathrm{open}}} \frac{\beta_i}{\gamma_i^\delta}, \tag{3.40}$$

which exactly matches the result [24, Cor. 1] derived under the same assumptions on $\gamma_1, \ldots, \gamma_{n_{\mathrm{open}}}$.

Further, [25, Rem. 6] shows that if in addition $n_{\mathrm{open}} = n_{\mathrm{tier}}$, then (3.40) is the probability of coverage under *arbitrary* i.i.d. fading on all links, not just Rayleigh fading. Similarly, for the case of arbitrary $\gamma_{i_1}, \ldots, \gamma_{i_k}$, (3.37) holds for arbitrary i.i.d. fading on all links if we assume no thermal noise, no min/max distance constraints, and the same path-loss exponent on all links.

Single tier: $k = 1$

For $k = 1, j_1 \equiv j \in \{1, \ldots, n_{\mathrm{open}}\}$, we obtain

$$\mathbb{P}\{\Gamma_j > \gamma\}$$

$$= (1 + \gamma) \sum_{n=1}^{\lceil \gamma^{-1} \rceil} \frac{(-1)^{n-1} [\beta_j \, \mathrm{sinc}(\delta)]^n}{n[1 - (n-1)\gamma]} \left[\frac{1}{\gamma} - (n-1)\right]^{\delta n} T_1(n), \quad j = 1, \ldots, n_{\mathrm{open}}$$

$$= F^{\mathrm{c}}(\gamma; \beta_j), \text{ say.} \tag{3.41}$$

Note that (3.41) is a finite sum, being a single term ($n = 1$) if $\gamma \geq 1$, two terms if $\gamma \geq 1/2$, etc. It has been seen from simulations [23] that the first two terms in the sum are sufficient to give good accuracy down to $\gamma \geq 1/4$, which is the lowest SINR threshold for coverage required in 3GPP [26]. Thus it is sufficient to know $T_1(1)$ (which is given by (3.39)), and $T_1(2)$, which is given by [23, App. E]

$$T_1(2) = \frac{2\sqrt{\pi}\delta^2 \Gamma(\delta)}{2^{2\delta}\Gamma\left(\delta + \frac{3}{2}\right)} \left(\frac{1-\gamma}{1+\gamma}\right)^2 \, _2F_1\left(\frac{1}{2}, 1; \delta + \frac{3}{2}; \left(\frac{1-\gamma}{1+\gamma}\right)^2\right),$$

where $\Gamma(z) = \int_0^\infty u^{z-1}\, e^{-u}\, du$ is the *Gamma function*.

Spectrum reuse vs. coverage

Now consider another multi-tier heterogeneous network with possibly different numbers n'_{open} and n'_{tier} of accessible and total tiers, respectively, and different densities $\{\lambda'_i\}$ and $\{P_i^{\text{tx}'}\}$, but the same δ and $\{K_i\}$ as before. Define $P'_i = K_i P_i^{\text{tx}'}$, $i = 1, \ldots, n'_{\text{tier}}$, and

$$\beta'_i = \frac{\lambda'_i(P'_i)^\delta}{\sum_{l=1}^{n'_{\text{tier}}} \lambda'_l(P'_l)^\delta}, \quad i = 1, \ldots, n'_{\text{open}}.$$

From (3.37), it can be shown that [23, eqn. (19)]

$$\mathbb{P}\left(\bigcup_{i=1}^{n'_{\text{open}}}\{\Gamma_i > \gamma\}\right) = F^{\text{c}}\left(\gamma; \sum_{i=1}^{n'_{\text{open}}} \beta'_i\right), \quad n'_{\text{open}} \geq \lceil \gamma^{-1} \rceil. \tag{3.42}$$

Thus the distribution of the actual SINR in the latter network when the serving BS is chosen for max-SINR connectivity without selection bias is identical to the distribution of the SINR at the user location in the former network from any accessible tier j, say, provided its "β" value β_j equals $\sum_{i=1}^{n'_{\text{open}}} \beta'_i$. In particular, if the original network had only one accessible (macrocellular, say) tier: $n_{\text{open}} = 1$, then adding more accessible tiers to obtain a total of n'_{open} accessible tiers *keeps the coverage probabilities unchanged for users in the network* so long as this "β" condition is met and there are sufficiently many accessible tiers to satisfy $n'_{\text{open}} \geq \lceil \gamma^{-1} \rceil$, where γ is the threshold SINR for coverage. We may therefore conclude that overlaying the original single accessible tier with more accessible tiers to create a heterogeneous network allows us to exploit spectrum reuse to increase system capacity with the number of tiers, while not affecting coverage.

3.7.3   Implications for system design

From (3.29) and (3.30), it follows that the joint complementary CDF of the SINR is completely described by the parameters N_0, κ, $\{\beta_i\}$, $\{\delta_i\}$, $\{m_i\}$, and $\{M_i\}$. Note also that $\{\delta_i\}$, $\{m_i\}$, and $\{M_i\}$ are all fixed for a given scenario, while N_0 is also fixed for a given class of receiver at the user terminal. Thus the adjustable parameters of the deployment, namely the set of densities $\{\lambda_i\}$ and transmit powers $\{P_i^{\text{tx}}\}$, *only affect the SINR distribution through the parameters* $\{\beta_i\}$. In particular, sets of $\{\lambda_i\}$, $\{P_i^{\text{tx}}\}$ that correspond to the same values of $\{\beta_i\}$ are equivalent in terms of the joint SINR distribution at an arbitrary user location, though they may represent very different heterogeneous networks. The network operator considering deploying such a heterogeneous network can choose from among these different sets of deployment parameters for a given set of $\{\beta_i\}$ on the basis of a utility function that accounts for revenue and the cost of deployment and operation of the network.

3.7.4 Marginal CCDF for different selection criteria for the serving BS

The (single) BS that *actually* serves the UE is chosen from among the candidate serving BSs from the accessible tiers according to some criterion. The SINR when the UE receives from this serving BS is the actual SINR at the UE. While the selection criterion for the candidate serving BS from each tier can be "nearest" or "strongest" (which is equivalent to maximum SINR at the UE), the selection criterion for the actual serving BS can also incorporate a *selection bias* in order to bias the selection of the serving BS toward candidate serving BSs from certain tiers as opposed to others. When applied to favor selection of a femtocell in a macro–femto HetNet, for example, this has the benefit of achieving greater offloading of traffic from the macrocellular network. Note that if the overall serving BS is chosen from among the candidate serving BSs from the tiers according to the "strongest" or "maximum SINR" criteria, the selection is not identical for the two criteria if selection bias is present. Further, it makes sense to select the overall serving BS according to the "nearest" criterion applied to the candidate serving BSs from the tiers only if (i) the candidate serving BSs are themselves chosen according to the "nearest" criterion from each tier, and (ii) we apply a selection bias among the tiers in order to account for the different transmit powers of the candidate serving BSs from the different tiers, and the different path-loss model parameters for the links to the UE from the candidate serving BSs.

Let the tier that actually serves the UE be denoted by the random variable I, taking values in $\{1, \ldots, n_{\text{open}}\}$. The actual SINR Γ at the UE is the SINR when receiving from the candidate serving BS of tier I:

$$\Gamma \equiv \Gamma_I.$$

We shall now derive the CCDF of Γ for several different criteria for the selection of candidate serving BSs and the serving BS from these candidate serving BSs.

With selection bias: serving BS is the "nearest"

As discussed above, the serving BS is not exactly the nearest candidate serving BS to the UE, because this would not account for the differences in path-loss models and transmit powers among the tiers of the HetNet. Instead, the tier whose candidate serving BS is chosen as the actual serving BS is given by $I = \arg\min_{1 \leq i \leq n_{\text{open}}} \tau_i R_i^{\alpha_i}$, where $R_1, \ldots, R_{n_{\text{open}}}$ are the distances of the candidate serving BSs from the UE, and the (non-random) bias term $\tau_i > 0$ accounts for the transmit power, the fixed part of the path loss, and any selection bias. This also models the case where serving BS selection is based on *averaged* (over fading) received power at the UE instead of *instantaneous* received power [27]. Recall that this selection criterion for the overall serving BS is only meaningful when paired with the "nearest" selection criterion for candidate serving BSs from the accessible tiers, i.e., when R_i is the distance from the UE to the nearest BS in tier i, $i = 1, \ldots, n_{\text{open}}$. The complementary CDF of the SINR is given by [28, Thm. 2]

$$\mathbb{P}\{\Gamma > \gamma\} = \sum_{i=1}^{n_{\text{open}}} e^{-m_i} \int_{m_i}^{M_i} \exp\left[-L\left(\gamma\left(\frac{u_i}{\pi v_i}\right)^{\frac{1}{\delta_i}}, u_1, \ldots, u_{n_{\text{open}}}\right)\right]$$

$$\times \exp\left[\sum_{j=1}^{n_{\text{open}}}\left(m_j - \max\{m_j, u_j\}\right)\right] du_i, \tag{3.43}$$

where $L(s, u_1, \ldots, u_n)$ is given by (3.25) with $k = n$, and $u_j = \lambda_j \pi\{(\tau_i/\tau_j)[u_i/(\lambda_i\pi)]^{\alpha_i/2}\}^{2/\alpha_j}$, $j = 1, \ldots, n_{\text{open}}$.

With selection bias: serving BS is the "strongest"

For this criterion to select the serving BS, the analysis is different for "nearest" and "strongest" criteria for selecting the candidate serving BS from each tier. It can be shown [28] that for the "nearest" criterion, the CCDF $\mathbb{P}\{\Gamma_i > \gamma, I = i\}$ conditioned on known distances to the UE from the candidate serving BSs in all accessible tiers is a sum of terms of the form $\mathbb{P}\{\mathbf{A}X > Z e_1^{(k)}\}$ over $k = 1, \ldots, n_{\text{open}}$, where for any k accessible tiers $j_1, \ldots, j_k$, X is the vector of received powers from the candidate serving BSs in these tiers, Z is the total noise plus interference power from all BSs other than these k candidate serving BSs, and

$$\mathbf{A} = \mathbf{I}_k + \left[(1 + \gamma^{-1})e_1^{(k)} - \boldsymbol{\theta}(j_1, \ldots, j_k)\right]\left(e_1^{(k)}\right)^{\top} - e_1^{(k)}\mathbf{1}_k^{\top},$$

where $\boldsymbol{\theta}(j_1, \ldots, j_k)$ is as defined in (3.44). It is also easy to verify (3.18) for this Z-matrix $\mathbf{A}$. Similarly, for the "strongest" criterion, the CCDF is also expressible as a sum of probabilities of the above form, where if X is of dimension $n = n_1 + \cdots + n_k$, then the first n_1 entries are the received power at the UE from n_1 BSs in tier j_1, etc., and Z is now the total noise plus received power at the UE from every BS in every tier. Following similar steps as in the proof of Theorems 3.2 and 3.3, we obtain the following results [28].

Theorem 3.4 *Suppose the overall serving BS is the candidate serving BS that is received strongest at the UE after applying bias τ_i to each accessible tier i, $i = 1, \ldots, n_{\text{open}}$, i.e., the serving BS is the candidate serving BS $B^*(I)$ from tier $I = \arg\max_{1 \le i \le n_{\text{open}}} \tau_i Y_{B^*(i)}$. Given any $\gamma > 0$, define $\rho = \gamma/(1 + \gamma)$. For any $k = 1, \ldots, n_{\text{open}}$, and any distinct $(j_1, \ldots, j_k) \in \{1, \ldots, n_{\text{open}}\}^k$, we define*

$$\boldsymbol{\theta}(j_1, \ldots, j_k) \equiv \left[1, \frac{\tau_{j_1}}{\tau_{j_2}}, \ldots, \frac{\tau_{j_1}}{\tau_{j_k}}\right]^{\top} = [\theta_1, \theta_2, \ldots, \theta_k]^{\top}, \textit{ say.} \tag{3.44}$$

Further, for a given $\boldsymbol{\theta} = [\theta_1, \theta_2, \dots, \theta_k]^\top \in \mathbb{R}_+^k$ and any $\boldsymbol{n} = [n_1, \dots, n_k]^\top \in \mathbb{N}^k$, we define

$$\zeta(\boldsymbol{n}) = \frac{1}{\rho^{-1} - \boldsymbol{n}^\top \boldsymbol{\theta}}, \quad \tilde{\boldsymbol{\theta}}(\boldsymbol{n}) = \Big[\underbrace{\theta_1, \dots, \theta_1}_{n_1}, \underbrace{\theta_2, \dots, \theta_2}_{n_2}, \dots, \underbrace{\theta_k, \dots, \theta_k}_{n_k} \Big]^\top .$$

Let Γ be the SINR at the UE when receiving from the serving BS.

1. *If the candidate serving BS from each tier is the BS in that tier that is nearest to the UE, then*

$$
\mathbb{P}\{\Gamma > \gamma\} = \sum_{k=1}^{n_{\text{open}}} (-1)^{k-1} \sum_{\substack{(j_1, \dots, j_k) \in \{1, \dots, n_{\text{open}}\}^k: \\ j_1 \ne j_l,\, l=2,\dots,k, \\ 1 \le j_2 < \cdots < j_k \le n_{\text{open}}}} [\zeta(\mathbf{1}_k)]_+
$$

$$
\times \int_{m_{j_1}}^{M_{j_1}} \cdots \int_{m_{j_k}}^{M_{j_k}} \mathrm{d}u_k \cdots \mathrm{d}u_1
$$

$$
\times \frac{\exp\left[-L\left(s_{\zeta(\mathbf{1}_k),\boldsymbol{\theta}(j_1,\dots,j_k)}(u_1, \dots, u_k), u_1, \dots, u_k \right) - \sum_{l=1}^{k} u_l \right]}{\prod_{l=1}^{k} \left[\mathbf{1}\{l > 1\} + s_{\zeta(\mathbf{1}_k),\boldsymbol{\theta}(j_1,\dots,j_k)}(u_1, \dots, u_k) \left(\frac{\pi \kappa \beta_{j_l}}{u_l} \right)^{1/\delta_{j_l}} \right]},
$$

$$\tag{3.45}$$

where we use the notation of (3.25) *and* (3.27).

2. *If the candidate serving BS from each tier is the BS in that tier that is received strongest at the UE, then*

$$
\mathbb{P}\{\Gamma > \gamma\} = \sum_{i=1}^{n_{\text{open}}} \sum_{k=1}^{n_{\text{open}}} \sum_{\substack{(j_1, \dots, j_k) \in \{1, \dots, n_{\text{open}}\}^k: \\ i \equiv j_1 \ne j_l,\, l=2,\dots,k, \\ 1 \le j_2 < \cdots < j_k \le n_{\text{open}}}} \sum_{\boldsymbol{n} = [n_1, \dots, n_k]^\top \in \mathbb{N}^k} \frac{(-1)^{n-1}}{n_1! \cdots n_k!} [\zeta(\boldsymbol{n})]_+
$$

$$
\times \int_{\tilde{m}_1}^{\tilde{M}_1} \cdots \int_{\tilde{m}_n}^{\tilde{M}_n} \mathrm{d}v_n \cdots \mathrm{d}v_1 \sum_{p=1}^{n_1} \frac{\exp\left[-\tilde{L}\left(\tilde{s}_{\zeta(\boldsymbol{n}),\tilde{\boldsymbol{\theta}}(\boldsymbol{n})}(v_1, \dots, v_n) \right) \right]}{\prod_{l=1}^{n} \left[\mathbf{1}\{l \ne p\} + \tilde{s}_{\zeta(\boldsymbol{n}),\tilde{\boldsymbol{\theta}}(\boldsymbol{n})}(v_1, \dots, v_n) \left(\frac{\pi \tilde{\kappa} \tilde{\beta}_l}{v_l} \right)^{1/\tilde{\delta}_l} \right]},
$$

$$\tag{3.46}$$

where we use the notation of (3.31), (3.32), (3.34), (3.35), *and* (3.36).

With selection bias: serving BS is "max-SINR"

The selection bias scheme is implemented by weighting the SINR from each candidate serving BS Γ_i from each accessible tier $i = 1, \dots, n_{\text{open}}$ by the bias $\tau_i \ge 1$, i.e., $I = \arg\max_{1 \le i \le n_{\text{open}}} \tau_i \Gamma_i$. The complementary CDF of the actual serving BS for the UE may be found from the joint PDF (3.5) of $\Gamma_1, \dots, \Gamma_{n_{\text{open}}}$ (for either the "nearest BS" or the "strongest BS" candidate serving BS selection criterion) as follows:

$$\mathbb{P}\{\Gamma > \gamma\} = 1 - \mathbb{P}\{\Gamma_I \le \gamma\}$$

$$= 1 - \sum_{i=1}^{n_{\text{open}}} \mathbb{P}\{\Gamma_i \le \gamma, I = i\}$$

$$= 1 - \sum_{i=1}^{n_{\text{open}}} \mathbb{P}\left\{\tau_j \Gamma_j \le \tau_i \Gamma_i \le \tau_i \gamma, j \in \{1, \dots, n_{\text{open}}\} \setminus \{i\}\right\}$$

$$= 1 - \sum_{i=1}^{n_{\text{open}}} \int_0^{\gamma} d\gamma_i$$

$$\times \int_0^{(\tau_i/\tau_1)\gamma_i} d\gamma_1 \cdots \int_0^{(\tau_i/\tau_{i-1})\gamma_i} d\gamma_{i-1} \int_0^{(\tau_i/\tau_{i+1})\gamma_i} d\gamma_{i+1} \cdots \int_0^{(\tau_i/\tau_{n_{\text{open}}})\gamma_i} d\gamma_{n_{\text{open}}}$$

$$\times f_{\Gamma_1, \dots, \Gamma_{i-1}, \Gamma_i, \Gamma_{i+1}, \dots, \Gamma_{n_{\text{open}}}}(\gamma_1, \dots, \gamma_{i-1}, \gamma_i, \gamma_{i+1}, \dots, \gamma_{n_{\text{open}}}). \tag{3.47}$$

No selection bias: serving BS is "strongest"

This is equivalent to saying that the BS that actually serves the UE is chosen as the candidate serving BS that is received with highest SINR at the UE. In other words, $I = \arg\max_{1 \le i \le n_{\text{open}}} \Gamma_i$ and the actual SINR at the UE, i.e., the SINR when receiving from the serving BS, is

$$\Gamma = \max_{1 \le i \le n_{\text{open}}} \Gamma_i.$$

Thus the complementary CDF of the SINR is

$$\mathbb{P}\{\Gamma > \gamma\} = \mathbb{P}\left(\bigcup_{i=1}^{n_{\text{open}}} \{\Gamma_i > \gamma\}\right)$$

$$= \sum_{k=1}^{n_{\text{open}}} (-1)^{k-1} \sum_{\substack{j_1, \dots, j_k: \\ 1 \le j_1 < \cdots < j_k \le n_{\text{open}}}} \mathbb{P}\{\Gamma_{j_1} > \gamma, \dots, \Gamma_{j_k} > \gamma\}, \tag{3.48}$$

with $\mathbb{P}\{\Gamma_{j_1} > \gamma, \dots, \Gamma_{j_k} > \gamma\}$ given by (3.29) and (3.30) for the "nearest BS" and "strongest BS" candidate serving BS selection criteria, respectively.

3.8 Application: camping probability in a macro–femto network

3.8.1 BS location model

We consider an LTE macrocellular deployment with an overlaid co-channel femtocellular layer. In our notation, we have $n_{\text{tier}} = 3$ (macros, OA femtos, and CSG femtos, respectively) and $n_{\text{open}} = 2$ (macros and OA femtos, respectively). The PPPs Φ_1, Φ_2, and Φ_3 for the macros, OA femtos, and CSG femtos, respectively, have intensities λ_1, λ_2, and λ_3. We may also define Φ' to be the PPP of the femto BS locations (both OA and CSG), with intensity $\lambda' = \lambda_2 + \lambda_3$. From the coloring theorem [18, p. 53], $\Phi' = \Phi_2 \cup \Phi_3$ for a model whereby each femto BS of the PPP Φ' is selected to operate either in OA mode with probability $p = \lambda_2/\lambda'$ or in CSG mode with probability $1 - p = \lambda_3/\lambda'$, independent of the other femto BSs in Φ'. Note that the analysis applies to a specific TTI, but is also applicable to the dynamic scenario wherein the BSs in the PPP Φ' independently get configured in OA mode with probability p and CSG mode with probability $1 - p$ in each transmission time interval (TTI).

3.8.2   Path-loss model

For camping, the quantity of interest is the reference signal received power (RSRP), i.e., the received power at the UE location from the pilots or *reference symbols* (to borrow LTE terminology) that are broadcast by each BS in each tier. Thus

$$P = K P_{\mathrm{RS}}, \quad P' = K' P'_{\mathrm{RS}}, \tag{3.49}$$

where P_{RS} and P'_{RS} are the RS transmit powers of the macro and femto (both OA and CSG) BSs, respectively.

We are interested in UE locations whose distance from the nearest macro BS exceeds some $r_{\min}$. Thus we have

$$d_{\min,1} = r_{\min}, \quad d_{\min,2} = d_{\min,3} = 0, \quad d_{\max,1} = d_{\max,2} = d_{\max,3} = \infty. \tag{3.50}$$

We also assume $\alpha_1 = \alpha_2 = \alpha_3 = \alpha$, say, and that the system is *interference limited*, i.e., $N_0 = 0$.

3.8.3   UE camping and outage criteria

In wireless system standards such as LTE, the camping criteria are expressed in terms of reference signal received quality (RSRQ), defined as the ratio of RSRP to the total carrier received power (including both the desired signal and interference and thermal noise), measured over the same band as the RSRP measurement. In other words, $\mathrm{RSRQ} = \mathrm{SINR}/(1 + \mathrm{SINR})$. We say that a UE can *camp on* the candidate serving macro [OA femto] BS at distance r_1 [r_2] if $\mathrm{RSRQ}_1(r_1) > \theta_1$ [$\mathrm{RSRQ}(r_2) > \theta_2$] for some threshold θ_1 [θ_2],[2] which are equivalent to the conditions $\Gamma_1 > \gamma_1$ [$\Gamma_2 > \gamma_2$], where $\gamma_1 = (1 - \theta_1)/\theta_1$ [$\gamma_2 = (1 - \theta_2)/\theta_2$]. In LTE, $\gamma_1 = -4\,\mathrm{dB}$ or higher [29]. We define the UE to be in *outage* at this location if it cannot camp on any access point (macro NodeB or femto BS operating under OA, as the case may be), i.e., if there is no access point such that $\Gamma_1 > \gamma_1$ or $\Gamma_2 > \gamma_2$ at this UE location. Since we assume that the candidate serving BS for a UE from any tier is the BS in that tier nearest to the UE location, all coverage/camping probabilities are calculated using (3.29).

---

[2] The LTE standard also requires $\mathrm{RSRP}_i(r_i) > P_{\min,i}$ for some $P_{\min,i}, i = 1, 2$, but these $P_{\min,i}$ are $-121\,\mathrm{dBm}$ or lower [29, Sec. 5.2.3.2] and are nearly always satisfied for any macro BS or OA femto BS at a UE location.

3.8.4 Probability that a UE can camp on a macro BS

The unconditional (area-wide) CCDF of reference symbol signal to interference ratio (SIR) that the UE receives from its nearest macro BS, i.e., the probability that the UE can camp on its nearest macro BS, is given by

$$
\mathbb{P}\{\Gamma_1 > \gamma\} = \frac{\exp\left(-m_1 \left\{1 + \gamma^\delta \left[G_\delta(\gamma^{-\delta}) + \beta/\mathrm{sinc}(\delta)\right]\right\}\right)}{1 + \gamma^\delta[G_\delta(\gamma^{-\delta}) + \beta/\mathrm{sinc}(\delta)]},
\tag{3.51}
$$

where $\delta = 2/\alpha$, $m_1 = \lambda_1 \pi r_{\min}^2$, and we slightly modify the definition of β in (3.20) to write

$$
\beta = \frac{\lambda'}{\lambda_1}\left(\frac{P'}{P}\right)^\delta = \frac{\lambda'}{\lambda_1}\left(\frac{K' P'_{\mathrm{RS}}}{K P_{\mathrm{RS}}}\right)^\delta.
\tag{3.52}
$$

Note from (3.51) that the probability that a UE can camp on the nearest macro BS is entirely described by the quantities m_1 and β defined in (3.52). Observe that m_1 is the mean number of macro BSs within a disk of radius $r_{\min}$. Further, β is a function of the intensity ratios λ'/λ_1 and the power ratios $P'_{\mathrm{RS}}/P_{\mathrm{RS}}$. For a given m_1, all choices of femto BS intensity λ' and transmit power P'_{RS} relative to the corresponding macro BS values that yield the same β will give the same UE probability of camping on a macro BS. Thus if the network operator chooses some values of m_1 and β (to satisfy a given macrocellular coverage requirement, say), then there is a variety of femto-cellular overlays that will have the same macrocellular coverage. An extreme example is a femto deployment giving the femto BSs the same transmit power capabilities as the macro BSs and the same density. In terms of macrocellular coverage, if $\alpha = 4$, say, this femtocellular deployment is equivalent to another deployment where the femto BSs have 20 dB less power than the macro BSs and ten times the density of the macro BSs.

3.8.5 Probability that a UE can camp on an OA femto BS

The probability that an arbitrary UE can camp on the nearest OA femto BS can be written after some manipulation as

$$
\mathbb{P}\{\Gamma_2 > \gamma\}
$$
$$
= \int_0^\infty \exp\left(-u\left\{1 + \gamma^\delta\left[G_\delta\left(\frac{1}{\gamma^\delta}\right) + \frac{\frac{1}{p} - 1}{\mathrm{sinc}(\delta)} + \frac{1}{p\beta}G_\delta\left(\frac{m_1 p \beta}{\gamma^\delta u}\right)\right]\right\}\right)\, du,
\tag{3.53}
$$

where $p = \lambda_2/\lambda' = \lambda_2/(\lambda_2 + \lambda_3)$ is the fraction of femto BSs in OA mode. Again, it is seen that for a given m_1 and fraction p of femto BSs that are in OA mode, any changes to the powers or intensities of the femto BS or macro BS processes that leaves β the same will lead to the same probability of coverage by an OA femto BS.

3.8.6   Probability that a UE can camp on a macro BS or an OA femto BS

For the case of no selection bias, i.e., setting the same threshold $\gamma < 1$ for Γ_1 and Γ_2, we can write

$$
\begin{aligned}
\mathbb{P}\{&\Gamma_1 > \gamma, \Gamma_2 > \gamma\} \\
&= \int_{r_{\min}}^{\infty} f_{R_1}(r_1) \int_0^{\infty} f_{R_2}(r_2)\, \mathbb{P}\{\Gamma_1 > \gamma, \Gamma_2 > \gamma \mid R_1 = r_1, R_2 = r_2\}\, \mathrm{d}r_2\, \mathrm{d}r_1 \\
&= \int_{m_1}^{\infty} \int_0^{\infty} \frac{p\beta[(1+\gamma)/(1-\gamma)]u\, e^{-u(1+p\beta v)}}{\left\{1 + [h(v^{-1})]^{-1/\delta}\right\}\left\{1 + [h(v)]^{-1/\delta}\right\}} \\
&\qquad \times \exp\left\{-u\left[\beta v\left(p\frac{G_\delta(h(v^{-1}))}{h(v^{-1})} + (1-p)\frac{1}{\operatorname{sinc}(\delta)h(v^{-1})}\right) + \frac{G_\delta(h(v))}{h(v)}\right]\right\}\, \mathrm{d}v\, \mathrm{d}u,
\end{aligned}
$$

$$(3.54)$$

where $u = \lambda_1 \pi r_1^2$, $v = \lambda' \pi r_2^2/(\beta u)$, and

$$
h(x) = \left[\frac{\gamma}{1-\gamma}\left(1 + x^{1/\delta}\right)\right]^{-\delta}. \tag{3.55}
$$

Finally, the probability that a UE can camp on a BS (macro or OA femto) is given by

$$
\mathbb{P}\{\Gamma_1 > \gamma \text{ or } \Gamma_2 > \gamma\} = \mathbb{P}\{\Gamma_1 > \gamma\} + \mathbb{P}\{\Gamma_2 > \gamma\} - \mathbb{P}\{\Gamma_1 > \gamma, \Gamma_2 > \gamma\}, \tag{3.56}
$$

where the terms on the right-hand side (RHS) are given by (3.51), (3.53), and (3.54), respectively. Note that the overall camping probability is also a function of m_1, β, p only since this is true of each term in (3.56).

3.8.7   Numerical results and discussion

We set $\alpha = 4$ and $\gamma_1 = \gamma_2 = \gamma = 0.4$ (i.e., $-4\,\mathrm{dB}$). We set $K' = 0.00063$ (i.e., $-32\,\mathrm{dB}$) and $K = 12.6$ (i.e., $11\,\mathrm{dB}$) from the outdoor-to-indoor propagation model [30, Sec. 20]. In Figure 3.1, we plot the overall camping probability for an arbitrary UE, given by (3.56), vs. p, the fraction of OA femto BSs, for several choices of $r_{\min}$ with $\lambda_1 = 0.1/\mathrm{km}^2$, $\lambda' = 10/\mathrm{km}^2$, $P_{\mathrm{RS}} = 33\,\mathrm{dBm}$, and $P'_{\mathrm{RS}} = 10\,\mathrm{dBm}$.

Note that once we choose the desired minimum distance to a macro BS, $r_{\min}$, to be some value in meters (m) or kilometers (km), a given value of m_1 on the plot corresponds to a macro BS density of $\lambda_1 = m_1/(\pi r_{\min}^2)$, in units of BSs/m^2 or BSs/km^2. As discussed earlier, a network operator considering a femtocellular overlay on an existing macrocellular deployment with intensity λ_1 can begin by choosing β to satisfy the macro BS camping requirement at distances $> r_{\min}$ from the nearest macro BS (i.e., for a given m_1). Further, many different combinations of femto BS power and intensity (relative to the macro BSs) could correspond to the same value of β. We see from (3.52) that for the deployment with the parameters above, the value of $\beta = 0.05$. However, the

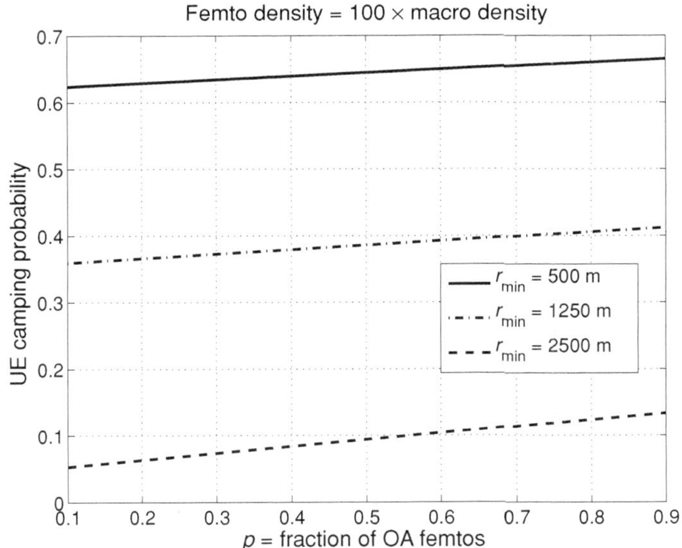

Figure 3.1 Plots of probability that UE at distance $\geq r_{min}$ from the nearest macro BS can camp on either that macro BS or the nearest OA femto BS, versus p, the mean number of macro BSs within distance r_{min} of the UE location, for three values of r_{min}. ©2012 IEEE. Reprinted, with permission, from [21].

same β could also correspond to the scenario $P'_{RS} = 23$ dBm and $\lambda' = 2.24/\text{km}^2$ (which is more characteristic of a macro–pico layout).

The optimal choice among these combinations of the dimensionless parameters m_1 and β involves factors such as the overall camping probability and some factors beyond the scope of this chapter such as the cost of femto and macro deployments. To illustrate the dependence of the overall camping probability (3.56) on p, we plot it versus m_1 for the same choices of β with $p = 0.1$ in Figure 3.2, and repeat with $p = 0.9$ in Figure 3.3. For completeness, we also include the macro BS camping probability (3.51) and the OA femto BS camping probability (3.53) versus m_1 for the same choices of β and p in Figures 3.2 and 3.3.

From Figures 3.2 and 3.3, we observe the general trend that for a given p and m_1, the OA femto BS camping probability increases with β, while the macro camping probability decreases with β. This causes the overall camping probability to decrease and then increase, but the increase is so gradual that the curve is almost flat with respect to m_1. Since for a given macro BS deployment, m_1 is equivalent to the minimum distance of the UE from a macro BS, this insensitivity of the overall camping probability to the minimum distance is desirable. We note that the overall camping probability gets flatter as p increases. For any camping probability curve, we observe that β needs to change by at least an order of magnitude to induce a significant change in the curve. Alternatively, plotting the camping probability vs. p in Figure 3.4, we note the increasing slope of the curves, i.e., greater sensitivity to p, as β increases.

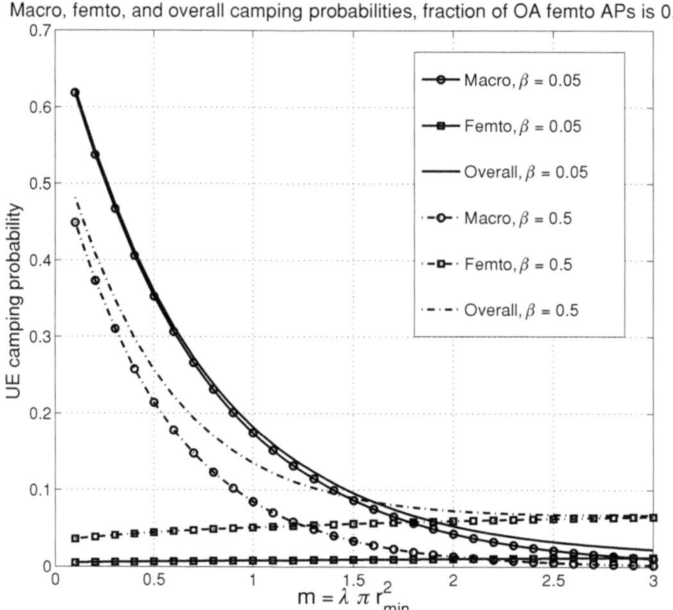

Figure 3.2 Plots of probability that UE at distance $\geq r_{\min}$ from the nearest macro BS can camp on that macro BS, the nearest OA femto BS, or either, versus m_1, the mean number of macro BSs within distance $r_{\min}$ of the UE location. We show the plots for $\beta = 0.05$ and 0.5. Here $p = 0.1$, meaning 10% of the femto BSs are in OA mode. ©2012 IEEE. Reprinted, with permission, from [21].

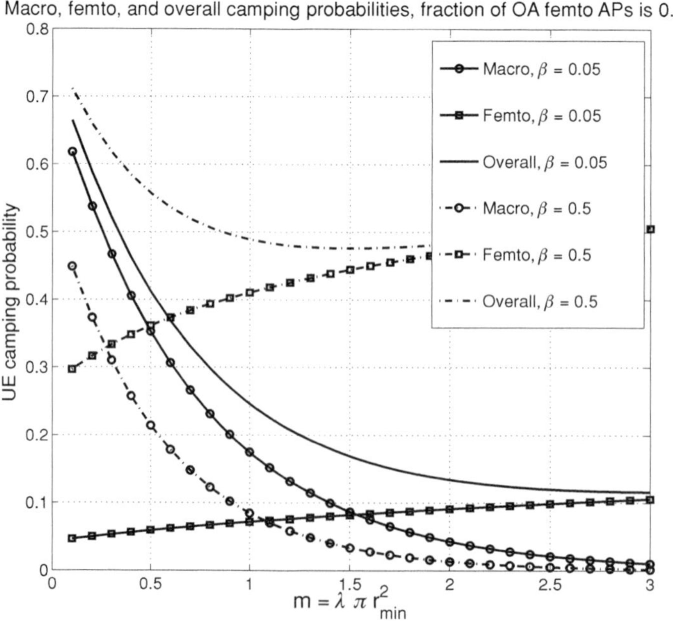

Figure 3.3 Same plots as in Figure 3.2, but with $p = 0.9$, corresponding to 90% of femto BSs operating in OA mode. ©2012 IEEE. Reprinted, with permission, from [21].

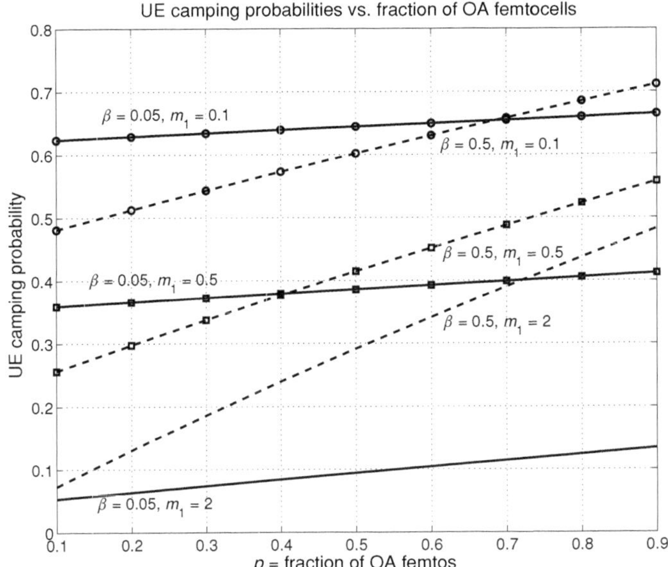

Figure 3.4 Probability that UE (at distance $\geq r_{\min}$ from the nearest macro BS, where $r_{\min}$ is such that m_1 is 0.1, 0.5, or 2), can camp on either that macro BS or the nearest OA femto BS, versus p, for $\beta = 0.05$ and 0.5. ©2012 IEEE. Reprinted, with permission, from [21].

3.9   Comparison between results for "regular" and PPP layouts

In this section we compare the SINR distributions obtained for "regular hexagonal" BS layouts with those obtained for PPP BS locations. The arguments are less rigorous and more heuristic than in the preceding sections, and are made for a single-tier network for the sake of simplicity.

3.9.1   Comparison of SIR distributions for PPP and regular BS layouts

Consider a single-tier network where the BSs are located on the standard "regular" hexagonal lattice with *inter-site distance* r_0. Assume that thermal noise is negligible and that the candidate serving BS (and therefore the actual serving BS, since this is a single tier) for any UE is the BS that is nearest to it. In the 19-cell wraparound model, we compute interference at any given UE location from the two "rings" of BSs around the serving BS, with the inner ring having 6 BSs and the outer ring having 12. Of these 18 interfering BSs, 6 each are located at distances of r_0, $2r_0$, and $\sqrt{3}r_0$, respectively, from the serving BS. Suppose r_0 is large enough to let us ignore the distance of the UE to the serving BS compared to the distance from the serving BS to the interfering BSs, so that we may assume that the UE is located at the serving BS for the purpose of computing the interference power. Then the total interference power from the 18 interfering BSs at the UE location is

$$Z_{\text{hex}} = \frac{1}{r_0^{\alpha}} \sum_{l=1}^{6} \left(\frac{H_{1,l}}{1} + \frac{H_{2,l}}{2^{\alpha}} + \frac{H_{3,l}}{3^{\alpha/2}} \right),$$

where $H_{i,1}, \ldots, H_{i,6}$ are the i.i.d. $\mathrm{Exp}(P)$ random variables representing the Rayleigh fading from the 6 interfering BSs at distances r_0 (for $i = 1$), $2r_0$ (for $i = 2$), and $\sqrt{3}r_0$ (for $i = 3$), respectively, from the serving BS. Thus the Laplace transform of Z_{hex} is

$$
\begin{aligned}
\mathcal{L}_{Z_{\text{hex}}}(s) &= \mathbb{E}\exp(-s Z_{\text{hex}}) \\
&= \left[\mathcal{L}_{H_{1,1}}\left(\frac{s}{r_0^\alpha}\right) \mathcal{L}_{H_{2,1}}\left(\frac{s}{(2r_0)^\alpha}\right) \mathcal{L}_{H_{3,1}}\left(\frac{s}{(\sqrt{3}r_0)^\alpha}\right) \right]^6 \\
&= \frac{1}{\left[\left(1 + \frac{sP}{r_0^\alpha}\right)\left(1 + \frac{sP}{r_0^\alpha}\frac{1}{2^\alpha}\right)\left(1 + \frac{sP}{r_0^\alpha}\frac{1}{3^{\alpha/2}}\right) \right]^6}.
\end{aligned}
\tag{3.57}
$$

Note that each hexagon has area $\sqrt{3}r_0^2/2$ and contains one BS. Thus the density of the BSs is $1/(\sqrt{3}r_0^2/2) = 2/(\sqrt{3}r_0^2)$.

Now suppose we use an alternative PPP location model for the BS locations, with intensity (density) $\lambda = 2/(\sqrt{3}r_0^2)$. Assume that we are at a UE location that is a maximum distance of $d_{\max} = r_0/M$ from the nearest BS (its serving BS), where $M \geq 2$. Then the total interference Z_{PPP} at the UE location from all BSs whose distance from this location is at least r_0/M has the following Laplace transform:

$$
\begin{aligned}
\mathcal{L}_{Z_{\text{PPP}}}(s) &= \exp\left[-2\pi\lambda \int_{r_0/M}^{\infty} \frac{r}{1 + r^\alpha/(sP)}\,dr \right] \\
&= \exp\left[-\pi\lambda(sP)^\delta G_\delta\left(\frac{(r_0/M)^2}{(sP)^\delta}\right) \right] \\
&= \exp\left[-\frac{2\pi}{\sqrt{3}}\left(\frac{sP}{r_0^\alpha}\right)^{2/\alpha} G_{2/\alpha}\left(\frac{1}{M^2(sP/r_0^\alpha)^{2/\alpha}}\right) \right].
\end{aligned}
\tag{3.58}
$$

Note that P/r_0^α is the received power from a BS at a location at a distance of r_0, the inter-site distance, from the transmitting BS, and should therefore be low, i.e., $P/r_0^\alpha \ll 1$. The Laplace transforms $\mathcal{L}_{Z_{\text{PPP}}}(s)$ and $\mathcal{L}_{Z_{\text{hex}}}(s)$ are plotted as functions of s for the example of $\alpha = 4$ and $P/r_0^2 = 0.01$ in Figure 3.5. Observe from Figure 3.5 that we have

$$
\mathcal{L}_{Z_{\text{PPP}}}(s) \leq \mathcal{L}_{Z_{\text{hex}}}(s) \text{ for all } s \geq 0,
$$

for this choice of α and P/r_0^α, and this property can be verified directly for other choices of α and small P/r_0^α by plotting the two functions. By definition [31] this corresponds to the following *Laplace transform ordering* of the random variables Z_{hex} and Z_{PPP}:

$$
Z_{\text{hex}} \leq_{\text{Lt}} Z_{\text{PPP}}.
$$

It is known [31] that the above Laplace transform ordering is equivalent to

$$
\mathbb{E}[g(Z_{\text{hex}})] \geq \mathbb{E}[g(Z_{\text{PPP}})]
$$

for all *completely monotone* (c.m.) functions $g : \mathbb{R}_+ \to \mathbb{R}$. For a UE at a given distance $R = r$ from the serving BS in either the regular hexagonal lattice scenario or the PPP

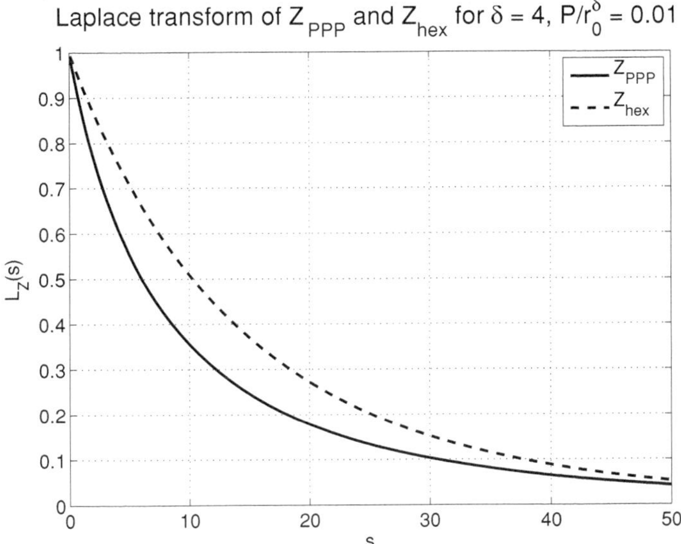

Figure 3.5 Plots of the Laplace transforms $\mathcal{L}_{Z_{\mathrm{PPP}}}(s)$ and $\mathcal{L}_{Z_{\mathrm{hex}}}(s)$ for $\alpha = 4$ and $P/r_0^\alpha = 0.01$.

scenario, the received power from the serving BS, U_{hex} and U_{PPP}, respectively, are identically distributed. Let us condition on this received power from the serving BS taking the value U (in either placement scenario for the BS locations). Then, for any $\gamma > 0$, the function

$$g(x) = 1\left\{\frac{U}{x} > \gamma\right\} = 1\left\{x < \frac{U}{\gamma}\right\}, \quad x > 0,$$

is trivially c.m. Thus from the above equivalence, we obtain

$$\mathbb{P}\{\Gamma_{\mathrm{hex}} > \gamma \mid U\} \geq \mathbb{P}\{\Gamma_{\mathrm{PPP}} > \gamma \mid U\} \text{ for all } \gamma > 0,$$

where $\Gamma_{\mathrm{hex}} = U/Z_{\mathrm{hex}}$ and $\Gamma_{\mathrm{PPP}} = U/Z_{\mathrm{PPP}}$ are, respectively, the SIRs at the UE for the two BS location scenarios. Taking another expectation over the distribution of U then yields

$$\mathbb{P}\{\Gamma_{\mathrm{hex}} > \gamma\} \geq \mathbb{P}\{\Gamma_{\mathrm{PPP}} > \gamma\} \text{ for all } \gamma > 0.$$

which agrees with earlier results obtained via simulation [7, Figure 6]. It is to be expected that the total interference power from PPP-located BSs will be larger than that in the regular lattice placement model, since the latter imposes a larger minimum distance from the interferers, and only adds 18 interference values instead of the infinitely many in the PPP placement model.

From direct calculation, we have the mean (averaged over the i.i.d. Rayleigh fading) interference power at the UE location due to these 18 interfering BSs to be

$$\bar{Z}_{\mathrm{hex}} = \frac{6P}{r_0^\alpha}\left(1 + \frac{1}{2^\alpha} + \frac{1}{3^{\alpha/2}}\right).$$

Further, the mean (over i.i.d. Rayleigh fading and the random locations of interferers) interference from the PPP-located interfering BSs at the UE location is

$$\bar{Z}_{\text{PPP}} = -\frac{\mathrm{d}}{\mathrm{d}s}\mathcal{L}_{Z_{\text{PPP}}}(s)\Big|_{s=0} = \pi\lambda P \int_{(r_0/M)^2}^{\infty} u^{-\alpha/2}\,\mathrm{d}u$$

$$= \frac{2\pi\lambda P}{\alpha-2}\left(\frac{M}{r_0}\right)^{\alpha-2} = \frac{2\pi M^{\alpha-2}}{3\sqrt{3}(\alpha-2)}\frac{6P}{r_0^{\alpha}}. \tag{3.59}$$

Thus the ratio of the mean interference power at the UE locations from the PPP and regular models is

$$10\log_{10}\left(\frac{\bar{Z}_{\text{PPP}}}{\bar{Z}_{\text{hex}}}\right) = 10(\alpha-2)\log_{10}M + 10\log_{10}\left[\frac{\frac{2\pi}{3\sqrt{3}}}{(\alpha-2)\left(1+\frac{1}{2^{\alpha}}+\frac{1}{3^{\alpha/2}}\right)}\right]\ \text{dB}.$$

For $\alpha = 4$ and $M = 2$, this ratio is approximately 3 dB. This is also consistent with simulation results in the literature, e.g., Figure 6 in [7], where a near-constant 3 dB difference between the complementary CDF curves of Z_{PPP} and Z_{hex} is seen. This permits a "calibration" of the analytical results obtained from the PPP BS location model versus the results from simulations for the regular hexagonal BS placement.

3.9.2 Comparison of achievable rates to UEs for the PPP and regular BS location models

Consider a single-tier network for now, and assume that the BSs transmit complex symbols from a finite constellation, e.g., a quadrature amplitude modulation (QAM) constellation such as quadrature phase shift keying (QPSK). In the regular hexagonal layout model for the BSs, the baseband equivalent complex total interference signal from the 18 interfering BSs can be approximated as additive circularly symmetric complex Gaussian noise using the central limit theorem (since the Gaussian limit distribution is quite accurate for a sum of 18 terms). Then the maximum rate on the downlink to any given UE can be calculated using the Shannon formula, as $\log_2(1 + \Gamma_{\text{hex}})$ bits/s/Hz.

Now, recall that all the analysis in this chapter has been for a "snapshot" of a network deployment, i.e., at one instant in time. For a given set of BS locations, this analysis may also be assumed to apply over a time duration (say T) short enough that the fading on the various links may be assumed not to vary over that duration. For such a given deployment of BSs (drawn from the PPP), the expressions derived in Section 3.7.4 are the *area-averaged* distributions of SINR under various association criteria. In other words, if we were to sample a single instance of the BS deployment at a very large number of locations, this would be the empirical CDF of the SINRs at those locations. Now, given a snapshot of the BS deployment, the same argument as for the regular hexagonal layout says that the maximum achievable rate to any UE whose SINR is Γ is $\log_2(1 + \Gamma)$ bits/s/Hz. Note that this is the achievable rate if the channels (i.e., locations of the BSs, and the fades on the links) do not vary over the duration of the transmission, i.e., the transmission time is short relative to the time duration T described above. Subject to this assumption, therefore, $\mathbb{P}\{\log_2(1 + \Gamma) > x\} = \mathbb{P}\{\Gamma > 2^x - 1\}$, where $\mathbb{P}\{\Gamma > \gamma\}$ is given by the expressions derived in 3.7.4, is the *area-averaged* distribution of the achievable rates to UEs under the corresponding association criteria. In particular, $\mathbb{E}[\log_2(1 + \Gamma)]$ is the area-averaged *mean downlink rate*.

It is important to understand, however, that notwithstanding its resemblance to the formula for *ergodic rate*, $\mathbb{E}[\log_2(1 + \Gamma)]$ is *not* the ergodic rate to any single UE. The reason is that the definition of ergodic rate assumes that transmissions occur over such a long time as to encounter every possible state of the channel, and this is exactly the opposite of the assumption discussed above for the calculation of the distribution of Γ. The ergodic rate calculation applies, for example, to a scenario where a given UE is moving over a space-filling curve trajectory through the deployment region, traveling fast enough that on the link between the UE and any BS, we may assume independent block fading over successive intervals of duration T. Unfortunately, when averaged over all PPP locations of the BSs, i.e., when averaged over all states of all channels of the system, the baseband equivalent complex total interference signal from all BSs is not Gaussian but symmetric alpha-stable [9]. This has also been verified by simulation in a single-tier network [32]. Further, this distribution has "alpha" < 2, hence it has no first or second moment. This implies that the achievable rate on the downlink to any UE over a transmission duration that is much longer than T cannot even be computed analytically from the Shannon formula using Lapidoth's results [33]. However, some progress can still be made, as follows.

Note that the baseband equivalent complex total interference signal from all BSs at distances greater than, say, r_0/M from the UE (using the notation of the previous section) can be treated as the sum of two terms: (i) the complex total interference signal from all BSs at distances between r_0/M and $\sqrt{3}r_0$, say (to match the distance of the furthest of the 18 interfering BSs in the regular hexagonal placement model), from the UE; and (ii) the complex total interference signal from all BSs at distances greater than $\sqrt{3}r_0$ from the UE. Owing to the PPP location model for the BSs and the assumption of independent fading on all links, these two complex signals are independent.

The number N of interfering BSs within distances r_0/M and $\sqrt{3}r_0$ of the UE location is a Poisson random variable with mean (and variance) equal to $\pi \lambda r_0^2(3 - M^{-2}) = 2\pi[\sqrt{3} - 1/(\sqrt{3}M^2)]$, where $\lambda = 2/(\sqrt{3}r_0^2)$ is chosen so as to "match" a reference regular BS placement with inter-site distance r_0. Since N will take a value larger than, say, $n = 6\pi[\sqrt{3} - 1/(\sqrt{3}M^2)]$ (i.e., 3 times this mean) with very low probability, the complex total interference signal from these N BSs can be approximated by a sum of at most n terms, which in turn can be approximated by a complex Gaussian from the central limit theorem.

Further, the total interference *power* Z_{PPP}, i.e., the squared magnitude of the complex total interference signal, can be written as the sum of two terms: $Z_{\text{PPP},a}$, due to the N BSs at distances between r_0/M and $\sqrt{3}r_0$ from the UE, and $Z_{\text{PPP},b}$, due to all BSs at distances greater than $\sqrt{3}r_0$ from the UE. The Laplace transforms of $Z_{\text{PPP},a}$ and $Z_{\text{PPP},b}$

are, respectively, given by

$$\mathcal{L}_{Z_{\mathrm{PPP},a}}(s) = \exp\left[-2\pi\lambda \int_{r_0/M}^{\sqrt{3}r_0} \frac{r}{1 + r^\alpha/(sP)}\, dr\right],$$

$$\mathcal{L}_{Z_{\mathrm{PPP},b}}(s) = \exp\left[-2\pi\lambda \int_{\sqrt{3}r_0}^{\infty} \frac{r}{1 + r^\alpha/(sP)}\, dr\right],$$

and the ratio of the mean values of these two powers is

$$\theta = \frac{\mathbb{E}Z_{\mathrm{PPP},b}}{\mathbb{E}Z_{\mathrm{PPP},a}} = \frac{-\dfrac{d}{ds}\mathcal{L}_{Z_{\mathrm{PPP},b}}(s)\big|_{s=0}}{-\dfrac{d}{ds}\mathcal{L}_{Z_{\mathrm{PPP},a}}(s)\big|_{s=0}} = \frac{1}{(\sqrt{3}M)^\alpha - 1} \ll 1.$$

In other words, the interference power contribution from BSs beyond $\sqrt{3}r_0$ is *dominated* by the contribution from the nearby BSs, and from the discussion above, the complex total interference signal from the nearby BSs is approximately Gaussian. It follows from Theorem 1 in [34] that the maximum rate on the downlink to the given UE for the case of PPP-located interferers is that obtained from the Shannon formula applied to the interference from the nearby BSs, i.e., $\log_2(1 + \Gamma_{\mathrm{PPP},a})$ bits/s/Hz, plus a correction term that is $o(\theta)$. Note that the distribution of $\Gamma_{\mathrm{PPP},a}$ is just that of Γ_{PPP} with the assumptions that $d_{\min} = r_0/M$ and $d_{\max} = \sqrt{3}r_0$. Thus a lower bound on the ergodic rate is easily calculated analytically as $\bar{C}_{\mathrm{PPP},a} = \mathbb{E}[\log_2(1 + \Gamma_{\mathrm{PPP},a})]$. Further, $\bar{C}_{\mathrm{PPP},a}$ is also less than the ergodic rate $\bar{C}_{\mathrm{hex}} = \mathbb{E}[\log_2(1 + \Gamma_{\mathrm{hex}})]$ for the regular hexagonal BS layout, since, as discussed above, the interference power in the PPP BS layout exceeds that in the regular hexagonal BS layout.[3] Thus results on maximum downlink rates and ergodic rates to UEs obtained analytically from the PPP location model for BSs should be expected to be on the conservative side.

3.10   Conclusions

We present the case for increased use of stochastic models for the locations of BSs in multi-tier heterogeneous cellular networks. We show how recent results for the case where BS locations are points of a PPP can yield analytical results on the distribution of SINR at arbitrary user locations in the network. These results are either available in closed form, or require relatively straightforward numerical integration. The applications of such results to HetNet system design and deployment are illustrated for a three-tier example, which shows the key deployment parameters that affect coverage. We also

---

[3] Strictly speaking, it was seen and shown that interference power from PPP-located interferers at distance less than r_0/M over the entire plane exceeded the interference power from 18 BSs located at points of a hexagonal lattice, but it can be verified via simulation (and checked from calculation) that the same holds for the interference power from all PPP-located BSs located at distances between r_0/M and $\sqrt{3}r_0$, say.

indicate the differences, both qualitative and quantitative, for the SINR distribution at and the maximum rate to such users, between the PPP BS location model and the conventional regular hexagonal BS layout for the case of a single-tier network. We believe that the vast number of deployment scenarios for HetNets require the use of tools other than exhaustive simulation in order to gain insights into the operation of such systems, and the choice of deployment parameters to optimize their performance (including accounting for operating costs and expected revenues). The analytical methods and results described in this chapter provide a useful set of tools to get a quick overview of the performance of proposed HetNets, and also to obtain a good starting set of deployment parameters for which detailed simulations can then be run for optimization of system performance.

References

[1] L. Chen, W. Chen, B. Wang, X. Zhang, H. Chen, and D. Yang, "System-level simulation methodology and platform for mobile cellular systems," *IEEE Commun. Mag.*, vol. 49, no. 7, pp. 148–55, July 2011.

[2] R. Mallik, "A new statistical model of the complex Nakagami-m fading gain," *IEEE Trans. on Communications*, vol. 58, no. 9, pp. 2611–20, Sep. 2010.

[3] A. Papoulis, *Probability, Random Variables, and Stochastic Processes*, 3rd edn. McGraw-Hill, 1991.

[4] R. A. DeVore and G. G. Lorentz, *Constructive Approximation*. Springer, 1993.

[5] S. Atapattu, C. Tellambura, and H. Jiang, "A mixture gamma distribution to model the SNR of wireless channels," *IEEE Trans. Wireless Commun.*, vol. 10, no. 12, pp. 4193–203, Dec. 2011.

[6] J. Almhana, Z. Liu, V. Choulakian, and R. McGorman, "A recursive algorithm for gamma mixture models," in *Proc. IEEE Int. Conf. on Commun. (ICC)*, June 2006, pp. 197–202.

[7] J. G. Andrews, F. Baccelli, and R. K. Ganti, "A tractable approach to coverage and rate in cellular networks," *IEEE Trans. on Communications*, vol. 59, no. 11, pp. 3122–34, Nov. 2011.

[8] P. C. Pinto and M. Z. Win, "Communication in a Poisson field of interferers-Part II: Channel capacity and interference spectrum," *IEEE Trans. Wireless Commun.*, vol. 9, no. 7, pp. 2187–95, July 2010.

[9] K. Gulati, B. L. Evans, J. G. Andrews, and K. R. Tinsley, "Statistics of co-channel interference in a field of Poisson and Poisson–Poisson clustered interferers," *IEEE Trans. Signal Processing*, vol. 58, no. 10, pp. 6207–22, Dec. 2010.

[10] K. W. Sung, H. Haas, and S. McLaughlin, "A semianalytical PDF of downlink SINR for femtocell networks," *EURASIP J. Wireless Commun. Networking*, vol. 2010, no. 4, 2010.

[11] R. Pupala, L. J. Greenstein, and D. G. Daut, "Throughput analysis of interference limited MIMO-based cellular systems," *IEEE Trans. Wireless Commun.*, vol. 9, no. 6, pp. 1946–51, June 2010.

[12] V. M. Nguyen, F. Baccelli, L. Thomas, and C. S. Chen, "Best signal quality in cellular networks: asymptotic properties and applications to mobility management in small cell networks," *EURASIP J. Wireless Commun. Networking*, vol. 2010, no. 4, 2010.

[13] S. Mukherjee, "Analysis of UE outage probability and macro-cellular traffic offloading for WCDMA macro network with femto overlay under closed and open access," in *Proc. IEEE Int. Conf. on Commun. (ICC)*, June 2011.

[14] A. J. Ganesh and G. L. Torrisi, "Large deviations of the interference in a wireless communication model," *IEEE Trans. Inform. Theory*, vol. 54, no. 8, pp. 3505–17, Aug. 2008.

[15] B. Blaszczyszyn, M. K. Karray, and F. X. Klepper, "Impact of the geometry, path-loss exponent and random shadowing on the mean interference factor in wireless cellular networks," in *Proc. IEEE Wireless Mob. Netw. Conf. (WMNC)*, Oct. 2010.

[16] A. Berman and R. J. Plemmons, *Nonnegative Matrices in the Mathematical Sciences*. Society for Industrial and Applied Mathematics, 1994.

[17] A. M. Hunter, J. G. Andrews, and S. Weber, "Transmission capacity of ad hoc networks with spatial diversity," *IEEE Trans. Wireless Commun.*, vol. 7, no. 12, pp. 5058–71, Dec. 2008.

[18] J. F. C. Kingman, *Poisson Processes*. Oxford University Press, 1993.

[19] R. W. Heath, Jr and M. Kountouris, "Modeling heterogeneous network interference," in *Proc. Inform. Theory and Appl. Workshop (ITA)*, Feb. 2012.

[20] S. Mukherjee, "UE coverage in LTE macro network with mixed CSG and open access femto overlay," in *Proc. IEEE Int. Conf. on Commun. (ICC)*, June 2011.

[21] S. Mukherjee, "Downlink SINR distribution in a heterogeneous cellular wireless network," *IEEE J. Select. Areas Commun. (JSAC)*, vol. 30, no. 3, pp. 575–85, Apr. 2012.

[22] M. L. Kleptsyna, A. Le Breton, and M. Viot, "General formulas concerning Laplace transforms of quadratic forms for general Gaussian sequences," *Journal of Applied Mathematics and Stochastic Analysis*, vol. 15, no. 4, pp. 309–25, 2002.

[23] S. Mukherjee, "Downlink SINR distribution in a heterogeneous cellular wireless network with max-SINR connectivity," in *Proc. Allerton Conference on Communication, Control, and Computing*, Sep. 2011.

[24] H. S. Dhillon, R. K. Ganti, F. Baccelli, and J. G. Andrews, "Modeling and analysis of K-tier downlink heterogeneous cellular networks," *IEEE J. Sel. Areas Commun. (JSAC)*, vol. 30, no. 3, pp. 550–60, Apr. 2012.

[25] P. Madhusudhanan, J. G. Restrepo, Y. Liu, T. X. Brown, and K. R. Baker, "Multi-tier network performance analysis using a shotgun cellular system," in *Proc. IEEE Global Telecommun. Conf. (GLOBECOM)*, Dec. 2011, pp. 1–6.

[26] NTT DoCoMo, "Discussions on UE capability for dual-antenna receiver," 3GPP Standard Contribution (R4-070745), May 2007.

[27] H. S. Jo, Y. J. Sang, P. Xia, and J. G. Andrews, "Outage probability for heterogeneous cellular networks with biased cell association," in *Proc. IEEE Global Telecommun. Conf. (GLOBECOM)*, Dec. 2011.

[28] S. Mukherjee, "Downlink SINR distribution in a heterogeneous cellular wireless network with biased cell association," in *Proc. IEEE Int. Conf. on Commun. (ICC), Small Cells Workshop (SmallNets)*, June 2012.

[29] 3GPP TS 36.304 Release 9 V9.3.0 (2010-06), "Technical specification group radio access network; evolved universal terrestrial radio access (E-UTRA); user equipment (UE) procedures in idle mode (release 9)," June 2010.

[30] Femto Forum, "Interference management in OFDMA femto-cells," http://www.femtoforum.org, Mar. 2010.

[31] C. Tepedelenlioglu, A. Rajan, and Y. Zhang, "Applications of stochastic ordering to wireless communications," *IEEE Trans. Wireless Commun.*, vol. 10, no. 12, pp. 4249–57, Dec. 2011.

[32] M. Aljuaid and H. Yanikomeroglu, "Investigating the Gaussian convergence of the distribution of the aggregate interference power in large wireless networks," *IEEE Trans. Veh. Technol.*, vol. 59, no. 9, pp. 4418–24, Nov. 2010.

[33] A. Lapidoth, "Nearest neighbor decoding for additive non-gaussian noise channels," *IEEE Trans. Inform. Theory*, vol. 42, no. 5, pp. 1520–29, Sep. 1996.

[34] M. S. Pinsker, V. L. Prelov, and S. Verdu, "Sensitivity of channel capacity," *IEEE Trans. Inform. Theory*, vol. 41, no. 6, pp. 1877–88, Nov. 1995.

4 Interference modeling for cognitive femtocells

Alberto Rabbachin, Tony Q. S. Quek, Hyundong Shin, and Moe Z. Win

4.1 Introduction

The possibility of increasing coverage and capacity of radio cellular systems by deploying femtocell base stations is strongly dependent on the capability of avoiding interference with the macrocell network (macrocell users and macrocell base stations). A centralized radio resource allocation procedure controlling the spectrum allocation of the multi-tier network would ensure a perfect coexistence between the macrocell and the femtocell networks. However, such a centralized system is of very high complexity. Considering the distributed nature of the femtocell network, with femtocell base stations placed without planning, a distributed resource allocation system is therefore desirable. In the broader field of cognitive radio networks, frequency band sharing can be based on different approaches. For less dynamic systems, access to the spectrum can be based on the information provided by a central database where geographical information on the spectrum usage is stored. In this case, the cognitive device is aware of its geographical position and can decide which frequency bands can be used to establish a communication link. For the case of femtocells, since the spectrum usage from the macrocell system is quite dynamic and changes frequently, a spectrum allocation relying on database information is not feasible. As for distributed secondary networks, cognitive radio resource management (CRRM) is a solution for a femtocell user (FU) to detect and access the idle spectrum [1–4]. A way to implement the CRRM is to activate a dedicated signaling channel between the macrocell and femtocell networks. The busy channel assessment can be based on the detection of a preamble shared between the macrocell and femtocell networks. A simpler solution for CRRM is to base the channel idle/busy assessment on the energy sensing of the macrocell network radio signals [5–7]. Moreover, since FUs might not always be required to transmit at maximum power, the cognitive femtocell network can implement a more flexible protocol, similar to detect-and-avoid for ultra-wideband systems [8], in which the transmission power level of the femtocell mobile devices is based on the sensed power of the macrocell network signals. In this chapter, we consider a CRRM procedure based on spectrum sensing that enables the femtocell network to learn about the spectrum usage of the macrocell network.

Small Cell Networks: Deployment, PHY Techniques, and Resource Management, ed. Tony Q. S. Quek, Guillaume de la Roche, İsmail Güvenç, and Marios Kountouris. Published by Cambridge University Press. © Cambridge University Press 2013.

The CRRM is, however, challenging because of the uncertainty associated with the aggregate interference generated by the FUs operating into the macrocell network. In distributed networks, the uncertainty can be the result of the unknown number and location of interferers as well as of channel fading, shadowing, and other uncertain environment-dependent conditions [9, 10]. Therefore, it is crucial to incorporate such uncertainty in the statistical interference model in order to quantify the effect of femtocell interference on the macrocell network system performance. In the scenario defined by femtocell networks, it is of great importance to accurately model the aggregate interference generated by multiple active femtocell users in the network. Due to lack of coordination among transmitters, the behavior of the aggregate interference generated by an ad hoc network cannot be modeled using Gaussian distribution since some interferers are dominant and the central limit theorem cannot be applied. Statistical models for non-Gaussian aggregate interference are often based on Gaussian mixtures [11, 12] or on α-stable random variables [13–15]. For the wider context of cognitive networks, in [16], the moment expression for the aggregate interference generated by Poisson distributed nodes in an arbitrary area was derived assuming the typical unbounded path-loss model. The performance of the unbounded path-loss model versus a bounded path-loss model was compared in [17]. The unbounded path-loss model results in significant deviation from the realistic performance. For cognitive radio networks, the log-normal distribution was proposed to model the sum of all interferers' powers [18]. This log-normal approximation was also used for the aggregate interference at primary users without accounting for the channel uncertainty due to fading [19]. The optimal power control strategies for secondary users were determined in [20] based on the Poisson model for the primary network. In [21], the characteristic function and cumulants of the cognitive network interference at a primary user is derived. A novel distribution named truncated-stable distribution was used to develop the statistical model for cognitive network interference. The model is flexible enough to include the effect of power control, non-circular regions, and the presence of obstacles.

In this chapter, we present a statistical model for *per-dimension* (real or imaginary part) aggregate interference generated by FUs based on the work in [21], accounting for the sensing procedure, secondary spatial reuse protocol, spatial density of the secondary users, and environment-dependent conditions such as path-loss, shadowing, and channel fading.

4.2   Stochastic geometry

Much of the work cited in the previous section makes use of tools from stochastic geometry to model the behavior of the interference. Stochastic geometry is a branch of applied probability that allows us to accurately study random point processes (e.g., Poisson point processes) on the plane or in a higher dimension. Among others, stochastic geometry finds applications in biology [22], astronomy [23], and positron emission tomography [24]. The use of stochastic geometry has considerably increased in the field of wireless communications, especially as a means to analyze connectivity in large-scale wireless networks where multiple devices can be simultaneously active due to lack of coordination (decentralized network) or frequency reuse purposes [25]. Moreover, with stochastic geometric tools we can account for the performance of each single link in the network for the location of both transmitting and receiving nodes. By using stochastic geometry, one of the main goals of network designers has been to find closed-form expressions for the signal to interference plus noise ratio (SINR) accounting for the spatial node distribution, the channel propagation characteristics, and the path-loss coefficient. In the context of femtocells, where a centralized control unit is missing and the FU activity is autonomous, the femtocell network can be considered as an ad hoc network with several users being simultaneously active. Therefore, by using stochastic geometric tools, the spatial distribution of FUs can be taken into account when modeling the femtocell network interference. As for ad hoc networks, the activity of the FUs is not kept under control but is subject to many sources of uncertainty. By applying tools of stochastic geometry, the effect of node activity can be taken into account. Therefore stochastic geometry provides the ability to model the statistical behavior of the aggregate interference generated by simultaneously active FUs.

4.3   System model

In femtocell networks different spectrum sensing procedures may be implemented in order to limit both intra-tier interference and inter-tier interference. In this chapter the case of inter-tier interference is considered. However, the proposed interference model can be easily extended to the intra-tier case. For the inter-tier case, the interference generated by the femtocell network can affect both macrocell uplink and downlink channels. While the transmissions generated by the femtocell base stations may cause potential interference to the macrocell users, the transmissions of the femtocell users generate potential interference to macrocell base stations. In the following, we consider this scenario and the interference generated by femtocell users is analyzed. In femtocell cognitive networks, the FUs sense the macrocell downlink channels before transmission in order to select those channels in which no harmful interference is caused to the macrocell network uplink. Furthermore, we consider the FUs hooked to different femtocell base stations. As such, there exists the possibility that FUs can transmit at the same time, on the same macrocell uplink channel, and that the activity of each user is independent on the others.

4.3.1   Activity protocols of the femtocell network users

The activity of each FU depends on the strength of the received downlink signal transmitted by the macrocell base station (MBS). In the simplest case, in order to guarantee a low level of interference, the femtocell user can transmit only if the received power from the MBS is below a certain threshold. However, for a given maximum interference power at the MBS, the level of the threshold can be adjusted according to the FU transmitted power. Therefore, to improve the spectrum usage of the femtocell network, a more sophisticated protocol can be used and the transmission power of the FU can be set according to the received MBS signal power. In the following, the general case of the multiple-threshold protocol is presented in detail and the single-threshold and the full activity protocols are also considered as extreme cases of the multiple-threshold protocol.

The transmission power of the femtocell network users is set according to the detected power level of the MBS downlink signal. This procedure is similar to a power control; however, the transmission power is not set according to the femtocell receive signal to noise ratio (SNR), but such that a protection margin to the MBS is guaranteed. We consider $N - 1$ normalized threshold values $\zeta_1, \zeta_2, \ldots, \zeta_{N-1}$ in increasing order to identify N different classes (or sets) of active FUs, denoted by $\mathcal{A}_k$, $k = 1, 2, \ldots, N$. Let $\zeta_0 = 0$ and $\zeta_N = \infty$. Then, the kth active class $\mathcal{A}_k$ obeys the following activation rule:[1,2]

$$\mathbb{1}_{[\zeta_{k-1}, \zeta_k]} \left(\mathsf{R}_i^{-2b} \mathsf{Y}_i \right) \sim \mathsf{Bern} \left(\mu_{\mathsf{Y}_i}^{(\mathrm{pt})} \left(0, \mathsf{R}_i^{2b} \zeta_{k-1}, \mathsf{R}_i^{2b} \zeta_k \right) \right), \tag{4.2}$$

where $\zeta_k \triangleq \frac{\beta_k}{K P_{\mathrm{MBS}}}$, β_k is the kth threshold value used by the FU terminal, P_{MBS} is the transmitted power of the MBS, Y_i is the squared fading path gain of the channel from the MBS to the ith FU, K is the gain accounting for the loss in the near-field, R_i is the distance between the macrocell MBS and the ith FU, and b is the amplitude path-loss exponent. Note that the power of the received MBS signal at the FU in the class $\mathcal{A}_k$ is between $K P_{\mathrm{MBS}} \zeta_{k-1}$ and $K P_{\mathrm{MBS}} \zeta_k$.

For a single-threshold value, the FU terminal is either ON or OFF. Therefore, the activity of the FU can be represented by the Bernoulli random variable:

$$\mathbb{1}_{[0, \zeta]} \left(\mathsf{R}_i^{-2b} \mathsf{Y}_i \right) \sim \mathsf{Bern} \left(F_{\mathsf{Y}_i} \left(\mathsf{R}_i^{2b} \zeta \right) \right), \tag{4.3}$$

Note that for $\zeta \to \infty$ all FUs are active.

4.3.2   Interference model

The interference signal at the MBS receiver generated by the ith FUs can be written as

$$\mathsf{I}_i = \sqrt{P_{\mathrm{FU}}} \mathsf{R}_i^{-b} \mathsf{X}_i, \tag{4.4}$$

where P_{FU} is the FU signal power at the limit of the near-far region[3], R_i is the distance between the ith FU and the MBS and X_i is the *per-dimension* fading channel path gain of the channel from the ith FU to the MBS. Note that $\mathsf{X}_i = Re\{\mathsf{H}_i\}$, where H_i is the complex path gain of the channel from the ith femtocell user to the MBS. In the following, we

---

[1] The nth-order partial moment $\mu_{\mathsf{X}}^{(\mathrm{pt})}(n, l, u)$ of the random variable X, calculated within the interval $[l, u]$, can be expressed as $\mu_{\mathsf{X}}^{(\mathrm{pt})}(n, l, u) \triangleq \int_l^u x^n f_X(x) \, dx$

[2] The indicator function is defined as

$$\mathbb{1}_{[p,q]}(x) = \begin{cases} 1, & \text{if } p \le x \le q \\ 0, & \text{otherwise,} \end{cases} \tag{4.1}$$

[3] We consider the near-far region limit at 1 meter.

assume that X_is are i.i.d. with common probability density function (PDF) $f_X(\cdot)$ and that they are mutually independent of Y_is.

The network of simultaneously active FUs in a given frequency band can be considered as an ad hoc network since FUs are linked to different femtocell base stations with no coordination among each others. Therefore, it is reasonable to consider the FUs spatially scattered according to a homogeneous Poisson point process (PPP) in the two-dimensional plane $\mathbb{R}^2$, where the victim MBS is assumed to be located at the center of the region. Let $\mathcal{S} \subset \mathbb{Z}^+$ be the index set of FUs in a region $\mathcal{R} \subset \mathbb{R}^2$. Then the probability that k FUs lie inside $\mathcal{R}$ depends only on the total area $A_{\mathcal{R}}$ of the region, and is given by [26]

$$\mathbb{P}\{|\mathcal{S}| = k\} = \frac{(\lambda A_{\mathcal{R}})^k}{k!} e^{-\lambda A_{\mathcal{R}}}, \quad k = 0, 1, 2, \ldots \tag{4.5}$$

where λ is the spatial density (in nodes per unit area). Furthermore, we assume that the region $\mathcal{R}$ is constrained in the annulus prescribed by two radius $d_{\min}$ and $d_{\max}$, which are minimum and maximum distances from the MBS receiver, respectively.[4] This allows us to consider a scenario where the FUs are located within a limited (bounded) region. As will be clear in the next section, by considering the contribution of limited regions, the model can be used to accurately describe the aggregate interference in environments with different shapes, FU densities, and radio propagation parameters.

4.4 Femtocell network interference analysis

All the simultaneously active FUs in the same frequency band contribute to the interference received by the MBS. As shown in Section 4.3.2, the contribution to the aggregate interference of each active FU depends on the MBS signal strength received, on the distance, and on the fading of the link between the FU and the MBS. To characterize the femtocell network interference generated by FUs, the cumulants for the case of the multiple-threshold protocol are first derived in Section 4.4.1. Using these cumulant expressions, we develop the symmetric truncated-stable model for the femtocell network interference in Section 4.4.2.

4.4.1 Multiple-threshold protocol

Each FU belongs to one of the N classes defined by the set of threshold values $\{\zeta_0, \zeta_1, \ldots, \zeta_N\}$ according to the distance and to the channel fading that affects the received MBS signal power. For each class, a transmission power is set in order to ensure very low interference to the MBS. The network interference generated by the

---

[4] Note that R_i in (4.4) can be smaller than 1. Therefore, the received interference power can be larger than P_I but it is finite since $d_{\min} > 0$.

FUs in all N classes can be expressed as

$$\mathsf{I}_{\mathrm{mt}} = \sum_{k=1}^{N} \mathsf{I}_{\mathrm{mt},k}, \tag{4.6}$$

where $\mathsf{I}_{\mathrm{mt},k}$ is the normalized aggregate interference generated by the FUs active in the kth class. In order to statistically characterize I_{mt}, the cumulants of the per-zone network interference are derived.

The network interference generated by the $\mathbb{Z}k$ class of FUs can be written as

$$\mathsf{I}_{\mathrm{mt},k} = \sqrt{P_{\mathrm{FU},k}} \underbrace{\sum_{i \in \mathbb{Z}k} \mathsf{R}_i^{-b} \mathsf{X}_i}_{\mathsf{Z}_k(\mathcal{R})}, \tag{4.7}$$

where $P_{\mathrm{FU},k}$ is the transmitted power of the FUs in the kth active class $\mathbb{Z}k$ and

$$\mathbb{Z}k = \left\{ i \in \mathcal{S} : \mathbb{1}_{[\zeta_{k-1},\zeta_k]}\left(\mathsf{R}_i^{-2b}\mathsf{Y}_i\right) = 1 \right\}. \tag{4.8}$$

The N power levels $P_{\mathrm{FU},1}, P_{\mathrm{FU},2}, \ldots, P_{\mathrm{FU},N}$ are set in decreasing order such that users active in classes characterized by higher detected power level of the downlink MBS signal transmit with lower power. Note that the power levels in a real system will be set such that the interference power received by the MBS by each of the FU is with high probability below a given value (usually lower than the noise level). To do so, it is necessary to take into account the effect of fading on the channel between FU and MBS. A good example of how these power levels are set is given in the detect-and-avoid protocol that ultra-wideband systems need to implement in order to transmit to some frequency bands. In this case the transmission levels are set such that the interference power is 3 dB lower than the noise level for 95% of the time.

By using Campbell's theorem [27, Theorem 3.1], the characteristic function (CF) of $\mathsf{Z}_k(\mathcal{R})$ can be expressed as

$$\psi_{\mathsf{Z}_k(\mathcal{R})}(j\omega) = \exp\left(-2\pi\lambda \int_{\mathsf{X}} \int_{\mathsf{Y}} \int_{d_{\min}}^{d_{\max}} \left[1 - \exp\left(j\omega x r^{-b}\right)\right]\right.$$

$$\left. \times \mathbb{1}_{[\zeta_{k-1},\zeta_k]}\left(r^{-2b}y\right) f_{\mathsf{X}}(x) f_{\mathsf{Y}}(y) r\, dr\, dy\, dx\right). \tag{4.9}$$

Since all $\mathsf{Z}_k(\mathcal{R})$s are statistically independent and considering that, if the RVs X and Y are independent, then

$$\kappa_{\mathsf{X}+\mathsf{Y}}(n) = \kappa_{\mathsf{X}}(n) + \kappa_{\mathsf{Y}}(n), \forall n,$$

we obtain the nth cumulant of the femtocell network interference I_{mt} for the multiple-threshold protocol as follows

$$\kappa_{\mathsf{I}_{\mathrm{mt}}}(n) = \sum_{k=1}^{N} P_{\mathrm{FU},k}^{n/2} \kappa_{\mathsf{Z}_k(\mathcal{R})}(n), \tag{4.10}$$

where $\kappa_{\mathsf{Z}_k(\mathcal{R})}(n)$ are given by (4.18), (4.19), and (4.21) in Appendix 4 A for $k = 1$, $k = 2, 3, \ldots, N - 1$, and $k = N$, respectively [21]. The multiple-threshold protocol includes the single-threshold protocol. It is easy to show that the nth cumulant of the femtocell network interference can be expressed as

$$\kappa_{I_{st}}(n) = P_{FU}^{n/2} \kappa_{Z_1(\zeta;\mathcal{R})}(n).$$ (4.11)

Moreover, in the case of full activity ($\zeta \to \infty$), the nth cumulant of the femtocell network interference can be expressed as

$$\kappa_{I_{fa}}(n) = P_{FU}^{n/2} \kappa_{Z_1(\infty;\mathcal{R})}(n).$$ (4.12)

Remark 4.1 *Using the cumulant expressions (4.10), (4.11), and (4.12), we can characterize statistical properties (e.g., mean, variance, and other higher order statistics) of the femtocell network interference for different femtocell spatial reuse protocols. For example, the second-order cumulant can be used to measure the power of the femtocell network interference.*

4.4.2 Truncated-stable distribution model

The truncated-stable distributions are a relatively new class of distributions that follow from the class of stable distributions [15]. The reasons for choosing truncated-stable distributions instead of stable distributions are well explained in [21]. It can be briefly summarized that truncated-stable distributions have smoothed tails and finite moments. Therefore they offer a statistical tool to model the aggregate interference in more realistic scenarios.

The CF and the parameters of the truncated-stable distributions are given in the following as reported in [21]. The CF of a symmetric truncated-stable random variable $T \sim \mathcal{S}_t(\gamma', \alpha, g)$ is given by [28]

$$\psi_T(j\omega) = \exp\left(\gamma'\Gamma(-\alpha)\left[\frac{(g-j\omega)^\alpha}{2} + \frac{(g+j\omega)^\alpha}{2} - g^\alpha\right]\right),$$ (4.13)

where $\Gamma(\cdot)$ is the Euler gamma function, and γ', α, and g are the parameters associated with the truncated-stable distribution. The parameters γ' and α are akin to the dispersion and the characteristic exponent of the stable distribution, respectively. The parameter g is the argument of the exponential function used to smooth the tail of the stable distribution. The nth cumulant of the truncated-stable distribution can be obtained using (4.13) as

$$\kappa_T(n) = \begin{cases} \gamma'\Gamma(-\alpha)g^{\alpha-n}\prod_{i=0}^{n-1}(\alpha-i) & \text{for even } n \\ 0 & \text{for odd } n. \end{cases}$$ (4.14)

For given α, using (4.14), the parameters γ' and g can be expressed in terms of the first two non-zero cumulants, namely, the second- and fourth-order cumulants.

To model the femtocell network interference using the truncated-stable distribution, we first fix the characteristic exponent to $\alpha = 2/b$ [21]. Let I_A be the femtocell network interference. Then, we can model the femtocell network interference I_A as the symmetric truncated-stable random variable, i.e.,

$$I_A \sim \mathcal{S}_t\left(\gamma'_A, \alpha = 2/b, g_A\right),$$ (4.15)

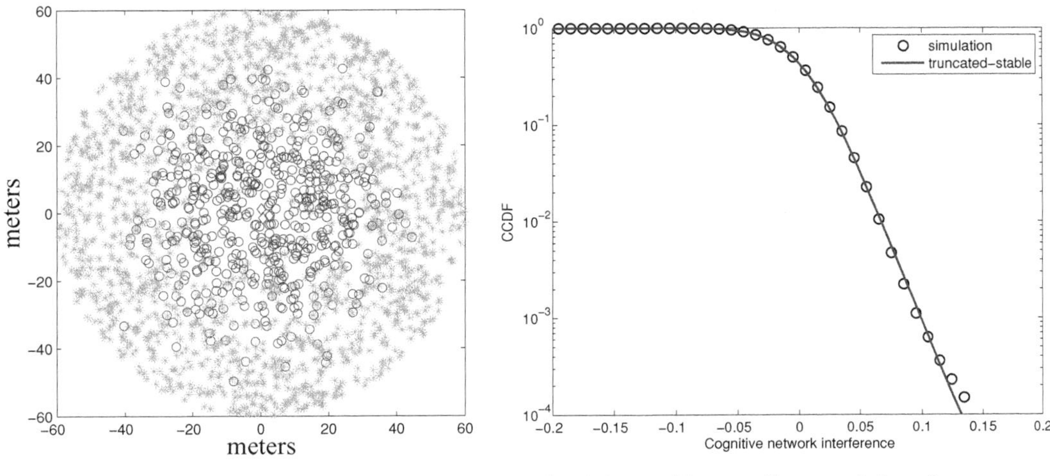

(a) Node displacements   (b) CCDF of femtocell network interference

Figure 4.1 (a) Node displacements (not only a single realization snapshot) and (b) network interference CCDF of a femtocell network with the multiple-threshold protocol for $d_{\min} = 1m$, $d_{\max} = 60m$, $\lambda = 0.1$ users$/m^2$, $N = 3$, $\zeta_1 = -42$ dBm, and $\zeta_2 = -20$ dBm. The transmitted power levels of the FUs are $P_{\mathrm{FU},1} = 0$ dBm for $\mathbb{Z}1$, $P_{\mathrm{FU},2} = -23.7$ dBm for $\mathbb{Z}2$, and $P_{\mathrm{FU},3} = -38.7$ dBm for $\mathbb{Z}3$. Both $|\mathsf{H}_i|$ and $\sqrt{\mathsf{Y}}$ follow a Nakagami distribution with shape parameter $m = 2$ and power parameter $\Omega = 1$. ©2011 IEEE. Reproduced, with permission, from [21].

where the dispersion and smoothing parameters γ'_{A} and g_{A} are given in terms of the second and fourth cumulants of I_{A} as

$$\gamma'_{\mathrm{A}} = \frac{\kappa_{\mathsf{I}_{\mathrm{A}}}(2)}{\Gamma(-\alpha)\,\alpha\,(\alpha - 1)\left[\frac{\kappa_{\mathsf{I}_{\mathrm{A}}}(2)(\alpha-2)(\alpha-3)}{\kappa_{\mathsf{I}_{\mathrm{A}}}(4)}\right]^{\frac{\alpha-2}{2}}}, \tag{4.16}$$

$$g_{\mathrm{A}} = \sqrt{\frac{\kappa_{\mathsf{I}_{\mathrm{A}}}(2)(\alpha - 2)(\alpha - 3)}{\kappa_{\mathsf{I}_{\mathrm{A}}}(4)}}. \tag{4.17}$$

Figure 4.1 shows the FUs' displacements and the complementary cumulative distribution function (CCDF) of the femtocell network interference generated by FUs that use the multiple-threshold protocol with $N = 3$. This figure shows that the proposed model provides a good approximation of the aggregate interference. It can also be noticed that due to the presence of fading the zones are overlapping.

4.5   Applications

As shown in [21], the proposed statistical model for the network interference is very flexible and can be easily adapted to account for both power control of the macrocell downlink and power control of the femtocell uplink. The downlink power control can be included in the model by evaluating the cumulants for the lth possible transmission power levels of MBS with $l \in [1, \ldots, L]$. The total cumulants of the femtocell interference is then given by the sum of the L cumulants and each contribution is weighted by the

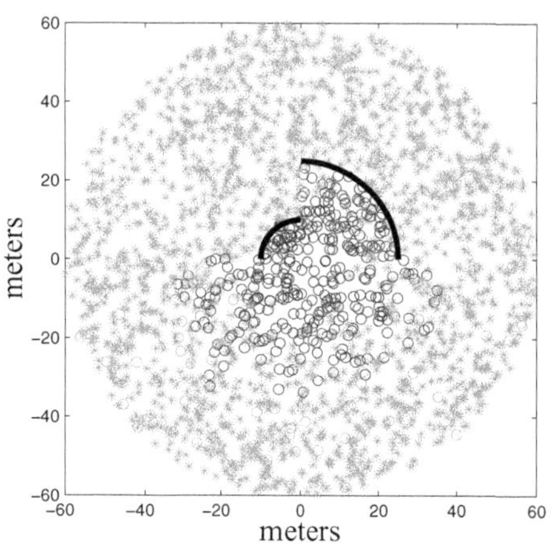

(a) Node displacements (not only a single realization snapshot)

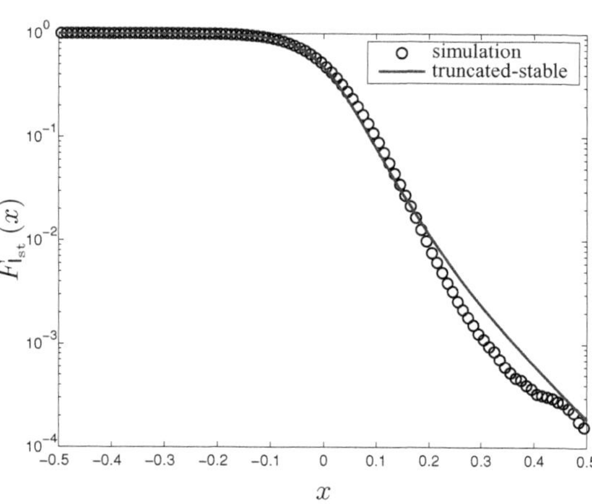

(b) CCDF of the femtocell network interference

Figure 4.2 (a) Node displacements (not only a single realization snapshot) of FUs and (b) CCDF of the femtocell network interference with the single-threshold protocol for $d_{\min} = 1$ meter, $d_{\max} = 60$ meters, $\lambda = 0.01$ users/m^2, $\zeta = -40$ dBm, $\theta_1 = \theta_2 = \pi/2$, and $\breve{\beta}_1 = \breve{\beta}_2 = 20$ dB. The two obstacles are present at 10 and 25 meters from the primary receiver, covering the angle of $\pi/2$, and causing additional attenuation of 20 dB. Both $|H_i|$ and $\sqrt{Y}$ follow a Nakagami distribution with shape parameter $m = 2$ and power parameter $\Omega = 1$. ©2011 IEEE. Reproduced, with permission, from [21].

probability that the base station transmits with that power. The uplink power control can be considered as a contribution to the fading affecting the interfering signal. Therefore, its contribution can be taken into account by simply weighting the total nth cumulant by the nth moment of the power control random variable.

Very often, aggregate interference models are not flexible enough to be applied in realistic environments where the shape of the area where the nodes are distributed and the obstacles present inside the area can play an important role. Implicitly, the model so far presented considers the polar coordinate system and places the MBS at the center of the region. Since the proposed model allows us to determine the aggregate interference generated by FUs confined in a limited circular region, the nth cumulant of the femtocell network interference generated by the FUs in a non-circular area of interest can be calculated by splitting the area into infinitesimal circular sections drawn from the MBS position, calculating the nth cumulant of the aggregate interference of each region, and summing the cumulants of all the regions together. Using this approach, we can also consider any position of the primary user within the area of interest. Moreover, the model allows us to include the effects of obstacles in the area of interest. Similarly to the approach used for a non-circular region, the presence of an obstacle can be taken into account by splitting the region of interest in circular sectors according to the position of the obstacles. In the calculation of the nth cumulant, for each circular sector behind the ith obstacle, the additional attenuation $\breve{\beta}_i$ due to the obstacle on both the macrocell uplink signal and the FU transmitted signals is taken into account. Two examples of the flexibility provided by the proposed models are given in Figures 4.2 and 4.3. In Figure 4.2

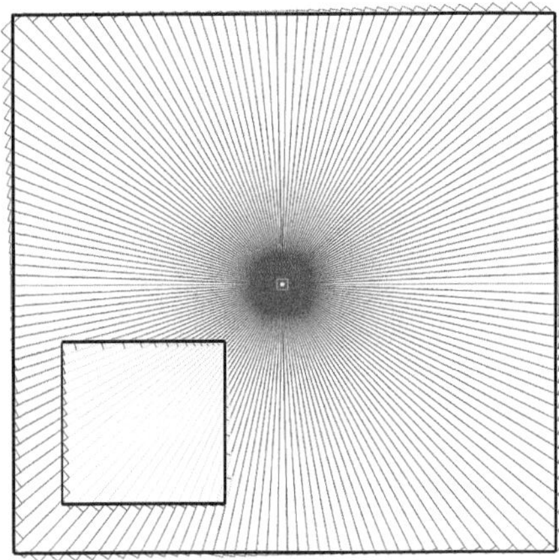

(a) Non-circular region

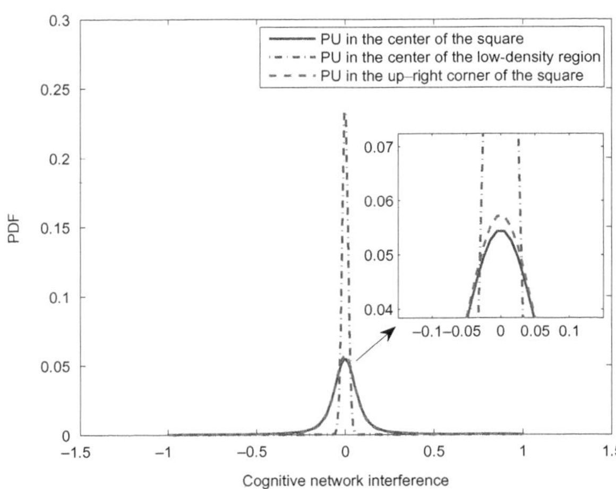

(b) PDF of the network interference in different positions

Figure 4.3 Femtocell interference in a non-circular region. ©2011 IEEE. Reprinted, with permission, from [21].

the FUs' displacement and the CCDF of the femtocell network interference are shown. From Figure 4.2(a), the effect of the obstacles on the FU activity can be noticed. Figure 4.2(b) shows that the proposed model offers a good approximation. Figure 4.3 shows how the interference in a non-circular region can be analyzed by also considering an area where there are no interferers. In Figure 4.3(b) the PDF of the femtocell interference is plotted for different positions of the MBS in the non-circular region. From this figure it can be noticed how the position of the MBS within the area of interest can affect the distribution of the aggregate interference.

Environments composed by different materials provide different path-loss exponents, which impact both the sensing procedure and the interfering signal. With the proposed model the effect of the exponent b on the mean aggregate interference power can

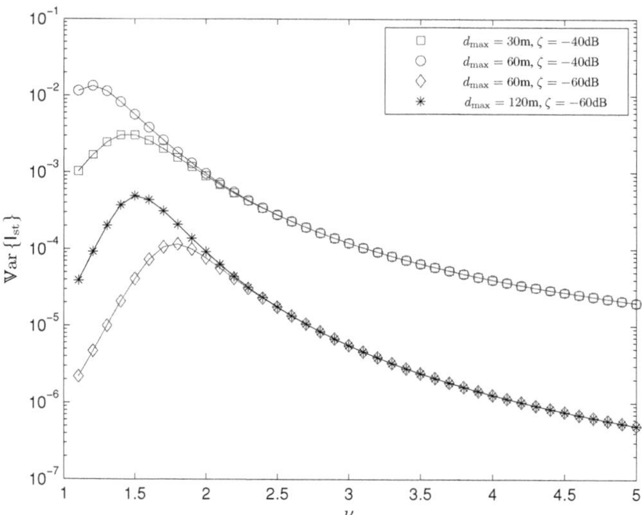

Figure 4.4 Variance of the femtocell network interference I_{st} for the single-threshold protocol as a function of the path-loss coefficient ν of the region $\mathcal{R}$ for $\lambda = 0.1$, $\zeta = -30$ dBm, $P_{FU} = 0$ dBm, and fading for primary and secondary links $\sqrt{Y} \sim$ Nakagami (2, 1) and $|H_i| \sim$ Nakagami (2, 1). ©2010 IEEE. Reprinted, with permission, from [29].

be easily addressed. As shown in Figure 4.4, the network interference generated by FUs exhibits a non-monotonic behavior with an increasing path-loss coefficient. Under certain condition of the Nakagami parameter m, R_f, and ζ, a higher path-loss coefficient may lead to higher activity of the nodes, which is not compensated by a reduction of the interference signal power.

Using the expression of the aggregate interference CF given in (4.13) allows us to calculate the effect of the interference on the MBS bit error probability and to evaluate the outage probability for a given MBS performance metric.

4.6   Conclusion

This chapter has presented a statistical model for the aggregate interference in cognitive femtocell networks. In particular the interference caused to the MBS from several active FUs has been investigated. The model accounts for the sensing procedure, the spatial node distribution, secondary spatial reuse protocol, and environment-dependent conditions, such as path loss, shadowing, and channel fading. The spatial reuse was first analyzed by considering the multiple-threshold protocol and then extended to the single-threshold and the no-threshold protocols. For these protocols, the characteristic function and the cumulant of the femtocell network interference at the MBS were derived. A statistical model based on truncated-stable distributions was proposed to model the interference generated by multiple FUs. In the chapter, it was shown how the proposed statistical model can be flexible and account for power control, shadowing, and the structure of the environment. The framework proposed in this chapter can be applied to carefully address the impact of femtocell networks deployment in specific areas. To conclude, it is worth mentioning that the proposed model can be easily adapted to model the interference generated by femtocell base stations to macrocell users or to model the intra-tier interference.

Appendix 4A Cumulants of $Z_k(\mathcal{R})$ ($k = 1, 2, \ldots, N$) for the multiple-threshold protocol

For the multiple-threshold protocol, the nth cumulant of $Z_k(\mathcal{R})$ can be expressed as[5]

- For $k = 1$

$$
\kappa_{Z_1(\mathcal{R})}(n) = \frac{2\pi\lambda\,\mathcal{M}\mathsf{X}n}{nb-2}\left[\left(d_{\min}^{2-nb} - d_{\max}^{2-nb}\right) F_\mathsf{Y}\left(d_{\min}^{2b}\zeta\right)\right.
$$
$$
\left. + \zeta^{\frac{nb-2}{2b}}\,\mu_\mathsf{Y}^{(\mathrm{pt})}\left(\frac{2-nb}{2b}, d_{\min}^{2b}\zeta, d_{\max}^{2b}\zeta\right) - d_{\max}^{2-nb}\,\mu_\mathsf{Y}^{(\mathrm{pt})}\left(0, d_{\min}^{2b}\zeta, d_{\max}^{2b}\zeta\right)\right].
$$

$$(4.18)$$

- For $k = 2, 3, \ldots, N-1$

$$
\kappa_{Z_k(\mathcal{R})}(n) = \frac{2\pi\lambda\,\mathcal{M}\mathsf{X}n}{nb-2}\left[d_{\min}^{2-nb}\,\mu_\mathsf{Y}^{(\mathrm{pt})}\left(0, d_{\min}^{2b}\zeta_{k-1}, \Delta_{\min}\right)\right.
$$
$$
- \zeta_{k-1}^{\frac{nb-2}{2b}}\,\mu_\mathsf{Y}^{(\mathrm{pt})}\left(\frac{2-nb}{2b}, d_{\min}^{2b}\zeta_{k-1}, \Delta_{\min}\right) + c_1\,\mu_\mathsf{Y}^{(\mathrm{pt})}\left(c_2, \Delta_{\min}, \Delta_{\max}\right)
$$
$$
\left. + \zeta_k^{\frac{nb-2}{2b}}\,\mu_\mathsf{Y}^{(\mathrm{pt})}\left(\frac{2-nb}{2b}, \Delta_{\max}, d_{\max}^{2b}\zeta_k\right) - d_{\max}^{2-nb}\,\mu_\mathsf{Y}^{(\mathrm{pt})}\left(0, \Delta_{\max}, d_{\max}^{2b}\zeta_k\right)\right],
$$

$$(4.19)$$

where $\Delta_{\min} = \min\{d_{\max}^{2b}\zeta_{k-1}, d_{\min}^{2b}\zeta_k\}$, $\Delta_{\max} = \max\{d_{\max}^{2b}\zeta_{k-1}, d_{\min}^{2b}\zeta_k\}$, and

$$
(c_1, c_2) = \begin{cases} \left(d_{\min}^{2-nb} - d_{\max}^{2-nb}, 0\right), & \text{if } d_{\min}^{2b}\zeta_k \geq d_{\max}^{2b}\zeta_{k-1} \\ \left(\zeta_k^{\frac{nb-2}{2b}} - \zeta_{k-1}^{\frac{nb-2}{2b}}, \frac{2-nb}{2b}\right), & \text{if } d_{\min}^{2b}\zeta_k < d_{\max}^{2b}\zeta_{k-1}. \end{cases}
$$

$$(4.20)$$

- For $k = N$

$$
\kappa_{Z_N(\mathcal{R})}(n) = \frac{2\pi\lambda\,\mathcal{M}\mathsf{X}n}{nb-2}\left[d_{\min}^{2-nb}\,\mu_\mathsf{Y}^{(\mathrm{pt})}\left(0, d_{\min}^{2b}\zeta_{k-1}, d_{\max}^{2b}\zeta_{k-1}\right)\right.
$$
$$
- \zeta_{k-1}^{\frac{nb-2}{2b}}\,\mu_\mathsf{Y}^{(\mathrm{pt})}\left(\frac{2-nb}{2b}, d_{\min}^{2b}\zeta_{k-1}, d_{\max}^{2b}\zeta_{k-1}\right)
$$
$$
\left. + \left(d_{\min}^{2-nb} - d_{\max}^{2-nb}\right) \bar{F}_\mathsf{Y}\left(d_{\max}^{2b}\zeta_{k-1}\right)\right].
$$

$$(4.21)$$

[5] $\mathcal{M}\mathsf{X}n$ is the nth moment of X: $\mathcal{M}\mathsf{X}n \triangleq \mathbb{E}\{\mathsf{X}^n\}$.

References

[1] J. Xiang, M. Y. Zhang, T. Skeie, and L. Xie, "Downlink spectrum sharing for cognitive radio femtocell networks," *IEEE Syst. J.*, vol. 4, no. 4, pp. 524–34, Dec. 2010.

[2] G. Gur, S. Bayhan, and F. Alagoz, "Cognitive femtocell networks: an overlay architecture for localized dynamic spectrum access," *IEEE Trans. Wireless Commun.*, vol. 17, no. 4, pp. 62–70, Aug. 2010.

[3] S.-Y. Lien, Y.-Y. Lin, and K.-C. Chen, "Cognitive and game-theoretical radio resource management for autonomous femtocells with QoS guarantees," *IEEE Trans. Wireless Commun.*, vol. 10, no. 7, pp. 2196–206, July 2011.

[4] S. Al-Rubaye, A. Al-Dulaimi, and J. Cosmas, "Cognitive femtocell," *IEEE Vehicular Technology Magazine*, vol. 6, no. 1, pp. 44–51, Mar. 2011.

[5] A. Ghasemi and E. S. Sousa, "Spectrum sensing in cognitive radio networks: requirements, challenges and design trade-offs," *IEEE Commun. Mag.*, vol. 46, no. 4, pp. 32–9, Apr. 2008.

[6] D. Cabric, S. M. Mishra, and R. W. Brodersen, "Implementation issues in spectrum sensing for cognitive radios," in *Proc. Asilomar Conf. on Signals, Systems, and Computers (ASILOMAR)*, Pacific, Grove, Nov. 2004, pp. 772–6.

[7] A. Sonnenschein and P. M. Fishman, "Radiometric detection of spread-spectrum signals in noise," *IEEE Trans. Aerosp. Electron. Syst.*, vol. 28, no. 3, pp. 654–60, July 1992.

[8] ECC, "Technical requirements for UWB DAA (detect and avoid) devices to ensure the protection of radiolocation services in the bands 3.1–3.4 GHz and 8.5–9 GHz and BWA terminals in the band 3.4–4.2 GHz," 2008.

[9] R. Tandra and A. Sahai, "SNR walls for signal detection," *IEEE J. Sel. Topics Signal Proc.*, vol. 2, no. 1, pp. 4–17, Feb. 2008.

[10] R. Tandra, S. M. Mishra, and A. Sahai, "What is a spectrum hole and what does it take to recognize one?" *IEEE Proceedings*, vol. 97, no. 5, pp. 824–48, May 2009.

[11] G. V. Trunk, "Further results on the detection of targets in non-Gaussian sea clutter," *IEEE Trans. Aerosp. Electron. Syst.*, vol. 7, no. 3, pp. 553–6, May 1971.

[12] G. V. Trunk and S. F. George, "Detection of targets in non-Gaussian sea clutter," *IEEE Trans. Aerosp. Electron. Syst.*, vol. 6, no. 5, pp. 620–8, Sep. 1970.

[13] C. L. Nikias and M. Shao, *Signal Processing with Alpha-Stable Distributions and Applications.* Wiley-Interscience, 1995.

[14] M. Shao and C. Nikias, "Signal processing with fractional lower order moments: stable processes and their applications," *IEEE Proceedings*, vol. 81, no. 7, pp. 986–1010, 1993.

[15] G. Samoradnitsky and M. Taqqu, *Stable Non-Gaussian Random Processes.* Chapman and Hall, 1994.

[16] E. Salbaroli and A. Zanella, "Interference analysis in a Poisson field of nodes of finite area," *IEEE Trans. Veh. Technol.*, vol. 58, no. 4, pp. 1776–83, May 2009.

[17] H. Inaltekin, M. Chiang, H. V. Poor, and S. B. Wicker, "The behavior of unbounded path-loss models and the effect of singularity on computed network characteristics," *IEEE J. Sel. Areas Commun. (JSAC)*, vol. 27, no. 7, pp. 1078–92, Sep. 2009.

[18] A. Ghasemi and E. S. Sousa, "Interference aggregation in spectrum-sensing cognitive wireless networks," *IEEE J. Sel. Topics Signal Proc.*, vol. 2, no. 1, pp. 41–56, Feb. 2008.

[19] R. Menon, R. M. Buehrer, and J. H. Reed, "On the impact of dynamic spectrum sharing techniques on legacy radio systems," *IEEE Trans. Wireless Commun.*, vol. 7, no. 11, pp. 4198–207, Nov. 2008.

[20] W. Ren, Q. Zhao, and A. Swami, "Power control in spectrum overlay networks: how to cross a multi-lane highway," *IEEE J. Sel. Areas Commun. (JSAC)*, vol. 27, no. 7, pp. 1283–96, Sep. 2009.

[21] A. Rabbachin, T. Q. S. Quek, H. Shin, and M. Z. Win, "Cognitive network interference," *IEEE J. Sel. Areas Commun. (JSAC)*, vol. 29, no. 2, pp. 480–93, Feb. 2011.

[22] M. Beil, F. Fleischer, S. Paschke, and V. Schmidt, "Statistical analysis of the three-dimensional structure of centromeric heterochromatin in interphase nuclei," *J. Micros.*, vol. 217, pp. 60–8, Jan. 2005.

[23] S. Chandrasekhar, "Stochastic problems in physics and astronomy," *Reviews of Modern Physics*, vol. 15, no. 1, pp. 1–89, Jan. 1943.

[24] M. Y. Vardi, L. Shepp, and L. Kaufman, "A statistical model for positron emission tomography," *J. Am. Statist. Assoc.*, vol. 80, no. 389, pp. 8–20, Mar. 1985.

[25] F. Baccelli and B. Błaszczyszyn, *Stochastic Geometry and Wireless Networks, Volume I – Theory*, ser. Foundations and Trends in Networking. NoW Publishers, 2009.

[26] J. F. Kingman, *Poisson Processes*. Oxford University Press, 1993.

[27] M. Z. Win, P. C. Pinto, and L. A. Shepp, "A mathematical theory of network interference and its applications," *IEEE Proceedings*, vol. 97, no. 2, pp. 205–30, Feb. 2009.

[28] P. Carr, H. Geman, D. B. Madan, and M. Yor, "The fine structure of asset returns: an empirical investigation," *J. Bus.*, vol. 75, no. 2, pp. 305–32, Apr. 2002.

[29] A. Rabbachin, T. Q. S. Quek and M. Z. Win, "Statistical modeling of cognitive network interference," Global Telecommunications Conference (GLOBECOM 2010), 2010 IEEE, pp. 1–6, 1–6 Dec. 2010.

5 Multiple antenna techniques in small cell networks

Salam Akoum, Marios Kountouris, and Robert W. Heath, Jr.

5.1 Introduction

Multiple input multiple output (MIMO) communication has been established both theoretically and practically as a means to increase data rates and improve reliability in wireless networks. While single input single output (SISO) wireless communication techniques rely on time domain or frequency domain processing to precode and decode the transmitted and received data signals, multiple antenna communication provides an extra spatial dimension to improve the wireless link performance in terms of error rate, coverage, and/or spectral efficiency.

As interest in MIMO communication has grown, upcoming cellular standards have embraced using multiple antennas at the base stations (BSs) and the mobile user terminals to increase the data rates and improve the performance of the radio link [1]. Multiple antennas are also being considered in small cell networks (SCNs) and femtocell networks as a means to improve coverage and manage interference [2, 3]. The development of MIMO techniques for two-tier networks needs to take into account the specific topology of the network, characterized by irregularity in terms of deployment, operation mode (closed access vs. open access), channel state information (CSI) availability, and backhaul connectivity. In this chapter, we provide an overview of MIMO communication techniques in two-tier networks. We present the state of the art in terms of MIMO precoding and coordination techniques to manage interference in heterogeneous networks. We illustrate the various gains and the associated challenges from using linear precoding with perfect and imperfect channel state information at the transmitter (CSIT) in femtocell networks and evaluate the potential role that multi-antenna communication is bound to play in two-tier networks.

5.2 Overview and benefits of MIMO communication

Depending on the availability of multiple antennas at the transmitter and the receiver, multiple antenna or space-time communication techniques can be classified into single

Small Cell Networks: Deployment, PHY Techniques, and Resource Management, ed. Tony Q. S. Quek, Guillaume de la Roche, İsmail Güvenç, and Marios Kountouris. Published by Cambridge University Press. © Cambridge University Press 2013.

input multiple output (SIMO) communication, multiple input single output (MISO) communication, or MIMO communication. Furthermore, multiple antenna communication can be divided into point-to-point or single user multiple input multiple output (SU-MIMO) communication and multi-user multiple input multiple output (MU-MIMO) communication. SU-MIMO can refer to the case when a BS serves one user terminal at each time and frequency resource block. When a BS uses its multiple antennas to serve several users simultaneously, an instance of MU-MIMO occurs.

Despite their wide range of application and their different classifications, the benefits of multi-antenna communication can be broadly summarized in terms of spatial diversity gain, array gain, spatial multiplexing gain, and interference reduction and avoidance. Note that, in general, it might not be possible to exploit all the aforementioned benefits simultaneously due to conflicting requirements on the spatial degrees of freedom. Using a combination of the benefits across a wireless network will, however, result in improved network capacity.

Spatial diversity gain corresponds to mitigating the effect of multi-path fading over wireless links. It is realized by using multiple antennas to transmit and/or receive multiple (ideally independent) copies of the transmitted signal over sufficiently decorrelated fading channels. As the number of independent copies increases, the probability of finding at least one channel that is not in deep fade increases, thereby improving the reliability of the reception. Spatial diversity gain is usually expressed in terms of an order. A MIMO channel with N_t transmit antennas and N_r receive antennas offers a maximum spatial diversity order of $N_t N_r$.

Array gain results from coherently combining the wireless signals at the receiver. Unlike the spatial diversity gain, it does not rely on the statistical decorrelation between the different fading channels. It improves the resistance to noise, and provides an increase in the coverage or the range of the wireless network. For the array gain, the received signal to noise ratio (SNR) still increases linearly with N_r, even if the channels are completely correlated.

MIMO communication offers a linear increase in data rates through *spatial multiplexing*. When the transmitter sends multiple independent data streams within the bandwidth of operation, and under suitable rich scattering channel conditions, the receiver can separate the data streams, resulting in a linear increase in the data rate. In general, for an $N_t \times N_r$ MIMO system, a $\min(N_t, N_r)$ increase in capacity can be achieved. Unlike code, time, or frequency division multiple access, MIMO multiplexing gain does not come at the cost of bandwidth expansion.

The MIMO spatial dimension can further be exploited for *interference reduction and avoidance*. For interference avoidance, multiple antennas can be used, through beamforming, to direct the transmitted energy toward the intended user, and to minimize the interference toward other users in the network. This increases the coverage and range of the wireless networks.

The basic MIMO techniques can be divided into single-user and multi-user methods. Depending on the availability of CSIT, they can be divided into two classes: open loop techniques, which do not require CSIT; and MIMO techniques that exploit CSIT, which are referred to as closed-loop MIMO methods. Depending on the multiplicity of antennas at the receiver, these methods can be further classified into multi-stream and single-stream transmissions. A detailed presentation of the vast spectrum of MIMO communication techniques can be found in [4, 5].

5.3 MIMO techniques in two-tier cellular networks

Cellular systems are inherently interference limited. If they were not, it would be possible to increase the spectral efficiency by lowering the frequency reuse or increasing the number of users per cell. Interference degrades the data rates and increases the outages, especially when it is not coordinated. In fact, the receiver on the downlink of a cellular system, for example, has to cope simultaneously with both spatial inter-stream interference and interference from neighboring BSs, which makes signal detection especially challenging. Most theoretical MIMO results are for high SNR regimes with idealized decoding; in practice MIMO techniques have to function in low signal to interference plus noise ratio (SINR) environments with low complexity receivers. In two-tier cellular networks, where the operator-deployed infrastructure is underlaid with small cell nodes, interference is the main performance bottleneck. Intra-cell interference arises from imperfect multi-user transmission within the same cell, whereas same-tier interference occurs from co-channel neighboring same-tier BSs. Cross-tier interference occurs from co-channel femtocell nodes on the cellular mobile user terminal or from co-channel macro base stations when the performance of the femtocell user terminal is under investigation. In addition to the increase in cross-talk between neighboring base stations, interference in two-tier networks is governed by the locations of the small cell nodes with respect to the macrocell user terminal and vice versa. The uncontrolled deployment of small cell nodes plays a major role in determining the effectiveness of various MIMO techniques.

Much of the prior research on femtocell networks has focused on single antenna interference management [6–10]. There is growing interest, however, in the application of multiple antenna methods to manage interference. In [11, 12], precoding techniques adopted in the 3rd Generation Partnership Project (3GPP) standards are applied at the femtocells to reduce interference and improve the throughput without taking into account the randomness in the base stations deployment. In [13], a beamforming strategy to minimize interference from the femtocells to the macrocell user is analyzed, assuming a reliable backhaul connection between the macrocell and the femtocells. In [14], a beam subset selection strategy with a maximum throughput scheduler is analyzed for a heterogeneous network with a femtocell underlay. In [15], a coordinated beamforming strategy is presented for the same setup. In [3], single-user beamforming is shown to achieve better coverage than multi-user beamforming on the downlink of a two-tier network, unless the latter is coupled with user scheduling [16]. The aforementioned results are derived assuming perfect CSI and do not consider the effect of channel uncertainty and same-tier interference on the performance of downlink beamforming. Channel uncertainty arises from feedback delay and quantization error. The effect of delay and channel quantization on the performance of limited feedback beamforming is investigated in the single cell literature [17–22]. An overview of limited feedback MIMO techniques in single-tier networks can be found in [23, 24].

As small cell networks are an instantiation of interference channels, multi-antenna techniques for small cell networks can be inspired by emerging technologies proposed for MIMO interference networks. Base station cooperation, also known as coordinated multi-point transmission (CoMP) in 3GPP LTE/ LTE-Advanced or network MIMO [25], exploits, instead of avoiding, inter-cell interference [26–28]. These techniques often rely on a reliable wired backhaul to exchange information between base stations and coordinate their transmissions. However, in femtocell-aided cellular networks, since the finite-rate backhaul is not always reliable and/or the information exchange and coordination between tiers is limited or absent, most of multi-cell MIMO techniques do not extend in a straightforward manner and many challenges exist. Recently, a new characterization of the interference channels has led to a transmission strategy known as interference alignment (IA) [29], which may lead to efficient precoding techniques in interference-limited networks.[1] Finally, several MIMO techniques proposed and analyzed in wireless ad hoc networks can also be employed in two-tier networks, especially in cases where stochastic geometry is used to model the network topology (see for instance [30–36] and references therein).

5.4　　System model

We consider a two-tier cellular system (see Figure 5.1), consisting of macrocells overlaid with a tier of low-power, short-range radio access nodes. Depending on the node density, the transmit power levels, and the operation characteristics, the second tier may be a relevant model for different types of small cells, e.g., femtocells, metrocells, picocells. In the remainder, we will assume closed-access femtocells but the results can be easily extended to open-access femtocells and picocells.

The macrocell base station (MBS) B_0 is located at the center of the macrocell and is equipped with N_{tc} antennas and provides service to a geographical region $\mathcal{C}$, assumed as a circular disk with radius R_c and area $|\mathcal{C}| = \pi R_c^2$. Each femtocell base station (FBS) is equipped with N_{tf} antennas. All user terminals in the network are equipped with a single receive antenna.

In a given time/frequency slot, each macrocell [resp. femtocell] employs its antennas for serving $1 \leq K_c \leq N_{tc}$ cellular [resp. $1 \leq K_f \leq N_{tf}$ indoor] users. When scheduling is considered, users are selected within a set of active users $\mathcal{U}_c$ [resp. $\mathcal{U}_f$] in each cell, and the set of served macrocell [resp. femtocell] users is denoted as $\mathcal{K}_c$ [resp. $\mathcal{K}_f$].

The users in each cell face interference from all the other macrocells and femtocells in the network. In our analysis, we focus on the downlink performance of the two-tier network, and we assume no coordination between the MBSs and the FBSs. The no coordination assumption is reasonable considering the uncontrolled and random deployment of femtocells. The uplink and the downlink transmissions of the two tiers are, however, synchronized.

---

[1] A downlink interference alignment scheme for cognitive femtocells is discussed in Chapter 6.

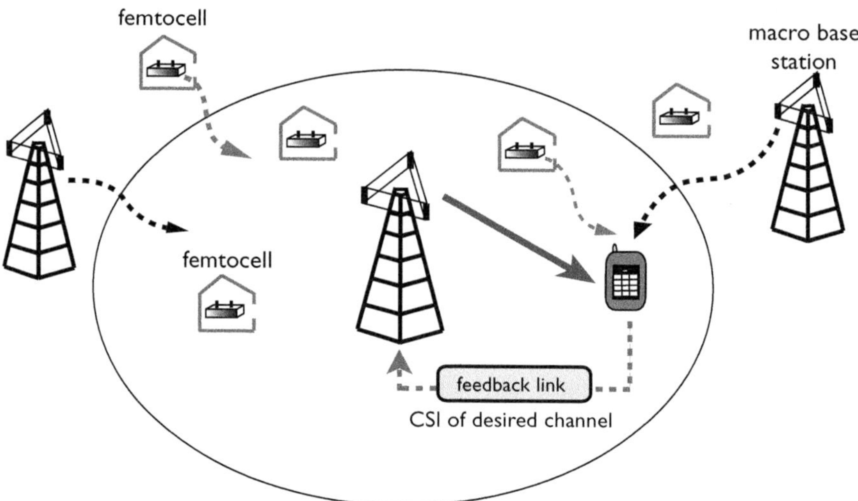

Figure 5.1 Mobile user M_0 experiences cross-tier interference from randomly distributed femtocells. It experiences same-tier interference from neighboring interfering macrocell base stations. M_0 feeds back its CSI to MBS B_0.

5.4.1   Network model

We use a stochastic geometry framework to model the distribution of the BS locations, which is a relevant model as FBS deployment is unplanned and varies over space and time. We analyze the performance inside a fixed macrocell (typical macrocell). The FBSs are located according to a spatial homogeneous Poisson point process (PPP) Π_f with intensity λ_f [37–39]. The MBS locations outside of the fixed cell are modeled according to a spatial homogeneous PPP Π_c with intensity λ_c, independent of the FBSs [40].

5.4.2   Channel model

The channel corresponding to the desired signal between the base station in the typical cell B_0 and a cellular user $k \in \{0, 1, \dots, K_c - 1\}$ is denoted by $\mathbf{h}_k \in \mathbb{C}^{N_{tc} \times 1}$, with its entries distributed as $\mathcal{CN}(0, 1)$. The direction of each vector channel is represented as $\tilde{\mathbf{h}}_k \triangleq \frac{\mathbf{h}_k}{\|\mathbf{h}_k\|}$. Designate $\tilde{\mathbf{H}} = [\tilde{\mathbf{h}}_0, \tilde{\mathbf{h}}_1, \dots, \tilde{\mathbf{h}}_{K_c-1}]^H \in \mathbb{C}^{K_c \times N_{tc}}$ as the concatenated matrix of channel directions. The interfering channel between the k-th cellular user and the i-th FBS F_i is denoted by $\mathbf{g}_{k,i,f} \in \mathbb{C}^{N_{tf} \times 1}$. The interfering channel between the k-th cellular user and the i-th interfering MBS B_i is denoted by $\mathbf{g}_{k,i,c} \in \mathbb{C}^{N_{tc} \times 1}$. When only one user is served, i.e., $K_c = K_f = 1$, the above channels for mobile user M_0 are denoted as $\mathbf{h}_0$, $\mathbf{g}_{i,f}$, and $\mathbf{g}_{i,c}$, respectively.

In this chapter, we assume linear precoding transmission with $\mathbf{V} \in \mathbb{C}^{N_{tc} \times K_c} = [\mathbf{v}_k]_{1 \le k \le K_c}$ denoting the precoding matrix of MBS B_0 and $\mathbf{W}_i = [\mathbf{w}_{k,i}]_{1 \le k \le K_f} \in \mathbb{C}^{N_{tf} \times K_f}$ denoting the precoding matrix used by each femtocell F_i to serve $1 \le K_f \le N_{tf}$ users.

The transmitted signals are subject to distance dependent path-loss effects. For analyzing the performance of MU-MIMO under perfect CSI, for tractability and exposition convenience, we consider the simplified power law path-loss model $\ell(r) = r^\alpha$ where $\alpha > 2$ is the path-loss exponent. With α_f [resp. α_c] we denote the path-loss exponent in the femtocell, both indoor and outdoor [resp. macrocell]. In the second part, where the performance of SU-MIMO with imperfect CSIT is investigated, we consider the non-singular path-loss model $\ell(r) = C \max(d_0, r)^\alpha$ where $d_0 > 0$ is the reference distance and $C > 0$ is a constant.

5.5   Zero-forcing beamforming in two-tier networks with perfect CSI

We investigate here the probability of coverage in the downlink of a two-tier network using zero-forcing (ZF) beamforming with perfect CSI at the central macrocell and the femtocells. We focus on ZF beamforming for analytical tractability and because it has low complexity, yet achieves the same multiplexing gain as higher complexity schemes such as dirty paper coding in MISO broadcast channels. For exposition convenience, interference from neighboring macrocell BSs is ignored.

Under ZF precoding transmission, the precoding matrix $\mathbf{V}$ used by B_0 is chosen as the normalized columns of the pseudoinverse $\tilde{\mathbf{H}}^H(\tilde{\mathbf{H}}\tilde{\mathbf{H}}^H)^{-1} \in \mathbb{C}^{N_{tc} \times K_c}$. Similarly, at each femtocell the columns of the precoding matrix $\mathbf{W}_j$ equal the normalized columns of $\tilde{\mathbf{F}}_j^{\,H}(\tilde{\mathbf{F}}_j\tilde{\mathbf{F}}_j^{\,H})^{-1} \in \mathbb{C}^{N_{tf} \times K_f}$, where the channel directions between F_j to its individual users are represented as $\tilde{\mathbf{f}}_j^H = [\tilde{\mathbf{f}}_{0,j},\ \tilde{\mathbf{f}}_{1,j},\ \ldots\ \tilde{\mathbf{f}}_{K_f-1,j}]$ and the entries of $\tilde{\mathbf{f}}_{k,j}$ are distributed as $\mathcal{CN}(0,1)$.

5.5.1   Coverage probability of femtocell users

Consider a reference femtocell F_0 at distance D from the macrocell B_0. During a given signaling interval, the received signal at femtocell user 0 at distance $d_{\max}$ w.r.t. F_0 is given as

$$y_0 = \underbrace{\sqrt{\frac{P_f}{K_f}}d_{\max}^{-\frac{\alpha_f}{2}}\mathbf{f}_0^H\mathbf{W}_0\mathbf{s}_{0,f}}_{\text{Desired signal}} + \underbrace{\sqrt{\frac{P_f}{K_f}}\sum_{F_i \in \Pi_f \backslash F_0}|X_{0,i}|^{-\frac{\alpha_f}{2}}\mathbf{f}_{i,f}^H\mathbf{W}_i\mathbf{s}_{i,f}}_{\text{Intra-tier interference}} + \underbrace{\sqrt{\frac{P_c}{K_c}}D^{-\frac{\alpha_c}{2}}\mathbf{f}_{0,c}^H\mathbf{V}\mathbf{s}_0}_{\text{Cross-tier interference}} + \mathbf{n}$$

where the vectors $\mathbf{s}_0 \in \mathbb{C}^{K_c \times 1}$ and $\mathbf{s}_{i,f} \in \mathbb{C}^{K_f \times 1}$ designate the normalized transmit data symbols for users in B_0 and F_i assuming equal power allocation, and $\mathbf{n}$ represents background noise. The term $\mathbf{f}_{0,c} \in \mathbb{C}^{N_{tc} \times 1}$ [resp. $\mathbf{f}_{i,f}$] designates the downlink vector channel from the interfering MBS B_0 [resp. interfering FBS F_i] to user 0 and $X_{0,i}$ denotes the location of i-th FBS. Neglecting receiver noise for analytical simplicity, the received signal to interference ratio (SIR) for user 0 is given as

$$\text{SIR}_f = \frac{\frac{P_f}{K_f}d_{\max}^{-\alpha_f}|\mathbf{f}_0^H\mathbf{w}_{0,0}|^2}{\frac{P_c}{K_c}D^{-\alpha_c}||\mathbf{f}_{0,c}^H\mathbf{V}||^2 + \frac{P_f}{K_f}\sum_{F_i \in \Pi_f \backslash F_0}||\mathbf{f}_{i,f}^H\mathbf{W}_i||^2|X_{0,i}|^{-\alpha_f}}. \tag{5.1}$$

User 0 can successfully decode its signal provided SIR_f is at least equal to its minimum SIR target T. We define

$$\mathcal{P}_f = \frac{P_c}{P_f} D^{-\alpha_c}, \quad \mathcal{Q}_f = d_{\max}^{\alpha_f} K_f, \tag{5.2}$$

and let

$$\kappa = \frac{\mathcal{P}_f \mathcal{Q}_f T}{K_c}. \tag{5.3}$$

The expression $\frac{\kappa}{\kappa+1} \in [0, 1)$ characterizes the relative strength of cellular interference. As κ increases (or $\frac{\kappa}{\kappa+1} \to 1$), user 0 experiences progressively poor coverage due higher cellular interference. Conversely, as $\kappa \to 0$, SIR_f is limited by interference from neighboring femtocells. Because of cellular interference, any femtocell user within $D \leq D_f$ meters of B_0 (D_f being the no-coverage femtocell radius) cannot satisfy its quality of service (QoS) requirement. As long as $D > D_f$, a femtocell user can tolerate interference from *both* cellular transmissions and hotspot transmissions. The no-coverage femtocell radius D_f can be computed using the following result.

Theorem 5.1 ([3]) *Any femtocell F_0 within $D < D_f$ meters from the macrocell B_0 cannot satisfy its QoS requirement ϵ, where D_f is given as*

$$D_f = \left[\frac{d_{\max}^{-\alpha_f}}{T} \frac{P_f/K_f}{P_c/K_c} \left(\frac{\mathcal{I}^{-1}(\epsilon; N_{tf} - K_f + 1, K_c)}{1 - \mathcal{I}^{-1}(\epsilon; N_{tf} - K_f + 1, K_c)} \right) \right]^{-1/\alpha_c}, \tag{5.4}$$

where the inverse function $\mathcal{I}^{-1}(y; a, b) \triangleq x$ is defined as the value of x for which $\mathcal{I}_x(a, b) = y$, with $\mathcal{I}_x(a, b)$ denoting the regularized incomplete beta function.

Remark 5.1 *Since $\mathcal{I}^{-1}(y; a, b)$ is monotonically increasing with a and monotonically decreasing with b for any $a, b \geq 0$, we can show that if SU-MISO beamforming is employed at the femtocells ($K_f = 1$), the no-coverage radius $D_{f,SU}$ is strictly smaller than the no-coverage radius $D_{f,MU}$ with multi-user beamforming transmission ($1 < K_f \leq N_{tf}$).*

We can also see that for $K_c = 1$, the reduction in the no-coverage radius using a SU-MISO transmission strategy at the femtocells relative to MU-MISO transmission to $K_f = N_{tf}$ users [resp. single-antenna transmission (SISO)] is given as

$$\frac{D_{f,SU}}{D_{f,MU}} = \left[\left(\frac{1 - \epsilon^{1/N_{tf}}}{\epsilon^{1/N_{tf}}} \right) \frac{\epsilon}{1 - \epsilon} \frac{1}{N_{tf}} \right]^{1/\alpha_c} \approx \left[\frac{\epsilon^{1-1/N_{tf}}}{N_{tf}} \right]^{1/\alpha_c}.$$

$$\frac{D_{f,SU}}{D_{f,SISO}} = \left[\left(\frac{1 - \epsilon^{1/N_{tf}}}{\epsilon^{1/N_{tf}}} \right) \frac{\epsilon}{1 - \epsilon} \right]^{1/\alpha_c} \approx \epsilon^{\frac{1}{\alpha_c}(1 - 1/N_{tf})}.$$

Furthermore, for fixed N_{tf}, K_f, and K_c, the no-coverage femtocell radius D_f in (5.4) scales as $(P_f/P_c)^{-1/\alpha_c}$. Decreasing D_f by a factor of k requires increasing P_f by $10\alpha_c \log_{10} k$ decibels. This suggests that a graph of D_f versus P_f/P_c is a straight line on a log-log scale with slope $-1/\alpha_c$.

We provide now the probability of coverage for a femtocell user, i.e., the probability that a randomly located femtocell user inside the femtocell has a target SIR greater than T, i.e., $p_{\mathrm{cov},\mathrm{f}}(\lambda_\mathrm{f}, \lambda_\mathrm{c}, T) = \mathbb{P}\left[\mathrm{SIR}_\mathrm{f} \geq T\right]$.

Theorem 5.2 ([3]) *The probability of coverage for a femtocell user in the presence of macrocell cross-tier interference and femtocell intra-tier interference is given by*

$$p_{cov,f}(\lambda_f, \lambda_c, T) = \sum_{k=0}^{N_{tf}-K_f} \frac{(-\mathcal{Q}_f T)^k}{k!} \sum_{j=0}^{k} \binom{k}{j} \frac{d^j}{d\theta^j} \left(1 + \frac{\mathcal{P}_f \theta}{K_c}\right)^{-K_c} \frac{d^{(k-j)}}{d\theta^{(k-j)}} e^{-\lambda_f C_f \theta^{\delta_f}}. \tag{5.5}$$

where $\delta_f = 2/\alpha_f$, $\theta = \mathcal{Q}_f T$,

$$C_f = \pi \delta_f K_f^{-\delta_f} \sum_{k=0}^{K_f-1} \binom{K_f}{k} B(k + \delta_f, K_f - k - \delta_f), \tag{5.6}$$

and $B(a, b) = \frac{\Gamma(a)\Gamma(b)}{\Gamma(a+b)}$ is the Euler beta function.

Based on the expression of the probability of coverage, we derive the maximum obtainable spatial reuse from multiple antenna femtocells when they share spectrum with cellular transmissions. Mathematically, the *maximum femtocell contention density* satisfying a QoS constraint is expressed as

$$\lambda_\mathrm{f}^*(D) = \arg\max \lambda_\mathrm{f}(D), \quad \text{subject to } \mathbb{P}(\mathrm{SIR}_\mathrm{f} \geq T) \geq 1 - \epsilon. \tag{5.7}$$

Theorem 5.3 ([3]) *In a two-tier network, the maximum femtocell contention density $\lambda_f^*(D)$ at distance D from the macrocell B_0, which satisfies (5.7) (in the small-ϵ regime) is given as*

$$\lambda_f^*(D) = \frac{1}{C_f(\mathcal{Q}_f \Gamma)^{\delta_f}} \left[\frac{\epsilon - \mathcal{I}_{\frac{\kappa}{\kappa+1}}(N_{tf} - K_f + 1, K_c)}{\frac{1}{\mathcal{G}_f} - \mathcal{I}_{\frac{\kappa}{\kappa+1}}(N_{tf} - K_f + 1, K_c)} \right] \tag{5.8}$$

where

$$\mathcal{G}_f = \left[1 + \frac{1}{(1+\kappa)^{K_c}} \sum_{j=0}^{N_{tf}-K_f-1} \left(\frac{\kappa}{\kappa+1}\right)^j \binom{K_c + j - 1}{j} \sum_{l=1}^{N_{tf}-K_f-j} \frac{1}{l!} \prod_{m=0}^{(l-1)} (m - \delta_f) \right]^{-1}.$$
$$\tag{5.9}$$

Note that whenever $K_\mathrm{f} = N_{\mathrm{tf}}$ we have that $\mathcal{G}_f = 1$.

Theorem 5.3 provides the maximum femtocell contention density at D considering both cross-tier cellular and hotspot interference from neighboring femtocells. Alternatively, given an average of λ_f transmitting femtocells per square meter, (5.8) can be inverted (numerically) to obtain the minimum D which guarantees that (5.7) is feasible.

A necessary condition for a positive femtocell contention density at distance D is $\mathcal{I}_{\frac{\kappa}{\kappa+1}}(N_{\mathrm{tf}} - K_\mathrm{f} + 1, K_\mathrm{c}) < \epsilon$, or $\kappa(D)$ in (5.3) is upper bounded as

$$\kappa \leq \frac{\mathcal{I}^{-1}(\epsilon; N_{\mathrm{tf}} - K_\mathrm{f} + 1, K_\mathrm{c})}{1 - \mathcal{I}^{-1}(\epsilon; N_{\mathrm{tf}} - K_\mathrm{f} + 1, K_\mathrm{c})}.$$

Violating the condition implies that the femtocells cannot guarantee reliable coverage because of cross-tier interference.

As $\kappa \to 0$ or $D^{-\alpha_c} \to 0$, the SIR at any femtocell located at D is primarily influenced by femtocell intra-tier interference. Consequently, $\lambda_f^*(D)$ in (5.8) approaches the limit $\check{\lambda}_f$ given by

$$\check{\lambda}_f = \lim_{\kappa \to 0} \lambda_f^*(D) = \frac{\epsilon\,\check{\mathcal{G}}_f}{\mathcal{C}_f(\mathcal{Q}_f T)^{\delta_f}}, \text{ where } \check{\mathcal{G}}_f = \lim_{\kappa \to 0} \mathcal{G}_f = \left[1 + \sum_{l=1}^{N_{tf}-K_f} \frac{1}{l!} \prod_{m=0}^{l-1}(m - \delta_f)\right]^{-1}.$$

$$(5.10)$$

The limit $\check{\mathcal{G}}_f$ determines the maximum contention density in the special case where macrocellular cross-tier interference can be neglected. In that case, we have that $\check{\mathcal{G}}_f$ and $\mathcal{C}_f$ scale as

$$\check{\mathcal{G}}_f \sim \Theta[(N_{tf} - K_f + 1)^{\delta_f}], \quad \mathcal{C}_f \mathcal{Q}_f^{\delta_f} \sim \Theta(K_f^{\delta_f}). \tag{5.11}$$

Moreover, $\forall \kappa \geq 0, \mathcal{G}_f \leq \check{\mathcal{G}}_f$ and $\mathcal{G}_f$ is bounded as $(N_{tf} - K_f + 1)^{\delta_f} \leq \check{\mathcal{G}}_f \leq \Gamma(1 - \delta_f)(N_{tf} - K_f + 1)^{\delta_f}$. In the case of multi-user transmission to $K_f = N_{tf}$ femtocell users, the femtocell area spectral efficiency (ASE) (in b/s/Hz/m^2), given as $(1 - \epsilon)K_f\check{\lambda}_f \log_2(1 + T)$ scales as $\Theta(N_{tf}^{1-\delta_f})$. With single-user transmission, the ASE scales as $\Theta(N_{tf}^{\delta_f})$. This implies that in path-loss environments with $\alpha_f < 4$, higher spatial reuse can be obtained (order-wise) provided that each FBS serves only one user. In contrast, multi-user femtocell transmission provides higher network-wide spatial reuse (order-wise) only when femtocell interference is significantly diminished ($\alpha_f > 4$).

5.5.2   Coverage probability of macrocell users

We now consider a cellular user M_0 at distance D from its MBS B_0, whose received signal is given by

$$y_0 = \sqrt{\frac{P_c}{K_c}} D^{-\frac{\alpha_c}{2}} \mathbf{h}_0^H \mathbf{V} \mathbf{s}_0 + \sqrt{\frac{P_f}{K_f}} \sum_{F_i \in \Pi_f} \mathbf{g}_{i,f}^H \mathbf{W}_i |X_{0,i}|^{-\frac{\alpha_f}{2}} \mathbf{s}_{i,f} + \mathbf{n} \tag{5.12}$$

where $|X_{0,i}|$ denotes the distance from the interfering femtocell F_i to user M_0. Neglecting the background noise, the probability of coverage of mobile user M_0 is given as

$$\mathbb{P}[\text{SIR}_c \geq T] = \mathbb{P}\left[\frac{|\mathbf{h}_0^H \mathbf{v}_0|^2}{\frac{1}{K_f}\sum_{F_i \in \Pi_f} ||\mathbf{g}_{i,f}^H \mathbf{W}_i||^2 |X_{0,i}|^{-\alpha_f}} \geq \mathcal{Q}_c T\right]. \tag{5.13}$$

where $\mathcal{Q}_c = K_c \frac{P_f}{P_c} D^{\alpha_c}$.

Using [30], the maximum femtocell contention density $\lambda_f^*(D)$ for which (5.13) satisfies the maximum outage probability constraint $\mathbb{P}(\text{SIR}_c \geq T) \geq 1 - \epsilon$ of a cellular user is given as

$$\lambda_f^*(D) = \frac{\epsilon\,\mathcal{G}_c}{\mathcal{C}_f(\mathcal{Q}_c T)^{\delta_f}}, \text{ where } \mathcal{G}_c = \left[1 + \sum_{j=1}^{N_{tc}-K_c} \frac{1}{j!} \prod_{k=0}^{j-1}(k - \delta_f)\right]^{-1}. \tag{5.14}$$

Note that $\mathcal{G}_c$ can be bounded as

$$(N_{tc} - K_c + 1)^{\delta_f} \leq \mathcal{G}_c \leq \Gamma(1 - \delta_f)\,(N_{tc} - K_c + 1)^{\delta_f}, \tag{5.15}$$

where $\Gamma(a, b)$ is the Euler gamma function.

Remark 5.2 *Since (5.14) varies as $\mathcal{G}_c/K_c^{\delta_f}$, approximating $\mathcal{G}_c$ by the upper bound in (5.15) shows that the maximum contention density for single-user beamforming denoted as $\lambda_{f,SU}^{*}(D)$ is proportional to $\Gamma(1 - \delta_f)N_{tc}^{\delta_f}$. With $1 < K_c < N_{tc}$ served users, the maximum femtocell contention density denoted as $\lambda_{f,MU}^{*}(D)$ is proportional to $\Gamma(1 - \delta_f)(N_{tc} - K_c + 1)^{\delta_f}/K_c^{\delta_f}$. Therefore, single-user beamforming increases the maximum femtocell density by a factor of $[N_{tc}K_c/(N_{tc} - K_c + 1)]^{\delta_f}$. With $K_c = N_{tc}$ users, we have that $\lambda_{f,MU}^{*}(D)$ is proportional to $N_{tc}^{-\delta_f}$, so that $\lambda_{f,SU}^{*}(D)/\lambda_{f,MU}^{*}(D)$ equals $\Gamma(1 - \delta_f)N_{tc}^{2\delta_f}$.*

Given an average of λ_f femtocells per square meter, inverting (5.14) yields the maximum distance up to which the cellular outage probability lies below ϵ. This cellular coverage radius D_c is given as

$$D_c = \left(\frac{P_c}{TK_cP_f}\right)^{1/\alpha_c} \left(\frac{\epsilon\mathcal{G}_c}{\lambda_f\mathcal{C}_f}\right)^{\frac{1}{\delta_f\alpha_c}}. \tag{5.16}$$

Remark 5.3 *Since D_c varies as $\left(P_c/P_f\right)^{1/\alpha_c}$, increasing the cellular coverage radius by a factor of k necessitates increasing P_c by $10\alpha_c \log_{10} k$ decibels relative to P_f.*

Since D_c is proportional to $(\frac{\mathcal{G}_c^{1/\delta_f}}{K_c})^{\frac{1}{\alpha_c}}$, with single-user transmission [resp. multi-user transmission to $K_c = N_{tc}$ users] at the macrocell and applying (5.15), the cellular coverage distance D_c scales with N_{tc} as $D_{c,SU} \sim \Theta(N_{tc}^{1/\alpha_c})$, $D_{c,MU} \sim \Theta(N_{tc}^{-1/\alpha_c})$. This suggests that single-user beamforming at the macrocell provides coverage improvement by a factor of N_{tc}^{2/α_c} (order-wise) relative to ZF beamforming.

5.5.3   Simulation results

Figure 5.3 shows that single-user transmission obtains a nearly $1.5\times$ reduction in the no-coverage femtocell radius D_f w.r.t. single-antenna transmission for $N_{tf} = 2$. Next, both Figures 5.2 and 5.3 show that single-user transmission reduces D_f by a factor of nearly $1.8\times$ relative to multi-user beamforming, indicating that single-user transmission significantly improves femtocell coverage. In Figures 5.3 and 5.4, the number of femtocells N_f increases from zero (at the no-coverage femtocell radius) to greater than 100 femtocells per cell-site within a few meters outside the no-coverage radius. This step-like transition from the cellular-limited to the femtocell-limited regime suggests that cross-tier cellular interference is the capacity-limiting factor even in densely populated femtocell networks, and interference between femtocells is negligible because of the proximity of home users to their FBSs.

Single-user transmission outperforms multi-user transmission even in the femtocell-limited regime with $\alpha_f = 3.8$. For example, with $N_{tf} = 2$ antennas, there is a nearly $1.7\times$ spatial reuse gain ($N_fK_f = 1080$ with single-user transmission versus $N_fK_f =$

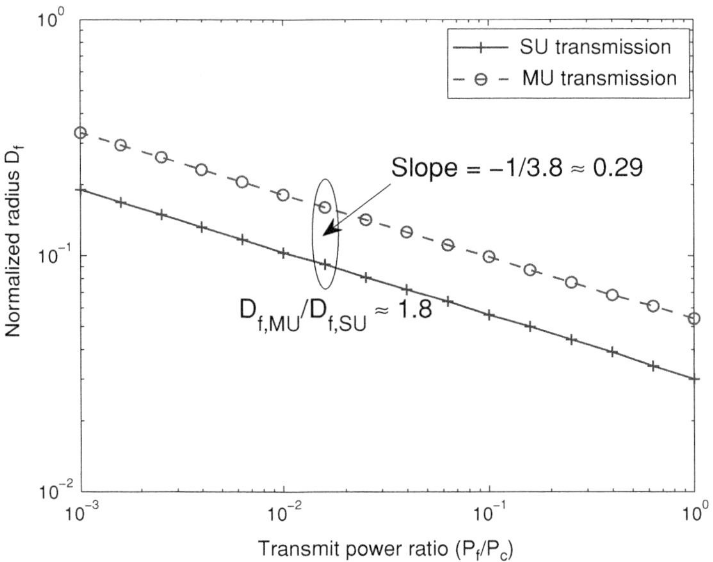

Figure 5.2 No-coverage femtocell radius for different values of $\frac{P_f}{P_c}$. ©2011 IEEE. Reprinted, with permission, from [16].

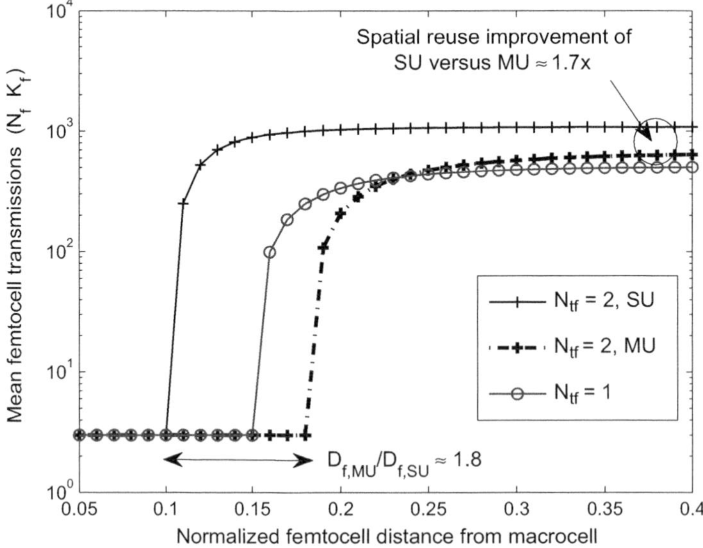

Figure 5.3 Maximum number of simultaneous femtocell transmissions $N_f K_f$ for different number of antennas and single-user versus multiple-user transmission per femtocell. ©2009 IEEE. Reprinted, with permission, from [3].

640 with multi-user transmission). In a scenario in which femtocell interference is *significantly diminished* (Figure 5.4 with $\alpha_f = 4.8$ and $N_{tf} = 3$ antennas), multi-user transmission to $K_f = 2$ femto users provides a marginally higher spatial reuse relative to single-user transmission. The conclusion is that achieving the multiplexing benefits

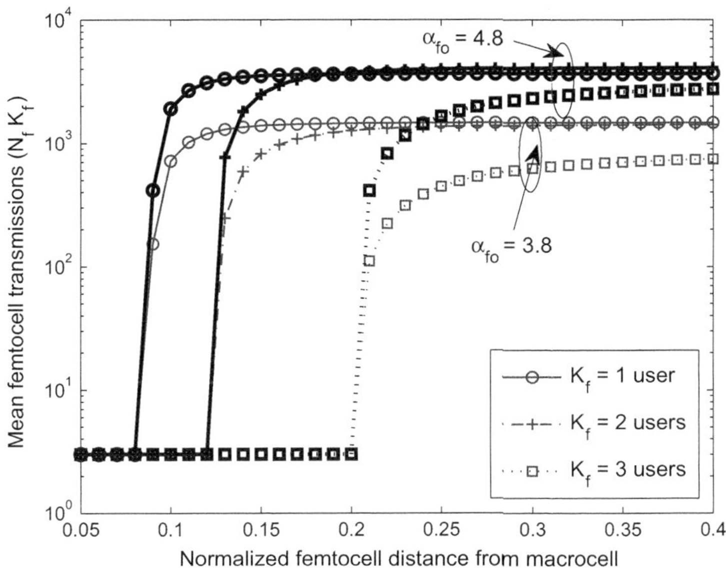

Figure 5.4 Maximum number of simultaneous femtocell transmissions $N_{\mathrm{f}}K_{\mathrm{f}}$ for different values of α_{f}. ©2009 IEEE. Reprinted, with permission, from [3].

of multi-user transmission requires relative isolation (or large α_{f}) among actively transmitting FBSs.

Figure 5.5 plots the maximum number of transmitting femtocells $N_{\mathrm{f}} = \pi \mathrm{R}_{\mathrm{c}}^2 \lambda_{\mathrm{f}}^*(D)$ as a function of the cellular user distance D. With $(P_{\mathrm{c}}/P_{\mathrm{f}})_{\mathrm{dB}} = 20$ and a desired $N_{\mathrm{f}} = 60$ femtocells/cell site, single-user transmission at the macrocell provides a normalized cellular coverage radius $D_{\mathrm{c}} \approx 0.35$. In contrast, the coverage provided by multi-user transmission is only $D_{\mathrm{c}} \approx 0.13$, resulting in a coverage loss of $2.7\times$ relative to single-user transmission. With single-user transmission and $(P_{\mathrm{c}}/P_{\mathrm{f}})_{\mathrm{dB}} = 0$, a cellular user at $D = 0.1$ can tolerate interference from nearly $N_{\mathrm{f}} = 62$ femtocells/cell site. In contrast, with multi-user transmission, N_{f} reduces to nearly 8 femtocells/cell site.

The preceding observations reveal that since the performance of multi-user transmission is significantly limited by residual femtocell interference, the macrocell should maximize cellular coverage by transmitting to only a single cellular user and that femtocells should adapt their transmit powers depending on their location in order to ensure reliable cellular coverage.

5.6   Zero-forcing beamforming with perfect CSI and user selection

In this section, we study the performance of ZF beamforming with perfect CSI where the MBS [resp. the FBSs] selects up to K_{c} [resp. K_{f}] out of U_{c} [resp. U_{f}] users at a time. The user selection is performed based on the desired receive signal power and not based on the instantaneous SINR as the aggregate interference (and the set of scheduled users)

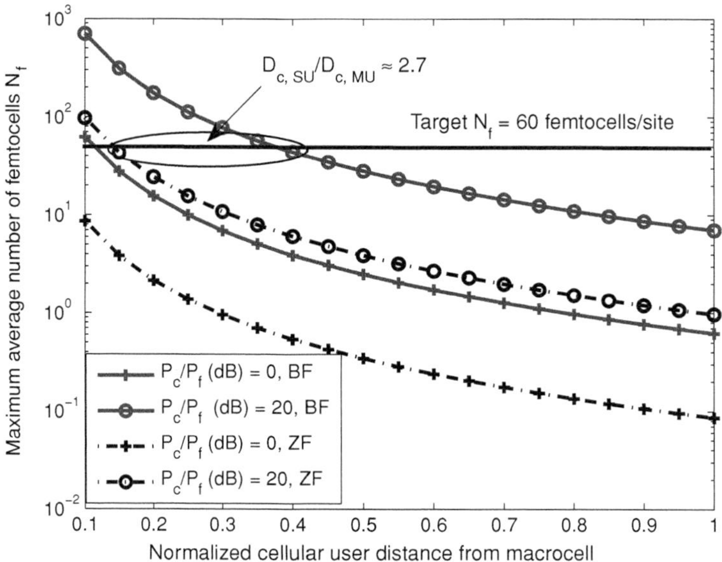

Figure 5.5 Maximum number of simultaneous femtocell transmissions satisfying outage probability constraint ϵ for a cellular user at different distances from the macrocell. ©2009 IEEE. Reprinted, with permission, from [3].

is not known during the feedback/scheduling stage. In the SU-MISO case, the channel quality indicator (CQI) fed back by the cellular and femtocells users takes on the form of the received SNR. In the MU-MISO case, the user group $\mathcal{K}$ is selected from the set of active users $\mathcal{U}$ as the one that maximizes the single-cell sum rate with no other cell interference, i.e.,

$$\mathcal{K} = \arg\max_{k \in \mathcal{U}} \sum_{k=1}^{K} \log_2\left(1 + \mathrm{CQI}_k\right), \tag{5.17}$$

where CQI_k is the received SNR for the k-th user under ZF beamforming. As the number of active users in each small cell is relatively low, exhaustive search is considered here.

Two different type of CQI metrics are investigated here

- *Channel-aware user selection (CUS)*: where only the effect of small-scale fading is considered,
- *Path-loss-aware user selection (PUS)*: where both fading and path-loss effects are taken into account into the CQI/rate calculations.

We provide, first, the following result for the user with the best channel when K_f users are served under CUS (the results are the same for macrocell users).

Theorem 5.4 ([16]) *The probability of coverage for a user having the largest effective channel gain selected among U_f active users and being at distance d_{max} from the FBS is* *given by*

$$p_{cov,f}^{(k^*)} = \sum_{u=1}^{U_f} (-1)^{u+1} \binom{U_f}{u} \sum_{t_1,t_2,\ldots,t_m} (-s)^{\mathcal{A}} \mathcal{X} \frac{\partial^{\mathcal{A}}}{\partial (us)^{\mathcal{A}}} \mathcal{L}_I(us) \qquad (5.18)$$

where the second summation is taken over all sequences of non-negative integer indexes t_1 through t_m such that the sum of all t_j is u, and $\mathcal{X} = \left(\frac{u!}{t_1! t_2! \cdots t_m!}\right) \left[\prod_{1 \le j \le m} ((j-1)!)^{t_j}\right]^{-1}$, $\mathcal{A} = \sum_{j=1}^{m}(j-1)t_j$, $m = N_{tf} - K_f + 1$, $s = T d_{max}^{\alpha_f}$, and $\mathcal{L}_I(\cdot) = \mathcal{L}_{I_{f,c}}(\cdot)\mathcal{L}_{I_{f,f}}(\cdot)$ denotes the Laplace transform of the aggregate interference.

In a SU-MISO beamforming system with $K_f = N_{tf}$, the above theorem takes on a simpler form, i.e.,

Lemma 5.1 *The probability of coverage for the femtocell user with the highest channel gain is given by*

$$p_{cov,f}^{(k^*)} = \sum_{u=1}^{U_f} (-1)^u \binom{U_f}{u} \frac{e^{-\lambda_f (us)^{2/\alpha_f} C_f}}{\left(1 + \frac{P_c D^{-\alpha_c}}{P_f} us\right)^{K_f}}. \qquad (5.19)$$

The coverage performance of ZF beamforming with user selection can be quantified using Theorem 5.4; however, the closed-form expression for the probability of coverage is rather involved. For that, we provide the following simple yet tight bounds:

Lemma 5.2 ([16]) *The probability of coverage of the best femtocell user under ZF beamforming satisfies*

$$\sum_{u=1}^{mU_f} \binom{mU_f}{u}(-1)^{u+1}\mathcal{L}_I(us) \le p_{cov,f}^{(k^*)} \le \sum_{u=1}^{mU_f} \binom{mU_f}{u}(-1)^{u+1}\mathcal{L}_I(vus) \qquad (5.20)$$

with $v = \Gamma(1+m)^{-\frac{1}{m}}$ and $m = N_{tf} - K_f + 1$.

In the case where the MBSs and the FBSs have knowledge of the distance-dependent path loss to their respective receivers, the user with the highest effective channel is the one having the largest product of channel gain (fading) and path-loss attenuation, which is now a random variable. However, since the cumulative distribution function (CDF) of the path-loss attenuation power is involved, deriving a closed-form expression for the probability of coverage of the selected user is hard to obtain and one can resort to results exploiting asymptotics and stochastic dominance.

5.6.1 Simulation results

We assess here the performance of ZF beamforming in a two-tier network with the following operating values, listed as [macro-parameter, femtocell-parameter]: cell radius [$R_c = 200, d_{max} = 20$] m, path-loss exponent [$\alpha_c = 4, \alpha_f = 3.5$], target SIR [$T_c = 0, T_f = 5$]

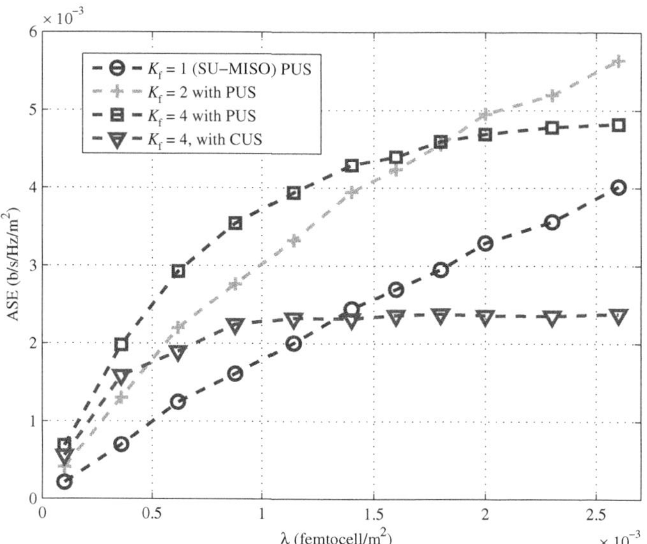

Figure 5.6 Area spectral efficiency vs. femtocell density with $U_f = U_c = 12$ active users per cell.

dB, transmit power $[P_c = 43, P_f = 13]$ dBm. Both tiers use $N_{tc} = N_{tf} = 4$ transmit antennas.

In Figure 5.6 we observe that PUS performs better than CUS, providing a gain of approximately 2×10^{-3} bit/s/Hz/m^2 for densities larger than 15 femtocells per hectare (ha). This is due to the increased multi-user diversity gains from the heavy tail behavior of the random distance (path-loss attenuation). We also see that the optimal number of streams K_f depends, among others, on the FBS density. For instance, for more than 17 FBS/ha, transmitting only two streams results in higher area spectral efficiency than using all transmit streams ($K_f = K_c = 4$). Note also that SU-MISO becomes optimal for very dense networks.

5.7   Beamforming in two-tier networks with imperfect CSI

We derive here the probability of coverage on the downlink of the macrocell network with quantized and delayed CSIT, first accounting for cross-tier interference from the femtocells, and then considering cross-tier interference and same-tier interference from the MBSs.

Since the edge of the typical cell is fixed and is not determined by interference levels, a guard region of radius R_g from the cell edge is imposed around the fixed cell in which no other transmitters can reside. In general, R_g is a parameter of the model; however, choosing $R_g = R_c/2$ was shown to provide a good performance relative to a hexagonal tessellation [41]. We compute the interference power from the interfering MBSs at a mobile user M_0 at a distance $0 < r < R_c$ from the typical base station B_0 located at the center of the cell. Since the exclusion distance around M_0 to the nearest interferer is not symmetric, the interference power outside a ball of radius $R_c + R_g - r$ around the mobile user is considered. This translation of the disk to be centered at M_0 removes the asymmetry in the interference field and results in an upper bound on the aggregate interference power. The exclusion region is denoted as $B(R_g + R_c - r)$.

The downlink performance in the typical cell is evaluated accounting for the MBSs interferers outside the guard region and femto stations within and outside the typical cell. The closed-access FBSs cause cross-tier interference to the macrocellular users and the interference from the femtocells is spatially invariant. To avoid the femtocells being located with the cellular user, we assume that the femtocells are located outside a ball of radius R_{cross} around the cellular user. R_{cross} is independent of r and the exclusion region due to cross-tier interference is denoted by $B(R_{cross})$.

For a narrowband flat-fading channel, the baseband discrete input–output relation for M_0 is given by

$$
y_0[n] = \sqrt{P_c}\, (\ell(r))^{-\frac{1}{2}}\, \mathbf{h}_0^H[n]\mathbf{v}_0[n - d]s_0[n]
$$

$$
+ \sqrt{P_f} \sum_{F_i \in \Pi_f / B_{cross}} (\ell(r_{i,f}))^{-\frac{1}{2}}\, \mathbf{g}_{i,f}^H[n]\mathbf{w}_{i,f}[n - d]s_{i,f}[n] \tag{5.21}
$$

$$
+ \sqrt{P_c} \sum_{B_i \in \Pi_c / B(R_c + R_g - r)} (\ell(r_{i,c}))^{-\frac{1}{2}}\, \mathbf{g}_{i,c}^H[n]\mathbf{w}_{i,c}[n - d]s_{i,c}[n] + v_0[n],
$$

$$
\tag{5.22}
$$

where $y_0[n] \in \mathbb{C}$ is the n-th received data sample at M_0. The vector $\mathbf{v}_0[n - d] \in \mathbb{C}^{N_{tc} \times 1}$ is the transmit beamforming vector at B_0 for M_0, calculated using delayed CSI. The vectors $\mathbf{w}_{i,f}[n - d] \in \mathbb{C}^{N_{tf} \times 1}$ and $\mathbf{w}_{i,c}[n - d] \in \mathbb{C}^{N_{tc} \times 1}$ are the transmit beamforming vectors used by femtocell F_i and the MBS B_i, respectively. The normalized symbol transmitted from B_0 intended for M_0 is given by s_0, and $s_{i,f}$, $s_{i,c}$ denote the normalized symbols transmitted by the F_i and the B_i to their users, respectively. Finally, $v_0[n] \in \mathbb{C}$ is additive white Gaussian noise (AWGN) at M_0 with variance σ^2.

5.7.1　Limited feedback beamforming and imperfect CSIT model

Since there is no coordination between the MBSs and the FBSs, limited feedback beamforming is employed on the downlink of the typical cell. For that, a single-cell, single-user limited feedback model is considered, in which the macrocell user M_0 quantizes the desired channel direction $\widetilde{\mathbf{h}}_0[n]$ by means of a unit norm vector codebook $\mathcal{V} = \{\mathbf{v}_1, \mathbf{v}_2, \cdots, \mathbf{v}_N\}$, of size $N = 2^B$ and known to both the transmitter and the receiver. The quantizer function $\mathcal{Q}$ at M_0 chooses the beamforming vector $\mathbf{v}_\ell$ that maximizes the inner product

$$
\mathbf{v}_\ell[n] = \mathcal{Q}\{\mathbf{h}_0[n]\} = \arg\max_{\mathbf{v}_k \in \mathcal{V}} |\widetilde{\mathbf{h}}_0^H[n]\mathbf{v}_k|^2 \quad 1 \le k \le N. \tag{5.23}
$$

The channel $\mathbf{h}_0[n]$ is then mapped to the index $I_n = \ell$ which is fed back to B_0 using B bits. The CSIT index received at B_0 at time n corresponds to the quantized channel index at time $n - d$, $I_{n-d} = m$. The beamforming vector $\mathbf{v}[n - d]$ corresponding to the index $I_{n-d} = m$ is then used to transmit to M_0 at time n.

To model the effect of delay in the feedback link, we use the Gauss–Markov autoregressive model, which has been shown in the literature to be reasonably accurate for small delays in the communication links [42, 43]. It has been widely used in the literature to model the effect of delay on the performance of wireless systems with limited feedback. Using the Gauss–Markov model, the current and delayed channel are related by

$$\mathbf{h}_0[n] = \eta \mathbf{h}_0[n - d] + \sqrt{1 - \eta^2}\,\mathbf{e}[n], \tag{5.24}$$

where $\mathbf{e}[n]$ is the channel error vector, uncorrelated with $\mathbf{h}_0[n - d]$. The entries of $\mathbf{e}[n]$ are distributed as $\mathcal{CN}(0, 1)$. The correlation coefficient for the desired channel is denoted by η. The Clarke's autocorrelation can, for example, be used to determine η as follows

$$\eta = \mathcal{J}_0\left(2\pi d f_{\mathrm{d}} \mathrm{T}_{\mathrm{s}}\right), \tag{5.25}$$

where $\mathcal{J}_0$ denotes the zero-th order Bessel function of the first kind, and f_{d} is the Doppler frequency.

We denote the aggregate cross-tier interference by

$$I_{\mathrm{f}} = \sum_{F_i \in \Pi_{\mathrm{f}}/B_{\mathrm{cross}}} \frac{|\mathbf{g}_{i,\mathrm{f}}^{\mathrm{H}}[n]\mathbf{w}_{i,\mathrm{f}}[n - d]|^2}{\ell(r_{i,\mathrm{f}})}$$

and the aggregate macrocell interference by

$$I_{\mathrm{c}}(r) = \sum_{B_i \in \Pi_{\mathrm{c}}/B(R_{\mathrm{c}}+R_{\mathrm{g}}-r)} \frac{|\mathbf{g}_{i,\mathrm{c}}^{\mathrm{H}}[n]\mathbf{w}_{i,\mathrm{c}}[n - d]|^2}{\ell(r_{i,\mathrm{c}})}.$$

The SINR estimated at M_0 is given by

$$\mathrm{SINR}[n, d] = \frac{|\mathbf{h}_0^{\mathrm{H}}[n]\mathbf{v}_0[n - d]|^2}{Q_{\mathrm{f}}(r)I_{\mathrm{f}} + Q_{\mathrm{c}}(r)I_{\mathrm{c}}(r) + \sigma_{\mathrm{c}}^2}, \tag{5.26}$$

where $Q_{\mathrm{f}}(r) = \frac{P_f \ell(r)}{P_{\mathrm{c}}}$ and $Q_{\mathrm{c}}(r) = \ell(r)$ are a function of the path-loss model, independent of the beamforming vectors, and $\sigma_{\mathrm{c}}^2(r) = \ell(r)\frac{\sigma^2}{P_{\mathrm{c}}}$. The interference functions I_{f} and $I_{\mathrm{c}}(r)$ are shot noise processes [39].

5.7.2   Analysis with cross-tier interference

For a given SINR threshold T, a key metric of interest, which serves as a building block for performance analysis, is the probability of coverage of M_0,

$$p_{\mathrm{cov,c}}(\lambda_{\mathrm{f}}, \lambda_{\mathrm{c}}, \alpha, T, d, r) = \mathbb{P}\left[\mathrm{SINR}[n, d] \geq T\right]. \tag{5.27}$$

The probability of coverage is the probability that M_0, randomly located inside the typical cell has a target SINR greater than T. It is evaluated as the complementary cumulative distribution function (CCDF) of the desired channel power $H_0 = |\mathbf{h}_0^H[n]\mathbf{v}_0[n - d]|^2$ given the aggregate cross-tier interference $I_f(r)$ and the receiver noise,

$$\mathbb{P}\left[\text{SINR}[n, d] \geq T\right] = \mathbb{P}\left[\left|\mathbf{h}_0^H[n]\mathbf{v}_0[n - d]\right|^2 \geq \left(Q_f(r)I_f + \sigma_c^2(r)\right) T\right]$$

$$= \mathbb{P}\left[H_0 \geq T\left(Q_f(r)I_f + \sigma_c^2(r)\right)\right]$$

$$= \mathbb{E}_{Q_f(r)I_f + \sigma_c^2(r)}\left\{\mathbb{P}\left[H_0 \geq T\left(Q_f(r)I_f + \sigma_c^2(r)\right)\right]\right\}. \quad (5.28)$$

To compute the probability of coverage, we need to derive an expression for the CCDF of the desired channel power under quantized and delayed feedback.

5.7.3 Distribution of the desired signal power

For the evaluation of the desired instantaneous channel power, we assume that for large values of η (low mobility), the terms with $\mathbf{e}[n]$ can be removed, as their coefficient $(1 - \eta^2)$ is usually small, i.e.,

$$\left|\mathbf{h}_0^H[n]\mathbf{v}_0[n - d]\right|^2 \approx \eta^2 \left|\mathbf{h}_0^H[n - d]\mathbf{v}_0[n - d]\right|^2. \quad (5.29)$$

This approximation is common in the literature to simplify the analysis [20, 44].

Furthermore, a quantization cell approximation (QCA) [45, 46] is used to approximate the performance of the MISO limited feedback system, which has been shown to closely model the performance of codebook design techniques such as random vector quantization (RVQ), and is used in the literature to analyze the performance of limited feedback systems [18–20].

Using the above assumptions, we obtain the following result.

Lemma 5.3 *The CDF of the effective desired channel power* $Z = \left|\mathbf{h}_0^H[n]\mathbf{v}_0[n - d]\right|^2$ *is given by*

$$F_Z(z) = 1 - 2^B \exp\left(-\frac{z}{2\eta^2}\right)$$

$$+ 2^B(1 - \delta)\exp\left(-\frac{z}{2\eta^2(1 - \delta)}\right) \sum_{i=0}^{N_{tc}-2} \sum_{\ell=0}^{i} \frac{\delta^i}{(i - \ell)!} \left(\frac{z}{2\eta^2(1 - \delta)}\right)^{i-\ell},$$

$$(5.30)$$

where $\delta = 2^{-\frac{B}{N_{tc}-1}}$.

Proof. See Appendix 5 A.   □

5.7.4 Probability of coverage

The probability of coverage follows from the desired signal CCDF by averaging over the interference power at the receiver and is given in the following theorem.

Theorem 5.5 *The probability of coverage for the macrocell user* M_0 *in the presence of femtocells cross-tier interference and thermal noise is given by*

$$p_{cov,c}(\lambda_f, \alpha, T, d, r) = \omega_3(T)\omega_1(T, \lambda_f) \exp\left(-\omega_1(T, \lambda_f)\right) \exp\left(-\frac{\sigma_c^2 T}{2\eta^2(1-\delta)}\right)$$

$$+ 2^B \exp\left(-\omega_2(T, \lambda_f)\right) \exp\left(-\frac{\sigma_c^2 T}{2\eta^2}\right)$$

$$- 2^B(1-2^{-B}) \exp\left(-\omega_1(T, \lambda_f)\right) \exp\left(-\frac{\sigma_c^2 T}{2\eta^2(1-\delta)}\right),$$

$$(5.31)$$

where

$$\omega_1(T, \lambda_f) = \lambda_f C_f \left(\frac{T Q_f}{2\eta^2(1-\delta)}\right)^{\frac{2}{\alpha}},$$

$$\omega_3(T) = 2^B(1-\delta)$$

$$\times \left[\sum_{i=1}^{N_{tc}-2} \sum_{\ell=0}^{i-1} \frac{\delta^i}{(i-\ell)!} \sum_{j=0}^{i-\ell} \binom{i-\ell}{j} \prod_{m=0}^{j-1} \left(\frac{2}{\alpha} - m\right)(-1)^j (Q_f)^{-j} \left(\frac{T\sigma_c^2}{2\eta^2(1-\delta)}\right)^{i-\ell-j}\right],$$

$$\omega_2(T, \lambda_f) = \lambda_f C_f \left(\frac{T Q_f}{2\eta^2}\right)^{\frac{2}{\alpha}}, \quad and \quad C_f = \frac{2\pi}{\alpha} \Gamma\left(\frac{2}{\alpha}\right) \Gamma\left(1 - \frac{2}{\alpha}\right). \quad (5.32)$$

Proof. See Appendix 5 B.   □

5.7.5   Maximum femtocells contention density

Using Theorem 5.5 and approximating the exponential terms by the first three terms of their Taylor series expansion around zero, i.e., $\exp(x) \approx 1 - x + x^2$ we can show that

Corollary 5.1 *The maximum femtocell density* $\lambda_f^*(r)$ *for which the probability of successful reception (coverage) satisfies the maximum outage probability constraint,*

$$\lambda_f^*(r) = \max \lambda_f(r) \ such \ that \ \mathbb{P}\left[\mathsf{SINR}[n, d] \geq T\right] \geq 1 - \epsilon, \quad (5.33)$$

is the largest root of the third order polynomial in λ_f, $P(\lambda_f)$, *such that* $P(\lambda_f) \geq 0$,

$$P(\lambda_f) = A_{\lambda_f} \lambda_f^3 + B_{\lambda_f} \lambda_f^2 + C_{\lambda_f} \lambda_f + D_{\lambda_f}, \quad (5.34)$$

where $A_\delta = \left(\frac{1}{1-\delta}\right)^{\frac{2}{\alpha_f}}$, $\quad A_{\lambda_f} = \left(C_f \left(\frac{\Omega Q_f}{2\eta^2}\right)^{\frac{2}{\alpha_f}}\right)^3 \left(\omega_3 \, A_\delta^3 \, e^{-\frac{\sigma_c^2 \Omega}{2\eta^2(1-\delta)}}\right),$

$$B_{\lambda_f} = \left(C_f \left(\frac{\Omega Q_f}{2\eta^2}\right)^{\frac{2}{\alpha_f}}\right)^2 \left(-\omega_3 \, A_\delta^2 \, e^{-\frac{\sigma_c^2 \Omega}{2\eta^2(1-\delta)}} + 2^B \, e^{-\frac{\sigma_c^2 \Omega}{2\eta^2}} - 2^B(1-2^{-B})A_\delta^2 e^{-\frac{\sigma_c^2 \Omega}{2\eta^2(1-\delta)}}\right),$$

$$C_{\lambda_f} = C_f \left(\frac{\Omega Q_f}{2\eta^2}\right)^{\frac{2}{\alpha_f}} \left(\omega_3 \, A_\delta \, e^{-\frac{\sigma_c^2 \Omega}{2\eta^2(1-\delta)}} - 2^B \, e^{-\frac{\sigma_c^2 \Omega}{2\eta^2}} + 2^B(1-2^{-B})A_\delta \, e^{-\frac{\sigma_c^2 \Omega}{2\eta^2(1-\delta)}}\right),$$

and $D_{\lambda_f} = 2^B \, e^{-\frac{\sigma_c^2 \Omega}{2\eta^2}} - 2^B(1-2^{-B}) \, e^{-\frac{\sigma_c^2 \Omega}{2\eta^2(1-\delta)}} - 1 + \epsilon.$

Calculating the maximum number of transmitting FBS per area provides an answer to the network provider about the expected coverage that the MBS can provide to its users at different locations inside the cell, given the average number of transmitting femtocells. To guarantee a certain quality of service requirement to the mobile users, the MBS may adjust its transmission technique to increase its coverage given the cross-tier interference. The network operator may further put a limit on the number of closed-access MBS that it can allow in certain locations so as not to hinder the macrocell network performance, through a certain pricing strategy or an enforced femtocell access-control strategy. Note that the maximum femtocell density increases as the feedback rate for the MISO limited feedback system increases, and caps at a maximum value close to that achieved by perfect CSIT.

5.7.6   Analysis with same-tier and cross-tier interference

In this section, we derive the probability of coverage for M_0 in the presence of both cross-tier interference from the femtocells and same-tier interference from neighboring MBSs. We assume that the system is interference limited, and that the received signal power is much higher than the noise power, thus thermal noise is neglected in our computations.

Theorem 5.6 *The probability of coverage for the macrocell user in the presence of same-tier inter-cell interference and cross-tier interference is upper bounded by*

$$
p_{cov,c}\left(\lambda_c, \lambda_f, \alpha, d, r\right) \leq 2^B \mathcal{L}_J\left(\frac{T}{2\eta^2}\right) - 2^B\left(1 - 2^{-B}\right)\mathcal{L}_J\left(\frac{T}{2\eta^2(1 - \delta)}\right)
$$

$$
+ 2^B\left(1 - \delta\right)\sum_{i=1}^{N_{tc}-2}\sum_{\ell=0}^{i-1}\frac{\delta^i}{(i - \ell)!}\left(-\frac{T}{2\eta^2(1 - \delta)}\right)^{i-\ell}\frac{\mathrm{d}^{i-\ell}}{\mathrm{d}\theta^{i-\ell}}\mathcal{L}_J\left(\theta\right)\Bigg|_{\theta=\frac{T}{2\eta^2(1-\delta)}} \tag{5.35}
$$

where

$$
\mathcal{L}_J(\theta) = \exp\left(-2\pi\theta\left(\frac{Q_f \mathrm{R}_{\mathrm{cross}}^{2-\alpha}\lambda_f}{C}\,{}_2F_1\left[1, 1 - \frac{2}{\alpha}, 2 - \frac{2}{\alpha}, -\frac{Q_f \mathrm{R}_{\mathrm{cross}}^{-\alpha}s}{C}\right]\right.\right.
$$

$$
\left.\left. + \left(\frac{Q_c(R_c + R_\mathrm{g} - r)^{2-\alpha}\lambda_c}{C}\,{}_2F_1\left[1, 1 - \frac{2}{\alpha}, 2 - \frac{2}{\alpha}, -\frac{Q_c(R_c + R_\mathrm{g} - r)^{-\alpha}s}{C}\right]\right)\right)\right).
$$

Proof. The proof follows that of Theorem 5.5, except that $\mathrm{d}\mathbb{P}(Q_f(r)I_f + \sigma_c^2 \leq s)$ is replaced by $\mathrm{d}\mathbb{P}(Q_f(r)I_f + Q_c(r)I_c \leq s)$ where $Q_c(r)I_c$ corresponds to the macrocell same-tier interference. $\qquad\square$

The bound on the probability of coverage in Theorem 5.6 is a function of the densities λ_f and λ_c and the relative powers Q_f and Q_c of the FBSs and the MBSs interferers. We observe from $\mathcal{L}_J$ the relative effect of the macrocell same-tier interference over that of cross-tier interference. Although the macrocell exclusion region $B(R_\mathrm{g} + R_\mathrm{c} - r)$ is greater than the exclusion region of the femtocells cross-tier interferers $B(R_{\mathrm{cross}}$, the MBSs transmit power P_c is higher than the transmit power of the FBSs, leading to a ratio

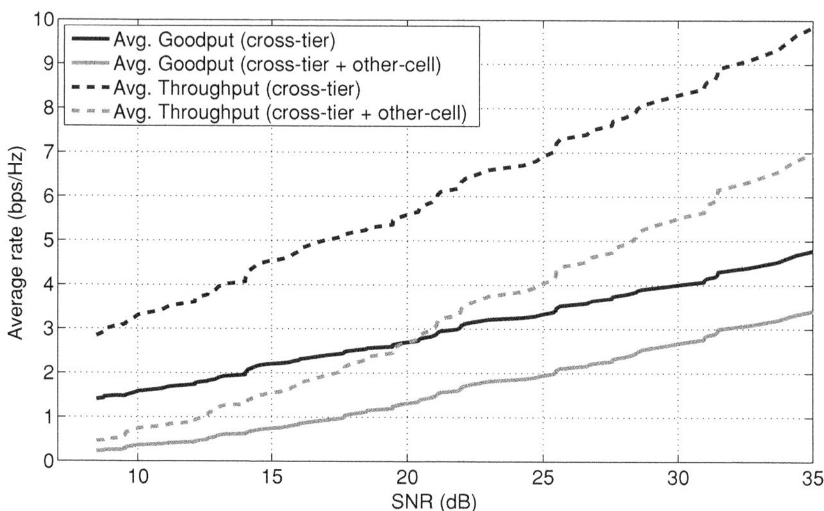

Figure 5.7 The average rate as a function of signal-to-noise ratio with and without other-cell interference for a limited feedback system with $N_{tc} = N_{tf} = 4$, and $N_{of} = 95$ femtocells per cell site, and $B = 5$ bits on the feedback channel.

$Q_f < Q_c$; the density of the MBSs interferers is also expected to be smaller than that of the femtocell cross-tier interferers $\lambda_c \leq \lambda_f$.

5.7.7   Simulation results

We consider a macrocell of radius $R_c = 300$ m and adopt a distance-based path-loss model corresponding to the IMT-2000 channel model [47] for outdoor and indoor path-loss. The Doppler spread is $f_d = \frac{vf_c}{c}$, where v is the relative velocity of the transmitter–receiver pair, f_c is the carrier frequency, and c is the speed of light. We assume fixed wall partition losses corresponding to indoor-to-outdoor and outdoor-to-indoor propagation, equal to 10 dB [47]. The outdoor and indoor-to-outdoor path-loss exponents are set to 3.8 and the carrier frequency is 2 GHz. The macrocell user terminal is uniformly distributed inside the fixed macrocell. For the limited feedback beamforming system, the quantization codebook is generated using RVQ.

Figure 5.7 compares the average goodput and throughput (rate) with femtocell cross-tier and macrocell same-tier interference for $\lambda_c = 5 \times 10^{-5}$ and $\lambda_f = 5 \times 10^{-4}, \frac{\lambda_f}{\lambda_c} = 10$. We observe from the figure that considering same-tier interference degrades the average goodput by 5 dB in SNR. The effect of the probability of outage is discernible for both cross-tier and other-tier interference. Taking other-cell interference into account is shown to degrade the performance by an average of 10 dB. This shows the importance of considering other-cell interference, when uncoordinated, in the analysis of heterogeneous networks.

In Figure 5.8, we plot the probability of outage for varying distances of M_0 from its BS. We observe that expectedly the probability of outage increases as the densities λ_f and

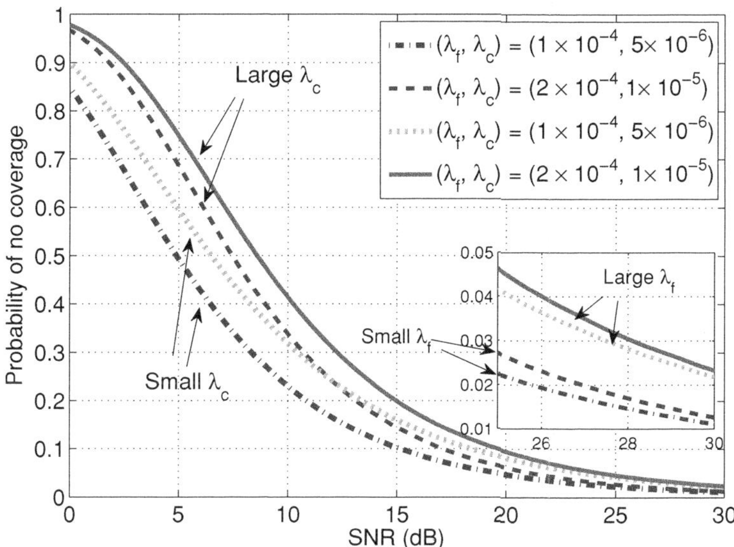

Figure 5.8 The probability of no coverage for varying densities of FBS and MBS interferers. We consider four cases, corresponding to (a) $(\lambda_f, \lambda_c) = (1 \times 10^{-4}, 5 \times 10^{-6})$, (b) $(\lambda_f, \lambda_c) = (2 \times 10^{-4}, 1 \times 10^{-5})$, (c) $(\lambda_f, \lambda_c) = (1 \times 10^{-4}, 5 \times 10^{-6})$, and (d) $(\lambda_f, \lambda_c) = (2 \times 10^{-4}, 1 \times 10^{-5})$.

λ_c increase. Furthermore, the probability of outage for cases (a) and (b) converge when the user is in the cell center, suggesting that the femtocells' interference becomes more prominent as the user moves toward the cell center. Similarly for cases (c) and (d) at the cell center. In contrast, when the user is close to the cell-edge, the macrocell interference dominates the sum interference term, and the effect of femtocells interference becomes more negligible. The probability of outage for cases (a) and (c) as well as that of cases (b) and (d) converge at the cell edge.

5.8　　Conclusions

In this chapter, we have investigated the performance of practical multiple antenna techniques for the downlink of a two-tier cellular network with other-cell interference and cross-tier interference from femtocells. We derived analytical expressions for the probability of coverage and the coverage zones in such a tiered architecture considering the number of antennas, the maximum tolerable outage probability accounting for path-loss and Rayleigh fading. In the case of perfect CSI, single-user beamforming in either tier is analytically shown to provide significantly superior coverage and spatial reuse, while performance of multi-user transmission (ZF beamforming) suffers from residual cross-tier interference. Multi-user transmission gains may appear when ZF beamforming is coupled with user scheduling. In the case of quantized and delayed CSIT, we derived the probability of coverage for a macrocell user as a function of the distance from the MBS, the temporal correlation of the channel, the density of the femtocell nodes, as well as the density of the other-cell interferers. We also showed that the density of transmitting femtocells increases as the rate on the feedback channel increases for the limited feedback beamforming system and caps at a maximum number of femtocells in accordance with perfect CSIT. Finally, we showed that the effect of other-cell interference is more prominent at the cell edge, which results in considerable average rate decrease.

Appendix 5A Proof of Lemma 5.3

Under QCA, for a given codebook $\mathcal{V}$, the quantization cell R_i is approximated as the union of two spherical caps with

$$R_i = \{\widetilde{\mathbf{h}} : |\widetilde{\mathbf{h}}^{\mathrm{H}}\mathbf{v}_i|^2 \geq 1 - \delta\}, \quad \text{with } \delta = 2^{-\frac{B}{N_{\mathrm{tc}}-1}}.$$

It was shown in [19], Lemma 2, that $|\mathbf{h}_0^{\mathrm{H}}[n]\mathbf{v}_0[n]|^2 = |\mathbf{h}_0[n]|^2\cos^2\left(\angle\left(\widetilde{\mathbf{h}}_0[n], \mathbf{v}_0[n]\right)\right)$ can be expressed as the sum of two independent chi-square random variables, $Z = X + Y$ where $X \sim \chi^2_{2(N_{\mathrm{tc}}-1)}(\eta^2(1-\delta))$ and $Y \sim \chi^2_2(\eta^2)$, respectively. The CDF in (5.30) is that of the sum of two independent chi-square random variables with $2(N_{\mathrm{tc}} - 1)$ and two degrees of freedom, and two different parameters, respectively. It follows from [48].

Appendix 5B Proof of Theorem 5.5

The probability of coverage is computed as

$$\mathbb{P}\left[\mathrm{SINR} \geq T\right] = \mathbb{P}\left[H_0 \geq T\left(Q_{\mathrm{f}}(r)I_{\mathrm{f}} + \sigma_{\mathrm{c}}^2\right)\right] = \int_0^\infty \mathbb{P}\left[H_0 \geq Ts\right] f_{Q_{\mathrm{f}}(r)I_{\mathrm{f}}+\sigma_{\mathrm{c}}^2}(s)ds$$

$$\overset{(a)}{=} \int_0^\infty \left[2^B \exp\left(-\frac{sT}{2\eta^2}\right) - 2^B(1-\delta)\exp\left(-\frac{sT}{2\eta^2(1-\delta)}\right)\right.$$

$$\left. \times \sum_{i=0}^{N_{\mathrm{tc}}-2}\sum_{\ell=0}^{i}\frac{\delta^i}{(i-\ell)!}\left(\frac{sT}{2\eta^2(1-\delta)}\right)^{i-\ell}\right] f_{Q_{\mathrm{f}}(r)I_{\mathrm{f}}+\sigma_{\mathrm{c}}^2}(s)ds$$

$$= \int_0^\infty \left[2^B \exp\left(-\frac{sT}{2\eta^2}\right) - 2^B(1-\delta)\exp\left(-\frac{sT}{2\eta^2(1-\delta)}\right)\frac{1-2^{-B}}{1-\delta}\right]$$

$$\times f_{Q_{\mathrm{f}}(r)I_{\mathrm{f}}+\sigma_{\mathrm{c}}^2}(s)ds \tag{5.36}$$

$$- \int_0^\infty \left[2^B(1-\delta)\exp\left(-\frac{sT}{2\eta^2(1-\delta)}\right)\sum_{i=1}^{N_{\mathrm{tc}}-2}\sum_{\ell=0}^{i-1}\frac{\delta^i}{(i-\ell)!}\left(\frac{sT}{2\eta^2(1-\delta)}\right)^{i-\ell}\right]$$

$$\times f_{Q_{\mathrm{f}}(r)I_{\mathrm{f}}+\sigma_{\mathrm{c}}^2}(s)ds \tag{5.37}$$

where step (a) follows from conditioning on $Q_{\mathrm{f}}(r)I_{\mathrm{f}} + \sigma_{\mathrm{c}}^2(r)$ and substituting for the CCDF of $H_0 = |\mathbf{h}_0^{\mathrm{H}}[n]\mathbf{v}_0[n-d]|^2$. The integral in (5.36) is expressed in terms of the Laplace transform of the interference term $Q_f(r)I_f$ and the Laplace transform of the noise term σ_c^2, i.e.,

$$\int_0^\infty \left[2^B \exp\left(-\frac{sT}{2\eta^2}\right) - 2^B(1-\delta)\exp\left(-\frac{sT}{2\eta^2(1-\delta)}\right) \sum_{i=0}^{N_{tc}-2} \delta^i \right] f_{Q_f(r)I_f+\sigma_c^2}(s)ds$$

$$= 2^B \mathbb{E}\left[e^{-(Q_f(r)I_f+\sigma_c^2)\theta}\right]\Bigg|_{\theta=\frac{T}{2\eta^2}} - \left(2^B\left(1-2^{-B}\right)\right)\mathbb{E}\left[e^{-(Q_f(r)I_f+\sigma_c^2)\theta}\right]\Bigg|_{\theta=\frac{T}{2\eta^2(1-\delta)}}$$

$$= 2^B \mathbb{E}\left[e^{-Q_f(r)I_f\theta}\right]\mathbb{E}\left[e^{-\sigma_c^2\theta}\right]\Bigg|_{\theta=\frac{T}{2\eta^2}} - \left(2^B\left(1-2^{-B}\right)\right)\mathbb{E}\left[e^{-Q_f(r)I_f\theta}\right]\mathbb{E}\left[e^{-\sigma_c^2\theta}\right]\Bigg|_{\theta=\frac{T}{2\eta^2(1-\delta)}}$$

$$= 2^B \mathcal{L}_{Q_f I_f}(\theta)\mathcal{L}_{\sigma_c^2}(\theta)\Bigg|_{\theta=\frac{T}{2\eta^2}} - \left(2^B\left(1-2^{-B}\right)\right)\mathcal{L}_{Q_f I_f}(\theta)\mathcal{L}_{\sigma_c^2}(\theta)\Bigg|_{\theta=\frac{T}{2\eta^2(1-\delta)}}.$$

The Laplace transform of the noise term σ_c^2 is given by $\mathcal{L}_{\sigma_c^2}(s) = e^{-s\sigma_c^2}$. The Laplace transform of the interference term $I_f = \displaystyle\sum_{F_i \in \Pi_f/B_{\text{cross}}} \frac{|\mathbf{g}_{i,f}^{H}[n]\mathbf{w}_{i,f}[n-d]|^2}{\ell(r_{i,f})} =$

$\displaystyle\sum_{F_i \in \Pi_f/B_{\text{cross}}} \frac{G_{i,f}}{\ell(r_{i,f})}$ is derived as

$$\mathcal{L}_{Q_f I_f}(s) = \mathbb{E}\left\{e^{-Q_f I_f s}\right\} = \mathbb{E}\left\{e^{-Q_f \sum_{F_i \in \Pi_f/B_{\text{cross}}} \frac{G_{i,f}}{\ell(r_{i,f})}}\right\}$$

$$= \mathbb{E}_{\Pi_f, G_{i,f}}\left\{\prod_{F_i \in \Pi_f/B_{\text{cross}}} e^{-sQ_f G_{i,f}/\ell(r_i,f)}\right\}$$

$$= \mathbb{E}_{\Pi_f}\left\{\prod_{F_i \in \Pi_f/B_{\text{cross}}} \mathbb{E}_{G_{i,f}}\left\{e^{-sQ_f G_{i,f}/\ell(r_i,f)}\right\}\right\}$$

$$= \exp\left(-2\pi\lambda_f \int_{R_{\text{cross}}}^\infty \left(1 - \mathbb{E}_{G_{i,f}}\left\{e^{-sQ_f G_{i,f}C^{-1}v^{-\alpha}}\right\}\right) vdv\right)$$

$$= \exp\left(-2\pi\lambda_f \int_{R_{\text{cross}}}^\infty \left(1 - \frac{1}{1+s\frac{Q_f}{C}v^{-\alpha}}\right) vdv\right)$$

$$= \exp\left[\frac{-2\pi\lambda_f C^{-1}Q_f(r)R_{\text{cross}}^{2-\alpha}}{\alpha-2} s\, {}_2F_1\left(1, 1-\frac{2}{\alpha}, 2-\frac{2}{\alpha}, -Q_f(r)C^{-1}R_{\text{cross}}^{-\alpha} s\right)\right].$$

$$(5.38)$$

The integral in (5.37) is evaluated as

$$
\int_0^\infty \left[2^B(1-\delta)\exp\left(-\frac{sT}{2\eta^2(1-\delta)}\right) \sum_{i=1}^{N_{tc}-2}\sum_{\ell=0}^{i-1}\frac{\delta^i}{(i-\ell)!}\left(\frac{sT}{2\eta^2(1-\delta)}\right)^{i-\ell} \right] f_{Q_f(r)I_f+\sigma_c^2}(s)\,ds
$$

$$
\overset{(a)}{=} 2^B(1-\delta)\sum_{i=1}^{N_{tc}-2}\sum_{\ell=0}^{i-1}\frac{\delta^i}{(i-\ell)!}\left(-\frac{T}{2\eta^2(1-\delta)}\right)^{i-\ell}\frac{d^{i-\ell}}{d\theta^{i-\ell}}\mathcal{L}_{I_f}(Q_f(r)\theta)\mathcal{L}_{\sigma_c^2}(\theta)\Bigg|_{\theta=\frac{T}{2\eta^2(1-\delta)}},
$$

where (a) follows from the identity $\mathcal{L}[x^k f(x)] = (-1)^k F^{(k)}(s)$ for the Laplace transform with $F^{(k)}(s)$ representing the k-th derivative of $F(s)$.

The $(i-\ell)$-th derivative of the product of the Laplace transforms $\mathcal{L}_{I_f}(Q_f(r)\theta)$ and $\mathcal{L}_{\sigma_c^2}(\theta)$ is computed using the Leibniz rule as

$$
\frac{d^{i-\ell}}{d\theta^{i-\ell}}\mathcal{L}_{I_f}(Q_f(r)\theta)\mathcal{L}_{\sigma_c^2}(\theta) = \sum_{j=0}^{i-\ell}\binom{i-\ell}{j}\frac{d^j}{d\theta_f^j}\mathcal{L}_{I_f}(\theta_f)\frac{d^{(i-\ell-j)}}{d\theta^{(i-\ell-j)}}\mathcal{L}_{\sigma_c^2}(\theta), \quad (5.39)
$$

where $\theta_f = Q_f(r)\theta$. The p-th derivative of $\mathcal{L}_{\sigma_c^2}(\theta)$ is given by $\frac{d^p}{d\theta^p}e^{-\sigma_c^2\theta} = (-1)^p(\sigma_c^2)^p e^{-\sigma_c^2\theta}$. The j-th derivative of the Laplace transform $\mathcal{L}_{I_f}(\theta)$ is given by taking the derivative of (5.38) in terms of the Gauss hypergeometric function.

For more tractable expressions of the probability of coverage in the presence of cross-tier interference, we assume that $R_{cross} \to d_0 \to 0$, the Laplace transform of the interference is then expressed as $\mathcal{L}_{Q_f I_f}(s) = \exp\left(-\lambda_f C_f Q_f^{\frac{2}{\alpha}} C^{-\frac{2}{\alpha}} s^{\frac{2}{\alpha}}\right)$, and $C_f = \pi\frac{2}{\alpha}\Gamma[\frac{2}{\alpha}]\Gamma[1-\frac{2}{\alpha}]$, and the jth derivative of $\mathcal{L}_{Q_f I_f}(s)$ is obtained by forming the first order Taylor expansion for the jth derivative and ignoring the higher order terms in $\lambda_f C_f \theta_f^{\frac{2}{\alpha}}$ [30],

$$
\frac{d^j}{d\theta^j}\exp\left(-\lambda_f C_f\theta_f^{\frac{2}{\alpha}}\right) = -\left[\lambda_f C_f\prod_{m=0}^{j-1}(\frac{2}{\alpha}-m)\,\theta_f^{\frac{2}{\alpha}-j}\exp\left(-\lambda_f C_f\theta_f^{\frac{2}{\alpha}}\right)\right] + \Theta(\lambda_f^2 C_f^2\theta_f^{2\frac{2}{\alpha}}).
$$

Plugging in the expressions for the derivatives of the Laplace transforms in (5.39), we have that

$$
2^B(1-\delta)\left(\sum_{i=1}^{N_{tc}-2}\sum_{\ell=0}^{i-1}\frac{\delta^i}{(i-\ell)!}\left(\frac{T\sigma_c^2}{2\eta^2(1-\delta)}\right)^{i-\ell}\sum_{j=0}^{i-\ell}\binom{i-\ell}{j}\right.
$$

$$
\times\left[-\prod_{m=0}^{j-1}(\frac{2}{\alpha}-m)\left(-Q_f(r)\,\theta\,\sigma_c^2\right)^{-j}\right]\Bigg)
$$

$$
\left.\times\left(\lambda_f C_f(Q_f(r)\theta)^{\frac{2}{\alpha}}\exp\left(-\lambda_f C_f(Q_f(r)\theta)^{\frac{2}{\alpha}}\right)\exp\left(-\sigma_c^2\theta\right)\right)\right)\Bigg|_{\theta=\frac{T}{2\eta^2(1-\delta)}}. \quad (5.40)
$$

Combining the two parts of the integration for the simplified path-loss model, the probability of coverage at M_0 is

$$
\mathbb{P}\left[\mathrm{SINR} \geq T\right] = 2^B e^{-\lambda_f C_f\left(\frac{TQ_f}{2\eta^2}\right)^{\frac{2}{\alpha}}} e^{-\frac{\sigma_c^2 T}{2\eta^2}} - \left(2^B(1-2^{-B})\right) e^{-\lambda_f C_f\left(\frac{TQ_f}{2\eta^2(1-\delta)}\right)^{\frac{2}{\alpha}}} e^{-\frac{\sigma_c^2 T}{2\eta^2(1-\delta)}}
$$

$$
+ 2^B(1-\delta)\left[\sum_{i=1}^{N_{tc}-2}\sum_{\ell=0}^{i-1}\frac{\delta^i}{(i-\ell)!}\left(\frac{T\sigma_c^2}{2\eta^2(1-\delta)}\right)^{i-\ell}\sum_{j=0}^{i-\ell}\binom{i-\ell}{j}\right.
$$

$$
\times \prod_{m=0}^{j-1}\left(\frac{2}{\alpha}-m\right)\left(\frac{-T\sigma_c^2 Q_f(r)}{2\eta^2(1-\delta)}\right)^{-j}\right]
$$

$$
\times \lambda_f C_f\left(Q_f(r)\frac{T}{2\eta^2(1-\delta)}\right)^{\frac{2}{\alpha}} \exp\left(-\lambda_f C_f\left(Q_f(r)\frac{T}{2\eta^2(1-\delta)}\right)^{\frac{2}{\alpha}}\right)
$$

$$
\times \exp\left(-\sigma_c^2\frac{T}{2\eta^2(1-\delta)}\right)
$$

$$
= \omega_3(T)\omega_1(T,\lambda_f)e^{-\omega_1(T,\lambda_f)}e^{-\frac{\sigma_c^2 T}{2\eta^2(1-\delta)}} + 2^B e^{-\omega_2(T,\lambda_f)}e^{-\frac{\sigma_c^2 T}{2\eta^2}}
$$

$$
- \left(2^B\left(1-2^{-B}\right)\right)e^{-\omega_1(T,\lambda_f)}e^{-\frac{\sigma_c^2 T}{2\eta^2(1-\delta)}}, \tag{5.41}
$$

where $\omega_1(T,\lambda_f)$, $\omega_2(T,\lambda_f)$, $\omega_3(T)$ and C_f are given in (5.32).

References

[1] J. Lee, J.-K. Han, and J. Zhang, "MIMO technologies in 3GPP LTE and LTE-Advanced," *EURASIP J. Wirel. Commun. Netw.*, vol. 2009, May 2009.

[2] J. G. Andrews, H. Claussen, M. Dohler, S. Rangan, and M. C. Reed, "Femtocells: past, present, and future," *IEEE J. Sel. Areas Commun. (JSAC)*, vol. 30, no. 3, pp. 497–508, Apr. 2012.

[3] V. Chandrasekhar, M. Kountouris, and J. G. Andrews, "Coverage in multi-antenna two-tier networks," *IEEE Trans. Wireless Commun.*, vol. 8, no. 10, pp. 5314–27, Oct. 2009.

[4] A. Paulraj, R. Nabar, and D. Gore, *Introduction to Space-time wireless Communications*. Cambridge: Cambridge University Press, 2003.

[5] H. Bolcskei, D. Gesbert, C. B. Papadias, and A.-J. van der Veen, *Space-Time Wireless Systems: From Array Processing to MIMO Communications*. Cambridge: Cambridge University Press, 2006.

[6] H. Claussen, "Performance of macro- and co-channel femtocells in a hierarchical cell structure," in *Proc. IEEE Int. Symp. Personal, Indoor, Mobile Radio Commun. (PIMRC)*, Athens, Greece, Sep. 2007, pp. 1–5.

[7] H. Claussen, L. T. W. Ho, and L. G. Samuel, "An overview of the femtocell concept," *Bell Labs Technical J.*, vol. 13, no. 1, pp. 221–46, Mar. 2008.

[8] H. S. Jo, P. Xia, and J. G. Andrews, "Downlink femtocell networks: open or closed?" *Proc. IEEE Int. Conf. on Commun. (ICC)*, Jun. 2011.

[9] D. López-Pérez, A. Valcarce, G. de la Roche, and J. Zhang, "OFDMA femtocells: a roadmap on interference avoidance," *IEEE Commun. Mag.*, vol. 47, no. 9, pp. 41–8, Sep. 2009.

[10] O. Simeone, E. Erkip, and S. Shamai, "Robust transmission and interference management for femtocells with unreliable network access," *IEEE J. Sel. Areas Commun. (JSAC) , Sp. Issue on Coop. Commun. on Cellular Networks.*, vol. 28, no. 9, pp. 1469–78, Dec. 2010.

[11] C. Jiang, L. J. Cimini Jr, and N. Himayat, "Interference mitigation with MIMO precoding in femtocellular systems," in *Annual Conference on Information Sciences and Systems (CISS)*, Mar. 2010, pp. 1–6.

[12] M. Husso, J. Hamalainen, R. Jantti, J. Li, E. Mutafungwa, R. Wichman, Z. Zheng, and A. Wyglinski, "Interference mitigation by practical transmit beamforming methods in closed femtocells," *EURASIP J. Wirel. Commun. Netw.*, vol. 2010, Apr. 2010.

[13] D.-C. Oh, H.-C. Lee, and Y.-H. Lee, "Power control and beamforming for femtocells in the presence of channel uncertainty," *IEEE Trans. Veh. Technol.*, vol. 60, no. 6, pp. 2545–54, Jul. 2011.

[14] S. Park, W. Seo, Y. Kim, S. Lim, and D. Hong, "Beam subset selection strategy for interference reduction in two-tier femtocell networks," *IEEE Trans. Wireless Commun.*, vol. 9, no. 11, pp. 3440–9, Nov. 2010.

[15] S. Park, W. Seo, S. Choi, and D. Hong, "A beamforming codebook restriction for cross-tier interference coordination in two-tier femtocell networks," *IEEE Trans. Veh. Technol.*, vol. 60, no. 4, pp. 1651–63, May 2011.

[16] D. Jaramillo-Ramirez, M. Kountouris, and E. Hardouin, "Downlink beamforming in multi-antenna two-tier networks with user selection," in *Proc. IEEE Global Telecommun. Conf. (GLOBECOM)*, Dec. 2011.

[17] Y. Isukapalli and B. D. Rao, "Finite rate feedback for spatially and temporally correlated MISO channels in the presence of estimation errors and feedback delay," in *Proc. IEEE Global Telecommun. Conf. (GLOBECOM)*, Nov. 26–30, 2007, pp. 2791–5.

[18] N. Jindal, "MIMO broadcast channels with partial side information," *IEEE Trans. Inf. Theory*, vol. 52, no. 11, pp. 5045–60, Nov. 2006.

[19] T. Yoo, N. Jindal, and A. Goldsmith, "Multi-antenna downlink channels with limited feedback and user selection," *IEEE J. Sel. Areas Commun. (JSAC)*, vol. 25, no. 7, pp. 1478–91, Sep. 2007.

[20] J. Zhang, M. Kountouris, J. G. Andrews and R. W. Heath Jr., "Multi-mode transmission for the mimo broadcast channel with imperfect channel state information," *IEEE Trans. Commun.*, vol. 59, no. 3, pp. 803–14, Mar. 2011.

[21] G. Caire, N. Jindal, M. Kobayashi, and N. Ravindran, "Multiuser MIMO achievable rates with downlink training and channel state feedback," *IEEE Trans. Inf. Theory*, vol. 56, no. 6, pp. 2845–66, Jun. 2010.

[22] S. Akoum and R. W. Heath Jr., "Limited feedback for temporally correlated MIMO channels with other cell interference," *IEEE Trans. Signal Process*, vol. 58, no. 10, pp. 5219–32, Oct. 2010.

[23] D. J. Love, R. W. Heath, Jr., V. K. N. Lau, D. Gesbert, B. Rao, and M. Andrews, "An overview of limited feedback in wireless communication systems," *IEEE J. Sel. Areas Commun. (JSAC)*, vol. 26, no. 8, pp. 1341–65, Oct. 2008.

[24] D. Gesbert, M. Kountouris, R. Heath, C.-B. Chae, and T. Salzer, "From single user to multiuser communications: shifting the MIMO paradigm," *IEEE Signal Process Mag*, vol. 24, no. 5, pp. 36–46, Sep. 2007.

[25] S. Venkatesan, A. Lozano, and R. Valenzuela, "Network MIMO: overcoming intercell interference in indoor wireless systems," in *Proc. Asilomar Conf. on Signals, Systems, and Computers (ASILOMAR)*, Pacific Grove, CA, US, Nov. 2007, pp. 83–7.

[26] S. Shamai (Shitz), O. Somekh, and B. Zaidel, "Multi-cell communications: an information theoretic perspective," *Joint Workshop on Communications and Coding (JWCC), Florence, Italy*, Oct. 2004.

[27] D. Gesbert, S. V. Hanly, H. Huang, S. Shamai, O. Simeone, and W. Yu, "Multi-cell MIMO cooperative networks: a new look at interference," *IEEE J. Sel. Areas Commun. (JSAC)*, vol. 28, no. 9, pp. 1380–408, Dec. 2010.

[28] P. Marsch and G. P. Fettweis, *Coordinated Multi-Point in Mobile Communications: From Theory to Practice.* Cambridge: Cambridge University Press, 2011.

[29] V. Cadambe and S. Jafar, "Interference alignment and degrees of freedom of the user interference channel," *IEEE Trans. Inf. Theory*, vol. 54, no. 8, pp. 3425–41, Aug. 2008.

[30] A. M. Hunter, J. G. Andrews, and S. Weber, "Transmission capacity of ad hoc networks with spatial diversity," *IEEE Trans. Wireless Commun.*, vol. 7, no. 12, pp. 5058–71, Dec. 2008.

[31] N. Jindal, J. G. Andrews, and S. Weber, "Multi-antenna communication in ad hoc networks: achieving MIMO gains with SIMO transmission," *IEEE Trans. Commun.*, Feb. 2011.

[32] K. Huang, J. G. Andrews, D. Guo, R. W. Heath Jr., and R. Berry, "Spatial interference cancellation for multiantenna mobile ad hoc networks," *IEEE Trans. Inf. Theory*, vol. 58, no. 3, pp. 1660–76, Mar. 2012.

[33] R. H. Y. Louie, M. R. McKay, and I. B. Collings, "Open-loop spatial multiplexing and diversity communications in ad hoc networks," *IEEE Trans. Inf. Theory*, vol. 57, no. 1, pp. 317–44, Jan. 2011.

[34] O. Ben-Sik-Ali, C. Cardinal, and F. Gagnon, "Performance of optimum combining in a Poisson field of interferers and Rayleigh fading channels," *IEEE Trans. Wireless Commun.*, vol. 9, no. 8, pp. 2461–7, Aug. 2010.

[35] M. Kountouris and J. G. Andrews, "Downlink SDMA with limited feedback in interference-limited wireless networks," *IEEE Trans. Wireless Commun.*, Aug. 2012.

[36] S. Weber and J. G. Andrews, *Transmission Capacity of Wireless Networks*, ser. Foundations and Trends in Networking. NoW Publishers, 2012.

[37] J. F. C. Kingman, *Poisson Processes.* Oxford University Press, 1993.

[38] D. Stoyan, W. Kendall, and J. Mecke, *Stochastic Geometry and Its Applications*, 2nd edn. John Wiley and Sons, 1996.

[39] F. Baccelli and B. Błaszczyszyn, *Stochastic Geometry and Wireless Networks, Volume I — Theory*, ser. Foundations and Trends in Networking. NoW Publishers, 2009.

[40] J. G. Andrews, F. Baccelli, and R. K. Ganti, "A tractable approach to coverage and rate in cellular networks," *IEEE Trans. Commun.*, vol. 59, no. 11, pp. 3122–34, Nov. 2011.

[41] R. W. Heath, Jr and M. Kountouris, "Modeling heterogeneous network interference," in *Proc. Inform. Theory and Appl. Workshop (ITA)*, Feb. 2012.

[42] C. C. Tan and N. Beaulieu, "On first-order Markov modeling for the Rayleigh fading channel," *IEEE Trans. Wireless Commun.*, vol. 48, pp. 2032–40, Dec. 2000.

[43] W. Turin, R. Jana, C. Martin, and J. Winters, "Modeling wireless channel fading," in *Proc. IEEE Vehicular Tech. Conf. (VTC)*, vol. 3, no. 4, Oct. 2001.

[44] R. Bhagavatula and R. Heath Jr, "Adaptive bit partitioning for multicell intercell interference nulling with delayed limited feedback," *IEEE Trans. Signal Process.*, vol. 59, no. 8, pp. 3824–36, Aug. 2011.

[45] K. K. Mukkavilli, A. Sabharwal, E. Erkip, and B. Aazhang, "On beamforming with finite rate feedback in multiple antenna systems," *IEEE Trans. Inf. Theory*, vol. 49, pp. 2562–79, Oct. 2003.

[46] S. Zhou, Z. Wang, and G. Giannakis, "Quantifying the power loss when transmit beamforming relies on finite-rate feedback," *IEEE Trans. Wireless Commun.*, vol. 4, no. 4, Jul. 2005.

[47] "Guidelines for evaluation of radio transmission technologies for IMT-2000," ITU Recommendation M.1225, Tech. Rep., 1997.

[48] M. K. Simon, *Probability Distribution involving Gaussian Random Variables. A Handbook for Engineers, Scientists and Mathematicians.* Springer, 2006.

6 Physical layer techniques for cognitive femtocells

Giuseppe Caire, Ansuman Adikhary, and Vasileios Ntranos

6.1 Breaking the cellular bottleneck

Consider a conventional macrocellular system with $N_{N,1}$ macrocell base stations (MBSs) with M antennas each, and $N_{U,1} = \sum_{j=1}^{N_{N,1}} N_{U,1,j}$ macrocell mobile stations (MMSs), equipped with a single antenna each. Assuming a time and frequency selective fading channel with coherence time τ_c and coherence bandwidth W_c [1–3], it is well known that, up to some close to unity multiplicative factor, the number of signal space dimensions undergoing the same fading channel state is given by $T \approx \tau_c W_c \gg 1$.[1] Therefore, it is safe to simplify the channel model and treat it as a *block-fading channel* where the frequency and time selective channel state is assumed to be piecewise constant over blocks that correspond to the timing of the time-frequency plane with $\tau_c \times W_c$ blocks, each spanning T signal dimensions [1]. This approximation is particularly meaningful when the system makes use of orthogonal frequency division multiplexing (OFDM), which is designed to exploit the locally linear time-invariant nature of the channel impulse response. In this chapter we will follow this block-fading approximation. If this approximation is removed and smooth variations in time and frequency are taken into account, an exact system performance analysis would be significantly more involved, since even the parallel channel decomposition operated by OFDM would not hold exactly. In fact, the block-fading model is a sort of realistic "best case" idealization, as far as the system design is concerned [4].

Next, we argue that the performance of a cellular system is fundamentally limited, no matter how many antennas per MBS and cooperation across the MBSs (network multiple input multiple output (MIMO)) we allow. Assume perfect cooperation between all the MBS antennas and all the MMSs. The system reduces to a giant MIMO point-to-point channel, the capacity of which behaves as

$$\mathcal{C}_{\text{coop, cell}}(\mathsf{SNR}) = M^*(1 - M^*/T) \times \log(\mathsf{SNR}) + O(1) \tag{6.1}$$

---

[1] The product $\tau_c W_c$ is much larger than 1 for *underspread* channels arising in wireless communications. For example, typical values from [3] yield $W_c \approx 200$ kHz and $\tau_c \approx 10$ ms, yielding $T \approx 2000$ symbols.

Small Cell Networks: Deployment, PHY Techniques, and Resource Management, ed. Tony Q. S. Quek, Guillaume de la Roche, İsmail Güvenç, and Marios Kountouris. Published by Cambridge University Press. © Cambridge University Press 2013.

where $M^* = \min\{MN_{N,1}, N_{U,1}, T/2\}$, and where $O(1)$ indicates a constant, independent of signal to noise ratio (SNR). The limitation due to T can be interpreted as the cost incurred by estimating the channel [5, 6]. This result shows that, even assuming ideal joint processing of all MMS (in the downlink) or of all MBSs (in the uplink), through a backbone network of infinite capacity [7], the system throughput is fundamentally limited by the ability of estimating the time and frequency selective channels. Therefore, even if we let $M \to \infty$ or have all MBSs cooperate, the throughput per user still decreases as $O(1/N_{U,1})$ [8–10]. Opportunistic scheduling and multi-user diversity, as implemented in the 3G so-called *high-data rate* uplink and downlink schemes [11], do not significantly improve this trend, yielding a per-user throughput that decreases as $O(\frac{\log \log N_{U,1}}{N_{U,1}})$.

In order to overcome the cellular dimensional bottleneck evidenced by (6.1), it is widely recognized that the single most valuable resource is *spatial reuse* [12]. In a cell of unit area and n users, under common distance-dependent path-loss models, it is possible to maintain a constant $O(1)$ throughput per user if source-destination pairs are at distance $O(1/\sqrt{n})$.[2] Hence, a possible solution consists of deploying a very dense cellular infrastructure, with base station density that grows *linearly* with the user density n. This solution, unfortunately, is not economically viable if these small cells are deployed and managed centrally, by the system operator. This is due to a number of obvious practical and economical reasons, including (for example) the cost of the maintenance of a very dense centrally planned infrastructure, and the non-trivial problem of finding cell sites for such a large number of base stations. Indeed, it is a matter of fact that people are not keen to allow the deployment of base stations in their own backyard, while almost everybody accepts without problems the installation of self-deployed wireless access points (e.g., wireless fidelity (WiFi)) in their homes.

User-deployed wireless local area networks (WLANs) (e.g., IEEE 802.11) achieve such dense spatial reuse in the unlicensed band. This solution has the advantage of providing very high data rates for short-range, mostly in-home, communication, but does not handle mobility as efficiently as cellular systems. The vertical handoff between macrocellular and WLAN is still not as efficient as it could be, essentially because both macrocellular and WLAN are "legacy" systems, developed independently: putting them together results in a sort of patchwork formed by pragmatic ad hoc solutions, rather than a coherently designed and carefully optimized system.

These considerations point naturally in the direction of an evolution of *licensed* cellular systems toward two-tier architectures, where a large number of user-deployed *femtocells* operate under a common *macrocell umbrella* [12, 14]. Femtocells have attracted significant attention both in cellular standards bodies such as the 3GPP [15–17] and academic research [12, 18], as reflected in the chapters of the present book. Some information theoretic literature [19] modeled femtocells as "primitive" relays [20], where a femtocell mobile station (FMS) signal is received both at the femtocell base station (FBS) and at the MBS, and the FBS serves as relay through the wired connection to the MBS. Beyond the interest in the information theoretic problem, this approach misses one of the main

---

[2] Notice that this does not violate the scaling laws of decode-and-forward networks [13], since the transport capacity in bit-meter per second is in the order of $n \times \frac{1}{\sqrt{n}} = \sqrt{n}$.

points about femtocells, i.e., the effect of offloading a large number of users (in fact, a linear fraction of users) from the MBS. Furthermore, since the wired link is typically provided by a digital subscriber line (DSL) connection, going through the Internet cloud possibly via some third-party Internet service provider (ISP), and eventually coming back to the wireless operator proprietary network through a special gateway, data transmission through the FBS is subject to throughput and delay uncertainty that makes the joint processing of the FBS and of the MBS signals basically infeasible.

A large body of literature on femtocells (see for example [12, 14, 21–25]) has focused on the cross-tier interference between the MBS and the femtocells. Another widely studied issue is open- vs. closed-access (see for example [26]), where closed access indicates the case where the use of FBSs is restricted to a closed subscriber group (CSG). Since femtocells are user deployed, an MMS in the premises of a femtocell may not have the permission to access the femtocell FBS and therefore may cause significant interference on the femtocell users when transmitting (at high power) to the MBS in the macrocell uplink (macro-UL). Similarly, such an MMS will suffer from significant cross-tier interference from the femtocell in the macrocell downlink (macro-DL). In order to address these issues, power control combined with interference cancellation and subband scheduling (exploiting the fractional frequency reuse power allocation pattern in different frequency bands) are studied in [14].

Most existing works on femtocells focus on continuous transmission of MBS and MMSs and on average signal to interference plus noise ratio (SINR) calculations, strongly inspired by today's 3G code division multiple access (CDMA) networks. These assumptions disregard the fact that the forthcoming generation of cellular systems (3GPP LTE and IEEE 802.16m [27, 28]) is based on OFDM/time division multiple access (TDMA), where *scheduling* is employed both in the macro-DL and in the macro-UL. It follows that MMSs are dynamically assigned to time-frequency slots in order to transmit (UL) or receive (DL), where this assignment changes from frame to frame. For example, if an MMS at some location in the cell is served on a given time-frequency slot in the macro-DL, all femtocells sufficiently far from it can reuse the same slot, provided that they control their transmit power in order to limit the interference caused to the MMS. In contrast, if an MMS transmits in a given time-frequency slot, only the femtocells in its vicinity will be affected by its interference. Therefore, each femtocell is impacted by strong uncoordinated interference from MMSs outside its CSG only for a small fraction of the time-frequency slots.

In order to exploit the implicit statistical multiplexing due to the macrocell UL and DL dynamic scheduling, in this chapter we advocate a "cognitive" approach, where the femtocells (both FBS and FMS) are given the ability of decoding the MBS control channel. We assume that location information is available as part of the protocol at negligible rate, since user mobility occurs at a very slow time-scale with respect to the signaling bandwidth. In this way, at the beginning of each allocation frame, each femtocell knows which MMSs will be scheduled on each time-frequency slot, including their transmit powers, rates, and geographic location in the cell. This common protocol information *implicitly* coordinates the femtocells, such that they can regulate their power and rate in order to cope with cross-tier interference. We shall see that with the implicit

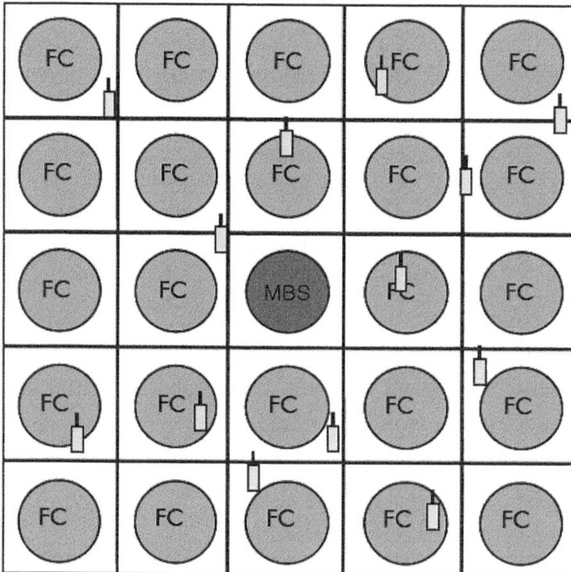

Figure 6.1 Reference layout of a macrocell containing one base station (MBS), macrocell users (MMSs), denoted in the picture by the cellphones, and femtocells arranged on a regular square grid. Each femtocell is formed by one femtocell base station (FBS) and by some number, larger than or equal to 1, of femtocell users (FMSs).

coordination obtained by the proposed cognitive approach, the traditional problems in femtocell systems studied in the conventional literature essentially disappear, and a very attractive throughput tradeoff region is achievable with almost elementary signal processing techniques.

This chapter is organized as follows. In Section 6.2, we present the basic system concept and assumptions. In Section 6.3, we consider the performance of the baseline system, including open vs. closed access, and opportunistic interference cancellation. Section 6.4 examines the application of the downlink interference alignment (DIA) idea [29] to the two-tier femtocell system, and Section 6.5 considers the use of multiple antenna in the FBSs. Conclusions are pointed out in Section 6.6.

6.2 Cognitive femtocells: system assumptions

Figure 6.1 shows a single macrocell ($N_{N,1} = 1$) containing $N_{N,2} \gg 1$ femtocells. For simplicity, we consider a square-shaped macrocell and assume that femtocells are circles placed on a square grid, motivated by the fact that in a typical suburban or low-density urban environment (e.g., think of Los Angeles, or Dallas), residential homes are essentially placed on a regular squared grid, separated by streets.

We consider a time division duplex (TDD) macrocell system, where macro-UL and macro-DL are accessed using OFDM/TDMA, and focus on a frequency-flat channel, corresponding to a set of subcarriers spanning a channel coherence bandwidth W_c.

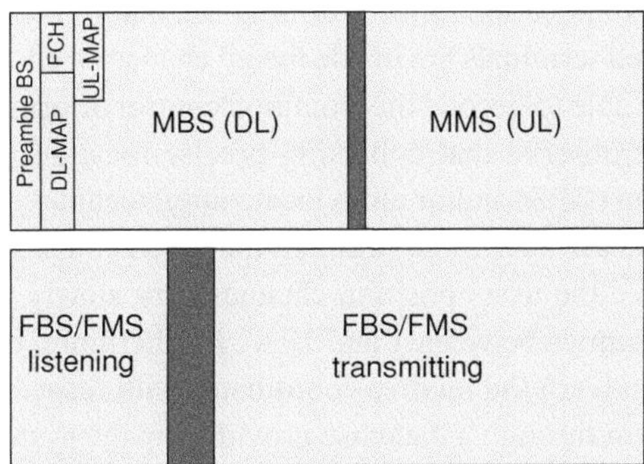

Figure 6.2 Frame structure for the proposed cognitive femtocell system. ©2011 IEEE. Reprinted, with permission, from [32].

Including frequency selectivity in our context is straightforward but requires some additional notation. Hence, for the sake of simplicity, here we focus on a single coherence bandwidth. We also hasten to say that most of the conclusions that hold for a TDD macrocell tier hold for a frequency division duplex (FDD) macrocell tier. In both cases, we consider femtocells operating in TDD, irrespectively of the duplexing mode used in the macrocell system. In the case of FDD macrocells, both the macro-UL and the macro-DL bands are used by the femtocells in TDD, for both the femto-UL and the femto-DL, depending on the cases.

The TDD macrocell and femtocell frame structures in the proposed cognitive femtocell architecture are shown, qualitatively, in Figure 6.2. These are closely reminiscent of the frame structure in the IEEE 802.16j standard for relays in 4th generation TDD systems [30]. The MBS frame is standard and includes the control channel (with UL and DL allocation maps), the macro-DL subframe and the macro-UL subframe, with a small guard interval in between. The femtocell frame structure is formed by two subframes. In the first one, overlapped with the MBS control channel, *all* femtocell terminals (both FBS and FMS) listen to the MBS control channel. After a guard interval, necessary to decode the control channel and acquire the allocation of the macrocell users on the macro-DL and macro-UL subframes, the femtocell terminals are active and transmit, using TDD, both in the uplink and in the downlink, depending on the case. As we will see, the various schemes presented in this chapter require different macro/femto UL/DL alignment. The important aspect to be stressed at this point is that we assume that all terminals in the femtocell can decode the base station (BS) control channel and exploit it as an *implicit common information for coordination*. In particular, we assume that the power, rate and location of the active macrocell users on each slot are known to all femtocells.

We hasten to say that this assumption is obviously optimistic. However, after careful consideration, we argue that it is not so far from what can be implemented in practice, through a careful system design. First, notice that by system design, all terminals in a macrocell, today, are controlled by the mBS, and decode the control information that defines the dynamic allocation of users, rates and powers over the frequency division multiple access (FDMA)/TDMA allocation "tiles" forming the macro-UL and macro-DL subframes. Since the femtocell terminals are inside the given macrocell, as any other macrocell terminal, they are also able to decode this common control information. Then, as already remarked before, we observe that collecting precise location information, through global positioning system (GPS) and/or radio localization techniques, for all the active macrocell users into a database accessible (through the wired connections) by the FBS should not be difficult, since the users position changes very slowly with respect to the signaling rate (typical dynamics between 1 and 10 s). Furthermore, the FBSs are placed in fixed positions, and can learn the relative coordinates with respect to all other femtocells and MBS in the system through a database provided by the system operator. This information changes with typical dynamics of the order of days, such that including positioning information as side information should not incur a large protocol overhead. Finally, we wish to point out that our assumption of perfect cognitive information about the macrocell users scheduling and positioning is *much less demanding* and significantly more realistic than the usual assumptions made in the information theoretic "overlay" cognitive radio model, where the cognitive side information consists of information messages of other users, "magically" and perfectly overheard at no extra cost by some cognitive terminal (see for example [31] and references therein).

Investigating to what extent the cognitive side information assumed here can be actually provided, with what accuracy, and at what costs in terms of protocol overhead, is certainly an interesting problem that, however, goes beyond the scope of this chapter. Here, we shall assume that this information is perfectly available, and examine the two-tier system performance in terms of the achievable throughput tradeoff region of the aggregate femtocell throughput versus the macrocell throughput. Notice that looking at the total system throughput is meaningless: the solution would be to shut down the MBS and forget the (few) outdoor MMSs, while serving only the (many) femtocell users. This, however, is not an acceptable solution in terms of fairness with respect to the macrocell users. Instead, we consider the achievable tradeoff region between macrocell and aggregate femtocell throughputs, leaving aside the problem of optimizing some cost function of these throughputs by operating at some desired point on the Pareto boundary of the tradeoff region. Specific cost functions corresponding to some desired notion *network utility* may take into account, for example, the relative priority of femtocell versus macrocell users, and might be dictated by fairness and revenues considerations, that go beyond the scope of this chapter.

6.2.1 Macrocell subsystem: DL/UL dynamic scheduling

In each macro-DL slot, the MBS schedules a single MMS. This can be done using round-robin, or some form of channel state driven scheduling. For example, in our numerical results we consider opportunistic proportional fair scheduling (PFS), in agreement with current *high data-rate* schemes such as evolution data only (EVDO) and high speed downlink packet access (HSDPA) [11]. Similarly, in each macro-UL slot, the MBS schedules a single MMS, where again PFS may be used, in agreement with high speed uplink packet access (HSUPA) [11].

The instantaneous channel gain between a given macrocell user u and the MBS, at each given time-frequency scheduling slot, is formed by the product of two components: (i) a distance dependent path-loss component; and (ii) a small-scale multi-path fading component. The typical duration over which the path loss can be considered constant in time is of the order of $10\,$s [3]. In contrast, the small-scale fading component has a coherence time of the order of $10\,$ms for low mobility users [3], i.e., at least three orders of magnitude faster than the path-loss component. Therefore, it is meaningful to consider the MMSs at fixed positions (i.e., we take a snapshot of the system for given user positions), while the performance is averaged over the small-scale fading components. It should be noticed that rates averaged with respect to the small-scale fading (the so-called "ergodic rates") here correspond to the long-term average throughputs achieved by the scheduler's dynamic rate allocation [4, 10, 33].

In these conditions, assuming that the users' fading gains are independent and identically distributed (i.i.d.) (e.g., Rayleigh fading for all users), PFS amounts *approximately* to serving the user with the largest small-scale fading component, irrespectively of its path-loss component [34]. In particular, letting $N_{\mathrm{U},1,1} = U$ denote the number of MMS in the system, with this approximation we have that each macrocell user is given an equal fraction $1/U$ of the time. This is the same as in round-robin, with the difference that now the small-scale fading statistics have changed, since for each scheduled user, its small-scale coefficient is the maximum of U independent coefficients. For the sake of simplicity, we will make this approximation throughout this chapter.

We assume that the MBS power, denoted by P_0, is equal to the aggregate power of all MMSs in the cell. This means that the MBS transmits (on the macro-DL) with constant maximum power P_0, while the MMSs transmit (in the macro-UL) with *instantaneous power* P_0 when they are scheduled, resulting in an average transmit power per MMS equal to P_0/U. Notice that we assume a fully loaded system, where all macrocell slots are used, both in the macro-DL and in the macro-UL.

6.2.2　Femtocell subsystem: power control

As mentioned earlier, femtocells operate in TDD with no constraint on aligning their UL and DL with the macro-UL and macro-DL slots (see Figure 6.2). We assume also that geographic location information is included in the scheduling protocol, such that the position of the scheduled MMS on any given allocation slot is known and each femtocell is aware of its own relative position in the cell.

Irrespectively of the number $N_{\mathrm{U},2,j}$ of users in femtocell $j = 1, \ldots, N_{\mathrm{N},2}$, we assume that each femtocell uses orthogonal intra-cell access and schedules one user per femto-UL or femto-DL slot. Therefore, as far as the femtocell throughput is concerned, it is sufficient to assume $N_{\mathrm{U},2,j} = 1$ for all j (we assume that no femtocells are idle). The case of $N_{\mathrm{U},2,j} > 1$ will be considered in Section 6.4, when treating the *Downlink IA* scheme.

The femtocells exploit the knowledge of the MBS control channel in order to implement a power control strategy that mitigates the cross-tier femto-to-macro interference. We consider a simple strategy based on *cross-tier interference temperature power control* – femtocells (both FBSs and FMSs) set their transmit power such that the average interference power at the macrocell receiver (either MBS or MMS) active on the given slot is not larger than some target level κ, and the transmitted power is not larger than a fixed peak power constraint P_1.

In the macro-DL slot, only the femtocells close to the scheduled MMS are forced to transmit at power significantly lower than their peak value P_1. This set of femtocells changes randomly from slot to slot due to the downlink scheduling of the MBS. Such randomization is beneficial, since it avoids cases where the *same* set of femtocells is permanently forbidden to use high transmit power in the macro-DL slot. In contrast, the femtocells located near the MBS are forced to use *permanently* a very small transmit power in the macro-UL slot, because the MBS receiver is active on all slots.[3] This means that the femtocells in the neighborhood of the MBS are inherently at a disadvantage with respect to the femtocells far from the MBS, since they can effectively transmit at high power only in the macro-DL slot. In Section 6.5 we shall alleviate this problem by using multiple antennas at the FMSs.

6.3   Baseline system performance

Our baseline system is based only on the above described power control strategy and may make use of opportunistic interference cancellation at the femtocells (both FBS and FMS), as explained later in this section. We consider a unit-side square cell $[-1/2, 1/2] \times [-1/2, 1/2]$ where the MBS is located at the origin 0, and $N_{N,2} = F^2$ femtocells are centered at points of coordinates $\left(\frac{2i-F+1}{2F}, \frac{2j-F+1}{2F} \right)$, for $i, j = 0, \ldots, F - 1$. The set of femtocell centers is denoted by $\mathcal{C}$. For example, in a typical low-density urban environment, single-family homes are spaced by $\approx 40\,\mathrm{m}$, corresponding to $F = 25$ for a macrocell of side equal to 1 km.

We consider a simplified geometry with disk-shaped femtocells of radius r_{fc}, shielded from the outdoor environment by walls. For two points $a, b \in \mathcal{C}$, the channel distance-dependent path gain is given by

$$g(a, b) = \frac{w^{n(a,b)}}{1 + ((d(a, b) - n(a, b)r_{\mathrm{fc}})/\delta)^{\alpha}} \tag{6.2}$$

where:

- $d(a, b)$ denotes the modulo-$\mathcal{C}$ distance between the centers of the femtocells containing points a and b. If point a (resp. b) is not contained in any femtocell, then the distance $d(a, b)$ is defined with respect to a (resp. b) itself. The modulo-$\mathcal{C}$ distance induces a torus topology on the square-shaped macrocell, that avoids boundary effects as far as the intra-tier femto-to-femto interference is concerned. In other words, every femtocell in the system sees the same average interference landscape from other femtocells, irrespective of its position.

---

[3] In contrast to most "cognitive radio" literature based on dynamic spectrum sensing and reusing time-frequency "holes," we assume a fully loaded macrocell system with no empty slots.

- The exponent $n(a, b)$ counts the number of walls between points a and b. In particular $n(a, b) = 0$ if both a and b are outdoor (not inside any femtocell), or they are in the same femtocell, $n(a, b) = 1$ if either a or b is indoor (inside a femtocell), and $n(a, b) = 2$ if a and b are indoor, in different femtocells.
- w denotes the wall absorption factor.
- δ is the path-loss "3 dB" distance threshold.
- α is the outdoor propagation exponent.

Notice that (6.2) yields that two points a, b inside the same femtocell have path gain equal to 1, i.e., we assume that indoor propagation yields no path loss. For homes of moderate size, especially when the FBS is located in the same room as the FMS(s), this is a realistic assumption.

When transmitter and receiver are inside the same femtocell, we assume no small-scale fading. Instead, for any two points a and b not located in the same femtocell, we model the corresponding small-scale fading power coefficient as the random variable $H(a, b) \sim$ exponential(1) (Rayleigh fading). For any set of distinct points pairs $(a_1, b_1), \ldots, (a_n, b_n)$, the joint fading coefficients distribution takes on the product form

$$\mathcal{P}\left(H(a_1, b_1) \leq h_1, \ldots, H(a_n, b_n) \leq h_n\right) = \prod_{i=1}^{n} F_H(h_i) \tag{6.3}$$

where $F_H(\mathsf{x}) = (1 - e^{-\mathsf{x}}) \, 1\{\mathsf{x} \geq 0\}$ denotes the fading gain marginal cumulative distribution function (CDF).

Closed subscriber group access is modeled by assuming that MMSs can be located *anywhere*, even inside a femtocell. In contrast, the case of open access (OA), is obtained by assuming that femtocells automatically "swallow" any macrocell user inside their radius. As a consequence, in the case of OA, MMSs are only located outdoors. This corresponds to the fact that, under OA, any macrocell user inside a femtocell gets controlled and served by the FBS, and therefore it disappears from the list of active macrocell users served by the MBS.

At this point, calculating the ergodic macrocell and aggregate femtocell throughput is just a matter of a rather tedious accounting of the interference levels seen at each receiver. For the sake of notation simplicity, in the following we let $\mathcal{U}$ indicate the set of MMSs locations, and identify users and femtocell indices with their spatial coordinates. Therefore, we shall refer to femtocell f as the femtocell whose center is $f \in \mathcal{C}$, and macrocell user u as the MMS located at $u \in \mathcal{U}$, where $|\mathcal{C}| = F^2$ and $|\mathcal{U}| = U$.

6.3.1 Macrocell throughput

Macro-UL:
The total interference power from all the femtocells to the MBS receiver in the UL slot is given by

$$I_{\text{bs}} = \sum_{f \in \mathcal{C}} \min\left\{\frac{\kappa}{g(f, 0)}, P_1\right\} g(f, 0) H(f, 0) \leq \kappa \sum_{f \in \mathcal{C}} H(f, 0) \tag{6.4}$$

Assuming a Gaussian random coding ensemble, the instantaneous rate for a macrocell user u, given that u is scheduled on the current UL slot, is given by

$$R_u = \log\left(1 + \mathcal{H}\frac{g(u, 0)P_0}{1 + I_{\text{bs}}}\right). \tag{6.5}$$

The statistics of the random variable $\mathcal{H}$ depend on what type of UL scheduling is used in the macrocell. In the case of round-robin scheduling, we have $\mathcal{H} = H(u, 0)$. In the case of PFS, under the approximation mentioned before, we have

$$\mathcal{H} = \max\{H(v, 0) : v \in \mathcal{U}\}.$$

In both cases, the statistics of $\mathcal{H}$ are independent of the user index u. In fact, it follows that the rates (6.5) depend on u only through the location-dependent path gain coefficients $g(u, 0)$.

Since each user is served for a fraction $1/U$ of the slots, summing over the whole macrocell user population we obtain the average UL sum throughput in the form

$$R_{\text{mc}}^{\text{UL}} = \frac{1}{U}\sum_{u \in \mathcal{U}}\mathbb{E}\left[\log\left(1 + \mathcal{H}\frac{g(u, 0)P_0}{1 + I_{\text{bs}}}\right)\right]. \tag{6.6}$$

The above formula captures the effect of CSG versus OA policy in the femtocells. In the OA policy case, the set of macrocell users locations $\mathcal{U}$ has empty intersection with the union of all femtocell disks. In contrast, in a CSG access policy case, the intersection of $\mathcal{U}$ with the union of the femtocell disks may be non-empty. In our numerical results, we assume uniformly distributed MMSs over the area not included in the femtocell disks, for OA, and uniformly distributed MMSs over the whole squared macrocell, for CSG. In the latter case, the probability that an MMS is inside a femtocell is given by $1/(\pi F^2 r_{\text{fc}}^2)$ (ratio between the macrocell unit area and the sum of all femtocell disk areas).

While the expectation in (6.6) can be given in closed form (see [35]), these are not much more suitable for calculation than Monte Carlo averaging, especially when the number of femtocells is large. A low-complexity alternative for fast closed form system performance evaluation can be obtained by using Jensen's inequality applied to I_{bs}. Namely, we have

$$\mathbb{E}[I_{\text{bs}}] \leq \sum_{f \in \mathcal{C}}\min\left\{\frac{\kappa}{g(f, 0)}, P_1\right\}g(f, 0) \leq \kappa F^2,$$

leading to the simple lower bound

$$R_{\text{mc}}^{\text{UL-lb}} = \frac{1}{U}\sum_{u \in \mathcal{U}}\mathbb{E}\left[\log\left(1 + \mathcal{H}\frac{g(u, 0)P_0}{1 + \kappa F^2}\right)\right]. \tag{6.7}$$

For both the round-robin and the PFS scheduling, the expectation with respect to $\mathcal{H}$ can be easily given in closed form, in terms of the exponential integral function $E_i(1, x) = \int_x^\infty \frac{e^{-t}}{t}dt$, for $x > 0$ (see for example [36]).

Macro-DL:
In this case, we let I_u denote the total interference power from all femtocells to the MMS located at u on the DL slot. This is given by

$$I_u = \sum_{f \in C} \min \left\{ \frac{\kappa}{g(f, u)}, P_1 \right\} g(f, u) H(f, u) \le \kappa \sum_{f \in C} H(f, u). \tag{6.8}$$

It follows that the instantaneous rate for a macrocell user u scheduled on the downlink is given by

$$R_u = \log \left(1 + \mathcal{H} \frac{g(u, 0) P_0}{1 + I_u} \right) \tag{6.9}$$

where the statistics of $\mathcal{H}$ depend on whether round-robin or PFS scheduling is used, as already discussed. Summing over the whole macrocell user population we obtain the average downlink sum throughput in the form

$$R_{mc}^{DL} = \frac{1}{U} \sum_{u \in \mathcal{U}} \mathbb{E} \left[\log \left(1 + \mathcal{H} \frac{g(u, 0) P_0}{1 + I_u} \right) \right]. \tag{6.10}$$

The same lower bound in (6.7) holds for this case too, since the statistics of the interference upper bounds in (6.4) and in (6.8) are identical.

6.3.2 Femtocell throughput

By assumption, femtocells can use the macro-UL and macro-DL slots in both directions. However, the interference landscape that each femtocell sees is different in the two cases. Therefore, we shall distinguish between the femtocell throughput in the macro-UL and in macro-DL slots. For both cases, easy to compute lower bounds analogous to (6.7) and closed form expressions for the expectations appearing in the throughput expressions can be obtained. Details are omitted for the sake of brevity.

Femtocell throughput in the macro-UL slot:
The intra-tier, femto-to-femto interference term for femtocell f is given by

$$I_f^{UL} = \sum_{f' \in C: f' \ne f} \min \left\{ \frac{\kappa}{g(f', 0)}, P_1 \right\} g(f, f') H(f, f') \le \kappa \sum_{f' \in C: f' \ne f} \frac{g(f, f')}{g(f', 0)} H(f, f'). \tag{6.11}$$

In addition, femtocell f receives interference from the macrocell user u, active on the uplink slot. Using the fact that macrocell users are scheduled for an equal fraction of time, the femtocell sum-throughput in the macro-UL slot is given by

$$R_{fc}^{UL} = \sum_{f \in C} \frac{1}{U} \sum_{u \in \mathcal{U}} \mathbb{E} \left[\log \left(1 + \frac{\min \left\{ \frac{\kappa}{g(f, 0)}, P_1 \right\}}{1 + I_f^{UL} + g(f, u) H(f, u) P_0} \right) \right]. \tag{6.12}$$

Notice that, in this case, irrespectively of the macrocell user u scheduled on the uplink, femtocells f located at short distance from the MBS are forced to use a small transmit power $\kappa / g(f, 0) \ll P_1$ in order to guarantee interference level κ to the MBS receiver.

Hence, the throughput of these femtocells in the macro-UL slot may be very small, depending on the wall loss factor w.

Femtocell throughput in the macro-DL slot:
The intra-tier, femto-to-femto interference term for femtocell f in this case depends on the macrocell user u scheduled on the downlink, and it is given by

$$I_f^{\mathrm{DL}}(u) = \sum_{f' \in \mathcal{C}: f' \neq f} \min\left\{ \frac{\kappa}{g(f', u)}, P_1 \right\} g(f, f') H(f, f')$$

$$\leq \kappa \sum_{f' \in \mathcal{C}: f' \neq f} \frac{g(f, f')}{g(f', u)} H(f, f'). \tag{6.13}$$

In addition, femtocell f receives interference from the MBS. Hence, the corresponding femtocell sum-throughput in the DL slot is given by

$$R_{\mathrm{fc}}^{\mathrm{DL}} = \sum_{f \in \mathcal{C}} \frac{1}{U} \sum_{u \in \mathcal{U}} \mathbb{E}\left[\log\left(1 + \frac{\min\left\{ \frac{\kappa}{g(f,u)}, P_1 \right\}}{1 + I_f^{\mathrm{DL}}(u) + g(f, 0) H(f, 0) P_0} \right) \right]. \tag{6.14}$$

Notice that, in this case, the transmit power of femtocell f depends on which macrocell user u is scheduled on the DL. Since macrocell users are given equal fractions of time, the time for which $\kappa/g(f, u)$ is less than the femtocell peak transmit power P_1 is equal to the fraction of macrocell users u such that $g(f, u) > \kappa/P_1$. For each f, and a given set of macrocell users positions $\mathcal{U}$, this defines the probability with which femtocell f is forced to back-off its transmit power by the power control rule. Interestingly, in this case the macro-DL scheduling induces a sort of statistical multiplexing such that each given femtocell f is formed to significantly back-off its transmit power only for a small fraction of time.

6.3.3   Interference cancellation at the femtocells

In this section we consider opportunistic successive interference cancellation (SIC) of the macro-to-femto cross-tier interference at the femtocells receivers (either FBS or FMS). The femtocell receiver has the option of decoding the macrocell signal first, by treating its own femtocell signal as additional noise, and then decoding its own desired signal after removing the macrocell interference. When this is not possible, the femtocell receiver still has the option of treating the macrocell interference as noise (as before). The decision of performing SIC is made at the femtocell receiver, and it depends on the received cross-tier interference power and rate, as detailed below. Since the instantaneous cross-tier interference power depends on the realization of the small-scale fading, which is not known in advance by the femtocell transmitter, it is not possible to allocate the transmission rate on a block-by-block basis. Nevertheless, the full benefit of *opportunistic* SIC can still be obtained by using rateless coding and incremental redundancy automatic repeat request (ARQ) [37] in the femtocells, analogously to what was done in [38] to handle unknown inter-cell interference in multi-cell systems. In this case, what matters is the average *mutual information* accumulated at the receiver per slot, irrespectively of slot-by-slot explicit rate adaptation. While these conclusions hold in the case of very large T (see [37, 38] for details), recent results on the performance of incremental redundancy schemes in the case of finite blocks (equivalently, for fixed and finite number of bits per information message) are presented in [39], and confirm that the large-T asymptotic behavior is closely attained for slots of practical size.

Opportunistic SIC is particularly useful when the femtocell receives the macrocell signal at high power, but the macrocell rate is small. In the macro-DL slot, this is typically the case of femtocells near the MBS, when the MBS schedules an MMS at the cell edge. In the macro-UL, this is typically the case of femtocells near a scheduled MMS located near the cell edge. In both cases, the macrocell rate is small due to the small path gain, while the macrocell signal is received by the femtocell at high power. Notice that this is also the most harmful case of cross-tier macro-to-femto interference, especially for a femtocell near the BS, operating in the macro-DL slot.

In the macro-UL slot, the rate region achievable at femtocell f receiver, using Gaussian i.i.d. independent code ensembles, SIC, and disregarding the macrocell message, is given by the union of the two regions $\mathcal{R}_{\mathrm{sic}} \cup \mathcal{R}_{\mathrm{su}}$ [40] given by:

$$
\mathcal{R}_{\mathrm{su}} =
\begin{cases}
R_f^{\mathrm{UL}} \leq \log\left(1 + \dfrac{\min\left\{\frac{\kappa}{g(f,0)}, P_1\right\}}{1+I_f^{\mathrm{UL}}}\right) \\[3mm]
R_u \leq \log\left(1 + \dfrac{g(f,u)H(f,u)P_0}{1+I_f^{\mathrm{UL}}+\min\left\{\frac{\kappa}{g(f,0)}, P_1\right\}}\right)
\end{cases}
\tag{6.15}
$$

and

$$
\mathcal{R}_{\mathrm{su}} = \left\{ R_f^{\mathrm{UL}} \leq \log\left(1 + \frac{\min\left\{\frac{\kappa}{g(f,0)}, P_1\right\}}{1 + I_f^{\mathrm{UL}} + g(f,u)H(f,u)P_0}\right) \right\},
\tag{6.16}
$$

corresponding to "single-user" decoding, i.e., treating the macrocell interference as noise. Noticing that the macrocell user instantaneous rate R_u is given by (6.5), we obtain the instantaneous femtocell rate with opportunistic SIC in the form

$$
R_f^{\mathrm{UL}} =
\begin{cases}
\log\left(1 + \dfrac{\min\left\{\frac{\kappa}{g(f,0)}, P_1\right\}}{1+I_f^{\mathrm{UL}}}\right), & \text{if } \mathcal{H}\frac{g(u,0)}{1+I_{\mathrm{bs}}} \leq H(f,u)\dfrac{g(f,u)}{1+I_f^{\mathrm{UL}}+\min\left\{\frac{\kappa}{g(f,0)}, P_1\right\}}. \\[3mm]
\log\left(1 + \dfrac{\min\left\{\frac{\kappa}{g(f,0)}, P_1\right\}}{1+I_f^{\mathrm{UL}}+g(f,u)H(f,u)P_0}\right) & \text{otherwise}
\end{cases}
\tag{6.17}
$$

Then, the sum throughput with opportunistic SIC is obtained by using (6.17) in the summation and averaging (with respect to the fading coefficients) as in (6.12). If joint decoding instead of SIC is used at the femtocell receiver, the achievable instantaneous rate can be improved [40]. However, here we restrict to SIC since it involves only single-user decoding steps and therefore it is more practical to be implemented, especially at the FMSs.

Following a similar argument, for the macro-DL slot we arrive at the femtocell instantaneous rate in the form

$$
R_f^{\mathrm{DL}} =
\begin{cases}
\log\left(1 + \dfrac{\min\left\{\frac{\kappa}{g(f,u)}, P_1\right\}}{1 + I_f^{\mathrm{DL}}(u)}\right), & \text{if } \mathcal{H}\dfrac{g(u,0)}{1+I_u} \le H(f,0)\dfrac{g(f,0)}{1+I_f^{\mathrm{DL}}(u)+\min\left\{\frac{\kappa}{g(f,u)}, P_1\right\}}. \\[2ex]
\log\left(1 + \dfrac{\min\left\{\frac{\kappa}{g(f,u)}, P_1\right\}}{1+I_f^{\mathrm{DL}}(u)+g(f,0)H(f,0)P_0}\right) & \text{otherwise}
\end{cases}
$$

$$\tag{6.18}$$

The sum throughput is obtained by using (6.18) in the summation and averaging (with respect to the fading coefficients) as in (6.14).

Notice that the multi-user diversity achieved through PFS scheduling in the macro-UL and macro-DL plays against the opportunity of performing SIC at the femtocells, since in this case $\mathcal{H}$ is distributed as the maximum of U independent fading gains. Without PFS, $\mathcal{H} = H(u, 0) \sim$ exponential(1), and the probability of performing SIC at femtocells is larger.

While evaluating the femtocell throughput with opportunistic SIC by Monte Carlo averaging is cumbersome but conceptually straightforward, by using Jensen's inequality and some closed form integration yields expressions that are easier to evaluate and offer significant simulation speed-up and insight in the effect of SIC. As an example, we focus on the macro-UL slot and rewrite (6.17) as

$$
R_f^{\mathrm{UL}} = \log\left(1 + \frac{\min\left\{\frac{\kappa}{g(f,u)}, P_1\right\}}{1 + I_f^{\mathrm{UL}} + g(f, u)H(f, u)P_0\, \mathbb{1}\{H(f, u) \le s(f, u)\mathcal{H}\}}\right) \tag{6.19}
$$

where we define $s(f, u) = \dfrac{g(u,0)(1+I_f^{\mathrm{UL}}+\min\left\{\frac{\kappa}{g(f,u)}, P_1\right\})}{g(f,u)(1+I_{\mathrm{bs}})}$. By averaging (6.19) for fixed f and u, and using Jensen's inequality, we can upper bound the resulting interference term at the denominator of the SINR in (6.19) as

$$
1 + \kappa \sum_{f' \in \mathcal{C}: f' \neq f} \frac{g(f, f')}{g(f', 0)} + g(f, u)P_0\mathbb{E}[H(f, u)\, \mathbb{1}\{H(f, u) \le s(f, u)\mathcal{H}\}].
$$

In order to evaluate the expectation in the last term, we can condition with respect to the random variable $s(f, u)$. Considering the case $\mathcal{H} \sim F_H(x)$ (no PFS in the macro-UL), we have

$$
\mathbb{E}[\,H(f, u)\, \mathbb{1}\{H(f, u) \le s\mathcal{H}\}|\, s(f, u) = s]
$$

$$
= \int_0^\infty \int_0^{sx} y\, dF_H(x)\, dF_H(y)
$$

$$
= \int_0^\infty \int_0^{sx} y\, e^{-x}e^{-y}\, dx\, dy
$$

$$
= \frac{s^2}{(s + 1)^2}. \tag{6.20}
$$

Table 6.1 Simulation parameters. ©2011 IEEE. Reprinted, with permission, from [32].

Parameter	Notation	Value
Macrocell side length	L	1000 m
Path-loss 3 dB distance	δ	50 m
Femtocell radius	r_{fc}	10 m
Distance between two femtocells	l	40 m
Path-loss exponent	α	3.5
Wall absorption loss	w	5 dB
Minimum SNR at cell edge	SNR_{min}	10 dB
Number of active MMSs	U	20

When PFS scheduling is used, $\mathcal{H}$ has probability density function (PDF) $F_{\mathcal{H}}(\mathsf{x}) = U(1 - e^{-\mathsf{x}})^{(U-1)} e^{-\mathsf{x}} 1\{\mathsf{x} \geq 0\}$. Repeating similar calculations, we obtain

$$\mathbb{E}\left[H(f, u) \, 1\{H(f, u) \leq s\mathcal{H}\} | s(f, u) = s \right] = \sum_{n=1}^{U} \binom{U}{n} (-1)^{n+1} \frac{s^2}{(s + n)^2}. \tag{6.21}$$

Finally, the results of (6.20) and (6.21) can be averaged with respect to $s(u, f)$ by Monte Carlo. As an alternative, for a large number of femtocells, we can approximate the terms I_f^{UL} and I_{bs} appearing in $s(f, u)$ with their mean value such that $s(f, u)$ becomes a deterministic quantity. Notice that the terms $s^2/(s + n)^2$, for $n = 1, \ldots, U$ appearing in (6.20) and in (6.21) are monotonically increasing between 0 and 1. Therefore, when $s(f, u)$ is small SIC is effective, while when $s(f, u)$ is large, SIC yields almost no gain. Similar bounds and approximations can be obtained for the macro-DL slot case.

6.3.4 Numerical results for the baseline system

The system parameters are listed in Table 6.1. We fixed the MBS power P_0 such that the received SNR (without interference) at the cell edge is 10 dB. Then, by changing the value of the interference temperature target value κ and letting P_1 be very large, we can sweep the boundary of the throughput tradeoff region achievable by treating interference as noise or using opportunistic SIC at the femtocells. At this point, it is worth remarking that for any set of achievable throughput points, the convex hull of such a set is also achievable, since we are considering time-averaged rates and therefore time sharing between different strategies is possible. If the region obtained as described above is non-convex, then the region can be enlarged by time sharing. In particular, if the region is included into the convex hull of the macro-only throughput (zero femtocell throughput, obtained by letting $\kappa = 0$) and of the femto-only throughput (zero macrocell throughput, obtained by letting $\kappa, P_1 \to \infty$), then the coding/decoding strategies analyzed before are dominated by simple orthogonalization, i.e., by reserving time-frequency slots for the two tiers. Instead, if the obtained region is convex, then orthogonalization of the two tiers is not convenient. Since the system is eventually interference limited, the shape of the tradeoff region depends on the power levels, and in particular on P_0 (indeed, κ and

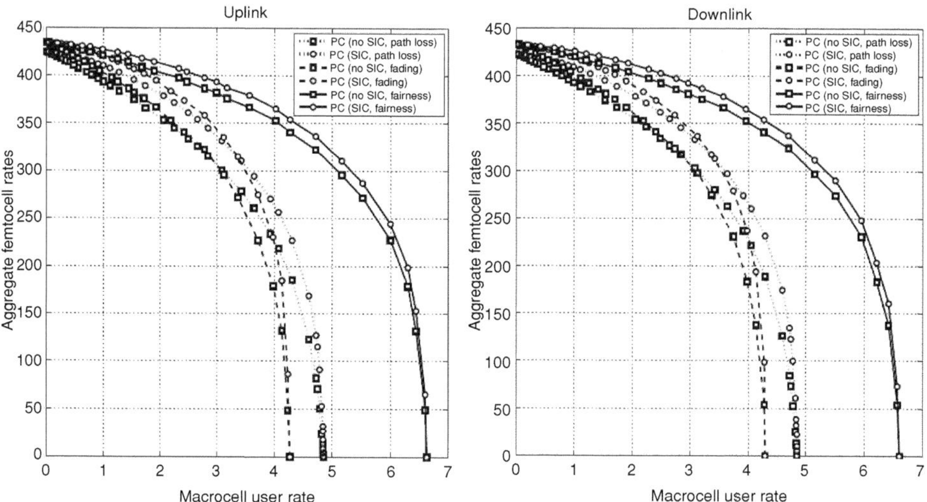

Figure 6.3 Femtocell sum throughput vs. macrocell throughput for different fading and macrocell scheduling scenarios, with and without opportunistic SIC, for the CSG case. ©2011 IEEE. Reprinted, with permission, from [32].

P_1 are given as fractions of P_0, expressed in dB). In our results, we have chosen a value of P_0 such that, at the cell edge, a macrocell user without any interference from the femtocell tier achieves an SNR of 10 dB (see Table 6.1).

In Figure 6.3 we show the throughput tradeoff region boundary for the case of no fading (path loss only), small-scale fading with round-robin macrocell scheduling, and small-scale fading with macrocell PFS, for the baseline cognitive femtocell system with interference temperature power control, denoted here as "PC," with and without opportunistic SIC at the femtocells. The results of Figure 6.3 correspond to CSG access, i.e., the positions of the MMSs are randomly generated over the whole macrocell area, and may fall inside some femtocells. A comparison between open vs. closed access, with and without SIC, is provided in Figure 6.4 for the case with PFS and Rayleigh fading.

We notice that in all cases the tradeoff regions are convex, thus indicating that the orthogonalization of the two tiers (as for example WiFi offloading would do) is not efficient in this range of transmit power. Also, we notice that SIC has a relatively minor effect in terms of aggregate femtocell throughput, also for the CSG access case. It is well known that, in the classical literature on conventional femtocell systems, macrocell users not belonging to the CSG of a femtocell may generate (resp., suffer from) very strong interference. This is regarded as one of the major limitation of femtocells in conventional two-tier systems (e.g., 3G-compatible systems). In contrast, we notice that with the relatively minor "cognitive" coordination assumed in this chapter, OA and CSG access yield virtually the same performance, even without using SIC.

While SIC does not yield appreciable benefits in terms of overall system sum through-put, it does provide very considerable performance improvements for femtocells close to the MBS, as illustrated in Figure 6.5. The charts in Figure 6.5 show the CDFs of the

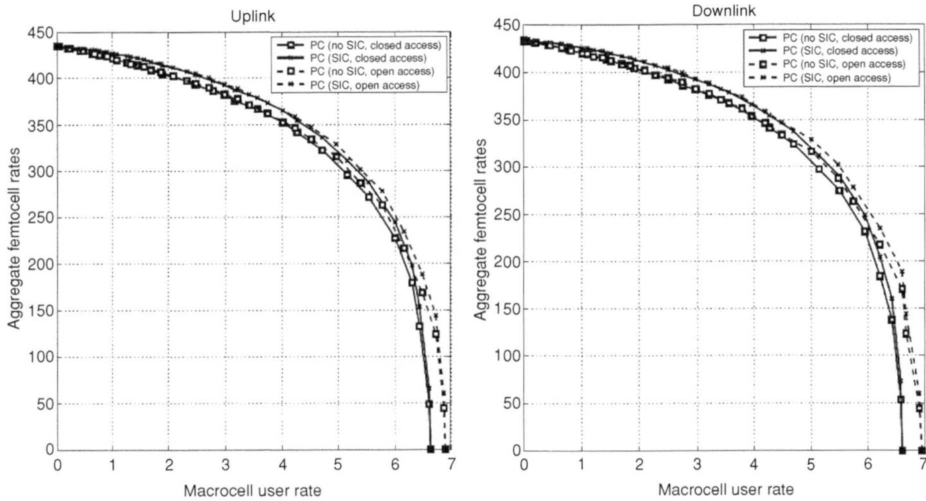

Figure 6.4 Femtocell sum throughput vs. macrocell throughput for PF scheduling with Rayleigh fading. OA vs. CSG, with and without SIC. ©2011 IEEE. Reprinted, with permission, from [32].

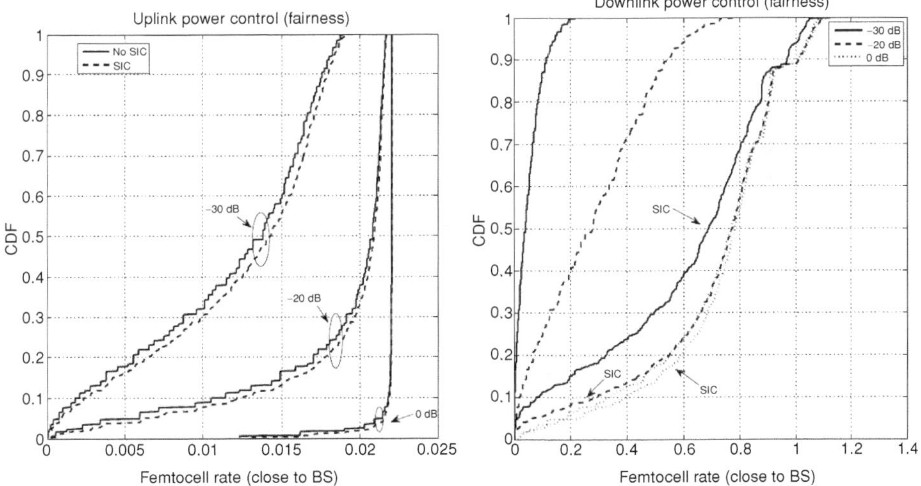

Figure 6.5 Throughput CDF for a femtocell close to the mBS.

femtocell ergodic rates for a particular femtocell located near the MBS. The ergodic rate CDF is obtained by generating random macrocell user positions $\mathcal{U}$ and, for each generation of $\mathcal{U}$, computing the ergodic achievable throughput according to the expressions obtained earlier. As anticipated before, femtocells near the MBS achieve very small rates on the macro-UL slot (notice the x-axis scale of the left chart in Figure 6.5, showing the throughput CDF on the macro-UL), since the interference temperature rule forces them to use a very small transmit power (in Figure 6.5, the dB label of each curve indicates $10 \log_{10}(\kappa / P_0)$). The only chance for these femtocells is to operate on the macro-DL slot. However, since they are close to the MBS they suffer from significant interference. Nevertheless, as shown by the right chart in Figure 6.5, using SIC allows us to achieve a throughput larger or equal to 0.6 bit/s/Hz with 80% probability, for interference temperature level equal to $-20\,\mathrm{dB}$ with respect to the MBS power P_0.

As an example, from Figure 6.3 with fading and PFS, we have seen that we can operate the system with macrocell tier spectral efficiency of ≈ 5.5 bit/s/Hz per km^2 and, at the same time, a total aggregate femtocell tier spectral efficiency of ≈ 300 bit/s/Hz per km^2. Although this is already well beyond what might be expected by other more conventional cellular techniques, even considering a large number of MBS antennas (e.g., see [8, 41]), the spectral efficiency per femtocell is just a modest 0.48 bit/s/Hz. This is due to the fact that the femtocell tier is dominated by interference, treated as noise. In the next sections, we examine two simple linear precoding schemes that can improve upon the baseline system performance.

6.4   Downlink interference alignment

In this section we adapt the scheme proposed in [29], known as *Downlink IA*, to the two-tier macro/femto network under consideration. The goal is to completely eliminate the cross-tier macro-to-femto interference on the macro-DL slot. In this way, the MBS transmit power can be increased arbitrarily without reducing the femtocell throughput. This is achieved by allowing for some redundancy in the signal set dimension, and therefore the scheme is efficient only if in each femtocell there are enough users to fill up these dimensions and recover this redundancy. While for the baseline system we assumed orthogonal access inside each femtocell, and therefore any number of users $N_{\mathrm{U},2,j} \geq 1$ in each femtocell j yields the same aggregate femtocell throughput, for the Downlink IA scheme the femtocell throughput depends critically on the number of users in each femtocell. For simplicity, we let $N_{\mathrm{U},2,j} = m$ for all j, and express the femtocell throughput as a function of m. While for the sake of comparison we insist on the system model defined in Section 6.2, we hasten to say that Downlink IA is more appropriate for a scenario comprising only *a few* tier-2 cells, each of which contains many users, as for example in the case of hotspots/picocells. For the Downlink IA scheme, we assume that the femto-DL and the macro-DL are aligned on the same slot.

Consider the block diagram in Figure 6.6. The mBS serves K MMSs on N signal dimensions using a fixed "tall" unitary precoding matrix $\mathbf{P} \in \mathbb{C}^{N \times K}$ with $K < N$, such that $\mathbf{P}^{\mathsf{H}}\mathbf{P} = \mathbf{I}_K$. For example, if these signal dimensions are obtained in the frequency domain (subcarriers), this corresponds to the well known multicarrier CDMA (MC-CDMA) with unitary spreading, where the columns of $\mathbf{P}$ are the frequency-domain spreading sequences used in the downlink (see for example [42, 43] and references therein). More generally, we consider that blocks of N signal dimensions, obtained in time and frequency, are used for precoding K macrocell users. The signal vector transmitted by the MBS is given by $\mathbf{x}_{\mathrm{bs}} = \mathbf{P}\mathbf{s}_{\mathrm{bs}}$, where $\mathbf{s}_{\mathrm{bs}}$ is the vector of coded symbols sent to the K MMSs on any given precoding block. Next, we consider femtocell f,

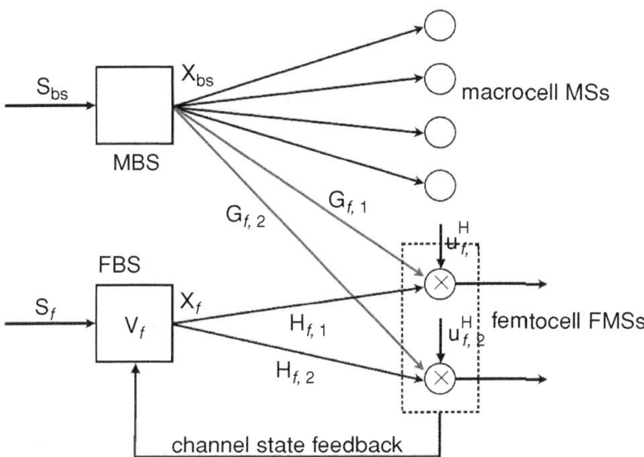

Figure 6.6 Downlink IA concept, including the MBS serving multiple MMSs, and one femtocell (indexed by f) serving multiple FMSs. ©2011 IEEE. Reprinted, with permission, from [32].

serving $1 \leq m \leq N$ femtocell users. The received signal at the i-th femtocell user over a precoding block is given by

$$\mathbf{y}_{f,i} = \sqrt{g(f, 0)}\mathbf{G}_{f,i}\mathbf{x}_{\text{bs}} + \mathbf{H}_{f,i}\mathbf{x}_f + \mathbf{z}_{f,i} \tag{6.22}$$

where $\mathbf{G}_{f,i}$ is a diagonal $N \times N$ matrix containing the fading coefficients over the N precoding signal dimensions, for the channel from the MBS to the i-th FMS in femtocell f, $\mathbf{H}_{f,i}$ is the diagonal $N \times N$ matrix for the downlink channel from the f-th FBS, $\mathbf{x}_f \in \mathbb{C}^{N \times 1}$ is the transmitted vector of the f-th fBS, and $\mathbf{z}_{f,i}$ is an i.i.d. Gaussian noise vector with components $\sim \mathcal{CN}(0, 1)$. Since $\mathbf{G}_{f,i}\mathbf{x}_{\text{bs}} \in \text{Span}\{\mathbf{G}_{f,i}\mathbf{P}\}$, for $K < N$ there exists a non-trivial orthogonal complement of the macrocell interference space at each i-th FMS receiver. We let $\mathbf{u}_{f,i}$ denote a unit-norm vector in such orthogonal complement. It follows that each i-th FMS receiver can project its signal vector onto $\mathbf{u}_{f,i}$, and obtain a scalar channel observation free from the cross-tier interference, given by

$$\tilde{y}_{f,i} = \mathbf{a}_{f,i}^{\mathsf{H}}\mathbf{x}_f + \tilde{w}_{f,i}. \tag{6.23}$$

where $\mathbf{a}_{f,i} = \mathbf{H}_{f,i}^{\mathsf{H}}\mathbf{u}_{f,i} \in \mathbb{C}^{N \times 1}$ is the equivalent channel vector and $\tilde{w}_{f,i} = \mathbf{u}_{f,i}^{\mathsf{H}}\mathbf{z}_{f,i}$ is also $\sim \mathcal{CN}(0, 1)$ is the equivalent channel noise (recall that $\mathbf{u}_{f,i}$ has unit norm). In order to serve simultaneously m femtocell users on N signal dimensions, each user i feeds back its channel vector $\mathbf{a}_{f,i}$, using standard TDD reciprocity.[4] Then, the FBS can use any suitable form of multi-user MIMO downlink precoding for the set of m femtocell users in the resulting cross-tier interference free vector broadcast channel given by (6.23), for $i = 1, \ldots, m$, where the $N \times m$ channel matrix

$$\mathbf{A}_f = \begin{bmatrix} \mathbf{a}_{f,1} & \mathbf{a}_{f,2} & \cdots & \mathbf{a}_{f,m} \end{bmatrix} \tag{6.24}$$

has rank m with probability 1.

---

[4] Notice that in this case operating the femtocells in TDD have a clear advantage over FDD.

At this point, any suitable scheme for the vector Gaussian broadcast channel with known channel matrix can be applied, depending on the desired transmitter complexity. For example, the FBS can use linear zero-forcing downlink precoding, and let $\mathbf{x}_f = \mathbf{V}_f \mathbf{s}_f$, where $\mathbf{V}_f$ is a unit-column normalized version of the pseudo-inverse of the equivalent channel matrix $\mathbf{A}_f^+ = \mathbf{A}_f(\mathbf{A}_f^H \mathbf{A}_f)^{-1}$ (see for example [10, 44]). Alternatives are provided by the regularized linear zero-forcing downlink precoding (see [45] and references therein), by various forms of *Dirty Paper Coding* (see [46–49]), by vector precoding (see [50]), or reduced lattice precoding (see [51]). In any case, the scheme achieves the alignment of the intra-femtocell multi-user interference with the inter-cell (macro-to-femto) interference, thus motivating the nickname of "downlink IA". Notice that if we had insisted on classical multicarrier CDMA (MC-CDMA), imposing a joint inter- and intra-cell zero-forcing interference constraint at each FMS yields $K + m \leq N$, instead of $K < N$ and $m \leq N$ achieved by this scheme.

Before entering the system performance analysis, a few remarks are in order. First, we notice that the MBS unitary precoding matrix $\mathbf{P}$ is fixed, and does not change from block to block. The fading channel coefficients in $\mathbf{G}_{f,i}$ change very slowly in time, since this is the channel from the MBS to the i-th user in femtocell f, and femtocell users are nomadic, typically moving slower than walking speed. Hence, the equivalent channel matrix $\mathbf{A}_f$ is very slowly varying in time, and can be accurately tracked by the FMS, even though the MBS schedules dynamically a different set of K macrocell users on each precoding block. Furthermore, under mild conditions on the joint distribution of the diagonal elements of the matrices $\mathbf{G}_{f,i}$ and $\mathbf{H}_{f,i}$ (e.g., in the case of independent Rayleigh fading, these elements are i.i.d. $\sim \mathcal{CN}(0, 1)$) and choosing $\mathbf{P}$ at random with uniform probability over the set of $N \times K$ unitary matrices, the probability that the equivalent channel matrix $\mathbf{A}_f$ has rank less than m is zero. Hence, the number of independent data streams that femtocell f can support on its DL slot, while being completely free from macrocell interference, is equal to m with probability 1. Finally, we notice that in [29] the scheme considers two mutually interfering cells, each of which makes use of a precoding matrix $\mathbf{P}$. However, in our case the femto-to-macro interference is caused by several femtocells surrounding the served MMSs. Then, in order to align interference from several FBSs it is necessary to exploit the knowledge of the cross-channel coefficients (from the FBSs to the MMSs). This is much more difficult to obtain, since the macrocell users may be mobile, and the MBS schedules them dynamically, so that these channel matrices changes at each slot. For this reason, we focus on the above described asymmetric scenario and just treat the femto-to-macro interference as noise.

6.4.1   Approximated asymptotic analysis

For finite N, K, m, the throughput tradeoff region of downlink IA is not amenable to simple analytic formulas and can be obtained by cumbersome Monte Carlo simulation. As an alternative, in this section we scale the system dimensions to infinity and obtain simple closed form formulas. It turns out that, as often happens in these cases, the asymptotic formulas yield very accurate approximation even for small system dimensions [52]. We let $\alpha_F = m/N$ and $\alpha_M = K/N$ denote the number of served users per dimension of the femtocell and macrocell precoders, respectively, and consider the system performance as $N \to \infty$ with fixed users/dimension ratios α_F and α_M. We assume that the group of K jointly precoded macrocell users is chosen independently on each macro-DL slot, with uniform probability over all $\binom{U}{K}$ possible groups. Since K users are served simultaneously on N dimensions, it is difficult to implement any form of power control as done in the baseline system. In fact, each femtocell should be able to regulate its power to guarantee a desired interference level to all K macrocell users at the same time. Instead, we consider a simplified system where femtocells transmit at constant power, owing to the fact that this system uses effectively some form of spreading and therefore it is more robust to interference. The MBS transmit power is given by

$$\frac{1}{N}\text{tr}\left(\mathbb{E}\left[\mathbf{P}\mathbf{s}_{bs}\mathbf{s}_{bs}^H\mathbf{P}^H\right]\right) = \frac{K}{N}P_{bs} = \alpha_M P_{bs} \tag{6.25}$$

where P_{bs} denotes the energy per symbol of the coded symbols (elements of $\mathbf{s}_{bs}$). Letting P_0 denote the MBS transmit power, we have $P_{bs} = P_0/\alpha_M$. Similarly, letting without loss of generality $\mathbf{x}_f = \mathbf{V}_f \mathbf{s}_f$ where $\mathbf{V}_f$ is a unit-norm columns downlink precoding matrix, the FBS transmit power is given by

$$\frac{1}{N}\text{tr}\left(\mathbb{E}\left[\mathbf{V}_f \mathbf{s}_f \mathbf{s}_f^H \mathbf{V}_f^H\right]\right) = \frac{m}{N}P_{fc} = \alpha_F P_{fc}, \tag{6.26}$$

where P_{fc} denotes the energy per symbol of the coded symbols (elements of $\mathbf{s}_f$). Letting P_1 denote the FBS transmit power, we have $P_{fc} = P_1/\alpha_F$.

From the macrocell point of view, approximating the cross-tier femto-to-macro interference on each macrocell user u from the surrounding femtocells as white noise, the system is completely analogous to the well studied downlink of MC-CDMA with unitary spreading. In this case, standard large-system analysis for various receivers is given in [42, 43]. In this chapter we focus on *linear* minimum mean squared error (MMSE) receivers at the MMSs, yielding the macrocell DL throughput approximation

$$R_{mc}^{DL} \approx \frac{\alpha_M}{U}\sum_{u \in \mathcal{U}}\log\left(1 + \frac{\eta_{mmse}(u)g(u,0)P_0/\alpha_M}{1 + I_u}\right) \tag{6.27}$$

where we let $I_u = P_1 \sum_{f \in \mathcal{C}} g(f, u)$ and where $\eta_{mmse}(u)$ is the multi-user efficiency for the MMSE receiver, solution of the fixed-point equation in the indeterminate η [42, 43]

$$\mathbb{E}\left[\frac{\mathcal{H}}{\alpha_M \tau_u \mathcal{H} + 1 + (1 - \alpha_M)\tau_u \eta}\right] = \frac{\eta}{1 + \tau_u \eta} \tag{6.28}$$

with $\mathcal{H} \sim \text{exponential}(1)$ and $\tau_u = \frac{g(u,0)P_0/\alpha_M}{1 + I_u}$. The throughput expression (6.27) is an approximation since we have ignored the fact that the cross-tier interference at user u from all the femtocells is colored due to the downlink precoding used by each FBS and we have summed over the macrocell users locations $\mathcal{U}$, although in the asymptotic large-system limit we let the number of served macrocell users K go to infinity.

In order to obtain an expression for the femtocell throughput, we make the assumption that the unit projection vectors $\mathbf{u}_{f,i}$ in the downlink IA scheme are random, independent, and uniformly distributed on the unit N-dimensional sphere. Assuming i.i.d. fading coefficients for all diagonal matrices $\mathbf{H}_{f,i}$, it follows that $\mathbf{A}_f$ conditionally on $\mathbf{u}_{f,1}, \ldots, \mathbf{u}_{f,m}$

is a matrix of independent elements, with asymptotic row and column regularity properties (see [53]). Hence, the empirical eigenvalue distribution of $\mathbf{A}_f^H \mathbf{A}_f$ converges almost surely to the Marcenko–Pastur distribution [43, 52]. Under these assumptions, we can obtain in closed form the femtocell throughput for linear or non-linear precoding (e.g., linear MMSE or dirty paper coding) for the vector broadcast channel with matrix $\mathbf{A}_f$. We let the received SNR for femtocell users be given by $\tau_f = \frac{P_1/\alpha_F}{1+I_f}$, where the inter-tier interference power is given by $I_f = P_1 \sum_{f' \in \mathcal{C}: f' \neq f} g(f, f')$.[5] Focusing again on linear schemes, in the numerical results of this section we consider the performance of *MMSE downlink beamforming* given by [43]

$$R_{\text{fc}}^{\text{DL}} = \alpha_F \sum_{f \in \mathcal{C}} \log \left(1 + \tau_f - \frac{\mathcal{F}(\tau_f, \alpha_F)}{4} \right), \tag{6.29}$$

where

$$Femtocell(x, z) = \left(\sqrt{x(1 + \sqrt{z})^2 + 1} - \sqrt{x(1 - \sqrt{z})^2 + 1} \right)^2.$$

This can be obtained by noticing that the channel seen at user i receiver of femtocell f is given by (6.23), where now the equivalent noise term $\tilde{w}_{f,i}$ has variance $1 + I_f$, due to the interference from the other femtocells. Then, by UL/DL duality [47, 54, 55], the SINR achieved with linear beamforming and total transmit energy $m P_{\text{fc}}$ coincides with that of the "dual UL" (virtual) multiple access channel with the same channel matrix $\mathbf{A}_f$, given by

$$\mathbf{r}_f = \mathbf{A}_f \mathbf{u}_f + \mathbf{z}_f,$$

with $\mathbf{z}_f$ i.i.d. with components $\sim \mathcal{CN}(0, 1 + I_f)$. Somehow arbitrarily, we choose equal symbol energy allocation in the dual UL and compute the corresponding SINRs that, in the large system limit, yield (6.29). We hasten to add that this is not the optimal (i.e., throughput maximizing) linear beamforming strategy. Since maximizing the sum throughput with linear beamforming is a well known non-convex problem whose iterative solution may converge to local maxima, and depends in general on the initialization [45, 49, 56, 57], here we prefer to provide the concise closed form expression (6.29) for the said, slightly suboptimal, MMSE linear beamforming scheme.

Figure 6.7 shows the system throughput tradeoff in the macro-DL slot when the MMSs make use of linear MMSE receivers and the femtocells use linear MMSE precoding, as explained above. The loading factors are set to $\alpha_M = \alpha_F = 0.95$. In this case, we considered an OA femtocell system (macrocell users located only outside each femtocell), and we provided in Figure 6.7 a comparison with the corresponding baseline scheme without SIC. We notice that as the BS power increases (SNR at the cell edge equal to 10, 20 and 30 dB), the downlink IA scheme outperforms the baseline scheme. Nevertheless, for 10 dB the baseline scheme performs better. The loss in the maximum sum-rate of the femtocells is due to the loading factor $\alpha_F < 1$ and to the noise enhancement of linear

---

[5] Notice that, for a symmetric system as considered here, I_f does not depend on f. However, we give the general expression that holds also for non-symmetric systems.

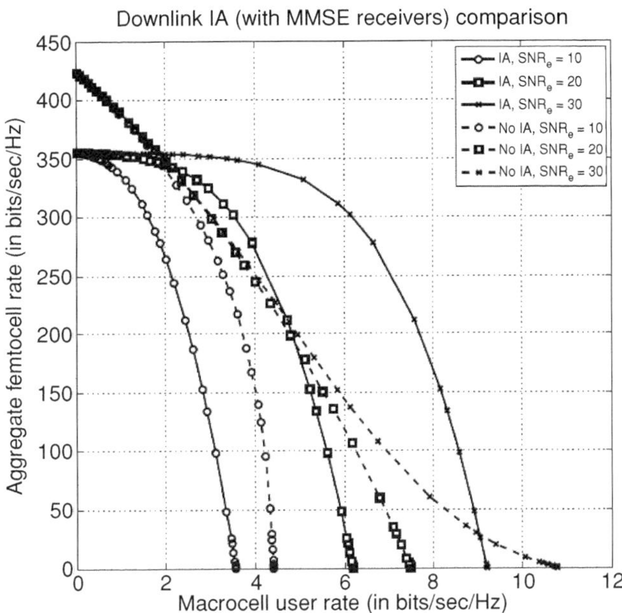

Figure 6.7 Comparison of downlink IA with the baseline power control strategy for increasing BS power and loading factors $\alpha_M = 0.95$, $\alpha_F = 0.95$. These results refer to OA where the baseline system uses round robin scheduling and no SIC. ©2011 IEEE. Reprinted, with permission, from [32].

precoding. Slightly better performance can be achieved by using non-linear dirty paper coding for the femtocell vector broadcast channel. It is worthwhile noticing also that in a typical residential femtocell the number of active users is small (typically 1 to 3). Therefore, the IA scheme should consider small precoding blocklength N, otherwise the ratio m/N is too small, leading to very poor spectral efficiency in the femtocells. In contrast, as mentioned at the beginning of this section, downlink IA may be attractive in the case where some picocell hotspots with many users (e.g., a train station, or a conference center) are located near an MBS. In this case, using downlink IA to "protect" the hotspot from the strong MBS interference in the macro-DL slot may yield significant performance improvements with respect to orthogonal access by bandwidth reservation.

6.5   Multiple antennas at the FBSs and reverse TDD

In this section we consider the use of multiple antennas at the FBSs, and show that a dramatic improvement of the femtocell throughput can be achieved with very simple signal processing. In fact, we shall use the multiple antennas at the FBSs to mitigate intra-tier and cross-tier interference, while insisting on simple orthogonal access inside each femtocell. Therefore, without loss of generality, we consider again the case

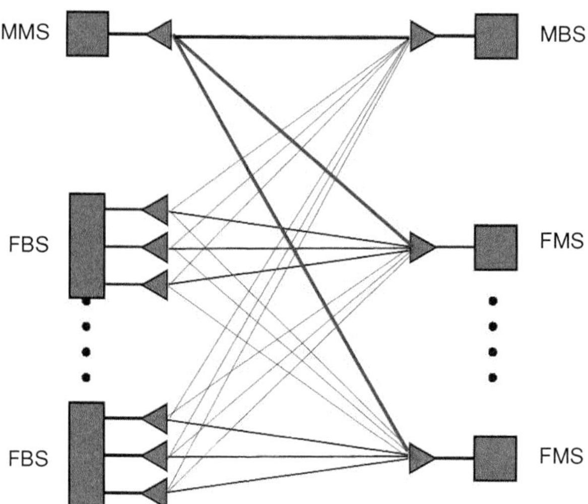

Figure 6.8 MISO (left to right) SIMO (right to left) interference channel for the two-tier network with multi-antenna FBSs. The thick gray lines indicate the links that determine the FMS transmit power, via the interference temperature power control. ©2011 IEEE. Reprinted, with permission, from [32].

of $N_{U,2,j} = 1$ user in each femtocell j. In this case, in contrast to what was done in the previous section, we propose *reverse TDD*, i.e., we align the macro-DL slot with the femto-UL slot and vice versa. Figure 6.8 shows the resulting channel model, where the macro-DL/femto-UL slot corresponds to transmitting from right to left, and the macro-UL/femto-DL slot corresponds to transmitting from left to right.

On the macro-DL/femto-UL slot we are in the presence of a single input multiple output (SIMO) interference channel. All transmitters (the MBS and the FMSs) have a single antenna and hence they are isotropic. In this case, the macrocell user's SINR during the macro-DL slot is imposed by the interference temperature power control at the femtocells. The received signal at the u-th mMS is given by

$$y_u = \sqrt{g(u, 0)}h_{u,0}x_{\text{bs}} + \sum_{f \in \mathcal{C}} \sqrt{g(f, u)}h_{f,u}x_f + z_u, \tag{6.30}$$

with SINR given by

$$\text{SINR}_u^{\text{DL}} = \mathcal{H}\frac{g(u, 0)P_0}{1 + I_u}, \tag{6.31}$$

where I_u is given by (6.8) with $H(f, u) = |h_{f,u}|^2$, and where $\mathcal{H} = |h_{u,0}|^2$, the statistics of which depend on the type of downlink scheduling used, as already discussed for the baseline system. The corresponding macro-DL throughput is given in (6.10). As already noticed, the interference temperature power control does not impact significantly the femtocell throughput since only the femtocells close to the scheduled macrocell user u in the current slot need to significantly reduce their transmit power. Furthermore, because of the MBS DL scheduling, the scheduled macrocell user changes at each slot, so that this effect is randomized over all the femtocells and no femtocell is forced to permanently use low transmit power.

The FBSs have M antennas each, and make use of a linear MMSE receiver, maximizing the SINR over all linear receivers. The received signal vector at FBS f is given by

$$\mathbf{y}_f = \mathbf{h}_{f,f} x_f + \sum_{f' \in \mathcal{C}: f' \neq f} \sqrt{g(f, f')} \mathbf{h}_{f,f'} x_{f'} + \sqrt{g(f, 0)} \mathbf{h}_{f,0} x_{\text{bs}} + \mathbf{z}_f. \tag{6.32}$$

The linear MMSE receiving vector for estimating the desired symbol x_f from $\mathbf{y}_f$ is given by (up to an irrelevant scaling factor)

$$\mathbf{u}_f = \mathbf{\Sigma}_f^{-1}(u) \mathbf{h}_{f,f} \tag{6.33}$$

where

$$\mathbf{\Sigma}_f(u) = \mathbf{I}_M + g(f, 0) P_0 \mathbf{h}_{f,0} \mathbf{h}_{f,0}^{\mathsf{H}} + \sum_{f' \in \mathcal{C}: f' \neq f} \min\left\{\frac{\kappa}{g(f', u)}, P_1\right\} g(f, f') \mathbf{h}_{f,f'} \mathbf{h}_{f,f'}^{\mathsf{H}}$$

is the interference plus noise covariance matrix at the f-th FBS receiver, which depends on the scheduled macrocell user u because of the power control. The receiver forms the scalar observation $\mathbf{u}_f^{\mathsf{H}} \mathbf{y}_f$, and the SINR of the resulting equivalent scalar output channel (treating residual interference as noise) is given by

$$\text{SINR}_f^{\text{simo}}(u) = \min\left\{\frac{\kappa}{g(f, u)}, P_1\right\} \mathbf{h}_{f,f}^{\mathsf{H}} \mathbf{\Sigma}_f^{-1}(u) \mathbf{h}_{f,f} \tag{6.34}$$

where the superscript "simo" indicates the SIMO interference channel nature of the channel model in the macro-DL/femto-UL slot (see Figure 6.8).

On the macro-UL/femto-DL slot we are in the presence of a multiple input single output (MISO) interference channel, given by

$$y_0 = \sqrt{g(u, 0)} h_{u,0}^* x_u + \sum_{f \in \mathcal{C}} \sqrt{g(0, f)} \mathbf{h}_{f,0}^{\mathsf{H}} x_f + z_0, \tag{6.35}$$

for the MBS receiver, and by the received signal at the FMS in femtocell f given by

$$y_f = \mathbf{h}_{f,f}^{\mathsf{H}} x_f + \sum_{f' \in \mathcal{C}: f' \neq f} \sqrt{g(f, f')} \mathbf{h}_{f,f'}^{\mathsf{H}} x_{f'} + \sqrt{g(u, f)} h_{u,f}^* x_u + z_f \tag{6.36}$$

for the f-th FMS receiver. Insisting on linear beamforming strategies, each FBS f sends the signal vector $x_f = \mathbf{w}_f s_f$, where $\mathbf{w}_f$ denotes a transmit beamforming vector and s_f is the corresponding coded symbol for the f-th femtocell user. From the well known UL-DL SINR duality for SIMO/MISO interference channels [58, 59], we have that by letting $\mathbf{w}_f = \mathbf{u}_f / \|\mathbf{u}_{vf}\|$, it is possible to achieve on the MISO channel defined by (6.35) and (6.36) the same SINRs achievable on the dual SIMO channel, given by (6.31) for the macrocell and by (6.34) for femtocell f. Furthermore, these SINRs are achieved with a power allocation with the same sum power of the dual channel, given by

$$P_{\text{tot}} = P_0 + \sum_{f \in \mathcal{C}} \min\left\{\frac{\kappa}{g(f, u)}, P_1\right\} \leq P_0 + F^2 P_1. \tag{6.37}$$

Notice that in the MISO direction, the sum power P_{tot} is allocated across the femtocells and the macrocell users in some way, that depends on the realization of the path loss and small-scale fading coefficients. With some extra effort it is possible to exploit a max–min duality result [55, 58] in order to impose a stricter *per-transmitter* power constraint. However, for the sake of simplicity, here we consider only the sum-power constraint, allowing for slightly optimistic results in the case the terminals that are strictly peak-power limited.

At this point, two issues remain to be addressed:

1. For given beamforming vectors $\mathbf{w}_f$, and given target SINRs $\gamma_0, \gamma_1, \ldots, \gamma_{F^2}$, the system needs to calculate the set of componentwise minimum transmit powers for the MISO channel that achieves the target SINRs, if these are feasible.
2. The linear MMSE receiving vectors (and consequently the transmit beamforming vectors) must be calculated at each FBS.

Both points are addressed in the following, yielding a simple and appealing physical layer protocol for cognitive femtocells with multiple antennas at the FBS.

Decentralized reverse link power allocation:
For all femtocells f, fix the unit-norm beamforming vectors $\mathbf{w}_f$, and the target SINRs $\gamma_0, \gamma_1, \ldots, \gamma_{F^2}$. In the macro-UL/femto-DL slot, we wish to solve the power allocation problem:

$$\text{minimize} \quad Q_u + \sum_{f \in \mathcal{C}} Q_f(u)$$

$$\text{subject to} \quad \text{SINR}_f^{\text{miso}}(u) \geq \gamma_f, \quad \forall \, f \in \mathcal{C}$$

$$\text{SINR}_u^{\text{UL}} \geq \gamma_0 \tag{6.38}$$

where, from (6.35), the macro-UL SINR is given by

$$\text{SINR}_u^{\text{UL}} = \frac{\mathcal{H}g(u, 0)Q_u}{1 + \sum_{f \in \mathcal{C}} g(0, f)|\mathbf{h}_{f,0}^{\mathsf{H}}\mathbf{w}_f|^2 Q_f(u)} \tag{6.39}$$

and, from (6.36), the femto-DL SINR is given by

$$\text{SINR}_f^{\text{miso}}(u) = \frac{|\mathbf{h}_{f,f}^{\mathsf{H}}\mathbf{w}_f^{\mathsf{H}}|^2 Q_f(u)}{1 + \sum_{f' \in \mathcal{C}: f' \neq f} g(f, f')|\mathbf{h}_{f,f'}^{\mathsf{H}}\mathbf{w}_{f'}|^2 Q_{f'}(u) + g(u, f)H(u, f)Q_u}. \tag{6.40}$$

As stated above, the UL-DL SINR duality for SIMO/MISO interference channels yields that, by choosing where $\mathbf{w}_f = \mathbf{u}_f/\|\mathbf{u}_f\|$ with $\mathbf{u}_f$ given by (6.33), there exist powers Q_u and $Q_f(u)$ for all $f \in \mathcal{C}$ such that $\gamma_0 = \text{SINR}_u^{\text{DL}}$ given by (6.31), and $\gamma_f = \text{SINR}_f^{\text{simo}}(u)$, given by (6.34) are achievable with the same (minimum) total power P_{tot} given by (6.37). Hence, we know that with this choice of the transmit beamforming vectors at the FBSs, problem (6.38) is feasible. Furthermore, we recognize that the "interference functions" given by

$$\mathcal{I}_{\text{bs}}(u) = \frac{1 + \sum_{f \in \mathcal{C}} g(0, f) |\mathbf{h}_{f,0}^{\mathsf{H}} \mathbf{w}_f|^2 Q_f(u)}{\mathcal{H}g(u, 0)}$$

and by

$$\mathcal{I}_f(u) = \frac{1 + \sum_{f' \in \mathcal{C}: f' \neq f} g(f, f') |\mathbf{h}_{f,f'}^{\mathsf{H}} \mathbf{w}_{f'}|^2 Q_{f'}(u) + g(u, f) H_{u,f} Q_u}{|\mathbf{h}_{f,f}^{\mathsf{H}} \mathbf{w}_f^{\mathsf{H}}|^2}$$

are *standard interference functions* according to the definition of [60]. Hence, for feasible SINRs, the iterative decentralized Yates' standard power control algorithm (also known as the Foschini–Miljanic algorithm), see [60, 61], can be applied in order to obtain the *componentwise* minimum powers Q_u, $Q_1(u), \ldots, Q_{F^2}(u)$ achieving the target SINRs. The componentwise minimum power allocation dominates any other power allocation achieving the same SINRs. Hence, it is the solution of (6.38). By the SINR duality, this must achieve the same sum-power P_{tot}.

For the sake of completeness, we give below the iteration of Yates' algorithm applied to our case. Let n denote the algorithm iteration index, and let $Q_u^{(0)}$, $Q_1^{(0)}(u), \ldots, Q_{F^2}^{(0)}(u)$ denote the initialization. For example, we can choose $Q_u^{(0)} = P_0$ and $Q_f^{(0)}(u) = P_f(u)$. Then, the algorithm power update iteration is given by

$$Q_u^{(n+1)} = \frac{\gamma_0 \left(1 + \sum_{f \in \mathcal{C}} g(0, f) |\mathbf{h}_{f,0}^{\mathsf{H}} \mathbf{w}_f|^2 Q_f^{(n)}(u) \right)}{\mathcal{H}g(u, 0)}$$

$$Q_f^{(n+1)}(u) = \frac{\gamma_f \left(1 + \sum_{f' \in \mathcal{C}: f' \neq f} g(f, f') |\mathbf{h}_{f,f'}^{\mathsf{H}} \mathbf{w}_{f'}|^2 Q_{f'}^{(n)}(u) + g(u, f) H_{u,f} Q_u^{(n)} \right)}{|\mathbf{h}_{f,f}^{\mathsf{H}} \mathbf{w}_f^{\mathsf{H}}|^2},$$

$$\forall f \in \mathcal{C}. \tag{6.41}$$

Figure 6.9 show the macro/femto throughput tradeoff region for system conditions similar to Figure 6.3 (in this case: OA femtocells, Rayleigh fading, with round-robin macrocell scheduling), for the case of $M = 5$ antennas at the FBSs. The macro-DL/femto-UL throughput curve shows a larger than three-fold increase in the femtocells aggregate throughput, for the same macrocell throughput. As far as the macro-UL/femto-DL throughput is concerned, notice that after some iterations of the distributed power control algorithm (6.41) the curve essentially coincides with the one in the reverse direction, as expected from UL–DL duality. It is also interesting to notice that even in the case of a single iteration, the throughput tradeoff is very competitive with respect to the baseline system. In practice, some good heuristic for power allocation in the reverse direction may even require no iteration at all. We leave this interesting and relevant topic for further investigation.

Decentralized estimation of the MMSE receiving vectors:
We use the matrix inversion lemma to rewrite $\mathbf{u}_f$ in the form

$$\mathbf{u}_f = (1 + \text{SINR}_f^{\text{simo}}(u)) \mathbf{K}_f^{-1} \mathbf{h}_{f,f}, \tag{6.42}$$

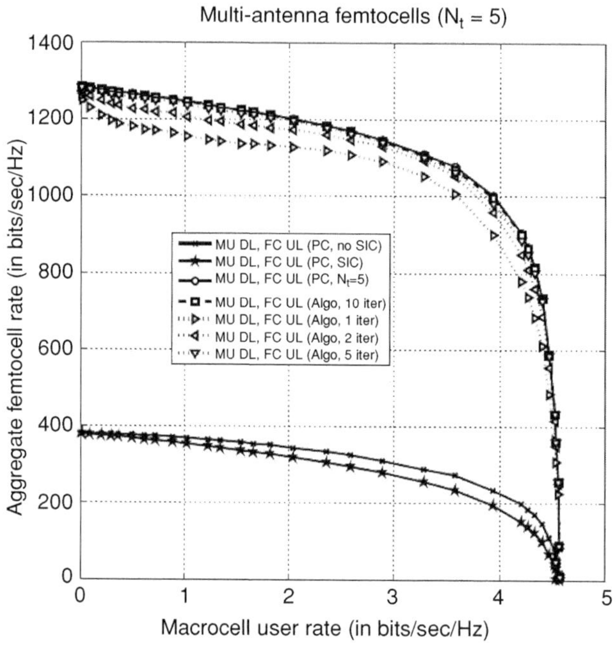

Figure 6.9 Throughput tradeoff for a multi-antenna FBS system ($M = 5$ antennas). These results are obtained under OA. ©2011 IEEE. Reprinted, with permission, from [32].

where

$$\mathbf{K}_f(u) = \mathbf{I}_M + g(f, 0)P_0\mathbf{h}_{f,0}\mathbf{h}_{f,0}^{\mathsf{H}} + \sum_{f' \in \mathcal{C}} \min\left\{\frac{\kappa}{g(f', u)}, P_1\right\} g(f, f')\mathbf{h}_{f,f'}\mathbf{h}_{f,f'}^{\mathsf{H}}$$

is the received signal covariance matrix at the f-th FBS receiver. Since the factor $1 + \mathsf{SINR}_f^{\mathrm{simo}}(u)$ in (6.42) is just a positive scaling, with no effect on the SINR, it is convenient to redefine $\mathbf{u}_f$ as the vector $\mathbf{K}_f^{-1}\mathbf{h}_{f,f}$. In order to calculate the MMSE receiving vectors, the signal covariance matrix $\mathbf{K}_f$ at each FBS receiver can be estimated using the sample covariance

$$\widehat{\mathbf{K}}_f = \frac{1}{N}\sum_{i=1}^{N} \mathbf{y}_f[i]\mathbf{y}_f^{\mathsf{H}}[i],$$

where N is the number of samples used for estimation, $\mathbf{y}_f[i]$ indicates the received signal vector at time (or signal dimension) i, and we assume $N \leq T$, such that all signal samples for estimation belong to the current femto-UL slot (and therefore they span the same coherence block). Notice that the estimation of the signal sample covariance matrix can take place over the whole slot and requires no training. Since $\mathbf{K}_f$ has dimension $M \times M$, where practical values of M (number of antennas per FBS) range from 4 to 10, it is safe to assume $T \gg M$, such that a very accurate estimate $\widehat{\mathbf{K}}_f \approx \mathbf{K}_f$ can be achieved. For simplicity, in our numerical results we assume that $\mathbf{K}_f$ is perfectly known.

In order to estimate the desired signal channel vector $\mathbf{h}_{f,f}$ at each FBS receiver, we assume that a training sequence of known symbols is included in each femto-UL slot. We consider $r+1$ orthogonal sequences of length $T_{\mathrm{tr}} \geq r+1$ symbols. Let $\boldsymbol{\Phi} = [\boldsymbol{\phi}_0, \ldots, \boldsymbol{\phi}_r] \in \mathbb{C}^{T_{\mathrm{tr}} \times (r+1)}$ denote the training matrix. Column $\boldsymbol{\phi}_0$ is used in the macro-DL training and has energy $\|\boldsymbol{\phi}_0\|^2 = T_{\mathrm{tr}} P_0$. The other r columns are distributed across the femtocells, in order to obtain a regular pilot reuse pattern with reuse factor r. The allocation of the training sequences to the femtocells is static, and can be implemented easily in a centralized fashion, e.g., by requiring that the FBSs get their training sequence allocation from a database, which can be updated at a very slow pace, every time a new femtocell is installed and joins the system. We assume that the FMSs transmit at constant power $P_{\mathrm{tr}} = \delta P_0$ during the femto-UL training phase. Since $\boldsymbol{\phi}_0^{\mathsf{H}} \boldsymbol{\phi}_\ell = 0$ for each $\ell = 1, \ldots, r$, no cross-tier interference exists during the training phase.

The relevant observation for the sake of channel estimation of femtocell f during the femto-UL training phase, after a suitable normalization, is given by

$$\mathbf{y}_f^{\mathrm{tr}} = \sum_{f' \in Femtocell_f} \sqrt{g(f, f')}\mathbf{h}_{f,f'} + \mathbf{z}_f^{\mathrm{tr}} \tag{6.43}$$

where the superscript "tr" stands for *training*, where $\mathbf{z}_f^{\mathrm{tr}}$ has i.i.d. components $\sim \mathcal{CN}(0, 1/(T_{\mathrm{tr}} P_{\mathrm{tr}}))$ and where $Femtocell_f$ denotes the set of femtocells sharing the same training sequence as femtocell f. The corresponding MMSE estimate is given by

$$\widehat{\mathbf{h}}_{f,f} = \frac{1}{\sum_{f' \in Femtocell_f} g(f, f') + 1/(T_{\mathrm{tr}} P_{\mathrm{tr}})} \mathbf{y}_f^{\mathrm{tr}}. \tag{6.44}$$

Eventually, the estimated beamforming vector at femtocell f is given by

$$\widehat{\mathbf{u}}_f = \widehat{\mathbf{K}}_f^{-1} \widehat{\mathbf{h}}_{f,f}$$

$$\approx \frac{1}{\sum_{f' \in Femtocell_f} g(f, f') + 1/(T_{\mathrm{tr}} P_{\mathrm{tr}})} \mathbf{K}_f^{-1} \mathbf{y}_f^{\mathrm{tr}}$$

$$\propto \underbrace{\mathbf{K}_f^{-1} \mathbf{h}_{f,f}}_{\mathbf{u}_f} + \mathbf{K}_f^{-1} \left(\underbrace{\sum_{f' \in Femtocell_f : f' \neq f} \sqrt{g(f, f')}\mathbf{h}_{f,f'} + \mathbf{z}_f^{\mathrm{tr}}}_{\text{pilot contamination}} \right). \tag{6.45}$$

The term denoted as "pilot contamination" contains the superposition of the interference channel vectors from femtocells f' sharing the same training sequence. From (6.45) we notice that if the pilot reuse factor r is sufficiently large, owing to the rapidly decreasing distance dependent path coefficients $g(f, f')$, the estimated beamforming vector is close to the optimal one, up to an irrelevant non-negative scaling factor. In the numerical results presented at the end of this section, in order to compute the macrocell and the femtocell aggregate throughputs we considered SINR values obtained by replacing the MMSE vectors $\mathbf{u}_f$ with their estimates $\widehat{\mathbf{u}}_f$ given by (6.45). Furthermore, the system throughput in the macro-DL/femto-UL slot is scaled by the training

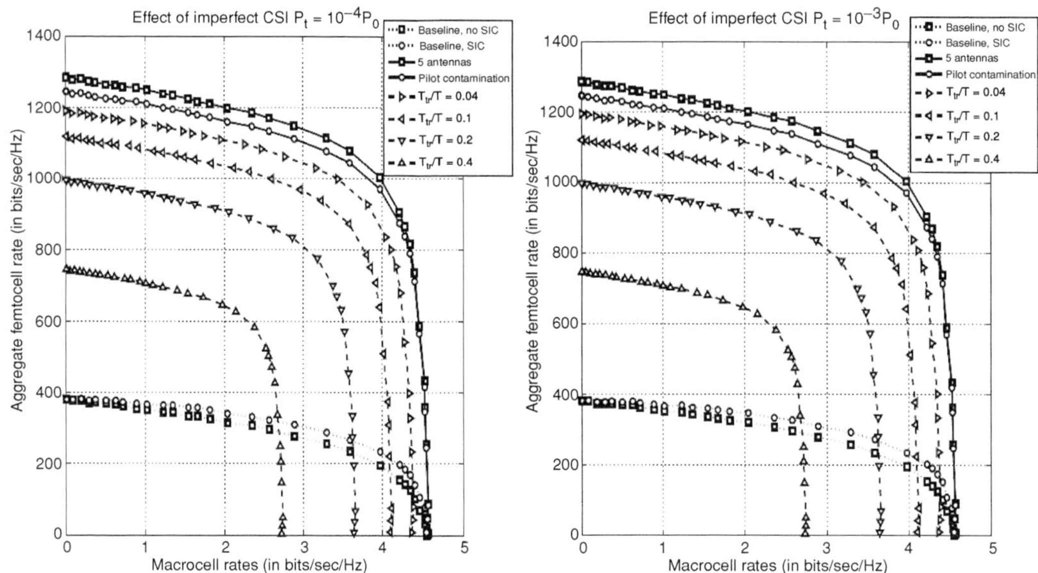

Figure 6.10 Throughput tradeoff for a multi-antenna FBS system ($M = 5$ antennas, OA, pilot reuse factor $r = 9$).

overhead factor $1 - \frac{T_{tr}}{T}$, accounting for the fact that T_{tr} dimensions per slot are used for training.

Figure 6.10 shows the tradeoff performance for the same system settings of Figure 6.9. Here, we focus on the macro-DL/femto-UL slot, where channel estimation takes place. The curve marked "pilot contamination" assumes $P_{tr} \to \infty$, and no training overhead. Therefore, the degradation with respect to the "perfect channel state information (CSI)" curve is uniquely due to the pilot contamination from femtocells reusing the same training sequence. Here, we considered a pilot reuse factor $r = 9$, i.e., the same training sequence is reused every three femtocells in both the South–North and East–West directions on the square grid of Figure 6.1. The other curves are obtained for different ratios T_{tr}/T, in the case of $P_{tr} = 10^{-4}P_0$ (left) and $P_{tr} = 10^{-3}P_0$ (right). We notice that the degradation due to noisy training is minimal, even for very small training power. However, the major effect of training is to reduce the throughput by the training overhead factor $(1 - T_{tr}/T)$. It follows that the smallest possible training dimension yields the best throughput. In the reverse direction, macro-UL/femto-DL, no training is required (apart from the standard amount of pilot symbols to enable coherent detection, which is neglected in our results), since the transmit beamforming vectors are normalized versions of the estimated receive beamforming vectors. As a result, the throughput performance in the reverse direction is basically indistinguishable from the "pilot contamination" curve, and therefore it is not shown here for the sake of chart readability.

It is also interesting to notice that UL–DL duality holds for any set of fixed beamforming vectors, not necessarily equal to the MMSE beamforming vectors. Therefore, the same distributed power control algorithm (6.41) can be used with $\widehat{\mathbf{w}}_f = \widehat{\mathbf{u}}_f / \|\widehat{\mathbf{u}}_f\|$

in lieu of $\mathbf{w}_f$. The results of Figure 6.10 are obtained under the assumption that the receivers can perfectly estimate their effective SINR, given by

$$\mathsf{SINR}_u^{\mathrm{UL}} = \frac{\mathcal{H}g(u,0)Q_u}{1 + \sum_{f\in\mathcal{C}} g(0,f)|\mathbf{h}_{f,0}^{\mathsf{H}}\widehat{\mathbf{w}}_f|^2 Q_f(u)}$$

for the MBS receiver in the macro-UL slot,[6] by

$$\mathsf{SINR}_f^{\mathrm{simo}}(u) = \frac{\min\left\{\frac{\kappa}{g(f,u)}, P_1\right\}|\widehat{\mathbf{u}}_f^{\mathsf{H}}\mathbf{h}_{f,f}|^2}{\widehat{\mathbf{u}}_f^{\mathsf{H}}\boldsymbol{\Sigma}_f(u)\widehat{\mathbf{u}}_f},$$

for the f-th fBS receiver (femto-UL slot), and by

$$\mathsf{SINR}_f^{\mathrm{miso}}(u) = \frac{|\mathbf{h}_{f,f}^{\mathsf{H}}\widehat{\mathbf{w}}_f^{\mathsf{H}}|^2 Q_f(u)}{1 + \sum_{f'\in\mathcal{C}:f'\neq f} g(f,f')|\mathbf{h}_{f,f'}^{\mathsf{H}}\widehat{\mathbf{w}}_{f'}|^2 Q_{f'}(u) + g(u,f)H(u,f)Q_u},$$

for the FMS receiver (femto-DL slot). Yates' algorithm can be applied in the form

$$Q_u^{(n+1)} = Q_u^{(n)}\frac{\gamma_0}{\mathsf{SINR}_u^{\mathrm{UL},n}}$$

$$Q_f^{(n+1)}(u) = Q_f^{(n)}(u)\frac{\gamma_f}{\mathsf{SINR}_f^{\mathrm{miso},n}(u)}, \qquad \forall\, f \in \mathcal{C} \tag{6.46}$$

where $\mathsf{SINR}_u^{\mathrm{UL},n}$ and $\mathsf{SINR}_f^{\mathrm{miso},n}(u)$ are the SINRs measured at the n-th iteration.

6.6   Conclusions

In this chapter we investigated an architecture for *cognitive femtocells*, assuming that the femtocell terminals (both FBSs and FMSs) can decode the control channel broadcasted by the MBS of the overlying macrocell system. The MBS control channel can be used to learn in advance which rate, power, and users will be scheduled on the macrocell macro-DL and in the macro-UL slots. This knowledge, combined with location information that can be provided through the wired network infrastructure and be updated at a relatively slow rate with respect to the signaling rate, provides *implicit* coordination information to the femtocells, allowing the implementation of tight *interference temperature* power control to keep the interference caused by the femtocells on the macrocell system below a specified threshold.

We studied the throughput tradeoff region achievable by very simple schemes, based on treating interference as noise, with possible opportunistic *successive interference cancellation* at the femtocell terminals in order to avoid strong interference from the macrocell system. We showed that SIC yields moderate benefits in terms of overall system throughput tradeoff, but it yields a very significant improvement of the spectral efficiency for femtocells located near the MBS. We also considered CSG vs. open access and showed that, thanks to the implicit coordination realized by the proposed

---

[6] In the macro-DL slot, no channel estimation and beamforming is involved and therefore $\mathsf{SINR}_u^{\mathrm{DL}}$ is given by (6.31).

architecture, the effect of closed access is minor, in contrast to the results of several existing studies that have considered more conventional femtocells systems, with the constraint of backward compatibility with existing 3G networks. In the considered approach, the key idea is that femtocells know *in advance* which MMS is going to be served in each slot, including its location. Since the macrocell system dynamically schedules the MMS, only the femtocells located near the active MMS are effectively constrained by the power control scheme, while the other femtocells can transmit at near-maximum power, without creating appreciable interference on the active macrocell user during the macro-DL slot. Similarly, on the macro-UL slot, only the femtocells near the active MMS suffer from significant interference.

Beyond the baseline system, we have investigated two potential improvements that go beyond treating interference as noise. The first scheme is based on a simplification of the so-called *downlink IA* scheme, based on unitary linear precoding for the macro-DL signal. This allows each femtocell to linearly null the cross-tier macro-to-femto interference in the macro-DL slot. This scheme requires that the femto-DL slot is aligned with the macro-DL slot. We provided an approximated closed-form analysis by letting the system dimensions become large. Although intuitively appealing, this scheme seems appropriate only for hotspot small cells covering areas with high user density. In fact, if only a small number (say 1 to 3) of users per femtocell are present, as typical in home-use, the dimensionality overhead due to precoding yields a net loss in the aggregate femtocell throughput.

The second, and much more attractive, improvement considers multiple antennas at the FBSs. In this case, we align the macro-DL slot with the femto-UL slot and vice versa. This yields a SIMO/MISO interference channel that, for the sake of simplicity and practicality, we address by using linear beamforming and UL–DL duality. The proposed scheme makes use of linear spatial MMSE beamforming vectors at the FBSs receivers (for the femto-UL) and of the same vectors (suitably normalized in order to have unit norm) at the FBSs transmitters (for the femto-DL). In the macro-DL/femto-UL slot, we use our proposed interference temperature power control scheme, based on the control channel and location side information, as in the baseline system. This guarantees a desired SINR for the macrocell users. Then, in the macro-UL/femto-DL slot (reverse direction) we use UL–DL duality and compute the transmit powers that realize the *same* SINRs in an iterative and distributed fashion. Numerical results show that with just a small number of antennas at the FBSs ($M = 5$ in our case) a very significant improvement of the femtocell throughput can be achieved without sacrificing the macrocell throughput, and that the iterative power control converges very quickly and yields good results even after a single iteration. We also discussed channel estimation and the beamforming calculation by using pilot symbols, and a pilot reuse scheme that mitigates the effect of *pilot contamination*, which is a major impairment in TDD systems. Our results show that also in the case of realistic channel estimation using pilot symbols, the system performs close to the ideal perfect CSI case.

As a general conclusion, it is interesting to notice that with relatively simple signal processing and power control algorithms the proposed system can achieve a femtocell aggregate spectral efficiency of ≈ 1000 bit/s/Hz per km^2, with macrocell spectral efficiency of 4 bit/s/Hz (see Figure 6.9). With 625 femtocells as in our test system, this corresponds to a modest 1.6 bit/s/Hz per femtocell, easily obtained by simple coded modulation (e.g., quadrature phase shift keying (QPSK) concatenated with a rate 4/5 binary coding). For a 4G system with 20 MHZ of total system bandwidth, this system would provide 32 Mb/s per residential femtocell (notice: this is average rate, not peak rate), which is competitive with respect to the rates offered by today's WLAN residential Internet connections. We also notice that a macrocell average spectral efficiency of 4 bit/s/Hz is in line with the rates expected by LTE/LTE-Advanced. The macrocell spectral efficiency can be significantly increased by considering multiple antennas also at the MBS, that were not considered here since the focus of this chapter is the femtocell system and its interaction with a baseline macrocell system. Most remarkably, the above performance can be achieved *in the same system bandwidth*, in contrast with today's combination of WLAN and cellular, using separate bandwidths. This yields a very significant improvement of the overall spectral efficiency (total bit/s/Hz) with respect to off-the-shelf WiFi offloading. Of course, in order to achieve the promised performance it is necessary to design the macrocell system in order to allow and facilitate the implicit coordination provided by the MBS control channel and by the availability of localization information. We believe that the results of this study provide a clear case for a system design that puts femtocells at the *heart* of system architecture, instead of considering them as an "afterthought."

References

[1] J. G. Proakis, *Digital Communications*. McGraw-Hill, 1987.

[2] A. Goldsmith, *Wireless Communications*. Cambridge: Cambridge University Press, 2005.

[3] D. N. C. Tse and P. Viswanath, *Fundamentals of Wireless Communication*, 1st edn. Cambridge: Cambridge University Press, 2005.

[4] H. Shirani-Mehr and G. Caire, "Channel state feedback schemes for multiuser MIMO-OFDM downlink," *IEEE Trans. Commun.*, vol. 57, no. 9, pp. 2713–23, Sep. 2009.

[5] T. Marzetta and B. Hochwald, "Capacity of a mobile multiple-antenna communication link in Rayleigh flat fading," *IEEE Trans. Inf. Theory*, vol. 45, no. 1, pp. 139–57, Jan. 2002.

[6] L. Zheng and D. Tse, "Communication on the Grassmann manifold: a geometric approach to the noncoherent multiple-antenna channel," *IEEE Trans. Inf. Theory*, vol. 48, no. 2, pp. 359–83, Feb. 2002.

[7] O. Somekh and S. Shamai, "Shannon-theoretic approach to a Gaussian cellular multiple-access channel with fading," *IEEE Trans. Inf. Theory*, vol. 46, no. 4, pp. 1401–25, Apr. 2002.

[8] T. L. Marzetta, "Noncooperative cellular wireless with unlimited numbers of base station antennas," *IEEE Trans. Wireless Commun.*, vol. 9, no. 11, pp. 3590–600, Nov. 2010.

[9] S. Ramprashad, G. Caire, and H. Papadopoulos, "Cellular and network MIMO architectures: MU-MIMO spectral efficiency and costs of channel state information," in *Proc. Asilomar Conf. on Signals, Systems, and Computers (ASILOMAR)*, Nov. 2010, pp. 1811–18.

[10] H. Huh, A. Tulino, and G. Caire, "Network MIMO with linear zero-forcing beamforming: large system analysis, impact of channel estimation and reduced-complexity scheduling," *IEEE Trans. Inf. Theory*, vol. 58, no. 5, pp. 2911–34, May 2012.

[11] H. Holma, A. Toskala, and J. Wiley, *HSDPA/HSUPA for UMTS: High Speed Radio Access for Mobile Communications*. John Wiley & Sons, Ltd., 2007.

[12] V. Chandrasekhar, J. G. Andrews, and A. Gatherer, "Femtocell networks: a survey," *IEEE Commun. Mag.*, vol. 46, no. 9, pp. 59–67, Sept 2008.

[13] P. Gupta and P. Kumar, "The capacity of wireless networks," *IEEE Trans. Inf. Theory*, vol. 46, no. 2, pp. 388–404, Feb. 2000.

[14] S. Rangan, "Femto-macro cellular interference control with subband scheduling and interference cancelation," in *Proc. IEEE Global Telecommun. Conf. (GLOBECOM) Workshops*, Dec. 2010.

[15] 3GPP, "UTRAN architecture for 3G Home Node B (HNB); Stage 2," *TS 25.467 (release 9)*, 2010.

[16] ——, "Service requirements for Home NodeBs (UMTS) and eNodeBs (LTE)," *TS 22.220 (release 9)*, 2010.

[17] ——, "3G Home Node B Study Item Technical Report," *TR 25.820 (Release 9)*, 2010.

[18] S. Yeh, S. Talwar, S. Lee, and H. Kim, "WiMAX femtocells: a perspective on network architecture, capacity, and coverage," *IEEE Commun. Mag.*, vol. 46, no. 10, pp. 58–65, 2008.

[19] O. Simeone, E. Erkip, and S. Shamai, "Robust communication against femtocell access failures," in *Proc. IEEE Inform. Theory Workshop (ITW)*, Oct. 2009, pp. 263–7.

[20] Y. Kim, "Capacity of a class of deterministic relay channels," *IEEE Trans. Inf. Theory*, vol. 54, no. 3, pp. 1328–29, Mar. 2008.

[21] Femto Forum, "Interference management in UMTS femtocells," White Paper, Dec. 2008. [Online]. Available: http://www.femtoforum.org/femto/Files/File/Interference Management in UMTS Femtocells.pdf

[22] V. Chandrasekhar, J. G. Andrews, T. Muharemovic, Z. Shen, and A. Gatherer, "Power control in two-tier femtocell networks," *IEEE Trans. Wireless Commun.*, vol. 8, no. 8, pp. 4316–28, Aug. 2009.

[23] V. Chandrasekhar and Z. Shen, "Optimal uplink power control in two-cell systems with rise-over-thermal constraints," *IEEE Commun. Lett.*, vol. 12, no. 3, pp. 173–5, Mar. 2008.

[24] Y. Tokgoz, F. Meshkati, Y. Zhou, M. Yavuz, and S. Nanda, "Uplink interference management for HSPA+ and 1xEVDO femtocells," in *Proc. IEEE Global Telecommun. Conf. (GLOBECOM)*, Dec. 2010, pp. 1–7.

[25] V. Chandrasekhar and J. G. Andrews, "Uplink capacity and interference avoidance for two-tier femtocell networks," *IEEE Trans. Wireless Commun.*, vol. 8, no. 7, pp. 3498–509, July 2009.

[26] P. Xia, V. Chandrasekhar, and J. Andrews, "Open vs closed access femtocells in the uplink," *Arxiv preprint arXiv:1002.2964*, 2010.

[27] "IEEE draft amendment standard for local and metropolitan area networks - part 16: Air interface for fixed and mobile broadband wireless access systems - advanced air interface," *IEEE P802.16m/D6 May 2010*, pp. 1 –932, 12 2010.

[28] 3GPP, "3GPP Specification Series: 36 series," Available: http://www.3gpp.org/ftp/Specs/html-info/36-series.htm.

[29] C. Suh, M. Ho, and D. Tse, "Downlink interference alignment," *IEEE Trans. Commun.*, vol. 59, no. 9, pp. 2616–2626, 2011.

[30] S. Peters and R. Heath, "The future of WiMAX: multihop relaying with IEEE 802.16 j," *IEEE Commun. Mag.*, vol. 47, no. 1, pp. 104–11, 2009.

[31] A. Goldsmith, S. Jafar, I. Maric, and S. Srinivasa, "Breaking spectrum gridlock with cognitive radios: an information theoretic perspective," *IEEE Proceedings*, vol. 97, no. 5, pp. 894–914, May 2009.

[32] A. Adhikary, V. Ntranos, and G. Caire, "Cognitive femtocells: breaking the spatial reuse barrier of cellular systems," in *Proc. Inf. Theory and Appl. Workshop (ITA)*, pp. 1–10, 2011.

[33] L. Georgiadis, M. Neely, M. Neely, and L. Tassiulas, *Resource Allocation and Cross-layer Control in Wireless Networks*. Foundations and Trends in Networking, Now Publisher, 2006.

[34] G. Caire, R. Muller, and R. Knopp, "Hard fairness versus proportional fairness in wireless communications: the single-cell case," *IEEE Trans. Inf. Theory*, vol. 53, no. 4, pp. 1366–85, Apr. 2007.

[35] E. Park and I. Lee, "Antenna placement for downlink distributed antenna systems with selection transmission," in *Proc. IEEE Veh. Technol. Conf. (VTC)*, May 2011, pp. 1–5.

[36] M. Simon and M. Alouini, *Digital Communication Over Fading Channels*. Wiley-IEEE Press, 2005.

[37] G. Caire and D. Tuninetti, "The throughput of hybrid-ARQ protocols for the Gaussian collision channel," *IEEE Trans. Inf. Theory*, vol. 47, no. 5, pp. 1971–88, July 2001.

[38] H. Shirani-Mehr, H. Papadopoulos, S. Ramprashad, and G. Caire, "Joint scheduling and ARQ for MU-MIMO downlink in the presence of inter-cell interference," *IEEE Trans. Commun.*, no. 99, pp. 1–12, 2010.

[39] Y. Polyanskiy, H. Poor, and S. Verdú, "Variable-length coding with feedback in the non-asymptotic regime," in *Proc. IEEE Int. Symp. Inform. Theory (ISIT)*, June 2010, pp. 231–5.

[40] F. Baccelli, A. E. Gamal, and D. Tse, "Interference networks with point-to-point codes," in *IEEE Trans. Inf. Theory*, vol. 57, no. 5, May 2011, pp. 2582–96.

[41] H. Huh, G. Caire, H. Papadopoulos, and S. Ramprashad, "Achieving massive MIMO spectral efficiency with a not-so-large number of antennas," *Arxiv preprint arXiv:1107.3862*, 2011.

[42] M. Debbah, W. Hachem, P. Loubaton, and M. de Courville, "MMSE analysis of certain large isometric random precoded systems," *IEEE Trans. Inf. Theory*, vol. 49, no. 5, pp. 1293–311, May 2003.

[43] A. M. Tulino and S. Verdú, *Random Matrix Theory and Wireless Communications*. Foundations and Trends in Communications and Information Theory, NOW Publishers Inc, 2004, vol. 1, no. 1.

[44] G. Caire, N. Jindal, M. Kobayashi, and N. Ravindran, "Multiuser MIMO achievable rates with downlink training and channel state feedback," *IEEE Trans. Inf. Theory*, vol. 56, no. 6, pp. 2845–66, June 2010.

[45] S. Wagner, R. Couillet, M. Debbah, and D. T. M. Slock, "Large system analysis of linear precoding in correlated MISO broadcast channels under limited feedback," *IEEE Trans. Inf. Theory*, vol. 58, no. 7, pp. 4509–37, July 2012.

[46] G. Caire and S. Shamai (Shitz), "On the achievable throughput of a multiantenna Gaussian broadcast channel," *IEEE Trans. Inf. Theory*, vol. 49, no. 7, pp. 1691–706, July 2003.

[47] S. Vishwanath, N. Jindal, and A. Goldsmith, "Duality, achievable rates, and sum-rate capacity of Gaussian MIMO broadcast channels," *IEEE Trans. Inf. Theory*, vol. 49, no. 10, pp. 2658–68, Oct. 2003.

[48] W. Yu and J. M. Cioffi, "Sum capacity of Gaussian vector broadcast channels," *IEEE Trans. Inf. Theory*, vol. 50, no. 9, pp. 1875–92, Sept. 2004.

[49] F. Boccardi, F. Tosato, and G. Caire, "Precoding schemes for the MIMO-GBC," in *2006 International Zurich Seminar on Communications*. IEEE, 2006, pp. 10–13.

[50] B. Hochwald, C. Peel, and A. Swindlehurst, "A vector-perturbation technique for near-capacity multiantenna multiuser communication part II: Perturbation," *IEEE Trans. Commun.*, vol. 53, no. 3, pp. 537–44, Mar. 2005.

[51] C. Windpassinger and R. Fischer, "Low-complexity near-maximum-likelihood detection and precoding for mimo systems using lattice reduction," in *Proc. IEEE Inf. Theory Workshop (ITW)*, Apr. 2003, pp. 345–8.

[52] R. Couillet and M. Debbah, *Random Matrix Methods for Wireless Communications*. Cambridge: Cambridge University Press, 2011.

[53] A. Tulino, A. Lozano, and S. Verdú, "Impact of antenna correlation on the capacity of multiantenna channels," *IEEE Trans. Inf. Theory*, vol. 51, no. 7, pp. 2491–509, July 2005.

[54] P. Viswanath and D. N. C. Tse, "Sum capacity of the vector Gaussian broadcast channel and uplink-downlink duality," *IEEE Trans. Inf. Theory*, vol. 49, no. 8, pp. 1912–21, Aug. 2003.

[55] L. Zhang, R. Zhang, Y.-C. Liang, Y. Xin, and H. V. Poor, "On Gaussian MIMO BC-MAC duality with multiple transmit covariance constraints," *IEEE Trans. Inf. Theory*, vol. 58, no. 4, pp. 2064–78, Apr. 2012.

[56] S. Christensen, R. Agarwal, E. Carvalho, and J. Cioffi, "Weighted sum-rate maximization using weighted MMSE for MIMO-BC beamforming design," *IEEE Trans. Wireless Commun.*, vol. 7, no. 12, pp. 4792–9, Dec. 2008.

[57] M. Stojnic, H. Vikalo, and B. Hassibi, "Rate maximization in multi-antenna broadcast channels with linear preprocessing," *IEEE Trans. Wireless Commun.*, vol. 5, no. 9, pp. 2338–42, Sep. 2006.

[58] F. Negro, I. Ghauri, and D. Slock, "On duality in the MISO interference channel," in *Proc. Asilomar Conf. on Signals, Systems, and Computers (ASILOMAR)*, Nov. 2010, pp. 2104–8.

[59] ——, "Beamforming for the underlay cognitive MISO interference channel via UL-DL duality," in *Proc. Int. Conf. on Cognitive Radio Oriented Wireless Networks and Communications (CROWNCOM)*, June 2010, pp. 1–5.

[60] R. D. Yates, "A framework for uplink power control in cellular radio systems," *IEEE J. Sel. Areas Commun. (JSAC)*, vol. 13, no. 7, pp. 1341–7, Sep. 1995.

[61] G. J. Foschini and Z. Miljanic, "A simple distributed autonomous power control algorithm and its convergence," *IEEE Trans. Veh. Technol.*, vol. 42, no. 4, pp. 641–6, Nov. 1993.

7 Femtocell coverage optimization

Holger Claussen, Lester Ho, Sam Samuel, and Florian Pivit

7.1 Introduction

The idea of having a cellular system deployed without planning is quite a challenging one. To have the resulting system work without any human involvement (except for the physical placement of cells) is even more challenging. The motivations for technology to head in this direction are numerous. Among these are the ever-present need to reduce the costs of operating a cellular network, and the need for the cellular network to keep pace with the increasing data demands placed upon it. The concept of a cellular network being able to deploy itself is not new and has been examined philosophically in [1]. The idea of femtocells, or more generally small cells, as a means to fill coverage gaps and to increase capacity is even older. Its roots can be traced back to [2]. There are, however, very good reasons why such deployments have not been successful in the past. The first is that as the size of the cells reduces, in order to increase the spatial frequency reuse, the deployment costs increase (due to the sheer number of additional cells). The second is that, once the cells are deployed, planning such a network manually can be complex and burdensome. If not carefully thought through, any insertion of a new cell into the topology can have detrimental rather than beneficial effects. These simple facts may often ruin the business case for the deployment of additional cells.

Pragmatically, then, in order to begin this particular journey toward what can be termed autonomic wireless cellular systems, we first need to find a deployment environment that allows us to examine some of the key problems. If we are successful in solving these problems then the opportunity presents itself to extend the applicability of these techniques to the wider aim indicated above. The context we choose to begin examining these issues is in the deployment of femtocells for enterprise and residential use. Here, we are fortunate that this deployment environment allows us to explore the issues in a

Parts of this chapter are based on "self-optimization of coverage for femtocell deployments," by H. Claussen, L. T. W. Ho and L. G. Samuel, which appeared in *Proc. Wireless Communications* [4], and "Femtocell coverage optimization using switched multi-element antennas," by H. Claussen and F. Pivit, in *Proc. IEEE International Conference on Communications* [5]. ©2008 and 2009 IEEE.

Small Cell Networks: Deployment, PHY Techniques, and Resource Management, ed. Tony Q. S. Quek, Guillaume de la Roche, İsmail Güvenç, and Marios Kountouris. Published by Cambridge University Press. © Cambridge University Press 2013.

semi-controlled way, where the propagation conditions for such deployments are favorable in the sense that some shielding from the macro network can be achieved.

When we examine the possibility of deploying a component of a cellular system automatically (in this case the base station) many engineering problems present themselves. There are the typical commercial engineering problems of fabricating a simple device that is cost optimized for the environment that it will be deployed into. Then there are cellular engineering problems such as ensuring that the solution does not adversely impact the already deployed macrocellular systems into which such a small cell will be deployed. The real problem, however, stems from the fact that the cells are deployed unplanned, meaning that the femtocell is placed into an existing radiating topology by the user. In the extreme we may have no knowledge of the cellular topology, and the physical location of the cell at all.

Indeed, often the primary driver in this type of deployment scenario is the position of the power socket or the available length of cabling to the device (Cat 5 or digital subscriber line (DSL) cable). In this context, such deployments will have to deal with:

- integration of a femto layer into the overall cellular system
- non-ideal cell placement (misplacement)
- prevalent user mobility
- prevalent traffic conditions.

The last three points can be summarized by the term *capture effects*. These are the effects that femtocells will have on mobile terminals that are moving through a macro coverage area, both in active and idle mode. In other words, the capture effects are the unnecessary mobility events triggered by transient mobile terminals moving to a femtocell. These effects have already been discussed in [3]. For successful femtocell deployments it is necessary to seek solutions to mitigate capture effects in a general sense. In [4], mobility event based coverage optimization methods were proposed to address this problem for femtocells with a single antenna. It was shown that a mobility event based self-optimization of coverage can both significantly reduce the total number of mobility events caused by femtocell deployments, and improve the indoor coverage for femtocells deployed in suitable locations compared to simpler methods that aim to achieve a constant cell radius. While the capture effects can be mitigated by such coverage optimizations, much more can be done when the antenna configuration of the femtocell is reconsidered. In [5], a novel multi-element antenna solution was proposed to increase the flexibility of the gain pattern in order to further reduce the core network mobility signaling over [4]. An additional consideration on the antenna configuration is the requirement for any solution to be of low cost. Therefore, a simple switched multi-element antenna system was considered. This allows the generation of a set of different gain patterns by switching between one or more of the available antennas. A prototype of the antenna system was built and the concept was confirmed by antenna pattern measurements in an anechoic chamber. The pattern measurements obtained were then used as input for a system-level simulation to evaluate the potential impact on the number of resulting mobility events and indoor coverage using a mobility event based self-optimization approach. Subsequently, in [6], a tunable multi-element antenna system was proposed, which results in improved flexibility and convergence speed over [5]. The material presented in this chapter summarizes and extends previous results in this area [4–6].

To some extent, some of the issues mentioned above are being addressed in the long term evolution (LTE)/system architecture evolution (SAE) evolution of the current 3rd Generation Partnership Project (3GPP) standard via the creation of equivalent paging areas [7]. However, pragmatically, until revisions of the current 3GPP standard occur or newer more targeted standards emerge that deal specifically with the nature of femtocell deployments, there will have to be concerted efforts on the issues of capture effects in practical femtocell deployments, since femtocells will be deployed for several years before these features are standardized.

The remainder of this chapter is organized as follows. We first discuss the problem statement for the effects caused by femtocellular coverage extending into public areas. We then examine the problem of pilot power and how this is linked to coverage optimization. Afterwards we examine how self-optimization techniques can be achieved through dynamic adjustment of the femtocell pilot power. This technique is initially explored through the context of simple single antenna femtocells. We then explore how this basic technique can be further enhanced by the addition of multiple antennas to the femtocell. This aspect is examined from both cost and operational perspectives. Subsequently, we analyze the performance of such systems. Finally, we explore the future evolution of these techniques.

7.2   Problem overview

The deployment of femtocells differs significantly from the deployment of the conventional underlay macrocells. Femtocells are deployed by users, and consequently little to no cell planning is taken into consideration during the deployment. This means that a femtocell may be deployed in unsuitable locations, where its pilot signals may radiate to areas outside of its intended area of coverage. This has the undesirable effect of triggering handovers from unwanted transient users who happen to be passing by the femtocell. Since the deployment of femtocells can be widespread, for example in dense urban areas, then this can cause a very large number of mobility events to occur. This, in turn, has an impact both on the network due to increased signaling load, and on the battery life of user terminals. In some cases, this is also exacerbated by the requirement, due to technical reasons, for the location area code used by a femtocell to be different from the one used by the macrocell underlay, and sometimes even to be different from those used by neighboring femtocells. The effect of this disparate allocation of location area codes is that the rise in unwanted mobility events, due to femtocell deployments, not only includes handover requests when the user terminal is in active mode during a call, but also cell relocation requests when the user terminal is in idle mode. Taking the example illustrated in Figure 7.1, a user terminal moving along a street where femtocells are deployed can move into and out of the coverage areas of multiple femtocells, triggering messages for mobility each time, even when the user is not making a call.

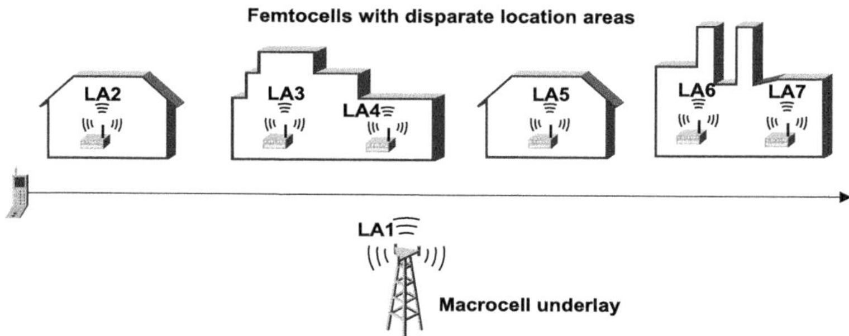

Figure 7.1 Femtocell deployment scenario and mobility problem. Used with permission from [6].

In order to prevent the detrimental effects described above, the femtocell needs to be able to auto-configure and self-optimize its coverage. Such capability is required not only for the reasons mentioned above but also for the very good reason that femtocells may start or stop radiating very dynamically, because they are controlled by the end user. Furthermore, the degree of self-optimization can be exacerbated by the need for closed, semi-closed (hybrid access) or open femtocell operation (i.e., the degree of public access to the femtocell). Here, we distinguish between auto-configuration and self-optimization schemes. The difference is that auto-configuration schemes provide an initial configuration, whereas self-optimization schemes optimize the current configuration during operation, for example, based on measurements or statistics collected over time.

In this chapter, two types of coverage auto-configuration methods are discussed: distance based and measurement based [4]. Both perform similarly, but the measurement-based approach is preferable since it requires much less information about the macrocell network. Three types of coverage self-optimization methods are discussed based on either a single antenna, a switched multi-element antenna system, or a tunable multi-element antenna system. The self-optimization methods in this chapter use information on mobility events and user equipment (UE) measurement reports. Depending on the scenario (e.g., public or restricted access to femtocells), such events can be handovers or idle mode (location area update) events, both successful or rejected (e.g., a mobile might be rejected by a femtocell using a closed-access model). All self-optimization methods are able to adapt to different types, shapes, and sizes of indoor environments.

The single antenna coverage self-optimization solution can reduce resulting mobility events significantly by up to 80% in the scenarios considered compared to using coverage auto-configuration methods only [4]. The method has a relatively short convergence time and is easily applicable in existing femtocells since it can be implemented via software and does not add any hardware costs.

The switched multi-element antenna coverage self-optimization solution uses multiple antennas connected to a switch. This allows the selection of one or two antennas with different patterns to provide additional gain pattern flexibility and allow better adaptation of femtocell coverage to the shape of the indoor environment compared to that of a single antenna solution [5]. The additional flexibility gained by the switched multi-element antenna solution results in further improvement, on average a 20% decrease in resulting mobility events and a 20% increase in indoor coverage compared to the best single antenna solution in the considered scenario. Moreover, due to the simplicity of the approach, the additional hardware costs for antennas and switch are relatively low. Further, the small footprint of the antennas means that they can easily be accommodated into existing femtocell designs.

The tunable multi-element antenna coverage self-optimization solution uses multiple patch antennas connected with attenuators, which allow the gain of each lobe to be individually adjusted [6]. This allows a further flexibility increase in terms of achievable antenna gain patterns. The proposed method achieves performance similar to the switched multi-element antenna solution, which suggests that the flexibility of the switched solution is usually sufficient. Moreover, the proposed tunable multi-element antenna solution has an advantage of faster convergence compared to the switched multi-element antenna solution, since each antenna lobe can be adjusted individually in each optimization step. However, a disadvantage of this solution is its increased size due to the use of four patch antennas.

In terms of deployment environments, the most challenging ones are those where buildings are closely located to busy thoroughfares (roads, pedestrian area, etc.), and where there is no wall separation between the femtocell and these busy outside spaces. Dense multi-story buildings can also be problematic since potential neighbor cells can be affected in more dimensions, but the passing traffic is usually lower than in the first example. In multi-story buildings, interference caused by other femtocells can cause coverage problems when a restricted access model is adopted. This can be largely resolved by moving to a public access model, or by providing separation in frequency. In a real network, typically large variations of traffic are encountered dependent on the time of the day. This is not considered in the analysis presented here, but can be taken into account by optimizing power levels independently for different times of the day using any of the self-optimization methods described herein.

7.3   Initial configuration of coverage

When a femtocell is first deployed, it needs to initialize its coverage. Here we can differentiate between two deployment scenarios: co-channel and dedicated channel deployment. While the concept is fairly similar in both scenarios, the difference is that for co-channel operation the cell boundary is typically at the locations where the femtocell pilot power is equal to the macrocell pilot power. Therefore, the achieved coverage is dependent on the femtocell pilot power, the macrocell pilot power, and the location of the femtocell within the macrocell. For dedicated channel deployments the cell boundary is at locations where the received pilot SNR exceeds the camping or handover threshold. Since the noise level is more or less constant, the coverage is only dependent on the femtocell pilot power and the camping or handover threshold, but independent of the femtocell location. In this section, different approaches for the initial auto-configuration of the pilot power are presented.

7.3.1   Co-channel operation

For co-channel operation of femtocells on the same frequency band as existing macrocellular networks, the following pilot power auto-configuration schemes are considered: distance based and measurement based.

Distance-based pilot power configuration. The femtocell pilot power is configured such that it is received on average with equal strength as the pilot power received from the strongest macrocell at a defined target cell radius of r, subject to its maximum power $P_{\text{pilot,max}}$ [4]. The initial femtocell pilot power can be calculated in decibels as

$$P_{\text{fBS,pilot}} = \min\left(P_{\text{mBS,pilot}} + G_{\text{mBS}} - L_{\text{mBS}} + L_{\text{fBS}}(r), P_{\text{pilot,max}}\right), \tag{7.1}$$

where $P_{\text{mBS,pilot}}$ is the pilot power transmitted by the macrocell, G_{mBS} is the macrocell antenna gain in direction of the femtocell, L_{mBS} denotes the path loss between the macrocell and the femtocell, and $L_{\text{fBS}}(r)$ is the estimated path loss from the femtocell to a UE at the target femtocell radius r. The path loss between a transmitter and a receiver separated by a distance d can be modeled as

$$L(d) = L_1 + L_2 \times 10\log_{10}(d) + L_{\text{wall}}. \tag{7.2}$$

For free-space loss L_1 and L_2 are 38.5 dB and 2 dB, respectively, for typical urban path loss L_1 and L_2 are assumed to be 28 dB and 3.5 dB, respectively, and L_{wall} accounts for additional wall losses. This achieves a roughly constant cell range independent of the distance to the macrocell in the co-channel hierarchical cell structure considered. Note that variations caused by the shadow fading losses are unpredictable and cannot be taken into account using the path-loss model. While this approach is suitable for the initial power configuration for femtocell deployments, it has the disadvantage that a large amount of information on the macrocellular network, such as cell locations, power levels, antenna orientation and gain, is required. The approach also relies on a reliable path-loss model for L_{mBS}. An example of the power configuration is shown in Figure 7.2, which shows the configured femtocell transmit power values (f) and (g), the resulting received power values from the macrocell (a) and (b), from the femtocell (d) and (e), as a function of the distance from the macrocell, and the resulting femtocell coverage range.

Measurement-based pilot power configuration. The femtocell pilot power is configured using the same principle as above, but with the difference that the received power $P_{\text{mBS,Rx-pilot}}$ is not estimated using a path-loss model, but measured at the femtocell. Then the initial femtocell pilot power can be calculated in decibels as

$$P_{\text{fBS,pilot}} = \min\left(P_{\text{mBS,Rx-pilot}} + L_{\text{fBS}}(r), P_{\text{pilot,max}}\right). \tag{7.3}$$

The path loss $L_{\text{fBS}}(r)$ from the femtocell to a UE at the target femtocell radius r can be modeled as described in (7.2). As with the distance based auto-configuration, this approach achieves a roughly constant cell range independent of the distance to the macrocell in a co-channel hierarchical cell structure. Note that variations are caused by the varying shadow fading losses that have an impact on $P_{\text{mBS,Rx-pilot}}$, but are unknown to the femtocell. This approach has the advantage that no information on the macrocell network is required. However, since the received signal from the macrocell $P_{\text{mBS,Rx-pilot}}$

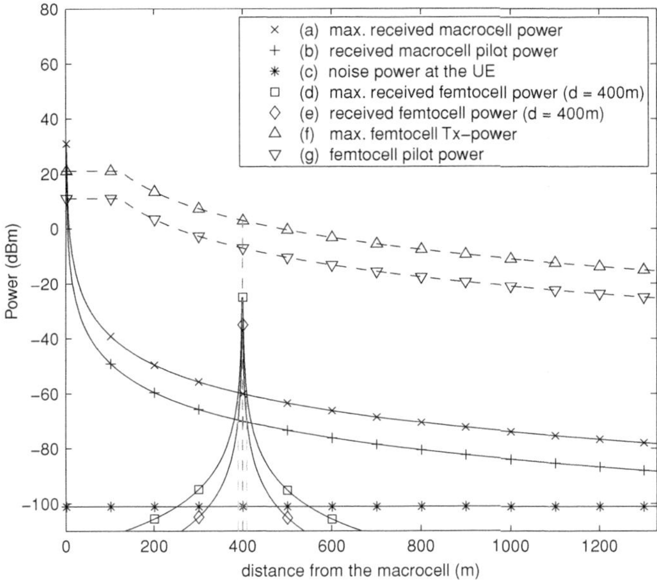

Figure 7.2 Distance-based pilot power configuration example.

depends highly on the deployment location and the resulting indoor wall losses, this causes some variability in the achieved cell range.

Effects of maximum transmit power on the coverage. For co-channel operation the maximum femtocell power is important since it has an effect on the size of the area within the macrocell where the femtocell can achieve its target cell coverage. This effect is also shown in Figure 7.2, curves (f) and (g), where the maximum configured power and pilot power remain flat at distances up to approximately 100 m from the macrocell. In this area the femtocell cannot achieve its target cell radius due to the power limit of 20 dBm in this example. While the area in which the coverage is not achievable is only a very small fraction of the overall coverage area of the macrocell, and shielding effects of walls usually help as well, the dimensioning of the maximum femtocell power should not be overlooked.

7.3.2   Dedicated channel operation

When femtocells are deployed on a dedicated carrier, the initial configuration of coverage becomes very simple since it is independent of the location of the femtocell with respect to the macrocell. The required pilot power for a target cell radius of r can be calculated in decibels as

$$P_{\text{fBS,pilot}} = \min\left(P_{\text{n}} + \rho_{\text{HO}} + L_{\text{fBS}}(r), P_{\text{pilot,max}}\right), \tag{7.4}$$

where P_{n} is the noise power and ρ_{HO} is the signal to noise ratio that is required to trigger a handover or camping event (i.e., idle mode cell relocation).

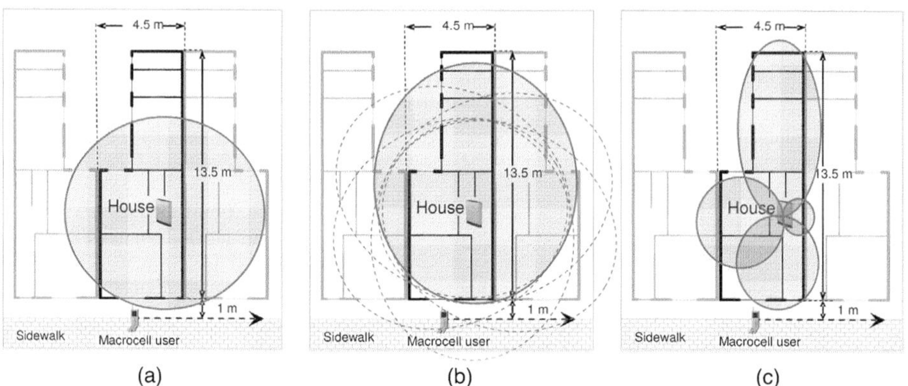

Figure 7.3 Coverage optimization approaches with different antenna configurations. Used with permission from [6].

7.4   Self-optimization of coverage

While the auto-configuration approaches in the previous section provide some useful starting points, it is necessary to fine tune the coverage to the deployment location within a building (e.g., large house or small house) and the traffic (e.g., if deployed next to a busy street or in the countryside). The objective of the self-optimization process is both to minimize the resulting mobility events and to maximize the indoor coverage.

In this section different mobility event based approaches for the self-optimization of coverage are presented to address this problem for femtocells with different antenna configurations. The benefits in coverage flexibility obtained by multi-element antenna solutions are illustrated in Figure 7.3 in a conceptual way (without wall effects) when mobility events of passing users are prevented. Figure 7.3(a) shows a conventional single antenna solution; Figure 7.3(b) shows additional flexibility gained through a switched multi-element antenna solution; and Figure 7.3(c) shows a solution with individually tunable antennas.

7.4.1   Coverage optimization for single antenna femtocells

For femtocells with a single antenna, the only way to increase coverage or reduce mobility events is by increasing or decreasing the pilot power (or handover bias). This case is illustrated in Figure 7.3(a). A mobility event based coverage self-optimization method for femtocells with a single antenna is described below.

Minimization of mobility events of passing users. This optimization method takes only unwanted mobility events from transient users into account that briefly hand over to the femtocell and hand back immediately after passing the house. The objective is to maximize the indoor coverage under the constraint of limiting the mobility events from such passing users. Starting from a measurement-based pilot auto-configuration, the femtocell counts mobility events and classifies them into wanted and unwanted events

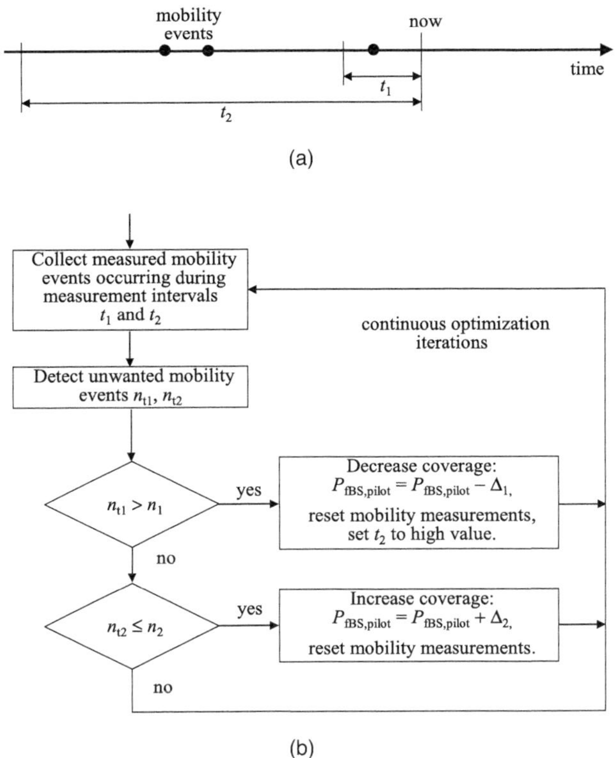

Figure 7.4 Overview of measurement intervals (a); and flowchart of mobility event based coverage optimization (b) that minimizes mobility events of passing users. ©2008 IEEE. Reprinted, with permission, from [4].

over time, dependent on whether the mobile is registered at the femtocell, and on the time a UE spends in the cell before moving back to the macrocell. If the number of unwanted mobility events of passing users n_{t1} exceeds a predefined value of n_1 events per timeframe t_1, the femtocell reduces its pilot power by a step Δ_1. The femtocell then starts a new mobility event count for the updated configuration. If the number of unwanted events n_{t2} is smaller or equal to a predefined acceptable value of n_2 events for a timeframe t_2, the femtocell increases its pilot power by a step Δ_2, to provide improved indoor coverage and starts a new mobility event count for the updated configuration.

The measurement intervals and the optimization process are illustrated in Figure 7.4. In this chapter it is assumed that all mobility events of passing users shall be prevented, therefore $n_1 = n_2 = 0$. The measurement interval t_1 is equal to the optimization iteration time, and for any occurring event the femtocell pilot power is reduced by $\Delta_1 = 3$ dB. In order to speed up the cell radius increase after the initial auto-configuration, t_2 is set to a low value of 120 seconds until a first unwanted mobility event is detected. Then t_2 is increased significantly to 6 hours so that the coverage is dominated by the decreases caused by unwanted mobility events. The step size Δ_2 for the pilot power increase was selected to be 0.3 dB. While the selected parameters perform well in the considered scenario, automatic optimization of these parameters is an item for future studies. Such optimized parameters would have to result in a fast convergence to the point with the best tradeoff between femtocell coverage and increase in core network signaling for the operator.

7.4.2   Coverage optimization using switched multi-element antennas

In current designs for home-based wireless equipment like digital enhanced cordless telecommunications (DECT) phones, wireless local area network (WLAN) modems, and WLAN hubs as well as in femtocell base stations the antenna system is of a static nature. Usually these installations use simple dipoles or printed circuit board (PCB) antennas with low gain and rather erratic and fixed patterns. These installations are not able to adapt in any way to their environment, except by increasing or decreasing their transmit power or changing handover parameters (see Figure 7.3(a)). Therefore, a femtocell base station deployed in an unsuitable location would have to reduce its output power to such an extent in order to limit the number of mobility events caused by outdoor users, that the coverage inside the building is compromised [4]. This is especially unfortunate if the antenna pattern is erratic (which is the case for most PCB and dipole antennas in consumer grade equipment) and if, for example, a stronger beam of the pattern points in the direction of the bypassing mobiles and a rather low-gain section of the pattern points toward the inside of the home. In this case it might happen that, in order to reduce the outdoor coverage, the output power of the femtocell would have to be reduced to such an extent that the signal strength inside the building may get too weak to provide sufficient coverage. It is obvious that with a single antenna it is difficult to adapt the femtocell coverage to its environment and only a limited reduction of mobility events can be achieved without compromising the coverage within the femtocell. On the other hand, if it were possible for the femtocell to choose from a certain number of diverse antenna patterns, then a significant improvement of the situation is possible: if the patterns show enough spatial diversity, then, for example, a pattern with a rather lower gain pointing to the outside and a rather higher gain pointing to the inside of the building could be chosen (see Figure 7.3(b)). This makes it possible to reduce the exposure outside of the building, and at the same time maintain sufficient coverage inside.

Multi-element antenna concept. Since a low-cost solution for such a multi-antenna system is essential, a simple switching network is proposed. This allows switching between multiple antennas and therefore allows us to choose between several fixed antenna patterns. For this scheme to work, it is required that the patterns of the antennas are diverse and show a reasonable directivity, so that either good coverage or low interference in each horizontal direction can be achieved. The antennas should then be arranged in a way that their patterns do not overlap too much, so that not too many antennas will be needed to cover the complete horizontal direction. The number of antennas needs to be chosen such that a reasonably large variety of pattern and coverage scenarios can be realized. Figure 7.5 shows a simplified illustration of the switching network with four antennas and four different patterns. By choosing the right pattern or

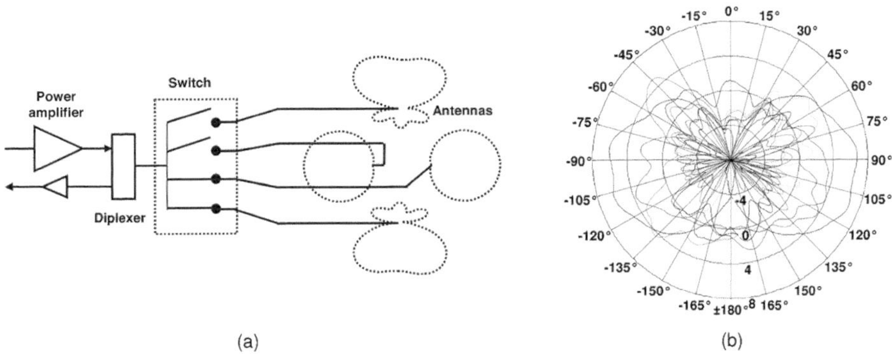

Figure 7.5 Antenna circuitry for switched multi-element antenna solution (a); and resulting measured antenna patterns (b). ©2009 IEEE. Reprinted, with permission, from [5].

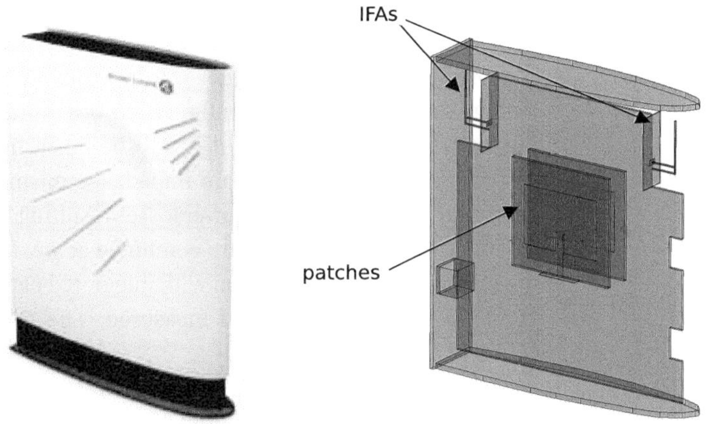

Figure 7.6 Physical design study of a femtocell and schematic of patch and inverted-F antenna location within the housing. Used with permission from [6].

pattern combination it will be possible to provide coverage within the building where it is required, as well as reducing the interference and number of mobility events. In order to choose the right pattern or pattern combination, an optimization process has to be performed. This process needs to be executable by the femtocell base station itself without further knowledge or coordination with the macro environment.

Antenna design. Since the available space inside the volume of a femtocell base station is limited, only a few types of antennas are suitable. In Figure 7.6 an example for a consumer-grade femto base station is shown; the picture shows the upright positioned housing and the placements of two patch and two inverted-F antennas (IFAs) within the femtocell. The size of the base station is approximately 170 mm×150 mm×35 mm ($h \times d \times w$). It contains a single main circuit board (MCB), which is located approximately in the center of the housing. The used antennas have to be of low cost and must be easy to integrate into the available volume of the housing. The patch antennas were chosen for their directional pattern, flat form factor, and the moderate implementation costs. They can easily be integrated into the existing base station by placing them parallel to the MCB. The IFAs were also chosen for their pattern, low cost, and ease of integration into the circuit board as a printed structure. With these four antennas the above stated requirements can be fulfilled: the antennas offer good diversity, in the shown orientation they cover the complete azimuth, and they offer good directivity at the same time. The pattern of the patch antenna shows a very good front-to-back ratio, such that two back-to-back placed antennas, one on either side of the MCB, provides two very diverse antenna patterns. The IFA antenna also shows a reasonable front-to-back ratio and, if placed in the upper corners of the MCB, generates a pattern, which complements the pattern of the patches. In addition, a combination of both IFA and patch antenna can be chosen.

Switch design. To make use of the different patterns, it is required to be able to switch between those antennas. In order to keep the cost and complexity of the switching circuitry as low as possible, a simple arrangement of four parallel switches is considered as shown in Figure 7.5. This allows the use of either one of the four antennas at a time, or any antenna combination thereof in parallel. This means no beamforming in the classical term (meaning the use of an array configuration of several elements with amplitude and phase control) is used. Of course, it has to be taken into account that if several antennas are operated in parallel, a certain impedance mismatch will occur, resulting in a lower efficiency. In order to keep the power losses resulting from impedance mismatch acceptable, no more than two antennas are combined at any time.

Prototype. A prototype of a femtocell base station with integrated patch and inverted-F antennas according to Figure 7.6 was built and measured. The measured antenna patterns are shown in Figure 7.5.

Optimization approach. The optimum pattern and pilot power configuration can be identified by a global search over all possibilities. However, this would lead to a long convergence time, not well suited for real-life systems. Therefore a different solution is required that achieves the same objectives and comes as close as possible to the optimal performance, but is better suited for implementation. Such a solution is described as follows. First, the femtocell pilot power is initialized for all antenna patterns using the measurement based auto-configuration method described above, and an initial antenna pattern is selected that provides similar gain in all directions. In this example, the pattern resulting from a combination of the two IFAs would be suitable. Note that each pattern is associated with its own individual pilot power value. During operation, the femtocell collects the information on mobility events such as handovers and idle mode mobility. Mobility events are classified as wanted or unwanted events over time, dependent on whether the mobile is registered at the femtocell and on the time a UE has spent in the cell before moving back to the macrocell. In addition, the femtocell periodically collects and stores the coverage information for different antenna patterns through path-loss measurements. These measurements can be performed by the UE using standardized methods and by the femtocell itself using its own measurement functionality and information on transmit powers and antenna gain, or by a combination

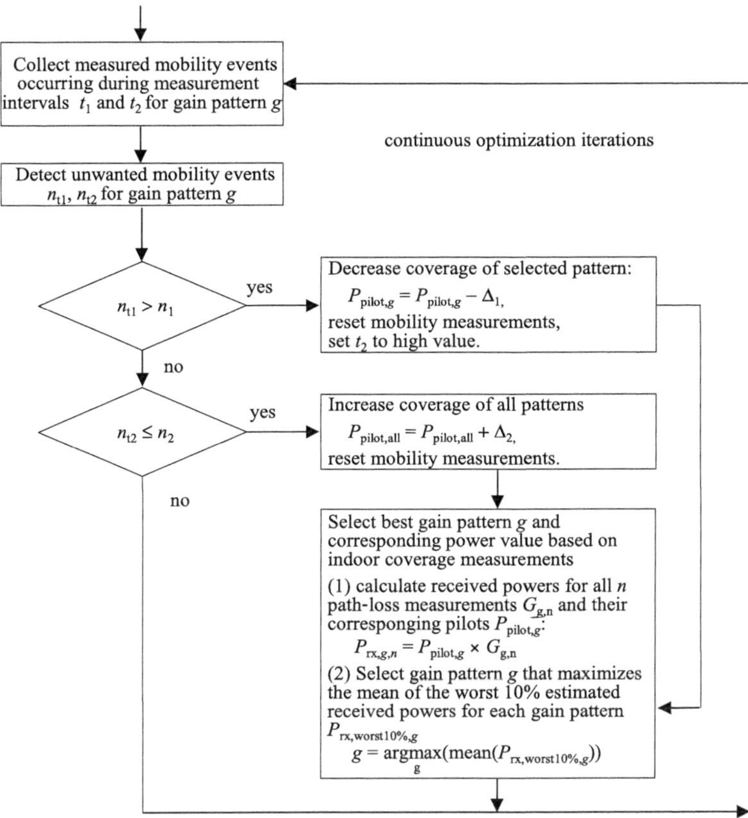

Figure 7.7 Flowchart of the switched multi-element antenna configuration process. ©2009 IEEE. Reprinted, with permission, from [5].

of both. The objective for the ongoing optimization is to maximize the indoor coverage under the constraint of limiting the mobility events from passing users. Figure 7.7 shows a flowchart of the method. If the number of unwanted mobility events of passing users n_{t1} exceeds a pre-defined value of n_1 events per timeframe t_1, the femtocell reduces its pilot power for the currently selected antenna pattern by a step Δ_1. The femtocell then starts a new mobility event count for the updated configuration. If the number of unwanted events n_{t2} is smaller or equal to a predefined acceptable value of n_2 events for a timeframe t_2, the femtocell increases its pilot power for the currently selected antenna pattern by a step Δ_2, to provide improved indoor coverage and starts a new mobility event count for the updated configuration. If the pilot power for the current antenna pattern changes, the pattern selection is re-evaluated, and the best pattern and pilot power combination is selected. For this, the estimated received powers at the measured points for the current power values for each antenna pattern are calculated. Then the pattern and power combination that maximizes the mean of the worst 10% of the estimated received powers is selected as the best new pattern and pilot power.

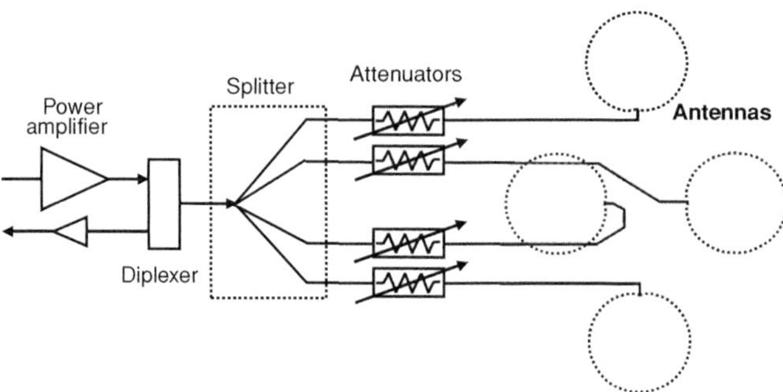

Figure 7.8 Antenna circuitry for tunable multi-element antenna solution. Used with permission from [6].

The parameters chosen for n_1, n_2, t_1, t_2, Δ_1, Δ_2 are equivalent to those for the similar implementable coverage optimization for a femtocell with a single antenna, as described above.

7.4.3 Coverage optimization using multi-element antennas with individually tunable attenuation

In the above-described method, it is possible to use more than one pattern at a time. But since only one transmitter is available, the power that is radiated by the two antennas cannot be individually controlled. In order to do so, each antenna could be connected to a separate transceiver, which would significantly increase the cost. Alternatively, the power level at each antenna can be tuned by individually adjustable attenuators, which is much more cost effective. This increases the flexibility for the coverage optimization even further compared to the previously discussed approaches (see Figure 7.3(c)). A schematic of the network with tunable attenuators is shown in Figure 7.8. The proposed self-optimization method for femtocells with gain-tunable multi-element antennas is described below.

Concept. In this section, an antenna configuration with four patch antennas, which are placed in a 90 degree orientation is considered as shown in Figure 7.8. This antenna configuration will further improve the coverage flexibility, since the maximum beam directions of each antenna fall into the same direction of the nulls of the remaining antennas. Therefore each direction can be either covered by an adjustable main beam or by a null.

Attenuator design. The use of attenuators will of course reduce the efficiency of the transceiver, but this is acceptable, considering that in most cases only a fraction of the overall power will need to be absorbed. For example, in a situation in which the coverage of one antenna causes the vast number of unwanted mobility events, and if this sector is turned off by absorbing all transmitted energy in that branch, then the overall power loss is only 25%. The impact on the receiver noise due to the additional attenuation is also negligible since the path loss between the mobile and the base station is low compared to the available power levels from the mobile and considering that the sectors in which

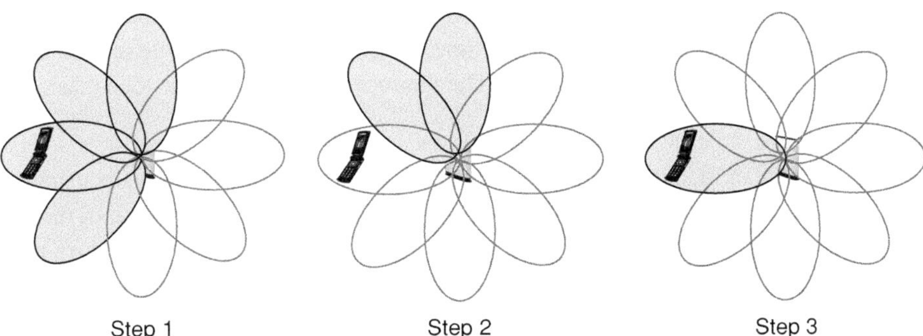

Figure 7.9 Example of serving lobe detection steps for a femtocell with eight lobes. Used with permission from [6].

a strong attenuation will be applied to the transmitted signal are exactly those sectors from which unwanted traffic is to be avoided.

Optimization approach. The optimum antenna and power configuration that minimizes the total number of mobility events while maximizing indoor coverage can be identified via a global search over all possibilities. Due to the high number of patterns resulting from the possibility to fine tune the gain for each lobe individually, and the time-consuming evaluation of each potential pattern, this approach is impractical. Alternatively, instead of jointly optimizing the coverage for all adjustable lobes, they can be optimized individually with the aim of preventing unwanted mobility events on each lobe. This is performed in a similar fashion as for the mobility event optimization for the single antenna femtocell case above, with the difference that instead of increasing or decreasing the pilot power, the gain for the serving lobe is modified. The serving lobe is defined as the one from which the mobile receives the strongest signal. As a result, one benefit over the switched pattern approach described above is a faster convergence of the solution, since only four attenuations need to be configured compared to ten different pilot power levels (one for each pattern). The identification of the serving lobe is performed by a series of gain modifications for each lobe and using the standardized reporting capability of the mobile to identify the correct lobe or beam. One example of this identification process for a femtocell with eight lobes is illustrated in Figure 7.9. These individual identification steps, taking both positive feedback (the UE reports measured power increase) and negative feedback (the UE does not report measured power increase) into account, are described below:

Step 1: The femtocell increases the gain for lobes 1 to 4 to identify in which sector of the coverage the connected UE is located. If the UE reports an increase in received pilot power, then the femtocell can conclude that the UE is covered by one of the four lobes.

Step 2: The femtocell narrows down the search by increasing the gain of two of the identified four lobes. If the UE is not covered by the two lobes it does not report an increased pilot power, so the femtocell can conclude that the mobile is not located in the two selected lobes, but in the other two remaining lobes.

Step 3: Finally the femtocell increases the gain of one of the remaining two lobes. In this example the correct lobe was chosen and the mobile reports back the received increase in pilot power. The femtocell has detected the lobe that serves the mobile best.

In an indoor environment the reflections of the signal via walls usually contribute to the received signal, which could lead to the case that an increase in received power can be reported by the mobile although this lobe with increased gain does not serve the UE best. As a result this can cause an unreliable identification of the best lobe. This can be resolved by basing the decision not only on whether an increased pilot is reported or not. In this case the gain would be increased sequentially for both choices in each step and the one with the highest reported signal increase is chosen. This doubles the number of gain modification steps required, but provides a more reliable result. One option to increase the detection speed is to use past statistics on where frequent handovers to the femtocell happen and use those to improve the guess. This can help to speed up the identification, for example for outside users passing the house, which would always be served by the same lobes.

7.5   Performance evaluation

The performance of the different self-optimization methods introduced in this chapter was analyzed using system level simulations. This section describes the simulation assumptions, the user mobility model, and summarizes the key results.

7.5.1   Simulation assumptions

System-level simulations were performed to derive both the number of mobility events for passing and indoor users, and indoor coverage as a function of the distance from the femtocell. A scenario with seven macrocells with three sectors each is considered, as shown in Figure 7.10. Femtocells are deployed randomly within the coverage area and reuse the same frequency as the macrocells in a hierarchical cell structure. An overview of the simulated scenario with different macrocell range and power can be found in [8]. Key simulation parameters including the propagation models for path loss, shadow fading, and antenna gain for the macrocell sectors assumed are shown in Table 7.1. The assumed transmission losses for the explicit house model as a function of the incident angle are shown in Figure 7.11 and are based on [9]. The assumed macrocell downlink transmit power is 20 watts (W) per sector and 10% of the total power is allocated to pilot channels. The cell range is calculated such that the received signal-to-noise ratio inside a house at the cell edge is 10 dB, with full coverage of the area, including indoor coverage at the cell, assuming an additional wall loss of 15 dB. As a result, the obtainable throughputs from the macrocell are mainly interference limited. The femtocell pilot powers are configured and optimized as described in the previous sections. It is assumed that a mobility event is triggered when the received pilot signal from a new cell is 4 dB

Table 7.1 Simulation parameters.

Path loss	Path loss is modeled as 38.5 dB $+ 20 \log_{10}(d) + L_{\text{walls}}$ dB for short distances with a transition to 28 dB $+ 35 \log_{10}(d) + L_{\text{walls}}$ dB in other cases, where d is the distance from the base station in meters. The loss L_{walls} is calculated based on the materials and the incidence angle of the walls/doors/windows in the signal path.
Shadow fading	Shadow fading is modeled as spatially correlated random process with log-normal distribution (8 dB standard deviation for the macrocell signal, 4 dB standard deviation for the femtocell signal spatial correlation $r(x) = e^{x/20}$ for distance x in meters [10]).
Receiver noise power	The receiver noise power is modeled as $10 \log_{10}(\text{kT} \times \text{NF} \times \text{W})$ where the effective noise bandwidth $W = 3.84 \times 10^6$ Hz, and $\text{kT} = 1.3804 \times 10^{-23} \times 290$ W/Hz. The noise figure at the UE is $\text{NF}_{[\text{dB}]} = 7$ dB.
Macrocell antenna gain	The macrocell antenna gain is calculated as $G(\Psi)_{[\text{dB}]} = G_{\max} - \min\left(12\left(\Psi/\alpha\right), \kappa\right), -\pi \leq \Psi \leq \pi$ with $\alpha = 70\pi/180$ (angle where gain pattern is 3 dB down from peak) $\kappa = 20$ dB (front-to-back ratio) $G_{\max} = 16$ dB (maximum antenna gain)

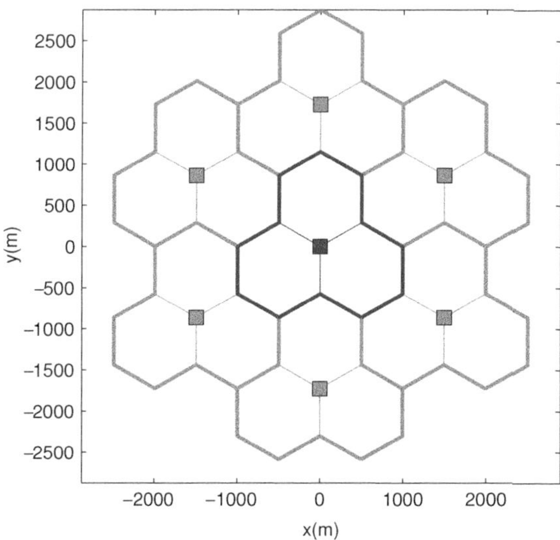

Figure 7.10 Simulated scenario. Used with permission from [6].

higher than the received pilot signal from the serving cell for a time of 500 ms. When a mobility event is triggered, the duration of the procedure is assumed to be 650 ms. Figure 7.11 illustrates the simulated area around a typical London terraced house as considered for this investigation and shows one example of the received pilot power levels within the coverage areas for both macrocell and femtocell. This house type was

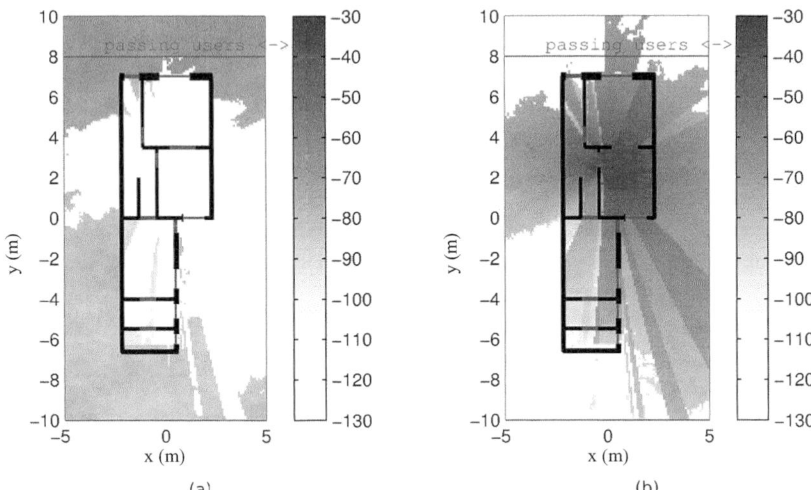

Figure 7.11 Coverage example for macrocell (a) and femtocell (b). ©2008 IEEE. Reprinted, with permission, from [4].

chosen as a worst-case example due to the narrow shape and its proximity and lack of wall separation to the footpath in front of the house.

7.5.2   Mobility model

To simulate the behavior of the femtocell user, an indoor user mobility model is used as described in more detail in [4]. The indoor model uses a collection of waypoints defined throughout the house model of the terraced house shown in Figure 7.11. The femtocell user moves from different rooms within the house and spends a certain amount of time at each waypoint in one room before moving on to another room at a speed of 1 m/s. Each room has a probability that it will be visited by the user, with the lounge and study being the rooms that are visited the most. In each room, the user will spend a certain amount of time at a waypoint. The amount of time the user spends at the waypoint is normally distributed. The probabilities of the user being in a room, and the mean time spent in the room were derived empirically from a short survey on user behavior during evening peak times (i.e., the period starting from when the user comes home from work to before going to sleep). To simulate the effect of underlay macrocell users, i.e., passers-by, an outdoor user model is used. The outdoor model simply moves the users along the sidewalk outside the house, whose location is shown in Figure 7.11. The users walk past the front of the house from one end of the sidewalk to the other in a straight line at a speed of 1 m/s. The entry point of the user (from the left or right side of the house) is randomly generated, as is the part of the sidewalk the user walks on. The inter-arrival times of pedestrians in a residential area were determined to be exponentially distributed using a separate large-scale outdoor pedestrian simulation. As such, the inter-arrival times used in the model are generated using an exponential distribution, with different means used for different scenarios.

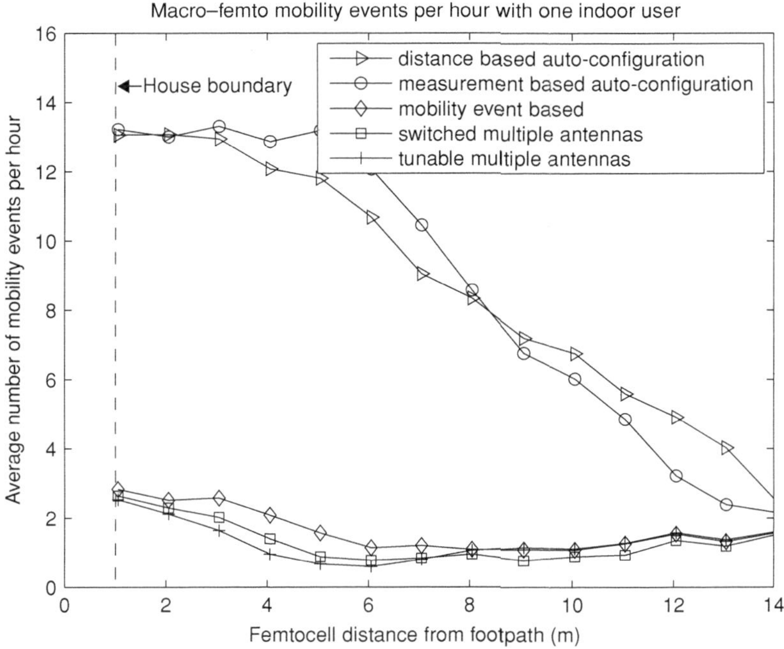

Figure 7.12 Average number of mobility events per hour for different femtocell locations. Adapted from [4–6].

7.5.3   Results

The proposed mobility event based coverage optimization methods are compared with the simpler auto-configuration schemes, and the resulting mobility events, indoor coverage, pilot powers, and gain pattern selection for the switched multi-element antenna solution are discussed.

Mobility events. Figure 7.12 shows a comparison of the resulting average number of mobility events per hour as a function of the femtocell distance from the footpath. The distance based single antenna auto-configuration method achieves performance similar to the simpler measurement based single antenna method. The mobility event based single antenna self-optimization method is able to significantly outperform both auto-configuration methods that do not take mobility events into account. The proposed switched multi-element antenna solution can reduce the number of mobility events further compared to a single element antenna solution with a similarly sophisticated coverage self-optimization. The improvement depends on where the femtocell is deployed in the house. In [6], it is shown that the performance of the implementable optimization approaches the optimal performance for this antenna configuration. The tunable multi-element solution with four patch antennas achieves an additionally improved performance for short distances from the footpath, but loses this advantage at larger distances due to its inability to focus the transmit power toward one direction without losses in the attenuators.

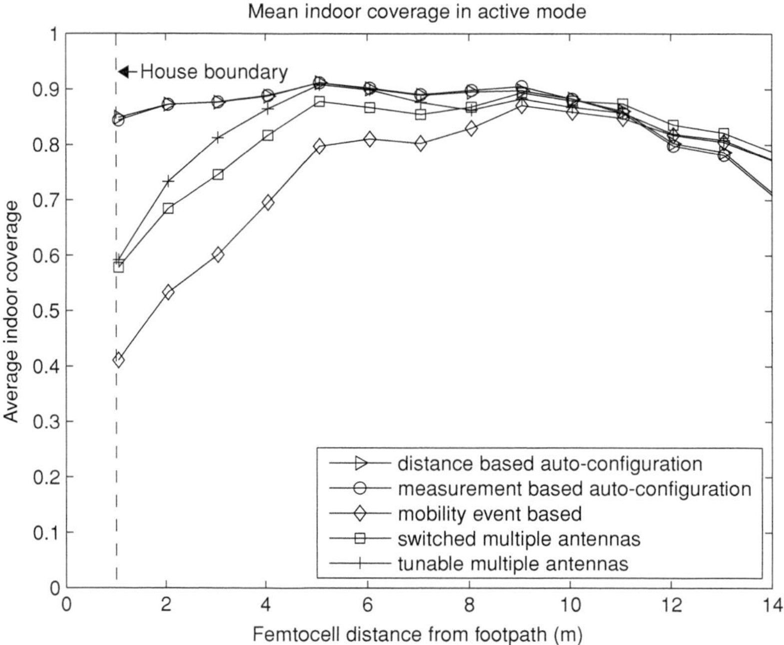

Figure 7.13 Average resulting indoor coverage for different femtocell locations. Adapted from [4–6].

Indoor coverage. Figure 7.13 compares results of indoor coverage, defined as the fraction of the area inside of the house where the pilot power of the femtocell is received stronger than the power of a macrocell, for the single antenna versus multiple-antenna solutions. The single antenna auto-configuration methods considered achieve very similar coverage. With a single antenna, the mobility event based self-optimization method achieves a high rate of coverage when femtocells are deployed in good locations, but a lower rate of coverage when deployed in unsuitable locations close to the footpath. This is a tradeoff resulting from the reduction in core network signaling, which is of higher priority. Note that the lower performance in unsuitable locations is not necessarily a disadvantage, since it would encourage the user to redeploy the femtocell (e.g., via short message service (SMS) notification), which will result in both better coverage for the user and a lower number of mobility events. For the multi-element antenna solutions, it is shown that in addition to the reduced number of mobility events discussed above, the indoor coverage can also be improved compared to a single antenna solution with a similar sophisticated coverage self-optimization. The most significant improvements in coverage are achieved when the femtocell is deployed in unsuitable locations close to the footpath, where the flexibility in terms of gain patterns allows better coverage of the rest of the house without increasing the number of unwanted mobility events and the associated core network signaling. In [6], it is shown that the performance of the implementable optimization approach for the switched multi-element antenna solution

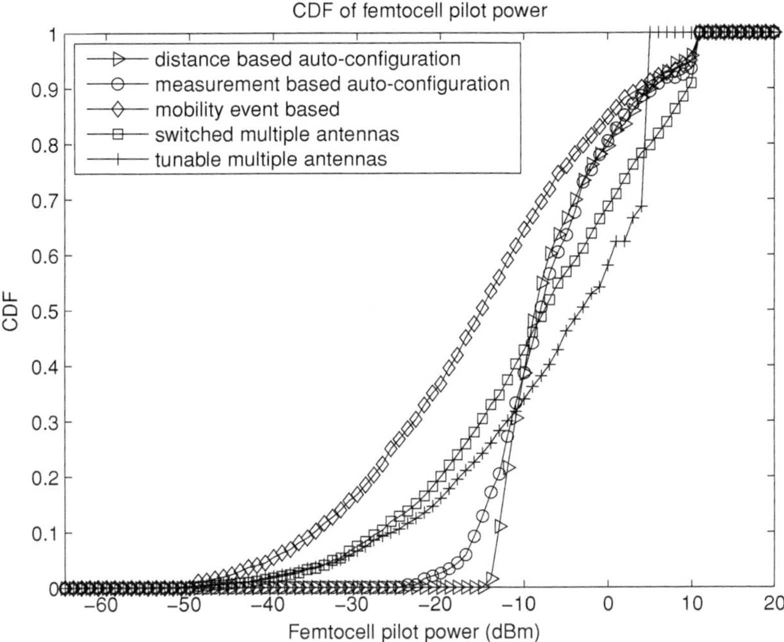

Figure 7.14 CDF of configured pilot powers. Adapted from [4–6].

approaches optimal performance. The tunable multi-element antenna solution results in slightly higher coverage for femtocells deployed close to the footpath due to the higher flexibility of the gain pattern compared with the switched pattern approach. However, with this solution, the coverage is slightly worse than that of the switched pattern solution for femtocell locations in the back of the house due to its inability to focus the transmit power toward one direction without losses in the attenuators. Note that since the model only considers transmission loss of walls, but neglects reflections on walls, floors, and ceilings, the coverage results are relatively conservative and slightly higher coverage can be expected in real deployments.

Pilot power. Figure 7.14 shows the cumulative distribution functions (CDFs) of the resulting pilot power for both scenarios. In general, all methods result in a large variation of pilot powers depending on the distance from the macrocell and the deployment location within the house. The mobility event based self-optimization methods show the largest variation in power. While in some cases they use a comparable power as the other auto-configuration methods in order to maximize the indoor coverage for the given environment, outdoor mobility, and deployment location, they also reduce the pilot power in some cases to −50 dBm. This happens to prevent excessive mobility events in cases where the user deploys the femtocell in an unsuitable location, for example, close to the window near the sidewalk. Current cellular standards such as for the Universal Mobile Telecommunication System (UMTS) have not been designed for femtocells and therefore specify much higher values for the minimum pilot power. For the multi-element

Table 7.2 Probability of antenna pattern use (• indicates activated antennas).

Pattern	1	2	3	4	5	6	7	8	9	10
IFA 1	•				•	•			•	
IFA 2		•					•	•	•	
Patch 1			•		•			•		•
Patch 2				•		•	•			•
Probability of use (%)	10.2	5.7	21.6	18.7	9.0	8.6	6.1	7.6	6.9	4.9

antenna solutions, the typically radiated pilot power can be around 10 dB higher than that of the single antenna solution for minimizing the total number of mobility events. This is enabled by the flexibility in terms of antenna gain of the proposed multi-element antenna solution and contributes to the better indoor coverage. The radiated power of the tunable antenna solution can be even higher, but the power per antenna is limited due to the fact that the total power is split equally among the antennas.

Pattern selection. Table 7.2 shows the probability of being optimal for each pattern in the proposed switched multi-element antenna solution. The simulations show that the patch antennas with their clean directional gain patterns and high spatial diversity seem to be most useful (patterns 3 and 4), since they are chosen far more often than the other patterns. Therefore, it can be given as a design rule that it is beneficial to achieve as diverse patterns as possible for switched multi-element antenna systems.

7.6   Coverage optimization of femtocell groups

The discussion so far has concentrated mainly on the deployment of femtocells working in isolation in residential scenarios. However, femtocells are also being developed for deployment in enterprise environments, such as in office buildings or shopping malls. In such scenarios, femtocells are deployed in groups in order to provide contiguous coverage over a larger area. Here, a collection of femtocells that are deployed together in order to jointly provide coverage is given the term "femtocell group." In many ways, the optimization of coverage for femtocell groups is similar to base stations in conventional cellular networks, as a femtocell group may be viewed as a cellular network, albeit covering a much smaller area. So unlike residential deployments, the coverage of a group of femtocells does not only take into consideration the interaction with the macrocell layer, but also with neighboring femtocells in the group. In this section, an overview of the aims and requirements of coverage optimization in femtocell group deployments is given.

Depending on the scenarios in which they are deployed, the optimization of coverage in femtocell groups is performed to achieve multiple objectives. One aspect of the coverage optimization is ensuring that the coverage of the femtocells is continuous in the intended area of coverage, without coverage gaps between the femtocells. There should also be enough overlap between the coverage of neighboring femtocells to ensure that handovers

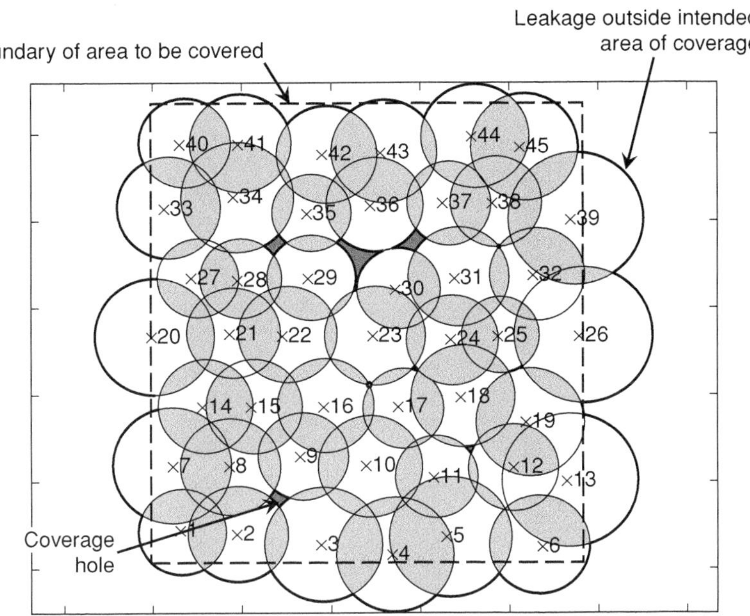

Figure 7.15 Illustration of joint coverage optimization problem.

between femtocells can be performed reliably. A second aspect to consider is to minimize the impact of the femtocell group on the co-channel macrocell network. The coverage should be set to reduce the leakage of coverage onto areas with a high number of macrocell users passing by, to reduce the amount of unnecessary mobility events (and hence signaling) from macrocell users. These two aspects are illustrated in Figure 7.15. The third aspect is the adjustment of the coverage size of individual femtocells according to their load. In many cases, the user traffic distribution in femtocell group deployment scenarios changes both with respect to location and time. Because of this, the coverage of a femtocell needs to be adjusted such that it does not capture too many users that cause it to become overloaded, and offloads users to less highly loaded neighboring femtocells. One example is illustrated in Figure 7.16 where one user of the center femtocell is offloaded to the left femtocell.

Optimizing the coverage of femtocell groups could be done through manual cell planning, where the femtocell coverage is set by engineers in order to satisfy the objectives above. Unlike residential femtocells where femtocells are user deployed, the deployment of enterprise femtocell groups is likely to be done by the network operators themselves, so performing manual cell planning is something that is possible in this case. However, manually optimizing the coverage, as with residential femtocells, is costly and time consuming. Changes to the environment and demand can also occur after deployment, which can have a significant impact to the performance of the network due to the relatively low transmit powers and capacities of femtocell base stations. Therefore the ability of femtocell groups to dynamically optimize their coverage would allow such deployments

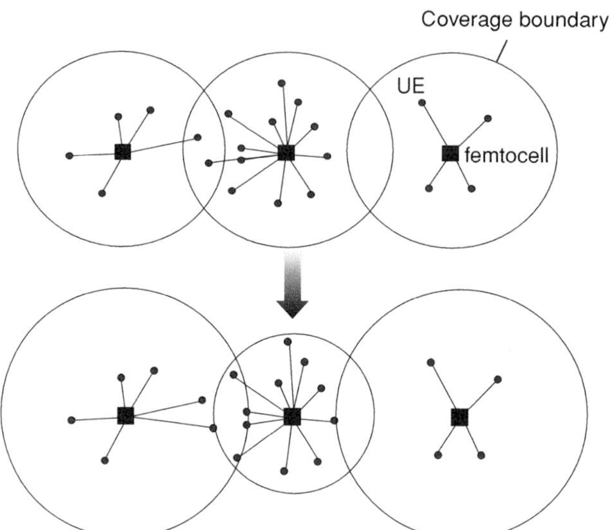

Figure 7.16 Illustration of adjustment of coverage to balance the load.

to be done without detailed cell planning, and allows adaptation to any changes to the environment, user demand, and topology of the femtocell network.

The dynamic optimization of femtocell groups requires information relating to coverage gaps, coverage leakage, and load to be collected by the femtocell network. Here, several examples of how this information can be obtained are given. As mentioned previously, the coverage of each femtocell should be adequate to ensure that there are no gaps in the coverage within the building. The ability of femtocells to detect gaps in coverage would therefore be required. This can be done by femtocells in many different ways, such as using measurement reports of received pilot powers from connected UEs or collecting statistics on handovers. A femtocell can suspect the existence of a coverage gap if the number of handovers toward macrocells is high. Obviously, there would be femtocells located at the exit/entry points of the building where handovers to and from the macrocell are expected to be high, relative to other femtocells. But techniques for identifying femtocells located at entry/exit points, and to distinguish handovers due to users exiting the building from those caused by coverage gaps can be easily used to overcome this. To detect leakage of coverage, the same techniques used for residential femtocells can be used, where mobility events from unregistered users are used to determine if the coverage of a femtocell is leaking onto areas of high macrocell user traffic. Collecting information on femtocell load is obviously straightforward, since this information is readily available locally at the femtocell.

The information on the femtocell's coverage gap, leakage, and load can then be used as inputs to an algorithm to calculate the adjustments to the coverage. The expected actions to be taken can be deduced quite intuitively. If a coverage gap is detected, the relevant femtocell should increase its coverage to cover the gaps. Conversely, if leakage of coverage is detected, the relevant femtocell should decrease its coverage to eliminate the leakage. The coverage optimization for load is triggered when a femtocell's load exceeds a set threshold, where the coverage of the femtocell is reduced in order to offload more users to neighboring femtocells. The load threshold used can be the number of active users, or the femtocell downlink or uplink capacity. The problem of femtocell group coverage optimization involves the consideration of different conflicting objectives. For example, there may be cases where reducing the coverage to perform load balancing or to reduce leakage may cause coverage gaps to occur. When designing the optimization algorithm, it would be necessary to place different priorities to each objective. These priorities would be determined by the requirements of the network operator, who may place more importance on minimizing coverage gaps above all other objectives.

One of the more complex aspects of femtocell group coverage optimization is ensuring that the adjustment of coverage is done jointly in conjunction with neighboring femtocells. When a femtocell reduces its coverage, its neighboring femtocell may need to increase its coverage so that the coverage overlap is maintained. In a distributed implementation (where femtocells act independently without coordination with neighboring femtocells), the reduction of coverage by a femtocell can create a coverage gap. However, once detected, the coverage gap would then be covered by neighboring femtocells. While this approach does not require any communication between femtocells, it does result in coverage gaps occurring temporarily before they are detected and remedied. In order to avoid this, communication between femtocells would be required so that neighboring femtocells can react in tandem, without delay, to any changes that a femtocell makes to its coverage. This exchange of status information between femtocells also allows the femtocell coverage optimization algorithm to make more informed decisions in order to avoid any potential oscillations that may occur. For example, if a femtocell knows that its neighbor is operating close to its capacity threshold, then it should not offload users to that neighbor as that would trigger said neighbor to attempt to offload users back. However, this exchange of information between femtocells would obviously introduce additional signaling overhead on the backhaul. Changes to the coverage of one femtocell may also trigger changes to neighboring femtocells that can "ripple" through to all other femtocells in the group. This not only further increases the signaling involved, but also increases the convergence time of a change. With this in mind, different steps have to be put into place to limit the "rippling" effect, for example by making sure that neighboring femtocells only react if the change in coverage made by the invoking femtocell is large enough to cause insufficient coverage overlaps or coverage gaps.

7.7　Conclusions

In this chapter, different auto-configuration and self-optimization methods were presented that aim to optimize the coverage of femtocells. Coverage adaptation methods were investigated that use information on mobility events of passing outdoor and indoor users to optimize the femtocell coverage in order to minimize the increase of core network mobility signaling. It was shown that for femtocells with a single antenna, mobility event based self-optimization of coverage can both significantly reduce the total number of mobility events caused by femtocell deployments and improve indoor coverage for femtocells deployed in suitable locations compared to simpler methods that aim to achieve a constant cell radius. In addition, two low-cost multi-element antenna solutions were presented to increase flexibility in terms of good deployment locations within the home where both good indoor coverage and a low number of resulting mobility events can be achieved. A prototype of the switched multi-element antenna solution was built and the resulting antenna patterns were measured. Two self-optimization methods were presented that optimize both power and pattern selection. Simulations were performed to compare the multi-element antenna solution based on the measured antenna patterns with a single antenna solution. It was shown that the flexibility gained due to the ability to select from multiple antenna patterns of the first solution allows shaping the femtocell coverage such that both the number of mobility events and the indoor coverage can be improved considerably. The second multi-element antenna solution uses four individually adjustable patch antennas, which further increases the flexibility of the achievable gain patterns. While the increased flexibility did not result in significant performance improvements compared to the switched multi-element antenna solution in the investigated scenario, it has some advantages in terms of convergence speed. Finally, the coverage optimization of groups of femtocells was discussed, which will be of great importance in the future, when enterprise or outdoor deployments of femtocells and other small cells will increase and become ubiquitous.

References

[1] H. Claussen, L. T. W. Ho, H. R. Karimi, F. J. Mullany, and L. G. Samuel, "I, base station: cognisant robots and future wireless access networks," in *Proc. IEEE Consum. Commun. and Networking Conf. (CCNC)*, Las Vegas, NV, Jan. 2006, pp. 595–9.

[2] A. C. Stocker, "Small-cell mobile phone systems," *IEEE Trans. Veh. Technol.*, vol. 33, no. 4, pp. 269–75, Nov. 1984.

[3] L. T. W. Ho and H. Claussen, "Effects of user-deployed, co-channel femtocells on the call drop probability in a residential scenario," in *Proc. IEEE Int. Symp. Personal, Indoor, Mobile Radio Commun. (PIMRC)*, Athens, Greece, Sep. 2007, pp. 1–5.

[4] H. Claussen, L. T. W. Ho, and L. G. Samuel, "Self-optimization of coverage for femtocell deployments," in *Proc. Wireless Telecommun. Symp. (WTS)*, Los Angeles, USA, Apr. 2008, pp. 278–85.

[5] H. Claussen and F. Pivit, "Femtocell coverage optimization using switched multi-element antennas," in *Proc. IEEE Int. Conf. on Commun. (ICC)*, Dresden, Germany, June 2009, pp. 1–6.

[6] H. Claussen, L. T. W. Ho, and F. Pivit, "Self-optimization of femtocell coverage to minimize the increase in core network mobility signalling," *Bell Labs Technical J.*, vol. 14, no. 2, pp. 155–84, Aug. 2009.

[7] 3GPP, "3GPP system architecture evolution: report on technical options and conclusions," 3rd Generation Partnership Project (3GPP), TR 23.882, July 2007.

[8] H. Claussen, "Performance of macro- and co-channel femtocells in a hierarchical cell structure," in *Proc. IEEE Int. Symp. Personal, Indoor, Mobile Radio Commun. (PIMRC)*, Athens, Greece, Sep. 2007, pp. 1–5.

[9] S. J. Fortune, D. M. Gay, B. W. Kernighan, O. Landron, R. A. Valenzuela, and M. H. Wright, "WISE design of indoor wireless systems: practical computation and optimization," *IEEE Comput. Sci. Eng.*, vol. 2, no. 1, pp. 58–68, Spring 1995.

[10] H. Claussen, "Efficient modelling of channel maps with correlated shadow fading in mobile radio systems," in *Proc. IEEE Int. Symp. Personal, Indoor, Mobile Radio Commun. (PIMRC)*, Berlin, Germany, Sep. 2005, pp. 512–16.

8   Random matrix methods for cooperation in small cell networks

Jakob Hoydis and Mérouane Debbah

8.1   Introduction

We are currently witnessing an exponentially increasing demand for wireless data services, which is mainly driven by the growing popularity of wireless modems, smartphones, and tablet personal computers (PCs). Not surprisingly, current cellular networks have already started reaching their capacity limits in densely populated areas and it is therefore necessary to assess which network architecture is most suited to carry the future data traffic. As additional spectrum resources are scarce, it seems inevitable that any future system architecture will rely to a significant extent on *network densification*, i.e., an increase of the number of antennas deployed per unit area. This can be achieved by either increasing the number of antennas per base station (BS) [1] or by deploying more BSs [2], or a combination of both [3]. More antennas per device lead to additional degrees of freedom, which can provide spatial multiplexing and diversity gains or can be used to cancel interference [4]. On the other hand, a denser deployment of BSs, such as femto or small cells [5], increases the spatial reuse of the radio spectrum. Although the network capacity would theoretically scale linearly with the BS density, dense networks suffer from increased inter-cell interference and user mobility becomes difficult to manage. The cooperation of multiple BSs, which jointly process user data from multiple cells [6] has shown its potential to counter inter-cell interference and to improve the cell edge coverage not only in theory [7] but also in practice [8]. Base station cooperation is therefore already considered as an essential feature of future cellular standards [9]. Apart from that, clusters of cooperating BSs forming *virtual cells* could also potentially reduce the amount of handover signaling between cells to enable user mobility in dense networks [10].

From a theoretical point of view, the performance analysis of dense networks is very challenging. This is not only due to the complexity of any meaningful channel model that includes path-loss, fading, and possibly line-of-sight components, but also due to practical limitations such as limited channel state information (CSI) and backhaul capacity [8], which must be accounted for. An additional degree of complexity appears when random deployments of BSs and random user distributions are considered [11]. The aim

Small Cell Networks: Deployment, PHY Techniques, and Resource Management, ed. Tony Q. S. Quek, Guillaume de la Roche, İsmail Güvenç, and Marios Kountouris. Published by Cambridge University Press. © Cambridge University Press 2013.

of large system analyses is to reduce the randomness, and hence the complexity, of such systems and to provide insight about the most relevant system parameters. For example, under the assumption that each BS is equipped with a very large number of antennas, the channels become essentially deterministic and the performance depends only on the statistical properties of the channel and not on a particular channel realization [12]. Similarly, if there is a large number of user terminals (UTs) in a network, the inter-cell interference becomes independent of the exact user locations and depends only on the statistical properties of the spatial point process underlying the user distribution [11]. In many cases, the performance of a system of realistic dimensions is very well approximated by its asymptotic performance. Thus, large system approximations can also be used for the optimization of certain network parameters, which would be intractable for finite dimensions, e.g. [13–15].

In this chapter, we will study the performance of BS cooperation in the uplink and downlink of dense small cell networks. Our analysis is based on the theory of large random matrices and relies on the assumption that the number of antennas per BS and the number of UTs per cell grow infinitely large at the same speed. We will derive asymptotically tight deterministic approximations of information-theoretic quantities, such as the mutual information and achievable rates with linear detectors and precoders. Those are shown by simulations to be accurate even for realistic system dimensions. Starting from a simple Rayleigh fading channel model where every device has full CSI and the locations of UTs and BSs are fixed, we gradually extend the results to account for imperfect CSI, Rician fading channels, and random user locations. Our aim is to provide the reader with a clear understanding of the applied methodology, which will allow him/her to carry out similar analyses in other scenarios of interest.

The rest of this chapter is organized as follows. In Section 8.2, we introduce the general system model. Section 8.3 reviews some basic lemmas of random matrix theory that are necessary to follow the proofs in the subsequent sections. Sections 8.4 and 8.5 deal with the performance analysis of the uplink and downlink with perfect and imperfect CSI, respectively. Rician channels are considered in Section 8.6, while random user locations are discussed in Section 8.7. The chapter is concluded in Section 8.8.

Notations: Boldface lower and upper case symbols represent vectors and matrices, respectively. $\mathbf{I}_N$ is the size-N identity matrix and $\mathrm{diag}(x_1, \ldots, x_N)$ is a diagonal matrix with elements x_i. Similarly, $\mathrm{diag}(\mathbf{X}_1, \ldots, \mathbf{X}_N)$ is a block-diagonal matrix with elements $\mathbf{X}_i$. The trace, transpose, and Hermitian transpose operators are denoted by $\mathrm{tr}\,(\cdot)$, $(\cdot)^{\mathrm{T}}$, and $(\cdot)^{\mathrm{H}}$, respectively. The spectral norm of a matrix $\mathbf{A}$ is denoted by $\|\mathbf{A}\|$. The notations $\Rightarrow$ and $\xrightarrow{\text{a.s.}}$ denote weak and almost sure convergence, respectively. We use $\mathcal{CN}\,(\mathbf{m}, \mathbf{R})$ to denote the circular symmetric complex Gaussian distribution with mean $\mathbf{m}$ and covariance matrix $\mathbf{R}$. We denote by $\mathbb{E}\,[\cdot|\cdot]$ the (conditional) expectation.

8.2   System model

We consider a small cell network consisting of B cells as schematically shown in Figure 8.1. For simplicity, we assume that there is one BS with N antennas and K

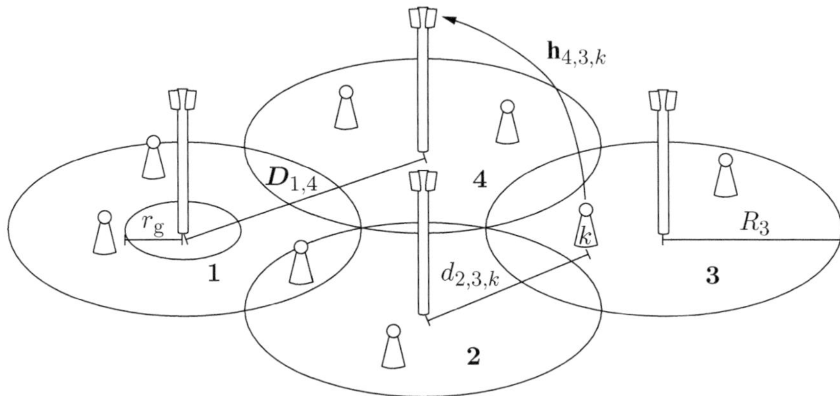

Figure 8.1 Schematic diagram of the system model with $B = 4$ BSs, each equipped with N antennas, and $K = 2$ single antenna UTs per cell.

single-antenna UTs in each cell. The extension to a different number of UTs in each cell and a different number of antennas at each BS is straightforward. We do not distinguish between macrocells and femtocells and presume that all BSs and UTs operate in the same frequency band. We consider a frequency-flat fading channel and denote by $\mathbf{h}_{i,j,k} \in \mathbb{C}^N$ the uplink channel transfer vector from the kth UT in cell j to the BS in cell i. Similarly, we use $\mathbf{h}_{i,j,k}^{\mathrm{H}}$ to denote the channel from BS i to UT k in cell j during downlink transmissions. The vectors $\mathbf{h}_{i,j,k}$ are modeled as

$$\mathbf{h}_{i,j,k} = \frac{1}{\sqrt{K d_{i,j,k}^{\alpha}}} \mathbf{w}_{i,j,k} \tag{8.1}$$

where $\mathbf{w}_{i,j,k} \sim \mathcal{CN}(\mathbf{0}, \mathbf{I}_N)$, $d_{i,j,k}$ denotes the distance between UT k in cell j and BS i and α is a path-loss exponent. Depending on the environment, α can take values in the range from 2 to 5. We assume a guard distance $r_{\mathrm{g}} > 0$ around each BS, so that $d_{i,j,k} \geq r_{\mathrm{g}}$ for all i, j, k. We define the aggregated channel transfer matrices $\mathbf{H}_{i,j} \in \mathbb{C}^{N \times K}$ from the UTs in cell j to BS i as

$$\mathbf{H}_{i,j} = \left[\mathbf{h}_{i,j,1} \cdots \mathbf{h}_{i,j,K}\right], \qquad 1 \leq i, j \leq B. \tag{8.2}$$

We assume that the BSs are allowed to cooperate by forming clusters or virtual cells, i.e., a cluster of BSs jointly processes the signals received on the uplink or transmitted on the downlink. Thus, a cluster of BSs can be regarded as a large distributed antenna array. For clarity of presentation, we ignore here the effects of limited backhaul capacity, delay, and synchronization errors and refer the interested reader to [6, 8]. We denote by $C \in [1, B]$ the number of clusters and denote by $\mathcal{C}_i = \{b_{i,1}, \ldots, b_{i,|\mathcal{C}_i|}\} \subset \{1, \ldots, B\}$, $i = 1, \ldots, C$, the set of BSs in cluster i, where $|\mathcal{A}|$ denotes the cardinality of set $\mathcal{A}$. Since each BS can only be assigned to a single cluster, the following holds: $\mathcal{C}_i \cap \mathcal{C}_j = \emptyset$ for $j \neq i$ and $\mathcal{C}_1 \cup \mathcal{C}_2 \cup \cdots \cup \mathcal{C}_C = \{1, \ldots, B\}$. Moreover, we have the two special cases $C = 1$ (full cooperation) and $C = B$ (no cooperation). For later use, we introduce the matrices $\mathbf{G}_{i,j} \in \mathbb{C}^{N_i \times K_j}$, defined as

$$
\mathbf{G}_{i,j} = \begin{pmatrix} \mathbf{H}_{b_{i,1},b_{j,1}} & \cdots & \mathbf{H}_{b_{i,1},b_{j,|\mathcal{C}_j|}} \\ \vdots & \ddots & \vdots \\ \mathbf{H}_{b_{i,|\mathcal{C}_i|},b_{j,1}} & \cdots & \mathbf{H}_{b_{i,|\mathcal{C}_i|},b_{j,|\mathcal{C}_j|}} \end{pmatrix}, \qquad 1 \le i, j \le C \tag{8.3}
$$

where $N_i = N|\mathcal{C}_i|$ is the total number of BS antennas in cluster i and $K_j = K|\mathcal{C}_j|$ is the total number of UTs in cluster j. The matrix $\mathbf{G}_{i,j}$ represents thus the aggregated channel between the UTs in the cells of cluster j and the BSs of cluster i. In order to simplify the notations later on, we rewrite the column vectors $\mathbf{g}_{i,j,k} \in \mathbb{C}^{N_i}$ of $\mathbf{G}_{i,j}$ as

$$
\mathbf{g}_{i,j,k} = \frac{1}{\sqrt{K}} \mathbf{R}_{i,j,k}^{\frac{1}{2}} \mathbf{u}_{i,j,k}, \qquad 1 \le k \le K_j \tag{8.4}
$$

where $\mathbf{u}_{i,j,k} \sim \mathcal{CN}\left(\mathbf{0}, \mathbf{I}_{N_i}\right)$ and $\mathbf{R}_{i,j,k} \in \mathbb{R}_+^{N_i \times N_i}$ is given as

$$
\mathbf{R}_{i,j,k} = \mathrm{diag}\left(d_{b_{i,s},b_{j,\lceil k/K \rceil},k-K(\lceil k/K \rceil-1)}^{-\alpha} \mathbf{I}_N, \; s = 1, \ldots, |\mathcal{C}_i| \right). \tag{8.5}
$$

Due to the guard distance r_g, all entries of $\mathbf{R}_{i,j,k}$ are bounded from above. This implies in particular $\|\mathbf{R}_{i,j,k}\| \le r_\mathrm{g}^{-\alpha}$ for all i, j, k; a property that will be of importance for the asymptotic analysis later on.

8.2.1   Parameters for numerical results

The numerical results in Sections 8.4, 8.5, and 8.6 are based on the same scenario. We consider a system with $B = 4$ BSs located at the positions $(-1, 0)$, $(0, -1)$, $(1, 0)$, and $(0, 1)$, respectively. We assume $N = 12$ antennas per BS and $K = 6$ UTs per cell. The UTs are placed on a circle of radius $2/3$ around each BS with an angular separation of $\pi/3$. We consider three cases: no cooperation ($C = 4$), full cooperation ($C = 1$), and clusters of size $C = 2$. Since the setup is perfectly symmetric, it is sufficient to consider the two cluster assignments $\mathcal{C}_1 = \{1, 2\}$, $\mathcal{C}_2 = \{3, 4\}$ and $\mathcal{C}_1 = \{1, 3\}$, $\mathcal{C}_2 = \{2, 4\}$. We further assume a path-loss exponent $\alpha = 3.7$ and compute average statistics from all cells by Monte Carlo simulations, which are compared to our analytical results.

In Section 8.7, we adopt a different model and assume that the UTs in cell i are not fixed, but randomly uniformly distributed in a disk of radius R_i centered at BS i (see Figure 8.1). We consider a system with $B = 4$ BSs located at the positions $(-1, 0)$, $(0, -1)$, $(1, 0)$, and $(0, 1)$, respectively. We assume $N = 12$ antennas per BS and $K = 6$ UTs per cell without BS cooperation ($C = 4$). We denote by $D_{i,j}$ the distance between BS i and j and assume a guard distance $r_\mathrm{g} = 0.1$ around each BS and radius $R_i = 1$ for all i. The path-loss exponent is assumed to be $\alpha = 3.7$. Average statistics over different user distributions and channel realizations from all cells are then obtained by Monte Carlo simulations and compared to our analytical results.

8.3 Background on random matrix theory

This section provides some background on random matrix theory and should enable the reader to follow the derivations in the subsequent sections. It might be skipped at a first read and used as a reference whenever needed later on. Due to space limitations and for the sake of simplicity the treatment is at many points superficial and we refer the interested reader to [16] for a thorough introduction to random matrix theory in wireless communications.

The main goal of our large system analysis based on random matrix theory is to provide deterministic approximations of information-theoretic quantities, such as the mutual information or achievable rates with linear detection or precoding, which are intractable to compute for finite system dimensions. These approximations, from now on called deterministic equivalents (DEs), become increasingly tight as the system dimensions increase. Surprisingly, DEs are often very accurate for realistic system sizes and can hence provide valuable insight about the most important system parameters without the need for simulations. Moreover, DEs can be used to simplify optimization problems that would be otherwise intractable. The justification of this approach is that if a certain parameter is optimal for an infinitely large system, it is likely to be close to optimal for a large finite-sized system. In several cases, this approach has been shown to yield very close to optimal results for realistic system dimensions, see e.g. [13–15].

In the following, we will use "$N \to \infty$" to denote that $N, K \to \infty$, such that $0 < \liminf \frac{K}{N} \le \limsup \frac{K}{N} < \infty$. A special case of this condition is $\lim K/N \to c \in (0, \infty)$. We will use "$a_N \asymp b_N$" to denote that $a_N - b_N \xrightarrow[N\to\infty]{\text{a.s.}} 0$ for two infinite series of complex-valued (random) numbers a_N and b_N. In more mathematical terms, we define a DE as follows:

Definition 8.1 (Deterministic equivalent) *Let a_N be an infinite sequence of complex-valued random variables and b_N be an infinite sequence of complex numbers. We say that b_N is a "deterministic equivalent" of a_N, if*

$$a_N - b_N \xrightarrow[N\to\infty]{\text{a.s.}} 0.$$

Sometimes, both a_N and b_N are deterministic and we will somewhat misuse the term DE here to denote that $a_N - b_N \to 0$ as $N \to \infty$. This is for example the case if a_N is the expected value of the elements of a random sequence. With the help of a simple example, we will now introduce the most important lemmas and theorems that are necessary to understand the proofs in the following sections. To this end, let us assume that we would like to find a DE $\bar{\gamma}_k$ of the quantity γ_k, defined as

$$\gamma_k = \mathbf{h}_k^{\mathrm{H}} \left(\mathbf{H}\mathbf{H}^{\mathrm{H}} + \rho \mathbf{I}_N \right)^{-1} \mathbf{h}_k \tag{8.6}$$

where $\rho > 0$ and $\mathbf{H} \in \mathbb{C}^{N \times K}$ has column vectors $\mathbf{h}_k = \frac{1}{\sqrt{K}} \mathbf{R}_k^{\frac{1}{2}} \mathbf{u}_k$, where $\mathbf{R}_k \in \mathbb{C}^{N \times N}$ are deterministic, and $\mathbf{u}_k \sim \mathcal{CN}(\mathbf{0}, \mathbf{I}_N)$. We assume that $\|\mathbf{R}_k\| < R_{\max}$ for all k. The quantity γ_k arises, for example, in the computation of the signal to interference plus noise ratio (SINR) at the output of the minimum mean squared error (MMSE) detector

[17]. The intuition behind the large system analysis is that, if the dimensions of $\mathbf{H}$ are sufficiently large, γ_k will behave with high probability like a deterministic quantity $\bar{\gamma}_k$. Although γ_k is random, its fluctuations around $\bar{\gamma}_k$ become very small and can be ignored. The first lemma allows us to write γ_k as the ratio of two quadratic forms.

Lemma 8.1 ([20, Eq. (2.2)]) *Let* $\mathbf{A} \in \mathbb{C}^{N \times N}$ *be Hermitian invertible. Then, for any vector* $\mathbf{x} \in \mathbb{C}^N$ *and any scalar* $\tau \in \mathbb{C}$ *such that* $\mathbf{A} + \tau \mathbf{x}\mathbf{x}^H$ *is invertible,*

$$\mathbf{x}^H \left(\mathbf{A} + \tau \mathbf{x}\mathbf{x}^H \right)^{-1} = \frac{\mathbf{x}^H \mathbf{A}^{-1}}{1 + \tau \mathbf{x}^H \mathbf{A}^{-1} \mathbf{x}}.$$

Since $\mathbf{H}\mathbf{H}^H = \sum_{j=1}^{K} \mathbf{h}_j \mathbf{h}_j^H$, we have

$$\mathbf{h}_k^H \left(\mathbf{H}\mathbf{H}^H + \rho \mathbf{I}_N \right)^{-1} \mathbf{h}_k = \frac{\mathbf{h}_k^H \left(\mathbf{H}\mathbf{H}^H - \mathbf{h}_k \mathbf{h}_k^H + \rho \mathbf{I}_N \right)^{-1} \mathbf{h}_k}{1 + \mathbf{h}_k^H \left(\mathbf{H}\mathbf{H}^H - \mathbf{h}_k \mathbf{h}_k^H + \rho \mathbf{I}_N \right)^{-1} \mathbf{h}_k}. \tag{8.7}$$

Note that the matrix $\left(\mathbf{H}\mathbf{H}^H - \mathbf{h}_k \mathbf{h}_k^H + \rho \mathbf{I}_N \right)^{-1}$ is now independent of $\mathbf{h}_k$. As a next step, we will seek to find DEs of the denominator and numerator of the right hand side (RHS) of (8.7), respectively. To this end, we will make use of the following lemma, which is essential for most of the results in this chapter:

Lemma 8.2 ([18, Lemma 2.7], [19, Lemma 4], [16, Theorem 3.7]) *Let* $\mathbf{A} \in \mathbb{C}^{N \times N}$ *be a random matrix with almost surely uniformly bounded spectral norm (with respect to* N). *Let* $\mathbf{x} \in \mathbb{C}^N$ *and* $\mathbf{y} \in \mathbb{C}^N$ *be two random vectors of independent and identically distributed (i.i.d.) entries with zero mean, variance* $1/K$, *and finite* 8*th order moment, both independent of* $\mathbf{A}$. *Then,*

$$(i) \quad \mathbf{x}^H \mathbf{A} \mathbf{x} - \frac{1}{K} \operatorname{tr} \mathbf{A} \xrightarrow[N \to \infty]{a.s.} 0 \quad \text{and} \quad (ii) \quad \mathbf{x}^H \mathbf{A} \mathbf{y} \xrightarrow[N \to \infty]{a.s.} 0.$$

Remark 8.1 *The assumption of finite* 8*th order moment of the entries of the vectors* $\mathbf{x}$ *and* $\mathbf{y}$ *in Lemma 8.2 is rather technical. For most distributions of practical relevance, such as* $\mathbf{X} \sim \mathcal{CN} \left(0, \frac{1}{K} \mathbf{I}_N \right)$, *this condition is satisfied.*

Applying Lemma 8.2 (i) to the numerator of (8.7), we obtain

$$\mathbf{h}_k^H \left(\mathbf{H}\mathbf{H}^H - \mathbf{h}_k \mathbf{h}_k^H + \rho \mathbf{I}_N \right)^{-1} \mathbf{h}_k = \frac{1}{K} \mathbf{u}_k^H \mathbf{R}_k^{\frac{1}{2}} \left(\mathbf{H}\mathbf{H}^H - \mathbf{h}_k \mathbf{h}_k^H + \rho \mathbf{I}_N \right)^{-1} \mathbf{R}_k^{\frac{1}{2}} \mathbf{u}_k$$

$$\asymp \frac{1}{K} \operatorname{tr} \mathbf{R}_k \left(\mathbf{H}\mathbf{H}^H - \mathbf{h}_k \mathbf{h}_k^H + \rho \mathbf{I}_N \right)^{-1}. \tag{8.8}$$

Here we have used the fact that the matrix $\mathbf{R}_k^{\frac{1}{2}} \left(\mathbf{H}\mathbf{H}^H - \mathbf{h}_k \mathbf{h}_k^H + \rho \mathbf{I}_N \right)^{-1} \mathbf{R}_k^{\frac{1}{2}}$ is independent of $\mathbf{u}_k$ and its spectral norm is upper bounded by $R_{\max}/\rho$. Part (ii) of Lemma 8.2 will be needed later to show that the scalar products $\mathbf{h}_j^H \mathbf{h}_k$, $k \neq j$, asymptotically vanish. We proceed by showing that the rank-1 perturbation $\mathbf{h}_k \mathbf{h}_k^H$ of $\mathbf{H}\mathbf{H}^H$ is asymptotically negligible in the trace of (8.8). For this, we need the following result:

Lemma 8.3 ([20]) *Let $z > 0$, $\mathbf{A} \in \mathbb{C}^{N \times N}$, $\mathbf{B} \in \mathbb{C}^{N \times N}$ with $\mathbf{B}$ Hermitian non-negative definite, and $\mathbf{v} \in \mathbb{C}^N$. Then,*

$$\left| \operatorname{tr} \mathbf{A} \left((\mathbf{B} + z\mathbf{I}_N)^{-1} - (\mathbf{B} + \mathbf{v}\mathbf{v}^H + z\mathbf{I}_N)^{-1} \right) \right| \leq \frac{\|\mathbf{A}\|}{z}.$$

If $\lim \sup_N \|\mathbf{A}\| < \infty$, *this implies in particular*

$$\left| \frac{1}{N} \operatorname{tr} \mathbf{A}(\mathbf{B} + z\mathbf{I}_N)^{-1} - \frac{1}{N} \operatorname{tr} \mathbf{A}(\mathbf{B} + \mathbf{v}\mathbf{v}^H + z\mathbf{I}_N)^{-1} \right| \xrightarrow[N \to \infty]{} 0.$$

As a consequence of Lemma 8.3 and (8.8), we have

$$\mathbf{h}_k^H \left(\mathbf{H}\mathbf{H}^H - \mathbf{h}_k\mathbf{h}_k^H + \rho\mathbf{I}_N \right)^{-1} \mathbf{h}_k \asymp \frac{1}{K} \operatorname{tr} \mathbf{R}_k \left(\mathbf{H}\mathbf{H}^H + \rho\mathbf{I}_N \right)^{-1}. \tag{8.9}$$

The reader will have noticed that the RHS of the last equation is still a random quantity and we need another result before we can establish a DE of γ_k.

Theorem 8.1 ([19, Theorem 1]) *Let $\mathbf{D} \in \mathbb{C}^{\beta_1 N \times \beta_1 N}$ and $\mathbf{S} \in \mathbb{C}^{\beta_1 N \times \beta_1 N}$ be Hermitian non-negative definite matrices and let $\mathbf{H} \in \mathbb{C}^{\beta_1 N \times \beta_2 K}$ be a random matrix with columns $\mathbf{h}_k = \frac{1}{\sqrt{K}} \mathbf{R}_k^{\frac{1}{2}} \mathbf{u}_k$, where $\beta_1, \beta_2 \in (0, \infty)$, $\mathbf{u}_k \in \mathbb{C}^N$ are random vectors of i.i.d. elements with zero mean, unit variance, and finite 8th order moment, and $\mathbf{R}_k \in \mathbb{C}^{\beta_1 N \times \beta_1 N}$ are deterministic covariance matrices. Assume that the spectral norms of $\mathbf{D}$, $\mathbf{S}$ and the matrices $\mathbf{R}_k$, $k = 1, \ldots, \beta_2 K$, are uniformly bounded with respect to N. Then, for any $\rho > 0$,*

$$\frac{1}{K} \operatorname{tr} \mathbf{D} \left(\mathbf{H}\mathbf{H}^H + \mathbf{S} + \rho\mathbf{I}_N \right)^{-1} - \frac{1}{K} \operatorname{tr} \mathbf{D}\mathbf{T}(\rho) \xrightarrow[N \to \infty]{a.s.} 0$$

where $\mathbf{T}(\rho) \in \mathbb{C}^{\beta_1 N \times \beta_1 N}$ is defined as

$$\mathbf{T}(\rho) = \left(\frac{1}{K} \sum_{k=1}^{\beta_2 K} \frac{\mathbf{R}_k}{1 + \delta_k(\rho)} + \mathbf{S} + \rho\mathbf{I}_{\beta_1 N} \right)^{-1}$$

and $\delta_1(\rho), \ldots, \delta_{\beta_2 K}(\rho)$ are the unique positive solution to the following set of coupled fixed-point equations

$$\delta_k(\rho) = \frac{1}{K} \operatorname{tr} \mathbf{R}_k \mathbf{T}(\rho), \qquad 1 \leq k \leq \beta_2 K. \tag{8.10}$$

A direct application of Theorem 8.1 to (8.9) for $\beta_1 = \beta_2 = 1$, $\mathbf{S} = \mathbf{0}$ yields

$$\mathbf{h}_k^H \left(\mathbf{H}\mathbf{H}^H - \mathbf{h}_k\mathbf{h}_k^H + \rho\mathbf{I}_N \right)^{-1} \mathbf{h}_k \asymp \frac{1}{K} \operatorname{tr} \mathbf{R}_k \mathbf{T}(\rho). \tag{8.11}$$

We would like to remark that $\mathbf{T}(\rho)$ in Theorem 8.1 can be very efficiently computed by a standard fixed-point algorithm that iteratively evaluates the equation (8.10) starting from some arbitrary initial values $\delta_1^{(0)}, \ldots, \delta_{\beta_2 K}^{(0)} > 0$. Since the numerator and denominator of the RHS of (8.7) differ only by the constant term 1, finding a DE of the denominator does not require any additional work. However, we need one more lemma to ensure that the ratio of both DEs is also a DE of γ_k:

Lemma 8.4 ([21, Lemma 1]) *Denote* a_N, $\bar{a}_N$, b_N, *and* $\bar{b}_N$ *four infinite sequences of complex random variables indexed by* N *and assume* $a_N \asymp \bar{a}_N$ *and* $b_N \asymp \bar{b}_N$. *If* $|a_N|$, $|\bar{b}_N|$ *and/or* $|\bar{a}_N|$,$|b_N|$ *are almost surely uniformly bounded above over* N, *then* $a_N b_N \asymp \bar{a}_N \bar{b}_N$. *Similarly, if* $|a_N|$, $|\bar{b}_N|^{-1}$ *and/or* $|\bar{a}_N|$,$|b_N|^{-1}$ *are almost surely uniformly bounded above over* N, *then* $a_N/b_N \asymp \bar{a}_N/\bar{b}_N$.

Since the DE of the numerator of γ_k satisfies $|\frac{1}{K}\operatorname{tr}\mathbf{R}_k\mathbf{T}(\rho)| \leq N R_{\max}/(K\rho)$ and the denominator of γ_k satisfies $|1 + \mathbf{h}_k^{\mathrm{H}}\left(\mathbf{H}\mathbf{H}^{\mathrm{H}} - \mathbf{h}_k\mathbf{h}_k^{\mathrm{H}} + \rho\mathbf{I}_N\right)^{-1}\mathbf{h}_k|^{-1} \geq 1$, it follows from Lemma 8.4 that

$$\gamma_k \asymp \bar{\gamma}_k \triangleq \frac{\frac{1}{K}\operatorname{tr}\mathbf{R}_k\mathbf{T}(\rho)}{1 + \frac{1}{K}\operatorname{tr}\mathbf{R}_k\mathbf{T}(\rho)}. \tag{8.12}$$

This concludes the derivation of a DE of γ_k. We would like to remark at this point that we have in addition to the almost sure convergence $\gamma_k - \bar{\gamma}_k \to 0$, also convergence in the mean, i.e.,

$$\mathbb{E}\left[|\gamma_k - \bar{\gamma}_k|\right] \xrightarrow[N\to\infty]{} 0. \tag{8.13}$$

Although not explicitly shown here, this result follows from the fact that the proofs of Lemma 8.2 and Theorem 8.1 are based on a combination of the first Borel–Cantelli lemma and the Markov inequality, which establish not only almost sure but also convergence in the 4th-order mean. For further details, we refer the reader to [16, Section 3.2.1]. As a consequence, we obtain the following auxiliary result

$$|\mathbb{E}\left[\log\left(1 + \gamma_k\right)\right] - \log\left(1 + \bar{\gamma}_k\right)| \leq \mathbb{E}\left[|\gamma_k - \bar{\gamma}_k|\right] \xrightarrow[N\to\infty]{} 0 \tag{8.14}$$

where the inequality is due to $|\mathbb{E}[a - b]| \leq \mathbb{E}[|a - b|]$ and $|\log(1 + a) - \log(1 + b)| \leq |a - b|$ for $a, b \geq 0$. Thus, we can also use the DE of γ_k to provide an asymptotically tight approximation of the associated ergodic rate $\mathbb{E}[\log(1 + \gamma_k)]$.

We will present now two more theorems of importance for the asymptotic performance analysis in the subsequent sections. The first can be used to provide a tight approximation of the normalized ergodic mutual information.

Theorem 8.2 ([22, Theorem 2.3]) *Under the same conditions as in Theorem 8.1,*

$$\mathbb{E}\left[\frac{1}{N}\log \det\left(\mathbf{I}_{\beta_1 N} + \rho\mathbf{S} + \rho\mathbf{H}\mathbf{H}^H\right)\right] - \bar{C}_N \xrightarrow[N\to\infty]{} 0$$

where

$$\bar{C}_N = \frac{1}{N}\log \det\left(\mathbf{I}_{\beta_1 N} + \rho\mathbf{S} + \frac{\rho}{K}\sum_{k=1}^{\beta_2 K}\frac{\mathbf{R}_k}{1 + \delta_k(1/\rho)}\right)$$

$$+ \frac{1}{N}\sum_{k=1}^{\beta_2 K}\log\left(1 + \delta_k(1/\rho)\right) - \frac{\delta_k(1/\rho)}{1 + \delta_k(1/\rho)}$$

and $\delta_1(1/\rho), \ldots, \delta_{\beta_2 K}(1/\rho)$ *as given by Theorem 8.1.*

The next theorem will turn out to be very useful for the study of linear receivers and precoders with imperfect CSI (see Section 8.5).

Theorem 8.3 ([19]) *Let* $\boldsymbol{\Theta} \in \mathbb{C}^{\beta_1 N \times \beta_1 N}$, $\beta_1 \in (0, \infty)$, *be a Hermitian positive definite matrix with uniformly bounded spectral norm (with respect to N). Let the matrices* $\mathbf{D}$, $\mathbf{S}$, $\mathbf{H}$, *and* $\mathbf{R}_k$ *for all k be as defined in Theorem 8.1. Then, for any $\rho > 0$,*

$$\frac{1}{K} \operatorname{tr} \mathbf{D} \left(\mathbf{H}\mathbf{H}^H + \mathbf{S} + \rho \mathbf{I}_N \right)^{-1} \boldsymbol{\Theta} \left(\mathbf{H}\mathbf{H}^H + \mathbf{S} + \rho \mathbf{I}_N \right)^{-1} - \frac{1}{K} \operatorname{tr} \mathbf{D} \mathbf{T}'(\rho) \xrightarrow[N \to \infty]{a.s.} 0$$

where $\mathbf{T}'(\rho) \in \mathbb{C}^{\beta_1 N \times \beta_1 N}$ *is defined as*

$$\mathbf{T}'(\rho) = \mathbf{T}(\rho)\boldsymbol{\Theta}\mathbf{T}(\rho) + \mathbf{T}(\rho)\frac{1}{K} \sum_{k=1}^{\beta_2 K} \frac{\mathbf{R}_k \delta_k'(\rho)}{(1 + \delta_k(\rho))^2} \mathbf{T}(\rho)$$

with $\mathbf{T}(\rho), \delta_1(\rho), \ldots, \delta_{\beta_2 K}(\rho)$ *given by Theorem 8.1 and* $\boldsymbol{\delta}'(\rho) = \left[\delta_1'(\rho), \ldots, \delta_{\beta_2 K}'(\rho) \right]^T$ *defined as*

$$\boldsymbol{\delta}'(\rho) = (\mathbf{I}_K - \mathbf{J}(\rho))^{-1} \mathbf{v}(\rho)$$

where $\mathbf{J}(\rho) \in \mathbb{C}^{\beta_2 K \times \beta_2 K}$ *and* $\mathbf{v}(\rho) = [v_1(\rho), \ldots, v_{\beta_2 K}(\rho)]^T \in \mathbb{C}^K$ *satisfy*

$$[\mathbf{J}(\rho)]_{kl} = \frac{\frac{1}{K}\operatorname{tr} \mathbf{R}_k \mathbf{T}(\rho) \mathbf{R}_l \mathbf{T}(\rho)}{K (1 + \delta_l(\rho))^2}, \qquad 1 \le k, l \le \beta_2 K$$

$$v_k(\rho) = \frac{1}{K} \operatorname{tr} \mathbf{R}_k \mathbf{T}(\rho) \boldsymbol{\Theta} \mathbf{T}(\rho), \qquad 1 \le k \le \beta_2 K.$$

8.4   Perfect channel state information

8.4.1   Uplink: mutual information and performance of linear receivers

In this section, we will focus on the uplink when full CSI is available at all BSs and UTs. The uplink performance with imperfect CSI is considered later in Section 8.5.2. The received signal $\mathbf{y}_i \in \mathcal{C}^{N_i}$ at the BSs of cluster i at a given time instant reads

$$\mathbf{y}_i = \sqrt{\rho}\mathbf{G}_{i,i}\mathbf{x}_i + \sqrt{\rho} \sum_j \mathbf{G}_{i,j}\mathbf{x}_j + \mathbf{n}_i \tag{8.15}$$

where $\mathbf{G}_{i,j} \in \mathbb{C}^{N_i \times K_j}$ are defined in (8.3), $\mathbf{x}_i = [x_{i,1}, \ldots, x_{i,K_i}]^T \sim \mathcal{CN}\left(\mathbf{0}, \mathbf{I}_{K_i}\right)$, $i = 1, \ldots, C$, is the message-bearing transmit vector of the UTs in the cells of cluster i, $\mathbf{n}_i \sim \mathcal{CN}(\mathbf{0}, \mathbf{I}_{N_i})$ is a noise vector, and $\rho > 0$ is the average transmit power per UT. The normalized instantaneous mutual information $\mathcal{I}_i$ between $\mathbf{x}_i$ and $\mathbf{y}_i$ is given as [23]

$$
\mathcal{I}_i = \frac{1}{N} \log \det \left(\mathbf{I}_{N_i} + \left(\frac{1}{\rho} \mathbf{I}_{N_i} + \sum_{j \neq i} \mathbf{G}_{i,j} \mathbf{G}_{i,j}^{\mathrm{H}} \right)^{-1} \mathbf{G}_{i,i} \mathbf{G}_{i,i}^{\mathrm{H}} \right)
$$

$$
= \frac{1}{N} \log \det \left(\mathbf{I}_{N_i} + \rho \sum_j \mathbf{G}_{i,j} \mathbf{G}_{i,j}^{\mathrm{H}} \right) - \frac{1}{N} \log \det \left(\mathbf{I}_{N_i} + \rho \sum_{j \neq i} \mathbf{G}_{i,j} \mathbf{G}_{i,j}^{\mathrm{H}} \right).
$$

$$(8.16)$$

We have decided to normalize the mutual information by N instead of N_i to be independent of the cluster assignment. The normalized ergodic mutual information I_i is then simply defined as the expectation of $\mathcal{I}_i$ with respect to the matrices $\mathbf{G}_{i,j}$, i.e.,

$$
I_i = \mathbb{E}\left[\mathcal{I}_i\right]
$$

$$(8.17)$$

and we denote by $I = \sum_i I_i$ the sum of the normalized ergodic mutual information over all clusters. Note that both terms in (8.16) correspond to the random matrix model as assumed in Theorem 8.2 for $\mathbf{S} = 0$. Thus, we can directly apply the theorem to both terms individually to obtain a DE of I_i.

Theorem 8.4 (Mutual information with perfect CSI)

$$
I_i - \bar{I}_i \xrightarrow[N \to \infty]{} 0
$$

where $\bar{I}_i = |\mathcal{C}_i|(\bar{I}_{i,1} + \bar{I}_{i,2})$, *with* $\bar{I}_{i,1} = \bar{C}_N$ *as given by Theorem 8.2 for* $\mathbf{H} = [\mathbf{G}_{i,1} \cdots \mathbf{G}_{i,C}]$ *and* $\mathbf{S} = 0$ *and* $\bar{I}_{i,2} = \bar{C}_N$ *as given by Theorem 8.2 for* $\mathbf{H} = [\mathbf{G}_{i,1} \cdots \mathbf{G}_{i,i-1} \mathbf{G}_{i,i+1} \cdots \mathbf{G}_{i,C}]$ *and* $\mathbf{S} = 0$.

We would like to remark that the factor $|\mathcal{C}_i| = N_i/N$ is needed in the theorem to ensure the correct normalization $1/N$ instead of $1/N_i$. In order to validate Theorem 8.4, let us consider the simple scenario as detailed in Section 8.2.1. Figure 8.2 shows I as obtained by Monte Carlo simulations and its deterministic approximation $\bar{I} = \sum_i \bar{I}_i$ as given by Theorem 8.4 versus the transmit signal to noise ratio (SNR) ρ for four different assignments of BSs to clusters. Surprisingly, there is an almost perfect match between both results over the full range of ρ. This is a first example that demonstrates the accuracy of DEs for realistic system dimensions. As expected, full cooperation achieves the best performance while no cooperation achieves the worst. Under the first cluster assignment $\mathcal{C}_1 = \{1, 2\}$ and $\mathcal{C}_2 = \{3, 4\}$, the distance between the cooperative BSs is equal to $\sqrt{2}$. Under the second cluster assignment $\mathcal{C}_1 = \{1, 3\}$ and $\mathcal{C}_2 = \{2, 4\}$, this distance is equal to 2. Our results therefore indicate that cooperation between the nearest BSs achieves the largest gains.

Let us now consider the performance of linear detectors in the same setup. We assume that the BSs of cluster i obtain estimates $\hat{x}_{i,k}$ of the transmitted symbols of its UTs by computing the inner products between the linear receive filters $\mathbf{r}_{i,k} \in \mathbb{C}^{N_i}$ and the

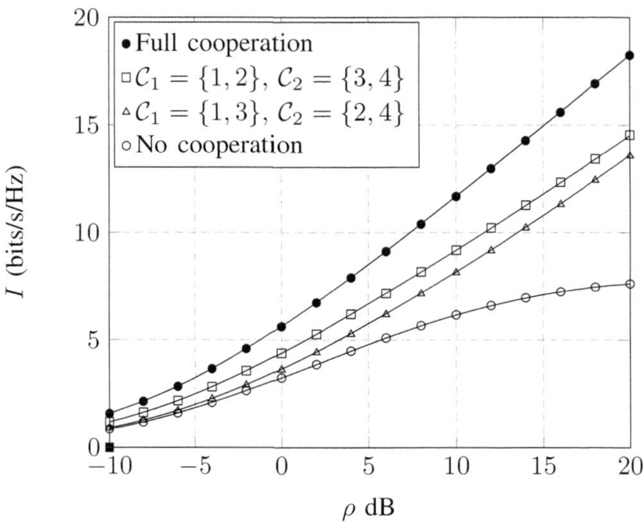

Figure 8.2 Sum of the normalized ergodic mutual information of all clusters I (markers) and its deterministic approximation $\bar{I}$ (solid lines) versus transmit SNR ρ under four different cluster assignments.

received signal vector $\mathbf{y}_i$, i.e., $\hat{x}_{i,k} = \mathbf{r}_{i,k}^{\mathsf{H}}\mathbf{y}_i$. The associated SINR $\gamma_{i,k}$ is given by

$$\gamma_{i,k} = \frac{\left|\mathbf{r}_{i,k}^{\mathsf{H}}\mathbf{g}_{i,i,k}\right|^2}{\mathbf{r}_{i,k}^{\mathsf{H}}\left(\frac{1}{\rho}\mathbf{I}_{N_i} + \sum_{(j,l)\neq(i,k)}\mathbf{g}_{i,j,l}\mathbf{g}_{i,j,l}^{\mathsf{H}}\right)\mathbf{r}_{i,k}} \tag{8.18}$$

and the associated ergodic rate $R_{i,k}$ is defined as

$$R_{i,k} = \mathbb{E}\left[\log\left(1 + \gamma_{i,k}\right)\right]. \tag{8.19}$$

We denote by $R_{\mathrm{sum},i}$ the normalized ergodic sum-rate of cluster i, given by

$$R_{\mathrm{sum},i} = \frac{1}{N}\sum_{k=1}^{K_i} R_{i,k} \tag{8.20}$$

and denote the normalized ergodic sum-rate over all clusters by $R_{\mathrm{sum}} = \sum_i R_{\mathrm{sum},i}$. Thus, R_{sum} can be seen as the counterpart to I for linear detectors. We will consider two receive filters $\mathbf{r}_{i,k}$ of practical interest. These are the matched filter (MF) $\mathbf{r}_{i,k}^{\mathrm{MF}}$ and the MMSE detector $\mathbf{r}_{i,k}^{\mathrm{MMSE}}$, which are respectively defined as [17]

$$\mathbf{r}_{i,k}^{\mathrm{MF}} = \mathbf{g}_{i,i,k} \tag{8.21}$$

$$\mathbf{r}_{i,k}^{\mathrm{MMSE}} = \left(\sum_j \mathbf{G}_{i,j}\mathbf{G}_{i,j}^{\mathsf{H}} + \frac{1}{\rho}\mathbf{I}_{N_i}\right)^{-1}\mathbf{g}_{i,i,k}. \tag{8.22}$$

The resulting SINR is respectively given for both detectors by

$$\gamma_{i,k}^{\mathrm{MF}} = \frac{\|\mathbf{g}_{i,i,k}\|^4}{\frac{\|\mathbf{g}_{i,i,k}\|^2}{\rho} + \sum_{(j,l)\neq(i,k)} \left|\mathbf{g}_{i,i,k}^{\mathrm{H}}\mathbf{g}_{i,j,l}\right|^2} \tag{8.23}$$

and

$$\gamma_{i,k}^{\mathrm{MMSE}} = \mathbf{g}_{i,i,k}^{\mathrm{H}} \left(\sum_j \mathbf{G}_{i,j}\mathbf{G}_{i,j}^{\mathrm{H}} - \mathbf{g}_{i,i,k}\mathbf{g}_{i,i,k}^{\mathrm{H}} + \frac{1}{\rho}\mathbf{I}_{N_i} \right)^{-1} \mathbf{g}_{i,i,k}. \tag{8.24}$$

With the techniques presented in Section 8.3, it is now easy to derive DEs of $\gamma_{i,k}^{\mathrm{MF}}$ and $\gamma_{i,k}^{\mathrm{MMSE}}$. These results are summarized in the following two theorems.

Theorem 8.5 (Matched filter with perfect CSI) $\gamma_{i,k}^{MF} - \bar{\gamma}_{i,k}^{MF} \xrightarrow[N\to\infty]{a.s.} 0$, *where*

$$\bar{\gamma}_{i,k}^{MF} = \frac{\left|\frac{1}{K}\operatorname{tr}\mathbf{R}_{i,i,k}\right|^2}{\frac{1}{\rho K}\operatorname{tr}\mathbf{R}_{i,i,k} + \sum_{(j,l)\neq(i,k)}\frac{1}{K^2}\operatorname{tr}\mathbf{R}_{i,i,k}\mathbf{R}_{i,j,l}}.$$

Proof. From Lemma 8.2, we have $\|\mathbf{g}_{i,i,k}\|^2 \asymp \frac{1}{K}\operatorname{tr}\mathbf{R}_{i,i,k}$ and $\left|\mathbf{g}_{i,i,k}^{\mathrm{H}}\mathbf{g}_{i,j,l}\right|^2 \asymp \frac{1}{K^2}\operatorname{tr}\mathbf{R}_{i,i,k}\mathbf{R}_{i,j,l}$ for $(j,l) \neq (i,k)$. Applying Lemma 8.4 concludes the proof. $\square$

Theorem 8.6 (MMSE detector with perfect CSI) $\gamma_{i,k}^{MMSE} - \bar{\gamma}_{i,k}^{MMSE} \xrightarrow[N\to\infty]{a.s.} 0$, *where*

$$\bar{\gamma}_{i,k}^{MMSE} = \frac{1}{K}\operatorname{tr}\mathbf{R}_{i,i,k}\mathbf{T}_i$$

and $\mathbf{T}_i = \mathbf{T}(1/\rho)$ *is given by Theorem 8.1 for* $\mathbf{H} = [\mathbf{G}_{i,1} \cdots \mathbf{G}_{i,C}]$, $\mathbf{S} = \mathbf{0}$.

Proof. The proof follows directly from Lemma 8.2, Lemma 8.3, and Theorem 8.1. $\square$

Due to (8.14), DEs of the ergodic rates $R_{i,k}$ with both receive filters (and also $R_{\mathrm{sum},i}$ and R_{sum}) can be directly obtained by replacing $\gamma_{i,k}$ in (8.19) by $\bar{\gamma}_{i,k}^{MF}$ and $\bar{\gamma}_{i,k}^{MMSE}$, respectively. Figure 8.3 depicts the achievable sum-rates R_{sum} for both detectors and their DEs as a function of ρ. Here, we only show results for the cluster assignment $\mathcal{C}_1 = \{1, 2\}$, $\mathcal{C}_2 = \{3, 4\}$. Although slightly less accurate as for the mutual information, the DEs also provide tight approximations of the system performance with linear detectors.

8.4.2 Downlink: performance of linear precoders

During downlink transmissions, the received baseband signal $\mathbf{y}_i = [y_{i,1}, \ldots, y_{i,K_i}]^{\mathrm{T}} \in \mathbb{C}^{K_i}$ at the UTs of cluster i at a given time instant reads

$$\mathbf{y}_i = \sqrt{\lambda_i}\mathbf{G}_{i,i}^{\mathrm{H}}\mathbf{F}_i\mathbf{x}_i + \sum_j \sqrt{\lambda_j}\mathbf{G}_{j,i}^{\mathrm{H}}\mathbf{F}_j\mathbf{x}_j + \mathbf{n}_i \tag{8.25}$$

where $\mathbf{x}_i = [x_{i,1}, \ldots, x_{i,K_i}]^{\mathrm{T}} \sim \mathcal{CN}\left(\mathbf{0}, \mathbf{I}_{K_i}\right)$, $i = 1, \ldots, C$, is the message-bearing signal vector for the UTs in the cells of cluster i, $\mathbf{F}_i \in \mathbb{C}^{N_i \times K_i}$ is a precoding matrix,

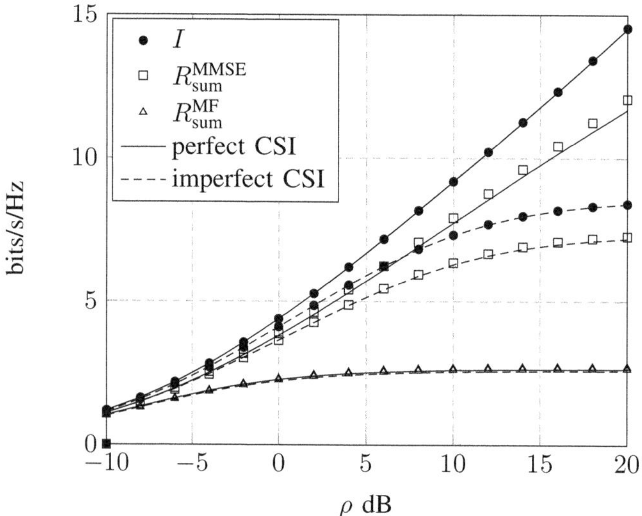

Figure 8.3 Uplink sum rate of all clusters R_{sum} (markers) and its deterministic approximation (solid lines) versus transmit SNR ρ for the cluster assignment $\mathcal{C}_1 = \{1, 2\}, \mathcal{C}_2 = \{3, 4\}$.

$\mathbf{n}_i \sim \mathcal{CN}\left(\mathbf{0}, \mathbf{I}_{K_i}\right)$ is a noise vector and λ_i is a power normalization factor, defined as

$$\lambda_i = \frac{|\mathcal{C}_i| K P}{\operatorname{tr} \mathbf{F}_i \mathbf{F}_i^{\mathrm{H}}} \tag{8.26}$$

where $P > 0$ is the average transmit power per BS per UT. Thus, the average transmit power of cluster i equals $\mathbb{E}[\lambda_i \mathbf{x}_i^{\mathrm{H}} \mathbf{F}_i^{\mathrm{H}} \mathbf{F}_i \mathbf{x}_i] = |\mathcal{C}_i| K P$. Note that the total transmit power scales linearly with the number of UTs. Without this assumption, the BSs would need to serve more and more UTs with the same total power and the individual rates would vanish as $K \to \infty$. For simplicity, we have imposed a sum-power constraint on each cluster and not an individual power constraint on each of the BSs. For more details on this topic, we refer the reader to the textbook [8]. Denote by $\mathbf{f}_{i,k} \in \mathbb{C}^{N_i}$ the kth column vector of $\mathbf{F}_i$. Then, the received SINR $\gamma_{i,k}$ related to symbol $x_{i,k}$ at UT k in the ith cluster can be expressed as

$$\gamma_{i,k} = \frac{\lambda_i \left|\mathbf{g}_{i,i,k}^{\mathrm{H}} \mathbf{f}_{i,k}\right|^2}{1 + \sum_{(j,l) \neq (i,k)} \lambda_j \left|\mathbf{g}_{j,i,k}^{\mathrm{H}} \mathbf{f}_{j,l}\right|^2}. \tag{8.27}$$

As for the uplink, we define the achievable rate of UT k in cell i as $R_{i,k} = \log(1 + \gamma_{i,k})$, denote by $R_{\mathrm{sum},i} = \frac{1}{N} \sum_k R_{i,k}$ the normalized sum-rate of cell i and define $R_{\mathrm{sum}} = \sum_i R_{\mathrm{sum},i}$. We consider two classical precoding techniques, eigenbeamforming (EBF) and regularized zero-forcing (RZF). Under EBF, the matrices $\mathbf{F}_i$ are given as

$$\mathbf{F}_{\mathrm{BF},i} = \mathbf{G}_{ii}. \tag{8.28}$$

Under RZF, the precoding matrices $\mathbf{F}_i$ are defined as

$$\mathbf{F}_{\mathrm{RZF},i} = \left(\mathbf{G}_{ii} \mathbf{G}_{ii}^{\mathrm{H}} + \mathbf{S}_i + \mu_i \mathbf{I}_{N_i}\right)^{-1} \mathbf{G}_{ii} \tag{8.29}$$

where $\mu_i > 0$ is a regularization parameter and $\mathbf{S}_i \in \mathbb{C}^{N_i \times N_i}$ is an arbitrary deterministic Hermitian non-negative matrix. Extensions to random $\mathbf{S}_i$, such as $\mathbf{S}_i = \sum_{j \neq i} \mathbf{G}_{ij} \mathbf{G}_{ij}^{\mathrm{H}}$ or a scaled version of this matrix, are also possible but not considered here. We denote by $\gamma_{i,k}^{\mathrm{BF}}$ and $\gamma_{i,k}^{\mathrm{RZF}}$ the SINR under EBF and RZF, respectively. The following theorems provide DEs of both quantities. We refer the reader to the proofs of Theorem 8.12 and 8.13 in Section 8.5.3, which provide DEs of achievable rates for both precoding strategies with imperfect CSI. With full CSI ($\tau = 0$), the theorems coincide.

Theorem 8.7 (Eigenbeamforming with perfect CSI) $\quad \gamma_{i,k}^{BF} - \bar{\gamma}_{i,k}^{BF} \xrightarrow[N \to \infty]{a.s.} 0$, where

$$\bar{\gamma}_{i,k}^{BF} = \frac{\bar{\lambda}_i \left| \frac{1}{K} \operatorname{tr} \mathbf{R}_{i,i,k} \right|^2}{1 + \sum_{(j,l) \neq (i,k)} \bar{\lambda}_j \frac{1}{K^2} \operatorname{tr} \mathbf{R}_{j,i,k} \mathbf{R}_{j,j,l}} \tag{8.30}$$

with

$$\bar{\lambda}_i = \frac{|\mathcal{C}_i| P}{\frac{1}{K} \sum_{k=1}^{K_i} \frac{1}{K} \operatorname{tr} \mathbf{R}_{i,i,k}}, \qquad 1 \leq i \leq C. \tag{8.31}$$

Theorem 8.8 (Regularized zero-forcing with perfect CSI) $\quad \gamma_{i,k}^{RZF} - \bar{\gamma}_{i,k}^{RZF} \xrightarrow[N \to \infty]{a.s.} 0$, where

$$\bar{\gamma}_{i,k}^{RZF} = \frac{\bar{\lambda}_i \delta_{i,k}^2}{(1 + \delta_{i,k})^2 + \frac{1}{K} \sum_{\substack{l=1 \\ l \neq k}}^{K_i} \frac{\bar{\lambda}_i \vartheta_{i,l,i,k}}{(1 + \delta_{i,l})^2} + \frac{1}{K} \sum_{\substack{j=1 \\ j \neq i}}^{C} \sum_{l=1}^{K_j} \frac{\bar{\lambda}_j \vartheta_{j,l,i,k} (1 + \delta_{i,k})^2}{(1 + \delta_{j,l})^2}}$$

with

$$\bar{\lambda}_i = |\mathcal{C}_i| P \left(\frac{1}{K} \sum_{l=1}^{K_i} \frac{\frac{1}{K} \operatorname{tr} \mathbf{R}_{i,i,l} \mathbf{T}_i'}{(1 + \delta_{i,l})^2} \right)^{-1}$$

$$\delta_{i,l} = \frac{1}{K} \operatorname{tr} \mathbf{R}_{i,i,l} \mathbf{T}_i$$

$$\vartheta_{j,l,i,k} = \frac{1}{K} \operatorname{tr} \mathbf{R}_{j,i,k} \mathbf{T}_{j,l}'$$

and where

- $\mathbf{T}_i = \mathbf{T}(\mu_i)$ *as given by Theorem 8.1 for* $\mathbf{H} = \mathbf{G}_{i,i}$ *and* $\mathbf{S} = \mathbf{S}_i$,
- $\mathbf{T}_i' = \mathbf{T}'(\mu_i)$ *as given by Theorem 8.3 for* $\mathbf{H} = \mathbf{G}_{i,i}$, $\mathbf{S} = \mathbf{S}_i$ *and* $\boldsymbol{\Theta} = \mathbf{I}_{N_i}$,
- $\mathbf{T}_{j,l}' = \mathbf{T}'(\mu_i)$ *as given by Theorem 8.3 for* $\mathbf{H} = \mathbf{G}_{j,j}$, $\mathbf{S} = \mathbf{S}_j$ *and* $\boldsymbol{\Theta} = \mathbf{R}_{j,j,l}$.

Based on the DEs of the SINR with both precoders and (8.14), we can compute DEs of the achievable rates by replacing $\gamma_{i,k}$ in $R_{i,k}$ by $\bar{\gamma}_{i,k}^{\mathrm{BF}}$ and $\bar{\gamma}_{i,k}^{\mathrm{RZF}}$, respectively. In Figure 8.4 we show the achievable sum-rates R_{sum} with both precoders as a function of the transmit power P for the setup as described in Section 8.2.1. We consider the cluster assignment $\mathcal{C}_1 = \{1, 2\}$, $\mathcal{C}_2 = \{3, 4\}$ and assume RZF with $\mu_i = 1/P$ and $\mathbf{S}_i = \mathbf{0}$ for all i. The deterministic approximations match the simulation results well over the full range of P. However, the approximations of RZF become less accurate at high values of P. At this point we would like to remind readers that our example scenario has rather

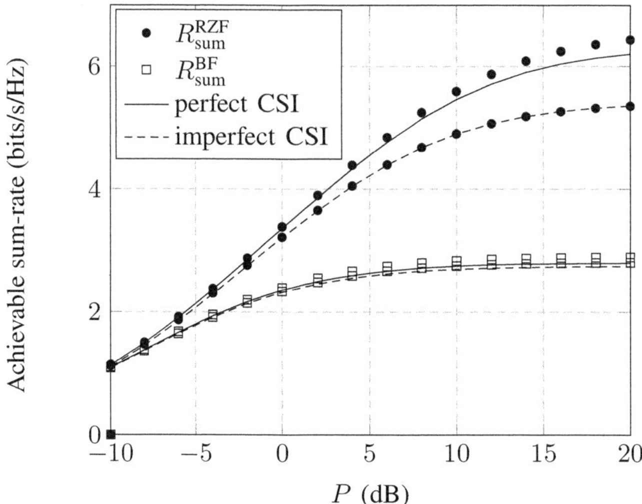

Figure 8.4 Downlink sum-rate of all clusters R_{sum} (markers) and deterministic approximations (solid lines) versus transmit power P for cluster assignment $\mathcal{C}_1 = \{1, 2\}, \mathcal{C}_2 = \{3, 4\}$.

small dimensions with only $N = 12$ antennas per BS and $K = 6$ UTs per cell. The approximations quickly become more accurate when N, K grow.

8.5 Imperfect channel state information

In this section, we will extend our previous results to the case where the BSs and the UTs have only estimates of their channels. We will first provide some background on how achievable rates with some classes of linear precoders and detectors under imperfect CSI can be obtained. Using the tools described in Section 8.3, we will then derive deterministic equivalents of these rates for the MF and MMSE detectors in the uplink, and EBF and RZF precoders in the downlink.

8.5.1 On achievable rates with imperfect channel state information

Consider the following multiple input multiple output (MIMO) channel with output $\mathbf{y}$:

$$\mathbf{y} = \mathbf{f}(\mathbf{x}, \mathcal{W})$$

where $\mathbf{f}$ is a deterministic function of the input $\mathbf{x} \sim \mathcal{CN}(\mathbf{0}, \mathbf{K}_{xx})$ and some set of possibly random parameters $\mathcal{W}$. For example, we could choose $\mathcal{W} = \{\mathbf{H}, \mathbf{n}\}$, where $\mathbf{H} \in \mathbb{C}^{N \times K}$ is a standard complex Gaussian matrix representing a fading channel and $\mathbf{n} \sim \mathcal{CN}(\mathbf{0}, \mathbf{I}_N)$ is noise, and consider the MIMO channel $\mathbf{f}(\mathbf{x}, \mathcal{W}) = \mathbf{Hx} + \mathbf{n}$. Under the assumption that the receiver has some side information $\hat{\mathcal{W}}$ about $\mathcal{W}$, we would like to compute the ergodic mutual information $I(\mathbf{y}; \mathbf{x}|\hat{\mathcal{W}})$ between $\mathbf{y}$ and $\mathbf{x}$ given $\hat{\mathcal{W}}$. The side information $\hat{\mathcal{W}}$ could correspond, for example, to some partial channel state information at the receiver. For most practical cases, $I(\mathbf{y}; \mathbf{x}|\hat{\mathcal{W}})$ is not known, but a lower bound and, thus, an achievable rate can be found according to the following steps [24, 25] (see also [26]):

$$I\left(\mathbf{y}; \mathbf{x}\,\middle|\,\hat{\mathcal{W}}\right) \overset{(a)}{=} h(\mathbf{x}) - h\left(\mathbf{x}\,\middle|\,\mathbf{y}, \hat{\mathcal{W}}\right)$$

$$\overset{(b)}{=} \log\det\left(\pi e \mathbf{K}_{xx}\right) - h\left(\mathbf{x} - \boldsymbol{\Psi}\mathbf{y}\,\middle|\,\mathbf{y}, \hat{\mathcal{W}}\right)$$

$$\overset{(c)}{\geq} \log\det\left(\pi e \mathbf{K}_{xx}\right) - h\left(\mathbf{x} - \boldsymbol{\Psi}\mathbf{y}\,\middle|\,\hat{\mathcal{W}}\right)$$

$$\overset{(d)}{\geq} \log\det\left(\pi e \mathbf{K}_{xx}\right) - \mathbb{E}\left[\log\det\left(\pi e \mathbb{E}\left[(\mathbf{x} - \boldsymbol{\Psi}\mathbf{y})(\mathbf{x} - \boldsymbol{\Psi}\mathbf{y})^{\mathrm{H}}\,\middle|\,\hat{\mathcal{W}}\right]\right)\right]$$

where (a) is the definition of the mutual information and we denote by $h(\cdot)$ the differential entropy, (b) follows since subtracting a known term does not change the entropy ($\boldsymbol{\Psi}$ is an arbitrary matrix that depends only on $\hat{\mathcal{W}}$), (c) follows since conditioning reduces the entropy, and (d) is obtained by upper bounding the entropy of $\mathbf{x} - \boldsymbol{\Psi}\mathbf{y}$ by the entropy of a complex Gaussian random variable with the same covariance matrix. Note that we have implicitly assumed $\mathbb{E}[\mathbf{y}|\hat{\mathcal{W}}] = \mathbf{0}$. Since the last lower bound holds for any choice of $\boldsymbol{\Psi}$, we can tighten this bound by choosing $\boldsymbol{\Psi} = \boldsymbol{\Psi}_{\mathrm{MMSE}}$ to be the linear MMSE estimator of $\mathbf{x}$ given $\mathbf{y}$ and $\hat{\mathcal{W}}$ [17]:

$$\mathbb{E}\left[(\mathbf{x} - \boldsymbol{\Psi}_{\mathrm{MMSE}}\mathbf{y})(\mathbf{x} - \boldsymbol{\Psi}_{\mathrm{MMSE}}\mathbf{y})^{\mathrm{H}}\,\middle|\,\hat{\mathcal{W}}\right] = \mathbf{K}_{xx} - \mathbf{K}_{xy}\mathbf{K}_{yy}^{-1}\mathbf{K}_{yx}$$

where $\mathbf{K}_{xy} = \mathbb{E}[\mathbf{x}\mathbf{y}^{\mathrm{H}}|\hat{\mathcal{W}}]$, $\mathbf{K}_{yx} = \mathbb{E}[\mathbf{y}\mathbf{x}^{\mathrm{H}}|\hat{\mathcal{W}}]$ and $\mathbf{K}_{yy} = \mathbb{E}[\mathbf{y}\mathbf{y}^{\mathrm{H}}|\hat{\mathcal{W}}]$. Thus,

$$I(\mathbf{y}; \mathbf{x}|\hat{\mathcal{W}}) \geq \log\det\left(\mathbf{K}_{xx}\right) - \mathbb{E}\left[\log\det\left(\mathbf{K}_{xx} - \mathbf{K}_{xy}\mathbf{K}_{yy}^{-1}\mathbf{K}_{yx}\right)\right]. \tag{8.32}$$

Consider now our exemplary MIMO channel $\mathbf{y} = \mathbf{f}(\mathbf{x}, \mathcal{W}) = \mathbf{H}\mathbf{x} + \mathbf{n}$ and assume that $\hat{\mathcal{W}} = \{\hat{\mathbf{H}}\}$ is an estimate of $\mathbf{H}$, satisfying $\mathbf{H} = \hat{\mathbf{H}} + \tilde{\mathbf{H}}$, where the estimation error $\tilde{\mathbf{H}}$ is independent of $\hat{\mathbf{H}}$. Using the above definitions, we have $\mathbf{K}_{xy} = \mathbf{K}_{xx}\hat{\mathbf{H}}^{\mathrm{H}}$, $\mathbf{K}_{yx} = \hat{\mathbf{H}}\mathbf{K}_{xx}$ and $\mathbf{K}_{yy} = \hat{\mathbf{H}}\mathbf{K}_{xx}\hat{\mathbf{H}}^{\mathrm{H}} + \mathbb{E}\left[\tilde{\mathbf{H}}\mathbf{K}_{xx}\tilde{\mathbf{H}}^{\mathrm{H}}\right] + \mathbf{I}_N$. From (8.32), we have after straightforward calculations

$$I(\mathbf{y}; \mathbf{x}|\hat{\mathcal{W}}) \geq \mathbb{E}\left[\log\det\left(\mathbf{I}_N + \left(\mathbf{I}_N + \mathbb{E}\left[\tilde{\mathbf{H}}\mathbf{K}_{xx}\tilde{\mathbf{H}}^{\mathrm{H}}\right]\right)^{-1}\hat{\mathbf{H}}\mathbf{K}_{xx}\hat{\mathbf{H}}^{\mathrm{H}}\right)\right].$$

Similarly, we can obtain achievable rates for different kinds of channels with varying degrees of channel state information. As another example, consider the channel $y = \mathbf{f}(x, \mathcal{W}) = \mathbf{r}^{\mathrm{H}}(\mathbf{h}_0 x + \mathbf{H}\mathbf{z} + \mathbf{n})$, where $\mathbf{r} \in \mathbb{C}^N$, $\mathbf{h}_0 \sim \mathcal{CN}(\mathbf{0}, \mathbf{I}_N)$, $x \sim \mathcal{CN}(0, \rho)$, $\mathbf{z} \sim \mathcal{CN}(\mathbf{0}, \rho\mathbf{I}_K)$ and $\mathbf{H}$, $\mathbf{n}$ as already defined above. Thus, we have $\mathcal{W} = \{\mathbf{r}, \mathbf{h}_0, \mathbf{H}, \mathbf{z}, \mathbf{n}\}$. Here, $\mathbf{r}$ can be seen as a linear receive filter and $\mathbf{H}\mathbf{z}$ as multi-user interference. Assume now that the receiver has estimates $\hat{\mathbf{H}}$ and $\hat{\mathbf{h}}_0$ of $\mathbf{H}$ and $\mathbf{h}_0$, respectively, and that $\tilde{\mathbf{H}}$ and $\tilde{\mathbf{h}}_0$ are the independent estimation errors, i.e., $\hat{\mathcal{W}} = \{\mathbf{r}, \hat{\mathbf{H}}, \hat{\mathbf{h}}_0\}$. From (8.32), one obtains after a few lines of calculus

$$I(y; x|\hat{\mathcal{W}}) \geq \mathbb{E}\left[\log\left(1 + \frac{\left|\mathbf{r}^{\mathrm{H}}\hat{\mathbf{h}}_0\right|^2}{\mathbf{r}^{\mathrm{H}}\left(\frac{1}{\rho}\mathbf{I}_N + \mathbb{E}\left[\tilde{\mathbf{H}}\tilde{\mathbf{H}}^{\mathrm{H}} + \tilde{\mathbf{h}}_0\tilde{\mathbf{h}}_0^{\mathrm{H}}\right] + \hat{\mathbf{H}}\hat{\mathbf{H}}^{\mathrm{H}}\right)\mathbf{r}}\right)\right]. \tag{8.33}$$

8.5.2　Uplink: achievable rates and performance of linear receivers

We assume now that the BSs have only estimates $\hat{\mathbf{G}}_{i,j}$ of the channel matrices $\mathbf{G}_{i,j}$. These could be either obtained through downlink training and feedback in frequency division duplex (FDD) systems [27] or through uplink training in time division duplex (TDD) systems [14, 26]. For the sake of clarity and ease of presentation, we assume that $\mathbf{g}_{i,j,k}$ can be decomposed into the channel estimate $\hat{\mathbf{g}}_{i,j,k}$ and the independent estimation error $\tilde{\mathbf{g}}_{i,j,k}$, such that

$$\mathbf{g}_{i,j,k} = \hat{\mathbf{g}}_{i,j,k} + \tilde{\mathbf{g}}_{i,j,k}. \tag{8.34}$$

For simplicity, we model both vectors as

$$\hat{\mathbf{g}}_{i,j,k} = \sqrt{\frac{1-\tau}{K}}\mathbf{R}_{i,j,k}^{\frac{1}{2}}\hat{\mathbf{u}}_{i,j,k}, \quad \tilde{\mathbf{g}}_{i,j,k} = \sqrt{\frac{\tau}{K}}\mathbf{R}_{i,j,k}^{\frac{1}{2}}\tilde{\mathbf{u}}_{i,j,k} \tag{8.35}$$

where $\hat{\mathbf{u}}_{i,j,k}, \tilde{\mathbf{u}}_{i,j,k} \sim \mathcal{CN}\left(\mathbf{0}, \mathbf{I}_{N_i}\right)$ and $\tau \in [0,1]$ reflects the accuracy of the channel estimates. In particular, for $\tau = 0$, we have perfect CSI and for $\tau = 1$, the channel is completely unknown. The extension to different parameters $\tau_{i,j,k}$ for each UT and cluster is straightforward. Channel estimates and errors could also be assumed to have different covariance matrices (see e.g., [12]). Let $\hat{\mathbf{G}}_{i,j} = \left[\hat{\mathbf{g}}_{i,j,1} \cdots \hat{\mathbf{g}}_{i,j,K_j}\right]$ and $\tilde{\mathbf{G}}_{i,j} = \left[\tilde{\mathbf{g}}_{i,j,1} \cdots \tilde{\mathbf{g}}_{i,j,K_j}\right]$. Then, under the new channel model, we can rewrite the received vector $\mathbf{y}_i$ at the BSs of cluster i as (cf. (8.15))

$$\mathbf{y}_i = \sqrt{\rho}\hat{\mathbf{G}}_{i,i}\mathbf{x}_i + \sqrt{\rho}\sum_{j\neq i}\hat{\mathbf{G}}_{i,j}\mathbf{x}_j + \sqrt{\rho}\sum_{j}\tilde{\mathbf{G}}_{i,j}\mathbf{x}_j + \mathbf{n}_i. \tag{8.36}$$

Applying (8.36) to (8.32) leads to the following lower bound $\hat{I}_i$ on the normalized ergodic mutual information $\frac{1}{N}I(\mathbf{y}_i; \mathbf{x}_i | \hat{\mathbf{G}}_{i,1}, \ldots, \hat{\mathbf{G}}_{i,C})$:

$$\hat{I}_i = \mathbb{E}\left[\frac{1}{N}\log\det\left(\mathbf{I}_{N_i} + \mathbf{\Theta}_i^{-1}\hat{\mathbf{G}}_{i,i}\hat{\mathbf{G}}_{i,i}^{\mathrm{H}}\right)\right]$$

$$= \mathbb{E}\left[\frac{1}{N}\log\det\left(\mathbf{I}_{N_i} + \rho\sum_{j}\tilde{\mathbf{Z}}_{i,j} + \rho\sum_{j}\hat{\mathbf{G}}_{i,j}\hat{\mathbf{G}}_{i,j}^{\mathrm{H}}\right)\right]$$

$$- \mathbb{E}\left[\frac{1}{N}\log\det\left(\mathbf{I}_{N_i} + \rho\sum_{j}\tilde{\mathbf{Z}}_{i,j} + \rho\sum_{j\neq i}\hat{\mathbf{G}}_{i,j}\hat{\mathbf{G}}_{i,j}^{\mathrm{H}}\right)\right] \tag{8.37}$$

where

$$\mathbf{\Theta}_i = \frac{1}{\rho}\mathbf{I}_{N_i} + \sum_{j\neq i}\hat{\mathbf{G}}_{i,j}\hat{\mathbf{G}}_{i,j}^{\mathrm{H}} + \sum_{j}\tilde{\mathbf{Z}}_{i,j} \tag{8.38}$$

$$\tilde{\mathbf{Z}}_{i,j} = \mathbb{E}\left[\tilde{\mathbf{G}}_{i,j}\tilde{\mathbf{G}}_{i,j}^{\mathrm{H}}\right] = \frac{\tau}{K}\sum_{k=1}^{K_j}\mathbf{R}_{i,j,k}. \tag{8.39}$$

The only difference with respect to the case of perfect CSI (8.17) is that all channels have been replaced by their estimates and the interference covariance matrix contains the additional term $\sum_j \tilde{\mathbf{Z}}_{i,j}$, which is due to channel estimation errors. We can now establish a DE $\hat{\bar{I}}_i$ of $\hat{I}_i$ by applying Theorem 8.2 to both terms in (8.37).

Theorem 8.9 (Mutual information lower bound with imperfect CSI)

$$\hat{I}_i - \hat{\bar{I}}_i \xrightarrow[N\to\infty]{} 0$$

where $\hat{\bar{I}}_i = |\mathcal{C}_i|(\hat{\bar{I}}_{i,1} - \hat{\bar{I}}_{i,2})$, *where* $\hat{\bar{I}}_{i,1} = \bar{C}_N$ *as given by Theorem 8.2 for* $\mathbf{H} = [\hat{\mathbf{G}}_{i,1} \cdots \hat{\mathbf{G}}_{i,C}]$ *and* $\mathbf{S} = \sum_j \tilde{\mathbf{Z}}_{i,j}$ *and* $\hat{\bar{I}}_{i,2} = \bar{C}_N$ *as given by Theorem 8.2 for* $\mathbf{H} = [\hat{\mathbf{G}}_{i,1} \cdots \hat{\mathbf{G}}_{i,i-1}\hat{\mathbf{G}}_{i,i+1} \cdots \hat{\mathbf{G}}_{i,C}]$ *and* $\mathbf{S} = \sum_j \tilde{\mathbf{Z}}_{i,j}$.

Next we will establish achievable rates for general linear detectors $\hat{\mathbf{r}}_{i,k} \in \mathbb{C}^{N_i}$ with imperfect CSI. Assume that the ith cluster estimates the signal of the kth UT in its cell based on the observation:

$$y_{i,k} = \hat{\mathbf{r}}_{i,k}^{\mathrm{H}}\mathbf{y}_i$$

$$= \sqrt{\rho}\hat{\mathbf{r}}_{i,k}^{\mathrm{H}}\left(\hat{\mathbf{g}}_{i,i,k}x_{i,k} + \sum_{l\neq k}\hat{\mathbf{g}}_{i,i,l}x_{i,l} + \sum_{j\neq i}\hat{\mathbf{G}}_{i,j}\mathbf{x}_j + \sum_j \tilde{\mathbf{G}}_{i,j}\mathbf{x}_j + \frac{1}{\sqrt{\rho}}\mathbf{n}_i\right). \quad (8.40)$$

Direct application of (8.32), leads to the following lower bound $\hat{R}_{i,k}$ of the ergodic mutual information $I(y_{i,k}; x_{i,k} | \hat{\mathbf{G}}_{i,1}, \ldots, \hat{\mathbf{G}}_{i,C})$:

$$\hat{R}_{i,k} = \mathbb{E}\left[\log\left(1 + \hat{\gamma}_{i,k}\right)\right] \quad (8.41)$$

where

$$\hat{\gamma}_{i,k} = \frac{\left|\hat{\mathbf{r}}_{i,k}^{\mathrm{H}}\hat{\mathbf{g}}_{i,i,k}\right|^2}{\hat{\mathbf{r}}_{i,k}^{\mathrm{H}}\left(\boldsymbol{\Theta}_i + \sum_{j\neq k}\hat{\mathbf{g}}_{i,i,j}\hat{\mathbf{g}}_{i,i,j}^{\mathrm{H}}\right)\hat{\mathbf{r}}_{i,k}} \quad (8.42)$$

where $\boldsymbol{\Theta}_i$ is defined in (8.38). Under MF and MMSE detection, the vectors $\hat{\mathbf{r}}_{i,k}$ are respectively given as

$$\hat{\mathbf{r}}_{i,k}^{\mathrm{MF}} = \hat{\mathbf{g}}_{i,i,k} \quad (8.43)$$

$$\hat{\mathbf{r}}_{i,k}^{\mathrm{MMSE}} = \left(\boldsymbol{\Theta}_i + \hat{\mathbf{G}}_{i,i}\hat{\mathbf{G}}_{i,i}^{\mathrm{H}}\right)^{-1}\hat{\mathbf{g}}_{i,i,k} \quad (8.44)$$

and we denote by $\hat{\gamma}_{i,k}^{\mathrm{MF}}$ and $\hat{\gamma}_{i,k}^{\mathrm{MMSE}}$ the related SINR. The next theorems provide DEs of the SINR for both detectors.

Theorem 8.10 (Matched filter with imperfect CSI) $\hat{\gamma}_{i,k}^{\mathrm{MF}} - \hat{\bar{\gamma}}_{i,k}^{\mathrm{MF}} \xrightarrow[N\to\infty]{a.s.} 0$, *where*

$$\hat{\bar{\gamma}}_{i,k}^{\mathrm{MF}} = \frac{(1-\tau)\left|\frac{1}{K}\mathrm{tr}\,\mathbf{R}_{i,i,k}\right|^2}{\frac{1}{\rho K}\mathrm{tr}\,\mathbf{R}_{i,i,k} + \frac{\tau}{K^2}\mathrm{tr}\,\mathbf{R}_{i,i,k}^2 + \sum_{(j,l)\neq(i,k)}\frac{1}{K^2}\mathrm{tr}\,\mathbf{R}_{i,i,k}\mathbf{R}_{i,j,l}}.$$

Proof. We apply Lemma 8.2 (i) to each term in (8.42) to obtain DEs of the denominator and numerator, respectively. We then use Lemma 8.4 to conclude the proof. $\square$

Theorem 8.11 (MMSE detector with imperfect CSI) $\hat{\gamma}_{i,k}^{MMSE} - \hat{\bar{\gamma}}_{i,k}^{MMSE} \xrightarrow[N\to\infty]{a.s.} 0$, *where*

$$\hat{\bar{\gamma}}_{i,k}^{MMSE} = \frac{1-\tau}{K} \operatorname{tr} \mathbf{R}_{i,i,k}\mathbf{T}_i \tag{8.45}$$

where $\mathbf{T}_i = \mathbf{T}(1/\rho)$ *is given by Theorem 8.1 for* $\mathbf{H} = [\hat{\mathbf{G}}_{i,1} \cdots \hat{\mathbf{G}}_{i,C}]$, $\mathbf{S} = \sum_j \tilde{\mathbf{Z}}_{i,j}$.

Proof. First notice that $\hat{\gamma}_{i,k}^{MMSE} = \hat{\mathbf{g}}_{i,i,k}^{H}(\boldsymbol{\Theta}_i + \sum_{j\neq k} \hat{\mathbf{g}}_{i,i,j}\hat{\mathbf{g}}_{i,i,j}^{H})^{-1}\hat{\mathbf{g}}_{i,i,k}$. Subsequent application of Lemma 8.2, Lemma 8.3, and Theorem 8.1 concludes the proof. $\qquad\square$

We show the achievable sum-rates $\hat{R}_{\text{sum}} = \sum_i \frac{1}{N} \sum_k \hat{R}_{i,k}$ with both detectors and the mutual information lower bound $\hat{I} = \sum_i \hat{I}_i$ as well as their deterministic approximations by Theorems 8.10, 8.11, and 8.9 in Figure 8.3. We have assumed the same system setup as for the numerical results of Section 8.4. The parameter τ was set to $\tau = 0.03$. As for the case of perfect CSI, the deterministic approximations match the simulations very well. Moreover, we can observe that imperfect CSI leads to a saturation of the achievable rates at high SNR. The impact of imperfect CSI on the performance of the matched filter is negligible. It becomes visible only for large values of τ.

8.5.3   Downlink: achievable rates with linear precoders

We assume now that the BS and the UTs in the ith cluster have an estimate $\hat{\mathbf{G}}_{i,i}$ of their local channel $\mathbf{G}_{i,i}$. The BSs compute their precoding matrices $\hat{\mathbf{F}}_i = [\hat{\mathbf{f}}_{i,1} \cdots \hat{\mathbf{f}}_{i,K_i}]$ as if the estimated channel was the real channel. The received signal $y_{i,k}$ at UT k in cell i is given as

$$y_{i,k} = \sqrt{\hat{\lambda}_i}\hat{\mathbf{g}}_{i,i,k}^{H}\hat{\mathbf{F}}_i\mathbf{x}_i + \sum_{j\neq i} \sqrt{\hat{\lambda}_j}\hat{\mathbf{g}}_{j,i,k}^{H}\hat{\mathbf{F}}_j\mathbf{x}_j + \sum_j \tilde{\mathbf{g}}_{j,i,k}^{H}\hat{\mathbf{F}}_j\mathbf{x}_j + \mathbf{n}_i. \tag{8.46}$$

Moreover, we assume that the UTs are aware of the precoding strategy and can hence compute the precoding matrices $\hat{\mathbf{F}}_i$ as well as the power normalization factors $\hat{\lambda}_i$ based on the knowledge of $\hat{\mathbf{G}}_{i,i}$. Under these assumptions, we can obtain the following lower bound $\hat{R}_{i,k}$ of the mutual information $I(y_{i,k}; x_{i,k} | \hat{\mathbf{G}}_{i,i,k})$ from (8.32):

$$\hat{R}_{i,k} = \mathbb{E}\left[\log\left(1 + \hat{\gamma}_{i,k}\right)\right] \tag{8.47}$$

where

$$\hat{\gamma}_{i,k} = \frac{\hat{\lambda}_i \left|\hat{\mathbf{g}}_{i,i,k}^{H}\hat{\mathbf{f}}_{i,k}\right|^2}{1 + \mathbb{E}\left[\hat{\lambda}_i \left|\tilde{\mathbf{g}}_{i,i,k}^{H}\hat{\mathbf{f}}_{i,k}\right|^2 + \sum_{(j,l)\neq(i,k)} \hat{\lambda}_j \left|\mathbf{g}_{j,i,k}^{H}\hat{\mathbf{f}}_{j,l}\right|^2 \Big| \hat{\mathbf{G}}_{i,i,k}\right]}. \tag{8.48}$$

Note that we could have also assumed that each UT only knows an estimate of its own channel $\hat{\mathbf{g}}_{i,i,k}$, the precoding vector $\hat{\mathbf{f}}_{i,k}$, and $\hat{\lambda}_i$. In this case, the conditional expectation in (8.48), would have been with respect to $\hat{\mathbf{g}}_{i,i,k}$, $\hat{\mathbf{f}}_{i,k}$ and $\hat{\lambda}_i$ instead of $\hat{\mathbf{G}}_{i,i}$. However, this assumption does not influence the asymptotic performance as we do not need to compute the conditional expectation explicitly. Thus, the same DEs hold under different assumptions on the CSI at the UTs. As for perfect CSI, the parameters $\hat{\lambda}_i = |\mathcal{C}_i| K P / \operatorname{tr} \hat{\mathbf{F}}_i \hat{\mathbf{F}}_i^{\mathrm{H}}$ are chosen to normalize the transmit power and the precoding matrices for EBF and RZF are respectively given as

$$\hat{\mathbf{F}}_{\mathrm{BF},i} = \hat{\mathbf{G}}_{i,i} \tag{8.49}$$

$$\hat{\mathbf{F}}_{\mathrm{RZF},i} = \left(\hat{\mathbf{G}}_{i,i} \hat{\mathbf{G}}_{i,i}^{\mathrm{H}} + \hat{\mathbf{S}}_i + \hat{\mu}_i \mathbf{I}_{N_i} \right)^{-1} \hat{\mathbf{G}}_{i,i}. \tag{8.50}$$

In the next two theorems, we provide DEs of the SINR $\hat{\gamma}_{i,k}$ for BF and RZF. Due to (8.14), these DEs can be directly used to provide DEs of the ergodic rates $\hat{R}_{i,k}$.

Theorem 8.12 (Eigenbeamforming with imperfect CSI) $\hat{\gamma}_{i,k}^{BF} - \hat{\bar{\gamma}}_{i,k}^{BF} \xrightarrow[N\to\infty]{a.s.} 0,$ *where*

$$\hat{\bar{\gamma}}_{i,k}^{BF} = \frac{(1-\tau)\hat{\bar{\lambda}}_i \left| \frac{1}{K} \operatorname{tr} \mathbf{R}_{i,i,k} \right|^2}{1 + \tau \hat{\bar{\lambda}}_i \frac{1}{K^2} \operatorname{tr} \mathbf{R}_{i,i,k}^2 + \sum_{(j,l)\neq(i,k)} \hat{\bar{\lambda}}_j \frac{1}{K^2} \operatorname{tr} \mathbf{R}_{j,i,k} \mathbf{R}_{j,j,l}}$$

with

$$\hat{\bar{\lambda}}_i = \frac{|\mathcal{C}_i| P}{\frac{(1-\tau)}{K} \sum_{k=1}^{K_i} \frac{1}{K} \operatorname{tr} \mathbf{R}_{i,i,k}}, \qquad 1 \le i \le C.$$

Proof. Let us first consider the power normalization factors $\hat{\lambda}_i$. As a simple consequence of Lemma 8.2, we have

$$\frac{1}{K} \operatorname{tr} \hat{\mathbf{F}}_{\mathrm{BF},i} \hat{\mathbf{F}}_{\mathrm{BF},i}^{\mathrm{H}} = \frac{1}{K} \sum_{k=1}^{K_i} \hat{\mathbf{g}}_{i,i,k}^{\mathrm{H}} \hat{\mathbf{g}}_{i,i,k} \asymp \frac{(1-\tau)}{K} \sum_{k=1}^{K_i} \frac{1}{K} \operatorname{tr} \mathbf{R}_{i,i,k}. \tag{8.51}$$

Thus, it follows from Lemma 8.4 that $\hat{\lambda}_i \asymp \hat{\bar{\lambda}}_i$ as given in the theorem. The terms inside the expectation in the denominator of $\hat{\gamma}_{i,k}$ converge almost surely to deterministic quantities, which can be easily found by applying Lemma 8.2 to the individual quadratic forms. In particular, we have for $(j,l) \neq (i,k)$

$$\left| \mathbf{g}_{j,i,k}^{\mathrm{H}} \hat{\mathbf{f}}_{\mathrm{BF},j,l} \right|^2 = \mathbf{g}_{j,i,k}^{\mathrm{H}} \hat{\mathbf{g}}_{j,j,l} \hat{\mathbf{g}}_{j,j,l}^{\mathrm{H}} \mathbf{g}_{j,i,k} \asymp \frac{1-\tau}{K^2} \operatorname{tr} \mathbf{R}_{j,i,k} \mathbf{R}_{j,j,l}. \tag{8.52}$$

Applying the same lemma to the remaining terms in denominator and numerator together with Lemma 8.4 concludes the proof. $\qquad\square$

Theorem 8.13 (RZF with imperfect CSI) $\hat{\gamma}_{i,k}^{RZF} - \hat{\bar{\gamma}}_{i,k}^{RZF} \xrightarrow[N\to\infty]{a.s.} 0,$ *where*

$$\hat{\bar{\gamma}}_{i,k}^{RZF} = \frac{\hat{\bar{\lambda}}_i \hat{\delta}_{i,k}^2}{\left(1 + \hat{\delta}_{i,k}\right)^2 + \tilde{v}_{i,k} + \frac{1-\tau}{K} \sum_{\substack{l=1 \\ l\neq k}}^{K_i} \frac{\hat{\bar{\lambda}}_i \hat{\vartheta}_{i,l,i,k}}{\left(1+\hat{\delta}_{i,l}\right)^2} + \frac{1-\tau}{K} \sum_{j\neq i} \sum_{l=1}^{K_j} \frac{\hat{\bar{\lambda}}_j \hat{\vartheta}_{j,l,i,k}\left(1+\hat{\delta}_{i,k}\right)^2}{\left(1+\hat{\delta}_{j,l}\right)^2}}$$

with

$$\hat{\bar{\lambda}}_i = |\mathcal{C}_i| P \left(\frac{1}{K} \sum_{l=1}^{K_i} \frac{\frac{1-\tau}{K} \operatorname{tr} \mathbf{R}_{i,i,l} \hat{\mathbf{T}}_i'}{\left(1 + \hat{\delta}_{i,l}\right)^2} \right)^{-1}$$

$$\tilde{v}_{i,k} = \frac{\tau \hat{\bar{\lambda}}_i \left(1 + \hat{\delta}_{i,k}\right)^2}{K} \sum_{l=1}^{K_i} \frac{\hat{\vartheta}_{i,li,k}}{\left(1 + \hat{\delta}_{i,l}\right)^2}$$

$$\hat{\delta}_{i,l} = \frac{1-\tau}{K} \operatorname{tr} \mathbf{R}_{i,i,l} \hat{\mathbf{T}}_i$$

$$\hat{\vartheta}_{j,l,i,k} = \frac{1-\tau}{K} \operatorname{tr} \mathbf{R}_{j,i,k} \hat{\mathbf{T}}_{j,l}'$$

and where

- $\hat{\mathbf{T}}_i = \mathbf{T}(\hat{\mu}_i)$ *as given by Theorem 8.1 for* $\mathbf{H} = \hat{\mathbf{G}}_{i,i}$, $\mathbf{S} = \hat{\mathbf{S}}_i$,
- $\hat{\mathbf{T}}_i' = \mathbf{T}'(\hat{\mu}_i)$ *as given by Theorem 8.3 for* $\mathbf{H} = \hat{\mathbf{G}}_{i,i}$, $\mathbf{S} = \hat{\mathbf{S}}_i$ *and* $\mathbf{\Theta} = \mathbf{I}_{N_i}$,
- $\hat{\mathbf{T}}_{j,l}' = \mathbf{T}'(\hat{\mu}_i)$ *as given by Theorem 8.3 for* $\mathbf{H} = \hat{\mathbf{G}}_{j,j}$, $\mathbf{S} = \hat{\mathbf{S}}_j$ *and* $\mathbf{\Theta} = \mathbf{R}_{j,j,l}$.

Proof. Let $\mathbf{\Delta}_j = \hat{\mathbf{G}}_{j,j} \hat{\mathbf{G}}_{j,j}^{\mathrm{H}} + \hat{\mathbf{S}}_j + \hat{\mu}_j \mathbf{I}_{N_j}$. Using the techniques described in Section 8.3, we can easily derive DEs of the power normalization terms $\hat{\lambda}_i$:

$$\frac{1}{K} \operatorname{tr} \hat{\mathbf{F}}_{\mathrm{RZF},i} \hat{\mathbf{F}}_{\mathrm{RZF},i}^{\mathrm{H}} = \frac{1}{K} \sum_{l=1}^{K_i} \hat{\mathbf{g}}_{i,i,l}^{\mathrm{H}} \mathbf{\Delta}_i^{-2} \hat{\mathbf{g}}_{i,i,l}$$

$$\overset{(a)}{=} \frac{1}{K} \sum_{l=1}^{K_i} \frac{\hat{\mathbf{g}}_{i,i,l}^{\mathrm{H}} (\mathbf{\Delta}_i - \hat{\mathbf{g}}_{i,i,l} \hat{\mathbf{g}}_{i,i,l}^{\mathrm{H}})^{-2} \hat{\mathbf{g}}_{i,i,l}}{\left(1 + \hat{\mathbf{g}}_{i,i,l}^{\mathrm{H}} (\mathbf{\Delta}_i - \hat{\mathbf{g}}_{i,i,l} \hat{\mathbf{g}}_{i,i,l}^{\mathrm{H}})^{-1} \hat{\mathbf{g}}_{i,i,l}\right)^2}$$

$$\overset{(b)}{\asymp} \frac{1}{K} \sum_{l=1}^{K_i} \frac{\frac{1-\tau}{K} \operatorname{tr} \mathbf{R}_{i,i,l} \mathbf{\Delta}_i^{-2}}{(1 + \frac{1-\tau}{K} \operatorname{tr} \mathbf{R}_{i,i,l} \mathbf{\Delta}_i^{-1})^2}$$

$$\overset{(c)}{\asymp} \frac{1}{K} \sum_{l=1}^{K_i} \frac{\frac{1-\tau}{K} \operatorname{tr} \mathbf{R}_{i,i,l} \hat{\mathbf{T}}_i'}{(1 + \frac{1-\tau}{K} \operatorname{tr} \mathbf{R}_{i,i,l} \hat{\mathbf{T}}_i)^2} \tag{8.53}$$

where (a) follows from applying Lemma 8.1 twice, (b) is a consequence of Lemmas 8.2, 8.3, and 8.4, and (c) follows from an application of Theorem 8.3 to the numerator, i.e., $\hat{\mathbf{T}}_i' = \mathbf{T}'(\hat{\mu}_i)$ as given by Theorem 8.3 for $\mathbf{H} = \hat{\mathbf{G}}_{i,i}$, $\mathbf{S} = \hat{\mathbf{S}}_i$, and $\mathbf{\Theta} = \mathbf{I}_{N_i}$, and an application of Theorem 8.1 to the denominator, i.e., $\hat{\mathbf{T}}_i = \mathbf{T}(\hat{\mu}_i)$ as given by Theorem 8.1 for $\mathbf{H} = \hat{\mathbf{G}}_{i,i}$, $\mathbf{S} = \hat{\mathbf{S}}_i$. From the same arguments, we have

$$\left| \hat{\mathbf{g}}_{i,i,k}^{\mathrm{H}} \hat{\mathbf{f}}_{\mathrm{RZF},i,k} \right|^2 = \left| \hat{\mathbf{g}}_{i,i,k}^{\mathrm{H}} \mathbf{\Delta}_i^{-1} \hat{\mathbf{g}}_{i,i,k} \right|^2 \asymp \frac{\hat{\delta}_{i,k}^2}{(1 + \hat{\delta}_{i,k})^2} \tag{8.54}$$

where $\hat{\delta}_{i,k} = \frac{1-\tau}{K}\operatorname{tr}\mathbf{R}_{i,i,k}\hat{\mathbf{T}}_i$. Consider now the terms $|\mathbf{g}_{j,i,k}^{\mathsf{H}}\hat{\mathbf{f}}_{\mathrm{RZF},j,l}|^2$ for $i \neq j$:

$$\left|\mathbf{g}_{j,i,k}^{\mathsf{H}}\hat{\mathbf{f}}_{\mathrm{RZF},j,l}\right|^2 = \mathbf{g}_{j,i,k}^{\mathsf{H}}\boldsymbol{\Delta}_j^{-1}\hat{\mathbf{g}}_{j,j,l}\hat{\mathbf{g}}_{j,j,l}^{\mathsf{H}}\boldsymbol{\Delta}_j^{-1}\mathbf{g}_{j,i,k}$$

$$\overset{(a)}{\asymp} \frac{1}{K}\hat{\mathbf{g}}_{j,j,l}^{\mathsf{H}}\boldsymbol{\Delta}_j^{-1}\mathbf{R}_{j,i,k}\boldsymbol{\Delta}^{-1}\hat{\mathbf{g}}_{j,j,l}$$

$$\overset{(b)}{\asymp} \frac{1}{K}\frac{\frac{1-\tau}{K}\operatorname{tr}\mathbf{R}_{j,i,k}\boldsymbol{\Delta}_j^{-1}\mathbf{R}_{j,j,l}\boldsymbol{\Delta}_j^{-1}}{(1+\hat{\delta}_{j,l})^2}$$

$$\overset{(c)}{\asymp} \frac{1}{K}\frac{\frac{1-\tau}{K}\operatorname{tr}\mathbf{R}_{j,i,k}\hat{\mathbf{T}}'_{j,l}}{(1+\hat{\delta}_{j,l})^2} \tag{8.55}$$

where (a) follows from Lemma 8.2, (b) follows from Lemmas 8.1, 8.2, 8.3, 8.4, and Theorem 8.1, and (c) is a consequence of Theorem 8.3 applied to the numerator, i.e., $\hat{\mathbf{T}}'_{j,l}$ is given by Theorem 8.3 for $\mathbf{H} = \hat{\mathbf{G}}_{j,j}$, $\mathbf{S} = \hat{\mathbf{S}}_j$, and $\boldsymbol{\Theta} = \mathbf{R}_{j,j,l}$. Denote by $\hat{\vartheta}_{j,l,i,k} = \frac{1-\tau}{K}\operatorname{tr}\mathbf{R}_{j,i,k}\hat{\mathbf{T}}'_{j,l}$. For $i = j$ but $k \neq l$, we obtain by the same arguments:

$$\left|\mathbf{g}_{i,i,k}^{\mathsf{H}}\hat{\mathbf{f}}_{\mathrm{RZF},i,l}\right|^2 = (\hat{\mathbf{g}}_{i,i,k} + \tilde{\mathbf{g}}_{i,i,k})^{\mathsf{H}}\boldsymbol{\Delta}_i^{-1}\hat{\mathbf{g}}_{i,i,l}\hat{\mathbf{g}}_{i,i,l}^{\mathsf{H}}\boldsymbol{\Delta}_i^{-1}(\hat{\mathbf{g}}_{i,i,k} + \tilde{\mathbf{g}}_{i,i,k})$$

$$\asymp \frac{1}{K}\frac{\hat{\vartheta}_{i,l,i,k}}{(1+\hat{\delta}_{i,k})^2(1+\hat{\delta}_{i,l})^2} + \frac{\tau}{K}\frac{\hat{\vartheta}_{i,l,i,k}}{(1+\hat{\delta}_{i,l})^2} \tag{8.56}$$

and also $|\tilde{\mathbf{g}}_{i,i,k}^{\mathsf{H}}\hat{\mathbf{f}}_{\mathrm{RZF},i,k}|^2 \asymp \frac{\tau}{K}\frac{\hat{\vartheta}_{i,k,i,k}}{(1+\hat{\delta}_{i,k})^2}$. Gathering all pieces and rearranging the terms concludes the proof. $\qquad\square$

Figure 8.4 shows the achievable sum-rates $\hat{R}_{\mathrm{sum}} = \frac{1}{N}\sum_i \sum_k \hat{R}_{i,k}$ with both precoders and the deterministic equivalent approximations by the last two theorems for $\tau = 0.03$ and the system setup as described in Section 8.2.1. We consider only the cluster assignment $\mathcal{C}_1 = \{1,2\}$, $\mathcal{C}_2 = \{3,4\}$. As in the previous examples, both results match very well.

8.6   Line-of-sight channels

In dense networks, it is likely that a UT is under line-of-sight conditions to one or more BSs [2]. This implies that the channel transfer matrices are not only composed of random fast fading but also deterministic or so-called *specular* components. An extension of our previous channel model to capture this behavior can be easily obtained by considering the new channel matrices $\bar{\mathbf{G}}_{ij} \in \mathbb{C}^{N_i \times K_j}$, defined as

$$\bar{\mathbf{G}}_{i,j} = \mathbf{G}_{i,j} + \mathbf{A}_{i,j} \tag{8.57}$$

where $\mathbf{G}_{i,j}$ is given by (8.4) and $\mathbf{A} \in \mathbb{C}^{N_i \times K_j}$ is a deterministic matrix. This channel model is also known as a Rician fading channel and has been studied for finite and infinite channel dimensions in several works [13, 28–30]. Most important for our applications is the following theorem:

Theorem 8.14 (Theorems 2.4 and 4.1 [30])　*Let* $\mathbf{H} = \mathbf{X} + \mathbf{A} \in \mathbb{C}^{\beta_1 N \times \beta_2 K}$, $\beta_1, \beta_2 \in (0, \infty)$, *be a random matrix where* $\mathbf{X}$ *has independent entries* x_{ij} *of zero mean, variance* σ_{ij}^2/K *and finite* $4 + \epsilon$ *order moment for some* $\epsilon > 0$, *and* $\mathbf{A}$ *is a deterministic matrix. Assume that the columns and rows of* $\mathbf{A}$ *have uniformly bounded Euclidean norm and that the* σ_{ij} *are uniformly bounded with respect to* N *and* K. *Then,*

$$\mathbb{E}\left[\frac{1}{N} \log \det \left(\mathbf{I}_N + \rho \mathbf{H}\mathbf{H}^H \right) \right] - \bar{V}_N \xrightarrow[N \to \infty]{} 0$$

where

$$\bar{V}_N = \frac{1}{N} \log \det \left(\rho \boldsymbol{\Psi}^{-1} + \mathbf{A} \tilde{\boldsymbol{\Psi}} \mathbf{A}^H \right) + \frac{1}{N} \log \det \left(\rho \tilde{\boldsymbol{\Psi}}^{-1} \right) - \frac{1}{\rho} \frac{1}{KN} \sum_{i,j} \sigma_{ij}^2 t_i \tilde{t}_j$$

where $\boldsymbol{\Psi} = diag(\psi_i)_{i=1}^{\beta_1 N}$ *and* $\tilde{\boldsymbol{\Psi}} = diag(\tilde{\psi}_j)_{j=1}^{\beta_2 K}$ *are defined as the unique positive solution to the set of* $\beta_1 N + \beta_2 K$ *implicit equations:*

$$\psi_i = \rho \left(1 + \frac{1}{K} \sum_{j=1}^{\beta_2 K} \sigma_{ij}^2 \tilde{t}_j \right)^{-1}, \qquad 1 \le i \le \beta_1 N$$

$$\tilde{\psi}_j = \rho \left(1 + \frac{1}{K} \sum_{i=1}^{\beta_1 N} \sigma_{ij}^2 t_i \right)^{-1}, \qquad 1 \le j \le \beta_2 K$$

where t_i *and* $\tilde{t}_j$ *are the* i-*th and* j-*th diagonal element of the matrices* $\mathbf{T} \in \mathbb{C}^{\beta_1 N \times \beta_1 N}$ *and* $\tilde{\mathbf{T}} \in \mathbb{C}^{\beta_2 K \times \beta_2 K}$, *respectively, which are defined as*

$$\mathbf{T} = \left(\boldsymbol{\Psi}^{-1} + \frac{1}{\rho} \mathbf{A} \tilde{\boldsymbol{\Psi}} \mathbf{A}^H \right)^{-1} \quad and \quad \tilde{\mathbf{T}}_j = \left(\tilde{\boldsymbol{\Psi}}^{-1} + \frac{1}{\rho} \mathbf{A}^H \boldsymbol{\Psi} \mathbf{A} \right)^{-1}.$$

We can directly apply the last theorem to provide a deterministic approximation of the normalized ergodic mutual information I_i^{Rice} of the uplink in Rician fading channels (defined as I_i in (8.17) for the new channel model in (8.57)) .

Theorem 8.15 (Mutual information for Rician fading channels)

$$I_i^{Rice} - \bar{I}_i^{Rice} \xrightarrow[N \to \infty]{} 0$$

where $\bar{I}_i^{Rice} = |\mathcal{C}_i|(\bar{I}_{i,1}^{Rice} - \bar{I}_{i,2}^{Rice})$, *where* $\bar{I}_{i,1}^{Rice} = \bar{V}_N$ *is given by Theorem 8.14 for* $\mathbf{H} = [\bar{\mathbf{G}}_{i,1} \cdots \bar{\mathbf{G}}_{i,C}]$ *and* $\bar{I}_{i,2}^{Rice} = \bar{V}_N$ *is given by Theorem 8.14 for* $\mathbf{H} = [\bar{\mathbf{G}}_{i,1} \cdots \bar{\mathbf{G}}_{i,i-1} \bar{\mathbf{G}}_{i,i+1} \cdots \bar{\mathbf{G}}_{i,C}]$.

We depict in Figure 8.5 the sum of the normalized ergodic mutual information of all clusters $I^{\text{Rice}} = \sum_i I_i^{\text{Rice}}$ and its deterministic approximation $\bar{I}^{\text{Rice}}$ based on Theorem 8.15 versus SNR ρ. The matrices $\mathbf{A}_{i,j}$ are drawn once at random according to the same distribution as the matrices $\mathbf{G}_{i,j}$, but then kept fixed during the simulations. The results are based on the setup as described in Section 8.2.1 and we compare the performance of all possible cluster assignments. The difference between the simulations and asymptotic approximations is negligible.

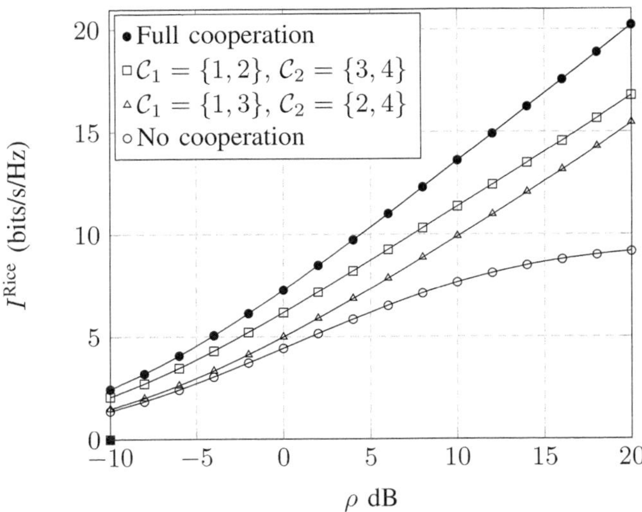

Figure 8.5 Sum of the normalized ergodic mutual information of all clusters I^{Rice} (markers) and its deterministic approximation $\bar{I}^{\mathrm{Rice}}$ (solid lines) versus SNR ρ.

The asymptotic analysis of Rician fading channels is so far more or less restricted to the capacity, see e.g., [13, 30], and results for linear receivers or precoders are rare. Exceptions are [31] and the very recently submitted work [32] (based on results in [33]), which treat the asymptotic performance of the MMSE receiver. However, due to space limitations, we will not discuss their results in detail.

8.7 Random user locations

In the preceding sections, we have tacitly assumed that the locations of the UTs are deterministic and, therefore, all of the previous results are only valid for a given snapshot of user locations. We will show in the following that we can also consider random user locations within the framework developed before. However, all results will apply only for the case of no cooperation between the BS, i.e., $C = B$. The study of cooperative systems with random user locations is very challenging and a topic of current research [34]. We would like to point out that our approach is fundamentally different from that in [11, 35], which considers an infinitely large network, where the distributions of both UTs and BSs are random and characterized by a spatial point process. We will assume from now on that the UTs in cell i are uniformly randomly distributed around the BS i within a disk of radius R_i, while keeping a guard distance of $r_{\mathrm{g}} > 0$ (see Figure 8.1). The distances $d_{i,i,k}$ can consequently be seen as i.i.d. random variables with density $f_{i,i}(x)$, given as

$$f_{i,i}(x) = \frac{2x}{R_i^2 - r_{\mathrm{g}}^2}, \qquad r_{\mathrm{g}} \le x \le R_i. \tag{8.58}$$

The distribution of the distances $d_{i,j,k}$, $j \neq i$, from the UTs to the neighboring BSs depends on both the radii R_j and the inter-BS distances $D_{i,j}$. One can show after some analysis that $d_{i,j,k}$, $j \neq i$ are i.i.d. random variables with density $f_{i,j}(x)$, which can be expressed as ($x \in [\max(D_{i,j} - R_j, 0), D_{i,j} + R_j]$)

$$f_{i,j}(x) = \frac{2x \left[g\left(r_g, D_{i,j}, x\right) - g\left(R_j, D_{i,j}, x\right) \right]}{\pi \left(R_j^2 - r_g^2 \right)} \tag{8.59}$$

where[1]

$$g(t, z, x) = \tan^{-1} \left(\frac{z^2 - t^2 + x^2}{\sqrt{2z^2 \left(t^2 + x^2\right) - \left(t^2 - x^2\right)^2 - z^4}} \right).$$

Under the above assumptions, we can rewrite the channel matrices $\mathbf{G}_{i,j}$ as

$$\mathbf{G}_{i,j} = \frac{1}{\sqrt{K}} \mathbf{U}_{i,j} \mathbf{L}_{i,j}^{\frac{1}{2}} \tag{8.60}$$

where $\mathbf{U}_{i,j} = [\mathbf{u}_{i,j,1} \cdots \mathbf{u}_{i,j,K}] \in \mathbb{C}^{N \times K}$, $\mathbf{L}_{i,j} = \mathrm{diag}(d_{i,j,k}^{-\alpha}, \ k = 1, \ldots, K)$ and α is a path-loss exponent. In contrast to the previous sections, $\mathbf{G}_{i,j}$ is now the product of two independent random matrices. As shown by the next theorem, we can also provide deterministic approximations of the mutual information in this case.

Theorem 8.16 ([20, Theorem 1.1], [36, Theorem 2])　*Let* $\mathbf{H} = \mathbf{X}\mathbf{T}^{\frac{1}{2}}$, *where* $\mathbf{X} \in \mathbb{C}^{\beta_1 N \times \beta_2 K}$, $\beta_1, \beta_2 \in (0, \infty)$, *has i.i.d. complex Gaussian entries with zero mean and variance* $1/K$ *and* $\mathbf{T} = \mathrm{diag}\left(t_j, \ j = 1, \ldots, \beta_2 K\right) \in \mathbb{R}_+^{\beta_2 K \times \beta_2 K}$ *is random. Let* $N, K \to \infty$ *such that* $K/N \to c \in (0, \infty)$ *and assume that the empirical distribution of* $\{t_1, \ldots, t_{\beta_2 K}\}$ *converges weakly and almost surely to a nonrandom probability distribution* F *with bounded support. Then, for any* $\rho > 0$,

$$\mathbb{E}\left[\frac{1}{N} \log \det \left(\mathbf{I}_{\beta_1 N} + \rho \mathbf{H}\mathbf{H}^H \right) \right] \xrightarrow[N \to \infty]{} \bar{J}$$

where

$$\bar{J} = \beta_1 \log(1 + \delta) + \beta_2 c \int \log \left(1 + \frac{\beta_1 \rho t}{c (1 + \delta)} \right) dF(t) - \frac{\beta_1 \delta}{1 + \delta}$$

and δ *is given as the unique positive solution to the following fixed-point equation:*

$$\delta = \beta_2 \int \frac{\rho}{t^{-1} + \frac{\beta_1 \rho}{c(1+\delta)}} dF(t). \tag{8.61}$$

Moreover,

$$\frac{1}{K} \mathrm{tr} \left(\mathbf{H}\mathbf{H}^H + \frac{1}{\rho} \mathbf{I}_{\beta_1 N} \right)^{-1} \xrightarrow[N \to \infty]{a.s.} \frac{\beta_1 \rho}{c (1 + \delta)}.$$

---

[1] We denote by $\tan^{-1}(u)$ the complex inverse tangent function, which takes the values $-\pi/2 \, \mathrm{sign}\{\mathrm{Im}\{u\}\}$, whenever $\mathrm{Re}\{u\} = 0$ and $|\mathrm{Im}\{u\}| > 1$.

Let us now define the matrices $\bar{\mathbf{G}}_i = \bar{\mathbf{U}}_i\bar{\mathbf{L}}_i^{\frac{1}{2}} \in \mathbb{C}^{N \times BK}$, where $\bar{\mathbf{U}}_i = \begin{bmatrix} \mathbf{U}_{i,1} \cdots \mathbf{U}_{i,B} \end{bmatrix}$ and $\bar{\mathbf{L}}_i = \mathrm{diag}\left(\mathbf{L}_{i,j},\ j = 1, \ldots, B\right)$, and $\tilde{\mathbf{G}}_{\bar{i}} = \bar{\mathbf{U}}_{\bar{i}}\bar{\mathbf{L}}_{\bar{i}}^{\frac{1}{2}} \in \mathbb{C}^{N \times (B-1)K}$, where $\bar{\mathbf{U}}_{\bar{i}} = \begin{bmatrix} \mathbf{U}_{i,1} \cdots \mathbf{U}_{i,i-1}\mathbf{U}_{i,i+1} \cdots \mathbf{U}_{i,B} \end{bmatrix}$ and $\bar{\mathbf{L}}_{\bar{i}} = \mathrm{diag}\left(\mathbf{L}_{i,j},\ j = 1, \ldots, i-1, i+1, \ldots, B\right)$. Then, the ergodic mutual information $I_{\mathrm{rnd},i}$ from (8.17) can be expressed as

$$I_{\mathrm{rnd},i} = \mathbb{E}\left[\frac{1}{N}\log\det\left(\mathbf{I}_N + \rho\bar{\mathbf{G}}_i\bar{\mathbf{G}}_i^{\mathsf{H}}\right) - \frac{1}{N}\log\det\left(\mathbf{I}_N + \rho\bar{\mathbf{G}}_{\bar{i}}\bar{\mathbf{G}}_{\bar{i}}^{\mathsf{H}}\right)\right] \qquad (8.62)$$

where the expectation is with respect to the matrices $\bar{\mathbf{G}}_i$ and $\bar{\mathbf{L}}_i$. Obviously, as $K \to \infty$, it follows from the strong law of large numbers that the empirical distributions of the diagonal elements of $\bar{\mathbf{L}}_i$ and $\bar{\mathbf{L}}_{\bar{i}}$ converge weakly to the distribution of the random variables $x^{-\alpha}$ and $y^{-\alpha}$, respectively, whose densities are given as

$$\bar{f}_i(x) = \frac{1}{B}\sum_{j=1}^{B} f_{i,j}(x) \qquad (8.63)$$

$$\bar{f}_{\bar{i}}(y) = \frac{1}{(B-1)}\sum_{j \neq i} f_{i,j}(y). \qquad (8.64)$$

We can now apply Theorem 8.16 to both terms in (8.62) to find the asymptotic normalized mutual information $\bar{I}_{\mathrm{rnd},i}$ as provided by the next theorem:

Theorem 8.17 (Mutual information with random user locations) *Let $N, K \to \infty$ such that $K/N \to c \in (0, \infty)$. Then,*

$$I_{rnd,i} \xrightarrow[N\to\infty]{} \bar{I}_{rnd,i} = \bar{J}_{i,1} - \bar{J}_{i,2}$$

where $\bar{J}_{i,1}$ is given by Theorem 8.16 for $\beta_1 = 1$, $\beta_2 = B$, $t = x^{-\alpha}$ and $dF(x) = \bar{f}_i(x)dx$, and $\bar{J}_{i,2}$ is given by Theorem 8.16 for $\beta_1 = 1$, $\beta_2 = (B-1)$, $t = x^{-\alpha}$, and $dF(x) = \bar{f}_{\bar{i}}(x)dx$.

Recall the definitions of the SINRs $\gamma_{i,k}^{\mathrm{MMSE}}$ (8.24) and $\gamma_{i,k}^{\mathrm{MF}}$ (8.23) with MMSE and MF detection and denote by $R_{\mathrm{sum,rnd},i}^{\mathrm{MMSE}}$ and $R_{\mathrm{sum,rnd},i}^{\mathrm{MF}}$ the associated ergodic sum-rates of cluster i (see (8.20)). Notice that the expectation of the ergodic sum-rates is now with respect to the channel matrices *and* the random user locations. Based on Theorem 8.16, we can also provide DEs of these quantities for the case of random user locations.

Theorem 8.18 (MMSE detection with random user locations) *Let $N, K \to \infty$ such that $K/N \to c \in (0, \infty)$. Then,*

$$(i) \qquad \gamma_{i,k}^{MMSE} - \frac{\rho}{d_{i,i,k}^{\alpha}c\,(1+\delta)} \xrightarrow[N\to\infty]{a.s.} 0$$

$$(ii) \qquad R_{sum,rnd,i}^{MMSE} \xrightarrow[N\to\infty]{} \bar{R}_{sum,rnd,i}^{MMSE} = c\int \log\left(1 + \frac{\rho}{x^{\alpha}c\,(1+\delta)}\right) f_{i,i}(x)dx$$

where δ is given as the unique positive solution to (8.61) for $\beta_1 = 1$, $\beta_2 = B$, $c = K/N$, $t = x^{-\alpha}$, and $dF(x) = \bar{f}_i(x)dx$.

Proof. By Lemmas 8.2, 8.3 and Theorem 8.16, we have

$$\gamma_{i,k}^{\mathrm{MMSE}} = \frac{d_{i,i,k}^{-\alpha}}{K} \mathbf{u}_{i,i,k}^{\mathrm{H}} \left(\bar{\mathbf{G}}_i \bar{\mathbf{G}}_i^{\mathrm{H}} - d_{i,i,k}^{-\alpha} \mathbf{u}_{i,i,k} \mathbf{u}_{i,i,k}^{\mathrm{H}} + \frac{1}{\rho} \mathbf{I}_N \right)^{-1} \mathbf{u}_{i,i,k}$$

$$\asymp d_{i,i,k}^{-\alpha} \frac{1}{K} \mathrm{tr} \left(\bar{\mathbf{G}}_i \bar{\mathbf{G}}_i^{\mathrm{H}} + \frac{1}{\rho} \mathbf{I}_N \right)^{-1}$$

$$\asymp \frac{\rho}{d_{i,i,k}^{\alpha} c(1+\delta)} \tag{8.65}$$

where δ is given as the unique positive solution to (8.61) for $\beta_1 = 1$, $\beta_2 = B$, $c = K/N$, $t = x^{-\alpha}$, and $dF(x) = \bar{f}_i(x)dx$. This proves part (i). The last result implies together with (8.14) that

$$\left| \mathbb{E}_{\bar{\mathbf{G}}_i} \left[\frac{1}{N} \sum_{k=1}^{K} \log\left(1 + \gamma_{i,k}^{\mathrm{MMSE}}\right) - \log\left(1 + \frac{d_{i,i,k}^{-\alpha}\rho}{c(1+\delta)}\right) \right] \right| \tag{8.66}$$

$$\leq \frac{1}{N} \sum_{k=1}^{K} \mathbb{E}_{\bar{\mathbf{G}}_i} \left[\left| \gamma_{i,k}^{\mathrm{MMSE}} - \frac{d_{i,i,k}^{-\alpha}\rho}{c(1+\delta)} \right| \right] \xrightarrow[N\to\infty]{} 0. \tag{8.67}$$

Since $\mathbb{E}_{\bar{\mathbf{G}}_i}\left[\left|\gamma_{i,k}^{\mathrm{MMSE}} - \frac{d_{i,i,k}^{-\alpha}\rho}{c(1+\delta)}\right|\right] \leq \mathbb{E}_{\bar{\mathbf{G}}_i}\left[|\gamma_{i,k}^{\mathrm{MMSE}}|\right] + \frac{d_{i,i,k}^{-\alpha}\rho}{c(1+\delta)} \leq \frac{2\rho}{cr_g^{\alpha}}$ it follows from the dominated convergence theorem [37] that

$$\mathbb{E}_{\bar{\mathbf{G}}_i,\bar{\mathbf{L}}_i} \left[\frac{1}{N} \sum_{k=1}^{K} \log\left(1 + \gamma_{i,k}^{\mathrm{MMSE}}\right) - \log\left(1 + \frac{d_{i,i,k}^{-\alpha}\rho}{c(1+\delta)}\right) \right] \xrightarrow[N\to\infty]{} 0. \tag{8.68}$$

To conclude the proof of part (ii), it is sufficient to realize that $\mathbb{E}\left[\log\left(1 + \frac{d_{i,i,k}^{-\alpha}\rho}{c(1+\delta)}\right)\right] = \int \log\left(1 + \frac{\rho}{x^{\alpha}c(1+\delta)}\right) f_{i,i}(x)dx.$ $\qquad\square$

Theorem 8.19 (Matched filter with random user locations) *Let $N, K \to \infty$ such that $K/N \to c \in (0, \infty)$. Then,*

$$(i) \qquad \gamma_{i,k}^{MF} - \frac{\frac{1}{c}d_{i,i,k}^{-\alpha}}{\frac{1}{\rho} + B \int x^{-\alpha} \bar{f}_i(x)dx} \xrightarrow[N\to\infty]{a.s.} 0$$

$$(ii) \qquad R_{sum,i}^{MF} \xrightarrow[N\to\infty]{} \bar{R}_{sum,rnd,i}^{MF} = c \int \log\left(1 + \frac{\frac{1}{c}y^{-\alpha}}{\frac{1}{\rho} + B \int x^{-\alpha} \bar{f}_i(x)dx}\right) f_{i,i}(y)dy.$$

Proof. Without cooperation, we have $\mathbf{R}_{i,j,k} = d_{i,j,k}^{-\alpha} \mathbf{I}_N$. Hence, by Theorem 8.5,

$$\gamma_{i,k}^{MF} \asymp \frac{\frac{1}{c}d_{i,i,k}^{-\alpha}}{\frac{1}{\rho} + B \frac{1}{KB} \sum_{j=1}^{B} \sum_{l=1}^{K} d_{i,j,l}^{-\alpha}}. \tag{8.69}$$

By the strong law of large numbers, we have

$$\frac{1}{KB} \sum_{j=1}^{B} \sum_{l=1}^{K} d_{i,j,l}^{-\alpha} \xrightarrow[N\to\infty]{a.s.} \int x^{-\alpha} \bar{f}_i(x)dx. \tag{8.70}$$

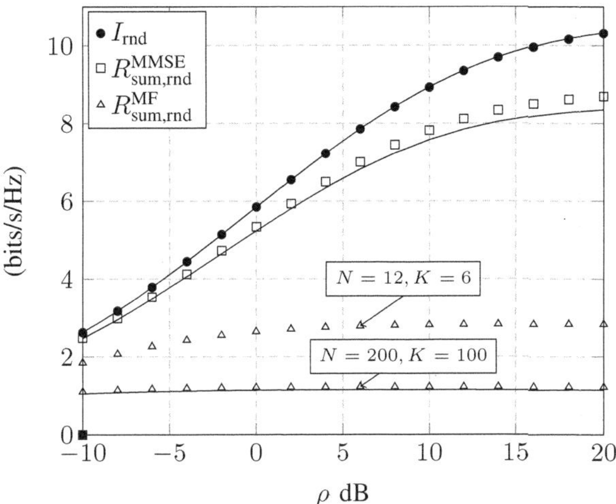

Figure 8.6 Sum of the normalized ergodic mutual information of all BSs I^{rnd} and sum-rates with MF and MMSE detection (markers) and their deterministic approximation $\bar{I}^{\mathrm{rnd}}$ (solid lines) versus transmit SNR ρ.

This proves part (i). Following the same arguments as in the proof of Theorem 8.18, we can show that

$$\mathbb{E}_{\bar{\mathbf{G}}_i, \bar{\mathbf{L}}_i} \left[\frac{1}{N} \sum_{k=1}^{K} \log\left(1 + \gamma_{i,k}^{\mathrm{MF}}\right) - \log\left(1 + \frac{\frac{1}{c} d_{i,i,k}^{-\alpha}}{\frac{1}{\rho} + B \int x^{-\alpha} \bar{f}_i(x) dx}\right) \right] \xrightarrow[N \to \infty]{} 0 \quad (8.71)$$

which concludes the proof of part (ii). $\qquad\square$

In order to validate the results of this section, we depict in Figure 8.6 $I_{\mathrm{rnd}} = \sum_i I_{\mathrm{rnd},i}$, $R_{\mathrm{sum,rnd}}^{\mathrm{MMSE}} = \sum_i R_{\mathrm{sum,rnd},i}^{\mathrm{MMSE}}$, $R_{\mathrm{sum,rnd}}^{\mathrm{MF}} = \sum_i R_{\mathrm{sum,rnd},i}^{\mathrm{MF}}$ and their asymptotic limits as provided by Theorems 8.16, 8.18, and 8.19 as a function of ρ for the simulation setup as described in Section 8.2.1. Even for a rather small system size ($N = 12$, $K = 6$), the match between the asymptotic and the simulation results is very good. One can observe that the matched filter requires much larger system dimensions $N = 200$, $K = 100$ to perform close to its asymptotic limit. In this section, we have only considered the uplink. However, analog results for the downlink can be obtained by similar methods.

8.8 Summary and conclusions

In this chapter, we have considered small cell networks with arbitrary clusters of cooperating BSs, which jointly process their user data. Starting from a simple system model, which accounts for path loss and fading, we have derived deterministic equivalents of the mutual information and achievable rates with linear detectors and precoders for uplink and downlink transmissions. We have then gradually extended these results to incorporate imperfect CSI and line-of-sight components in the channel. Finally, we have moved away from deterministic network topologies and considered random user locations. Our analysis is based on asymptotic results from large random matrix theory, which assumes that the number of antennas per BS and the number of UTs per cell grows infinitely large at the same speed. However, we have demonstrated by simulations that the asymptotic performance approximations are very tight for realistic system dimensions. Thus, the results might find useful applications for the optimization of certain system parameters, e.g., optimal power allocation, BS clustering and placement, scheduling, etc., which would be intractable based on an exact analysis. In the course of this chapter, we have also mentioned several open research problems. Among them count the asymptotic performance analysis of linear precoders and detectors for Rician fading channels and the study of cooperative systems with random user (and possibly BS) distributions. The last point would allow one, for example, to find the optimal BS clustering for a given statistical distribution of the UTs.

References

[1] T. L. Marzetta, "Noncooperative cellular wireless with unlimited numbers of base station antennas," *IEEE Trans. Wireless Commun.*, vol. 9, no. 11, pp. 3590–600, Nov. 2010.

[2] J. Hoydis, M. Kobayashi, and M. Debbah, "Green small-cell networks," *IEEE Veh. Technol. Mag.*, vol. 6, no. 1, pp. 37–43, Mar. 2011.

[3] S. A. Ramprashad, H. C. Papadopoulos, A. Benjebbour, Y. Kishiyama, N. Jindal, and G. Caire, "Cooperative cellular networks using multi-user MIMO: trade-offs, overheads, and interference control across architectures," *IEEE Commun. Mag.*, vol. 49, no. 5, pp. 70–7, May 2011.

[4] J. Andrews, "Interference cancellation for cellular systems: a contemporary overview," *IEEE Wireless Commun.*, vol. 12, no. 2, pp. 19–29, Apr. 2005.

[5] H. Claussen, "The future of small-cell networks," *IEEE COMMSOC MMTC E-Lett.*, pp. 32–6, Sep. 2010.

[6] D. Gesbert, S. V. Hanly, H. Huang, S. Shamai, O. Simeone, and W. Yu, "Multi-cell MIMO cooperative networks: a new look at interference," *IEEE J. Sel. Areas Commun. (JSAC)*, vol. 28, no. 9, pp. 1380–408, Dec. 2010.

[7] S. Venkatesan, A. Lozano, and R. Valenzuela, "Network MIMO: overcoming intercell interference in indoor wireless systems," in *Proc. Asilomar Conf. on Signals, Systems, and Computers (ASILOMAR)*, Pacific Grove, CA, US, Nov. 2007, pp. 83–7.

[8] P. Marsch and G. P. Fettweis, *Coordinated Multi-point in Mobile Communications: From Theory to Practice*. Cambridge: Cambridge University Press, 2011.

[9] S. Sesia, M. Baker, and I. Toufik, *LTE: The UMTS Long Term Evolution: From Theory to Practice*. Wiley-Blackwell, July 2011.

[10] P. Charriere, J. Brouet, and V. Kumar, "Optimum channel selection strategies for mobility management in high traffic TDMA-based networks with distributed coverage," in *Proc. IEEE Int. Conf. Personal Wireless Commun. (ICPWC)*, Mumbay, India, Dec. 1997.

[11] J. G. Andrews, F. Baccelli, and R. K. Ganti, "A tractable approach to coverage and rate in cellular networks," *IEEE Trans. Commun.*, vol. 59, no. 11, pp. 3122–34, Nov. 2011.

[12] J. Hoydis, S. ten Brink, and M. Debbah, "Massive MIMO: how many antennas do we need?" in *Proc. Allerton Conference on Communication, Control, and Computing*, Urbana-Champaign, Illinois, US, Sep. 2011, pp. 545–50.

[13] J. Dumont, W. Hachem, S. Lasaulce, P. Loubaton, and J. Najim, "On the capacity achieving covariance matrix for Rician MIMO channels: an asymptotic approach," *IEEE Trans. Inf. Theory*, vol. 56, no. 3, pp. 1048–69, Mar. 2010.

[14] J. Hoydis, M. Kobayashi, and M. Debbah, "Optimal channel training in uplink network MIMO systems," *IEEE Trans. Signal Process.*, vol. 59, no. 6, pp. 2824–33, June 2011.

[15] H. Huh, G. Caire, H. C. Papadopoulos, and S. A. Ramprashad, "Achieving massive MIMO spectral efficiency with a not-so-large number of antennas," *IEEE Trans. Wireless Commun.*, 2012.

[16] R. Couillet and M. Debbah, *Random Matrix Methods for Wireless Communications*. Cambridge: Cambridge University Press, 2011.

[17] S. M. Kay, *Fundamentals of Statistical Signal Processing: Estimation Theory*. Prentice-Hall, Inc. Upper Saddle River, NJ, USA, 1993.

[18] Z. D. Bai and J. W. Silverstein, "No eigenvalues outside the support of the limiting spectral distribution of large dimensional sample covariance matrices," *Ann. Probab.*, vol. 26, no. 1, pp. 316–45, Jan. 1998.

[19] S. Wagner, R. Couillet, M. Debbah, and D. T. M. Slock, "Large system analysis of linear precoding in correlated MISO broadcast channels under limited feedback," *IEEE Trans. Inf. Theory*, vol. 58, no. 7, pp. 4509–37, July 2012.

[20] J. W. Silverstein and Z. D. Bai, "On the empirical distribution of eigenvalues of a class of large dimensional random matrices," *Journal of Multivariate Analysis*, vol. 54, no. 2, pp. 175–92, 1995.

[21] M. J. M. Peacock, I. B. Collings, and M. L. Honig, "Eigenvalue distributions of sums and products of large random matrices via incremental matrix expansions," *IEEE Trans. Inf. Theory*, vol. 54, no. 5, pp. 2123–38, May 2008.

[22] S. Wagner, "MU-MIMO transmission and reception techniques for the next generation of cellular wireless standards (LTE-A)," Ph.D. dissertation, EURECOM, 2011.

[23] T. Cover and J. A. Thomas, *Elements of Information Theory*, 2nd edn. John Wiley & Sons, Inc., 2006.

[24] M. Medard, "The effect upon channel capacity in wireless communications of perfect and imperfect knowledge of the channel," *IEEE Trans. Inf. Theory*, vol. 46, no. 3, pp. 933–46, May 2000.

[25] T. Yoo and A. Goldsmith, "Capacity and power allocation for fading MIMO channels with channel estimation error," *IEEE Trans. Inf. Theory*, vol. 52, no. 5, pp. 2203–14, May 2006.

[26] B. Hassibi and B. M. Hochwald, "How much training is needed in multiple-antenna wireless links?" *IEEE Trans. Inf. Theory*, vol. 49, no. 4, pp. 951–63, Apr. 2003.

[27] G. Caire, N. Jindal, M. Kobayashi, and N. Ravindran, "Multiuser MIMO achievable rates with downlink training and channel state feedback," *IEEE Trans. Inf. Theory*, vol. 56, no. 6, pp. 2845–66, June 2010.

[28] D. Hoesli, Y.-H. Kim, and A. Lapidoth, "Monotonicity results for coherent MIMO Rician channels," *IEEE Trans. Inf. Theory*, vol. 51, no. 12, pp. 4334–9, Dec. 2005.

[29] M. Kang and M. Alouini, "Capacity of MIMO Rician channels," *IEEE Trans. Wireless Commun.*, vol. 5, no. 1, pp. 112–22, Jan. 2006.

[30] W. Hachem, P. Loubaton, and J. Najim, "Deterministic equivalents for certain functionals of large random matrices," *Ann. Appl. Probab.*, vol. 17, pp. 875–930, 2007.

[31] J. Dumont, P. Loubaton, S. Lasaulce, and M. Debbah, "On the asymptotic performance of MIMO correlated Rician channels," in *Proc. IEEE Int. Conf. on Acoustics, Speech, and Sig. Proc. (ICASSP)*, vol. 5, Mar. 2005, pp. 813–16.

[32] A. Kammoun, M. Kharouf, R. Couillet, J. Najim, and M. Debbah, "On the fluctuations of the SINR at the output of the Wiener filter for non centered channels: the non Gaussian case," in *Proc. IEEE Int. Conf. on Acoustics, Speech, and Sig. Proc. (ICASSP)*, Kyoto, Japan, Mar. 2012.

[33] W. Hachem, P. Loubaton, J. Najim, and P. Vallet, "On bilinear forms based on the resolvent of large random matrices," *Annales de l'IHP: Probability and Statistics*, 2011 [Online]. Available: http://arxiv.org/abs/1004.3848

[34] J. Hoydis, A. Müller, R. Couillet, and M. Debbah, "Analysis of multicell cooperation with random user locations via deterministic equivalents," in *Proc. Workshop on Spatial Stochastic Models for Wireless Networks (SPASWIN'12)*, Paderborn, Germany, May 2012.

[35] K. Huang and J. G. Andrews, "A stochastic-geometry approach to coverage in cellular networks with multi-cell cooperation," in *Proc. IEEE Global Telecommun. Conf. (GLOBECOM)*, Houston, Texas, US, Dec. 2011.

[36] R. Couillet, M. Debbah, and J. W. Silverstein, "A deterministic equivalent for the analysis of correlated MIMO multiple access channels," *IEEE Trans. Inf. Theory*, vol. 57, no. 6, pp. 3493–514, June 2011.

[37] P. Billingsley, *Probability and Measure*, 3rd edn. Wiley, 1995.

9 Mobility in small cell networks

Veeraruna Kavitha, Sreenath Ramanath, and Eitan Altman[*]

9.1 Introduction

Small cell networks (SCNs) made of portable pico and femto base stations (BSs) serve dense urban areas, commercial and office spaces, hotspots, etc. Their design and deployment pose many new challenges to the optimal system design. Managing mobile users deriving service from such SCNs is one of the key challenges. Furthermore, reducing cell size increases the frequency of handovers, which results in an increased number of call drops before completing the service. However, they also offer better communication rates to cell-edge users resulting in reduced service times. The study of such tradeoffs is an important topic while designing optimal systems.

In order to prevent large numbers of handovers that would result from reducing cell size, it has been proposed (for e.g., see [1, 2]) to group together a number of small cells (SCs) into one virtual macrocell. This helps to restrict the effort of preventing losses due to the handover only to those handovers that occur between SCs of the same virtual cell. In between the SCs some fast switching mechanisms are proposed such as frequency following mechanisms where the frequency used by a mobile follows it from one SC to the next. This requires reserving the same channel for a user in the entire macrocell.

In this chapter, we consider a large macrocell divided into a number of SCs and study the impact of mobility in such systems,[1] especially the effect of frequent handovers. We assume that an ongoing call is never dropped at the SC boundary within a virtual cell (in later sections, we relax this assumption and study handover at SC boundaries within a virtual cell). We further assume that BS switching (BSS) at any SC boundary requires some fixed amount of information (in terms of bytes) to be exchanged. Further, there is a possibility of calls being dropped at macrocell boundaries. We also assume that the active users cross macrocell boundaries at maximum once, i.e., the calls always end before reaching the second macrocell boundary. The handovers at the cell boundaries are

[*] The major part of this work was carried out in INRIA, Sophia Antipolis, France and UPAV, Avignon, France.
[1] Throughout the chapter, macrocell and virtual cells are used interchangeably.

Small Cell Networks: Deployment, PHY Techniques, and Resource Management, ed. Tony Q. S. Quek, Guillaume de la Roche, İsmail Güvenç, and Marios Kountouris. Published by Cambridge University Press. © Cambridge University Press 2013.

modeled using an independent Poisson process, which is a commonly made assumption (for example see [3, 4]).

The first goal of this chapter is to model the system so as to compute its performance measures. In this system, arrivals occur in space and each arrival utilizes the resources for a certain time (*service time*) that depends on the position and movement of the user and the placement of the serving BS(s). We are thus interested in developing tools, that take into account not only the instantaneous geometry but also the way it varies in time. The queues modeling these type of systems are referred to as spatial queues and have been used in various applications [5–8] and we use similar tools. The first important quantity that we compute is the moments of the *service time*. It is the time taken by the system to serve[2] the users, which in our case will be equal to the time the BS spends on a call. We derive expressions for these service times, during which the information is exchanged between the moving user and the set of appropriate BSs (which it encounters during its journey), using variable rate of transmission. We further simplify the expressions for service time under the following assumption: in a typical SCN, a moving user would have traversed across a number of cells before completing the call.

We model the macrocells by various types of queues and well known results from queuing theory (e.g., [9]) are used to obtain performance measures like expected waiting time, expected service time, drop or blocking probability, etc. One can use these results for preliminary dimensioning purposes in planning the SCN catering to pedestrian and vehicular mobility, typical of urban and suburban areas. We derive closed form expressions of useful performance metrics considering free space path loss, handover constraints, traffic type, etc. We also obtain closed form expressions for optimal cell size (for various performance metrics), when all the users move with velocity profiles with certain statistical characteristics. We make the following theoretical and/or simulation-based observations:

1. **Maximum possible velocity (Section 9.3.2, Theorem 9.2):** For any fixed transmission power P, there exists a maximum user velocity $V_{\lim}(P)$ and the users moving at speeds greater than $V_{\lim}(P)$ cannot successfully communicate with the BS.
2. **Larger cells for larger velocities (Section 9.3, Theorems 9.3, 9.4, and 9.5):** Given P, the optimal cell size increases with an increase in the highest velocity that the system has to support. This is true as long as the highest velocity is less than $V_{\lim}(P)$. However, the system cannot cater to velocities above the limit $V_{\lim}(P)$, even if one increases the cell size indefinitely.
3. **Insensitivity to application (Section 9.3, Theorems 9.3, 9.4, and 9.5):** The optimal cell size remains the same for non-elastic (NES) as well as elastic (ES) calls for large file sizes, when the rest of the parameters remain same.
4. **Two-dimensional Manhattan Grid (Section 9.3.9):** The one-dimensional results can be applied directly to a two-dimensional regular street grid (see e.g., [10, 11]).

---

[2] Throughout the chapter we use terms from queuing theory such as arrivals, service time, etc.

We also worked on systems where a small number of call drops can occur even at SC boundaries. Some of these results also considered the case of randomly wandering users with a finite number of transmission rates. Toward the end of this chapter, a brief discussion of these results is presented in Appendix 9A.

Several authors have examined the impact of mobility on the performance of wireless networks. The authors in [12] have shown that mobility in fact increases the capacity in ad hoc networks. In [13], the authors discuss the tradeoff between delay and achievable throughput in the presence of mobility in wireless ad hoc networks. Further, in [14–17], the authors discuss the impact of inter- and intra-cell mobility on capacity, flow level performance, tradeoff between throughput and fairness, etc. They show that mobility tends to increase the capacity with globally optimal as well as fair sharing policies, when the BSs interact. In conclusion, we see that the mobility (via multi-user diversity, opportunistic scheduling, etc.) can in fact improve the overall performance of the system. However, most of the work assumes that the handovers occur without extra cost. But in reality, each handover requires exchange of some information between the user and the new BS and has a risk of not finding free resources in the new cell. Our work mainly focuses on these issues, which become significant whenever the frequency of handovers increases, as in the case of SCNs.

To begin with, we consider the case with continual rates and with an umbrella cell concept, where a number of cells are grouped into a *virtual* macrocell and in which the calls are never dropped when a user crosses the SCs of the same virtual cell. The system model with an umbrella cell concept and continual rates is described in Section 9.2, while the service time is discussed in Section 9.3.1 and optimized in Section 9.3.5. The NES and ES calls are modeled by appropriate queuing models and performance measures are derived in Sections 9.3.7 and 9.3.8, respectively. The two-dimensional regular street grid is studied in Section 9.3.9, while numerical examples are provided in Section 9.4. Brief discussions of some initial results on a system with possible call drops at SC boundaries and with a decentralized call admission control are provided in Section 9.5. Some lengthy derivations are provided in two appendices (Appendix 9B and 9C) placed at the end of the chapter.

9.2   Umbrella concept and continual rates

We consider a macrocell, with a coverage area in $[-D, D]$, divided into a number of SCs of length L as shown in Figure 9.1. Each SC has a BS located at the center and all of these BSs communicate to a control unit (CU), which controls the entire system. *We assume that there is no interference between any pair of cells.*

Traffic types: Define the waiting time as the duration between the arrival of the call and the instant its service starts. We consider two types of traffic: ES and NES. The two types of calls result from two different types of applications. The NES traffic (multimedia streaming, voice calls, etc) is very sensitive to the waiting time. These callers can tolerate some errors in transmission (for example in voice calls the human ears cannot distinguish the errors when they are up to a certain perseverance limit) but

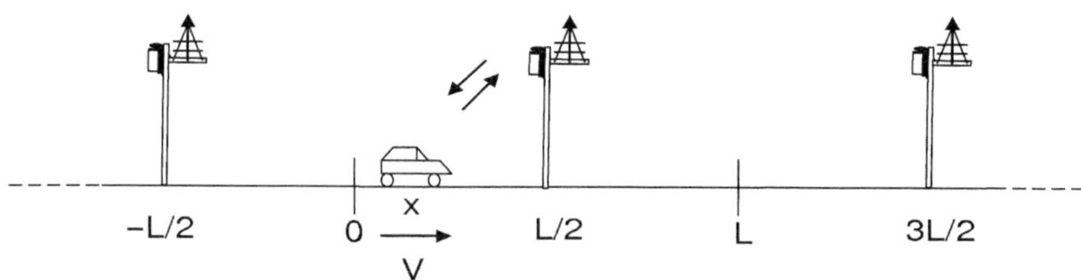

Figure 9.1 User moving with velocity V along a line. ©2010 IEEE. Reprinted, with permission, from [18].

are very impatient. They in fact drop the call if not picked up within a small waiting time. Elastic traffic (e.g., a data call) on the other hand can wait for some time before it is picked up, but is very sensitive to errors in transmission. In such calls, the natural performance metrics are either the expected waiting time or the expected sojourn time. The probability of a call being blocked, P_B and the drop probability, i.e., the probability that an ongoing call is dropped before completion, P_{Dr} are important performance measures for NES calls. Systems are designed with more stringent requirements on P_{Dr} than P_B. The two types of users are assumed to be served using independent sets of resources.

Arrivals: The two (ES, NES) arrivals are modeled by two independent Poisson arrivals with rates λ. Every arrival is associated with Marks (X, V, S): $X \in [-D, D]$ is the location of arrival, S is the file size requirement, and V is the user velocity, distributed respectively according to $P_{n,X}$, $P_{n,V}$, and $P_{n,S}$, with respective densities $f_{n,X}$, $f_{n,V}$, and $f_{n,S}$. Let $P_n := (P_{n,X}, P_{n,V}, P_{n,S})$. We *assume symmetry in both directions, i.e., that* $P_n((X, V, S) \in A) = P_n((X, V, S) \in -A)$ *for all Borel sets A.* Throughout the chapter we consider without loss of generality (w.l.g.) $V \geq 0$.

Handovers: In the major part of this research (except for Appendix 9A), we assume that handovers are completely successful at SC boundaries. However, we do not assume the same for macrocell boundaries, i.e., a crossover into a new macrocell results in a successful handover only if the new macrocell has free servers. We model each handover into a macrocell as a Poisson arrival, stochastically independent of the new call arrivals (as done for the example in [3, 4]). We further assume that there can be at maximum one handover, i.e., the calls get finished before reaching the second macrocell boundary. This simplifies the analysis to a good extent and is a valid assumption as the macrocells are typically large in dimension. We consider generalization of this assumption in our future work. Lastly, in Section 9.5 we obtain some initial results by relaxing the assumption that the handovers at SC boundaries are completely successful.

Radio conditions: The BS communicates with the mobiles using a wireless link, at a rate that depends upon the distance between the two. Since our primary focus is on mobility we implicitly consider SCs deployed outdoors, for example urban or suburban scenarios. Hence we can assume significant line-of-sight signal. Further, SCs being small in size, it will be sufficient to consider only the direct path for communication. A user located at x communicates with BS of cell m using unit transmit power (when receiver noise variance is 1) at rate[3] given by (with 1{.} representing an indicator function),

$$\bar{R}(x;m) := 1\left\{\left|x - \left(mL - \frac{L}{2}\right)\right| \le d_0\right\}$$

$$+ 1\left\{\left|x - \left(mL - \frac{L}{2}\right)\right| > d_0\right\}d_0^{\beta}\left|x - \left(mL - \frac{L}{2}\right)\right|^{-\beta}, \quad (9.1)$$

where $\beta \ge 1$ represents the path-loss factor and $d_0 > 0$ is a small distance up to which there is no propagation loss. The above model is valid for systems with low signal to noise ratios, wherein the rates are directly given by the SNRs.

9.3   System analysis and performance measures

The users are moving continuously with a fixed but random velocity. The macrocell can handle at maximum K parallel calls. Transmission always occurs at fixed power P. Since SCs are small in size, the movement of the users results in frequent handovers. The number of handovers will be quite large such that it would be complicated to design a reliable system without redundancy: we assume that every BS can also handle K parallel calls.[4] This ensures that, once a call is picked up it is not dropped at any SC boundary: when a user crosses over to a new BS, the new BS would at maximum be handling $K - 1$ calls and hence will have at least one free server. However, it is important to note that the maximum power used at any time in the system equals KP. We further assume that:

1. Every BS (at an SC boundary) requires fixed B_{h} bytes of information to be communicated (independent of the user's velocity), after which the user's service is resumed by the BS of the cell it just entered.
2. The user is served by the BS of the SC in which it is moving, as it is physically nearest to this BS.

In the following sections up to and excluding Section 9.5, all the results are derived under the assumption that no call drops ever happen at SC boundaries.

---

[3] The analysis will go through for any other rate functions, for example like $R(x;1) = (1 + (x - L/2)^2)^{-\beta/2}$ [19], $R(x;1) = \log\left[1 + (1 + (x - L/2)^2)^{-\beta/2}\right]$ [19], $R(x;1) = \log\left[1 + (d_0^{\beta}|x - L/2|^{-\beta}\mathcal{X}\{|x - L/2| > d_0\} + \mathcal{X}\{|x - L/2| \le d_0\})\right]$ etc. Some of the simplifications that we obtain in subsequent sections may not be possible with these rate functions. However, one can always conduct Monte Carlo simulations to obtain the required inferences.

[4] In practical systems, each BS will have M back-up servers to manage handovers. This means each BS can handle M parallel calls. In general M need not be equal to K; however, M has to be chosen large enough to ensure negligible call drops at SC boundaries, taking into consideration the large number of handovers associated with SCs. With this large enough M the system's behavior will be close to the system considered in this work (the case with $M = K$).

9.3.1   Time required for communicating S bytes (T_c)

We define by $T_c(S, X, V)$ the time required to communicate a packet of length S bytes to a user located at X (when the service starts) and moving with velocity V. If the user can communicate at a fixed rate r bytes/sec then the communication time would have been S/r. The maximum rate at which a user can communicate with the BS in cell m is given by (9.1). This position dependent is minimum when the user is at the cell edges and increases as the user moves toward the cell center. This poses a need to calculate the communication time considering the variable rates. The location of the user (under service) will change according to $X(t) = X + Vt$ (Figure 9.1). At time t, if the user is in cell m, i.e., if $X(t) \in [(m-1)L, mL]$, it communicates with the BS of the mth cell. Hence the user gets service at a time varying rate given by

$$R(t; X, V) := P\bar{R}(X(t); m) \text{ if } t \in \left[\frac{(m-1)L - X}{V}, \frac{mL - X}{V} \right].$$

Without loss of generality we consider the users, whose communication started in the first SC, i.e., with $X \in [0, L]$. The communication time T_c required by the user, i.e., the time required to communicate S bytes satisfies:

$$S = \int_0^{T_c} R(t; X, V)\mathrm{d}t. \tag{9.2}$$

Let
$$g(l) := \int_0^{l/V} P\bar{R}(Vt; 1)\mathrm{d}t = P \int_0^l \bar{R}(l'; 1)\frac{\mathrm{d}l'}{V}, \tag{9.3}$$

represent the number of bytes communicated while the mobile traverses interval $[0, l]$. Note that[5]

$$g(L) = \frac{Pd_0^\beta}{V} \int_0^{L/2-d_0} \left(\frac{L}{2} - l \right)^{-\beta} \mathrm{d}l + \frac{Pd_0^\beta}{V} \int_{L/2+d_0}^{L} \left(l - \frac{L}{2} \right)^{-\beta} \mathrm{d}l + \frac{2Pd_0}{V}$$

$$= \begin{cases} \frac{2Pd_0}{V(\beta-1)} \left(\beta - \left(\frac{L}{2d_0} \right)^{1-\beta} \right) & \text{when } \beta > 1 \\ \frac{2Pd_0}{V} (\log(L/2) - \log(d_0)) & \text{when } \beta = 1. \end{cases} \tag{9.4}$$

For any m, the number of bytes communicated as the user traverses through the mth SC (by change of variable $l = X + Vt - (m-1)L$) becomes

$$\int_{\frac{(m-1)L-X}{V}}^{\frac{mL-X}{V}} R(t; X, V)\mathrm{d}t = \int_{\frac{(m-1)L-X}{V}}^{\frac{mL-X}{V}} P\bar{R}(X + Vt; m)\mathrm{d}t$$

$$= \frac{P}{V} \int_0^L \bar{R}(l; 1)\mathrm{d}l = g(L) \tag{9.5}$$

and thus is independent of m. Out of this number, B_h number of bytes are dedicated for BSS. Hence, irrespective of the cell that the user traverses, $g(L) - B_h$ number of bytes are transmitted during the user's journey via one SC. Thus the communication time can have three components:

---

[5] Throughout the chapter, it is assumed that $L > 2d_0$, as it can be easily shown that the optimal cell size has to be greater than $2d_0$.

1. Time taken in the originated cell: $(L - X)/V$.
2. Time taken to travel N full cells, where (with $\lfloor t \rfloor$ representing the largest integer in t)

$$N = N(S, X, V) := \left\lfloor \frac{(S - (g(L) - g(X)))}{g(L) - B_h} \right\rfloor$$

represents the number of cells traveled during the communication of S bytes, and
3. Time taken in the cell in which the call terminates, i.e., time taken to communicate leftover bytes

$$S_l := S - (g(L) - g(X)) - N(g(L) - B_h).$$

Further, a call can be handled only if the number of bytes that can be communicated while traversing through a cell, $g(L)$, is greater than the number of bytes required for BSS B_h. From (9.2), the communication time $T_c(S, X, V)$ can be calculated as:

$$T_c(S, X, V) = \begin{cases} \frac{1}{V} \arg\inf_{l \in (X, L]} \{(g(l) - g(X)) \geq S\} \\ \qquad\qquad\qquad\qquad \text{if } S < (g(L) - g(X)) \\ \infty \qquad\qquad\qquad\qquad \text{if } B_h > g(L) \\ \frac{L-X}{V} + N\frac{L}{V} + \frac{1}{V} \arg\inf_{l \in (0, L]} \{g(l) - B_h \geq S_l\} \\ \qquad\qquad\qquad\qquad \text{otherwise.} \end{cases}$$

From (9.4), g is a continuous and monotonically increasing function, so g^{-1} exists and thus:

Theorem 9.1 *The time to communicate S bytes with a user initially located at X and moving with velocity V is,*

$$T_c(S, X, V) = \begin{cases} \frac{g^{-1}(S+g(X);V)-X}{V} & \text{if } S < (g(L) - g(X)) \\ \infty & \text{if } B_h > g(L) \\ \frac{(L-X)+NL+g^{-1}(S_l+B_h;V)}{V} & \text{otherwise} \end{cases}$$

where $g^{-1}(s; v)$ $\qquad\qquad\qquad\qquad\qquad\qquad\qquad\qquad\qquad$ (9.6)

$$= \begin{cases} \frac{L}{2}\left(1 - e^{-\frac{sv}{Pd_0}}\right), & \text{if } \beta = 1 \ \frac{sv}{Pd_0} < \log\left(\frac{L}{2d_0}\right) \\[2mm] \frac{L}{2} + \frac{2d_0^2 e^{-2}}{L} e^{\frac{sv}{Pd_0}}, & \text{if } \beta = 1 \ \frac{sv}{Pd_0} > \log\left(\frac{L}{2d_0}\right) + 2 \\[2mm] \frac{L}{2} + \frac{sv}{P} - d_0 \log\left(\frac{L}{2d_0}\right) - d_0, & \text{if } \beta = 1 \ \text{otherwise} \\[2mm] \frac{L}{2} - \left(\frac{sv(\beta-1)}{Pd_0^\beta} + \left(\frac{L}{2}\right)^{-\beta+1}\right)^{\beta-1}, & \text{if } \beta > 1 \ \frac{sv(\beta-1)}{Pd_0^\beta} < d_0^{-\beta+1} - \left(\frac{L}{2}\right)^{-\beta+1} \\[2mm] \frac{L}{2} + \left(\frac{2\beta d_0^{-\beta+1}}{\beta-1} - \frac{sv(\beta-1)}{Pd_0^\beta} - \left(\frac{L}{2}\right)^{-\beta+1}\right)^{\beta-1}, & \text{if } \beta > 1 \ \frac{sv(\beta-1)}{Pd_0^\beta} > \frac{(\beta+1)d_0^{-\beta+1}}{\beta-1} - \left(\frac{L}{2}\right)^{-\beta+1} \\[2mm] \frac{L}{2} + \frac{d_0^\beta}{\beta-1}\left(\frac{sv(\beta-1)}{Pd_0^\beta} + \left(\frac{L}{2}\right)^{-\beta+1} - d_0^{-\beta+1}\beta\right), & \text{if } \beta > 1 \ \text{otherwise.} \end{cases}$$

Approximation: In SC based systems, a user traverses a large number of SCs while receiving service. Hence the communication time can be approximated by the product

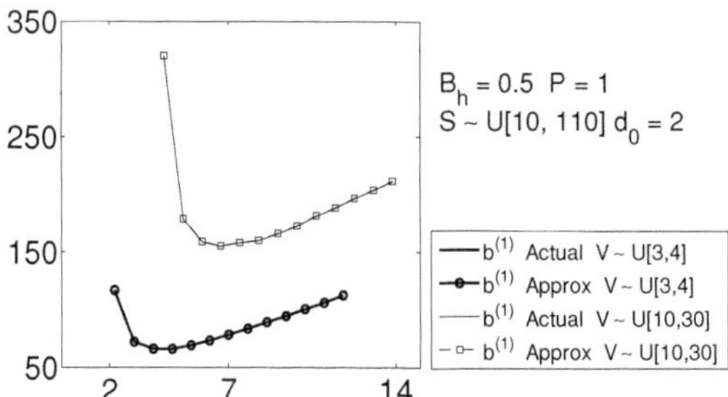

Figure 9.2 Approximation of communication time T_c. x-axis: cell size L; y-axis: expected communication time $b^{(1)}$.

of the number of cells, $S/(g(L) - B_h)$, and the time taken for traversing each cell L/V:

$$T_c(S, X, V) \approx \frac{S}{g(L) - B_h} \frac{L}{V} \quad \text{when } g(L) > B_h. \tag{9.7}$$

In Figure 9.2, we show the validity of this approximation. We plot the expected value of actual communication time (given in Theorem 9.1) and the expected value of the approximation (9.7), for two different velocity profiles. As expected the approximation matches well with the actual communication time B_h, in fact for all velocity profiles (one can hardly distinguish the two lines in the figure).

Remark 9.1 *If a call is originated at position X and moves with velocity $V \neq 0$ then, in case of ES applications, the call will be picked up at a position $X_s = X + VW$, where W is the waiting time. It is difficult to estimate X_s (as W is unknown) and the time taken to communicate S bytes (T_c) actually depends upon X_s but not on X. However, with the above approximation, $T_c(S, X_s, V) = T_c(S, V)$, i.e., T_c does not depend upon the location of the user when its communication started.*

9.3.2   Maximum velocity handled by the system

The communication time, T_c, is finite only if the number of bytes transfered $g(L)$ per cell is strictly greater than the bytes required for BSS, B_h. The system cannot handle velocities for which the communication times are infinite and hence we obtain (proof in Appendix B.1):

Theorem 9.2 *When the path-loss factor $\beta = 1$, for any transmit power P, the system can handle all velocity profiles. When $\beta > 1$, there exists a bound $V_{\lim}(P) < \infty$ (increasing linearly with P) on the maximum velocity that can be handled by the system, where*

$$V_{\lim}(P) := \frac{2Pd_0\beta}{\beta - 1}. \tag{9.8}$$

Henceforth we consider only the velocity profiles that satisfy:

$$P_{n,V}(V < V_{\max}) = 1, \quad \text{where } V_{\max} < V_{\lim}(P). \tag{9.9}$$

9.3.3   Service time: the time of the macrocell spent for user's service

The user reaches the boundary of the macrocell starting from a point X in time:

$$T_\partial(X, V) := \frac{D - X}{V}.$$ (9.10)

The macrocell has to serve the user either until all its S bytes are transmitted (which takes time T_c) or until the user reaches the boundary. Thus, the overall service time requirement of the user in a macrocell is $T_D := \min\{(D - X_s)/V, T_c(V, S)\}$, where X_s (defined in Remark 9.1) is the user position when its service starts. By Remark 9.1, T_c does not depend upon X_s. For NES applications $X_s = X$ is the position of arrival. For ES applications, it is difficult to estimate X_s, instead we approximate X_s with X, i.e., $T_D(X, V, S) \approx \min\{T_\partial(X, V), T_c(V, S)\}$. The error in this approximation is given (with W representing the waiting time) by:

$$E_{rr} = W 1_{\{T_c > \frac{D-X}{V}\}} + \left(T_c - \frac{D - X_s}{V}\right) 1_{\{\frac{D-X_s}{V} < T_c < \frac{D-X}{V}\}}$$

and so for any k,

$$\mathbb{E}[E_{rr}^k] \leq \mathbb{E}\left[W^k 1_{\{T_c > \frac{D-X}{V} - W\}}\right].$$

The error is small either whenever the number of servers is large (so waiting times are small) or when the macrocell is large in size (which is usually the case).

9.3.4   Macro handovers

Handovers are modeled as Poisson arrivals (as done in [3, 4]). In this subsection we derive the other characteristics of the handover calls.

Distribution of handover call marks (X, S)**:** In general, handover densities will be different from the new call densities $f_{n,X}$, $f_{n,S}$, and $f_{n,V}$. As the users move in either direction with equal probability, $P_{h,X}$ the position of handover arrival occurs either at $-D$ or at D with half probability. If handover occurs at $-D$ the corresponding velocity will be positive, which is the case we consider without loss of generality. We assume $f_{n,S}$ is exponential, i.e., $f_{n,S}(s) = \mu e^{-\mu s}$ for some $\mu > 0$, in which case by memoryless property, $f_{h,S} = f_{n,S}$.

Rate of handovers: Let

$$v(v, L) := \frac{\eta(L) - v B_h}{L} = v \frac{g(L) - B_h}{L}.$$ (9.11)

Then, from (9.7) and (9.10),

$$P_{ho} := \mathbb{P}(T_\partial < T_c) = \mathbb{E}_{n,X,V}\left[e^{-\mu v(V,L)(D-X)/V}\right]$$ (9.12)

gives the probability that a call is not completed in one macrocell. This precisely represents that fraction of new arrivals that gets converted into handover calls. So

$$\lambda_{ho} := \lambda P_{ho}$$

is the rate at which handovers occur.

Speed of handover arrival: A handover call arrives at $X = -D$ with velocity $v > 0$ only if a new call with velocity v is not completed before reaching the boundary. Here we use the assumption that handover occurs at maximum at one macrocell boundary, i.e., a handover call is not converted to another handover. Let

$$P_{ho,v} := \mathbb{P}(T_\partial < T_c | v) \;=\; \mathbb{E}_{n,X}\left[e^{-\mu v(v,L)(D-X)/v}\right] \tag{9.13}$$

represent the conditional probability given $V = v$. Then, the handover speed distribution is given by

$$f_{h,V}(v) = \frac{f_{n,V}(v)P_{ho,v}}{P_{ho}}.$$

9.3.5   Moments of service time

Under assumption (9.9), the kth moment of T_D exists (whenever the corresponding for S and V^{-1} exist) and equals, $b^{(k)} := \mathbb{E}_{X,V,S}[(T_D(X, V, S))^k]$, where $\mathbb{E}_{X,V,S}$ is the expectation with respect to the (new call and handover call) joint distribution,

$$P_{X,V,S} := \frac{\lambda(P_{n,X}, P_{n,V}, P_{n,S}) + \lambda_{ho}(P_{h,X,V}, P_{h,S})}{\lambda + \lambda_{ho}}.$$

In Appendix 9B.2, we obtain (recall $\mathbb{E}_n$ is the expectation with respect to P_n, the new call distribution):

$$b^{(k)} = \mathbb{E}_n\left[\frac{T_D(x, v, s)^k + T_D(-D, v, s)^k P_{ho,v}}{1 + P_{ho}}\right] \text{ and} \tag{9.14}$$

$$\frac{db^{(k)}}{dL} = \mathbb{E}_n\left[\frac{d}{dL}\left(\frac{T_D(x, v, s)^k + T_D(-D, v, s)^k P_{ho,v}}{1 + P_{ho}}\right)\right]. \tag{9.15}$$

9.3.6   Cell size: optimizing the moments of the service time

The number of bytes that can be communicated in a cell increases with the increase in cell size: from (9.4) g is continuous and monotonically increasing with respect to L. For any given velocity there exists a minimum cell size (the smallest cell size at which one can transmit more than B_h bytes per cell), below which successful communication is not possible. When cell size is closer to this smallest one, the number of useful bytes transmitted per cell ($g(L) - B_h$) is very small and hence it takes more time to transmit S bytes, i.e., the communication time T_c will be large. As the cell size increases from this smallest size, the communication time T_c starts reducing. However, after some point, due to path loss, the number of bytes per cell starts saturating and hence the gain in terms of useful bytes transmitted per cell will be small in comparison with the extra time taken to traverse each cell, resulting in increasing the communication time again. *Thus there exists an optimal cell size for every fixed velocity.* One can extrapolate similar things even for random velocity. We derive the optimal cell size $L^*_{b^{(1)}}$ and relate the same to the optimizer of more interesting performance measures for ES and NES calls in the subsequent subsections.

We define $L^*_v(v) := \arg \max_L v(v, L)$, the maximizer of the function v given by (9.11). In Appendix 9B.3, we show that there exists a unique $L^*_v(v) > 0$ for every velocity v. In Appendix 9B.4, we further show that for fixed velocities (i.e., when $V \equiv \bar{v}$), the derivatives of all the (existing) k-th moments of the service time vanish only at $L^*_v(\bar{v})$. Thus *for fixed velocities, all the (existing) moments of the service time have unique and common minima*:

$$L^*_{b^{(k)}} := \arg \min_L b^{(k)} = \arg \max_L v(\bar{v}, L) = L^*_v(\bar{v}) \quad \text{for all } k.$$

The common optimizer L^* (for example for $\beta > 1$) satisfies

$$\frac{\partial v(\bar{v}, L)}{\partial L}\bigg|_{L=L^*} = 0 \quad \text{or} \quad 2P\left(\frac{L^*}{d_0}\right)^{-\beta} L^* - \eta(L^*) + \bar{v} B_{\mathrm{h}} = 0 \quad (9.16)$$

and therefore we have the following theorem.

Theorem 9.3 *For a fixed velocity profile, i.e., $P_{n,V}(V = \bar{v}) = 1$ the optimal cell size for the expected service time is,*

$$L^*_{b^{(1)}} = L^*_v = \begin{cases} 2\left(\dfrac{2Pd_0^\beta \frac{\beta}{\beta-1}}{2Pd_0 \frac{\beta}{\beta-1} - \bar{v} B_h}\right)^{\frac{1}{\beta-1}} & \text{when } \beta > 1 \\ 2d_0 e^{\frac{\bar{v} B_h}{2Pd_0}} & \text{when } \beta = 1. \end{cases} \quad (9.17)$$

Further, if the k-th moment of the service time exists then $L^*_{b^{(k)}} = L^*_{b^{(1)}}$.

For velocity profiles with small variance, the optimizers of all the moments of the service time will be approximately equal. Hence when $P_{n,V}$ has small variance with mean $\bar{v}$ then $L^*_{b^{(1)}}$ is close to $L^*_v(\bar{v})$. *From the above it is clear that $L^*_{b^{(1)}}$ increases when the mean $\bar{v}$ increases.*

Having obtained the service time, we now turn our attention to model various configurations of the macrocell with appropriate queues to further obtain their performance measures.

9.3.7 Elastic calls: average waiting time

Each macrocell can handle at maximum of K parallel calls. The control unit (CU) of the macrocell keeps a record of the users entered into the system and serves them in first in first out (FIFO) order via the BSs of the SCs. When a new user initiates a call, it is immediately picked up if there are less than K active calls in the system. If not, the user will have to wait. Its service will start at the time: (i) when one of the active K users finishes their service and exits, (ii) if there are no other waiting users that arrived before it. The BS nearest to the user, at the time of its service start, will initiate the call. Hence, after its call is served (by the macrocell under consideration) as discussed in Section 9.3.1 either till its service is over or till it reaches the macrocell boundary. When it reaches the boundary the call will be transferred to the next macrocell as a handover call, and the handover call is treated by the new cell in a similar way to that of a new call. Thus each macrocell can be modeled by a $M/G/K$ queue with service time T_D and Poisson arrivals at rate $\lambda + \lambda_{\text{ho}}$. This queue has been analyzed to a good extent and the system is stable only if ([20])

$$\rho := \frac{(\lambda + \lambda_{\text{ho}})b^{(1)}}{K} < 1.$$

For stable queues, the expected waiting time of a randomly arrived user can be approximated by [20]:

$$\mathbb{E}[W]_K = \left(\frac{b^{(2)}}{2(b^{(1)})^2}\right)\left(\frac{b^{(1)}}{K(1-\rho)}\right)\left(\frac{(K\rho)^K}{K!}\right)\pi_0; \tag{9.18}$$

$$\pi_0^{-1} = \frac{(K\rho)^K}{K!} + (1-\rho)\sum_{i=0}^{K-1}\frac{(K\rho)^i}{i!},$$

where $b^{(1)}$, $b^{(2)}$ are given by (9.14). If the system is unstable the number of waiting users grows toward infinity and thus one should consider only the cell sizes L with $\rho < 1$. Hence, the optimal size, which minimizes (9.18) is

$$L_{\text{ES}}^* := \arg\min_{\{L:\rho<1\}} \mathbb{E}[W]_K.$$

We saw in the previous section that the optimizer of $b^{(2)}$ is same as that of $b^{(1)}$ for fixed velocities and will be close to each other for smaller velocity variances. In a similar way the same thing is true for ρ, i.e., for fixed velocities,

$$L_\rho^* := \arg\min_{\{L:\rho<1\}} \rho = L_{b^{(1)}}^*.$$

The expected waiting time (9.18) is continuously differentiable in $b^{(1)}$, $b^{(2)}$, and ρ. Thus (minimizer of (9.18) is a zero of its derivative and $E[W]_K$ depends upon L only via $b^{(1)}$, $b^{(2)}$, ρ), we have the following theorem.

Theorem 9.4 *Optimal cell size for a system with ES traffic, with $P_{n,V}(V = \bar{v}) = 1$ is,*

$$L_{\text{ES}}^* = \arg\min_{\{L:\rho<1\}} \mathbb{E}[W]_k = L_{b^{(1)}}^*. \tag{9.19}$$

So, $L_{b^{(1)}}^*$ minimizes both expected waiting time and expected service time. Also, it minimizes the expected sojourn time, as it is the sum of expected waiting and service times.

Also from (9.18), when $L_{b^{(1)}}^*$ and $L_{b^{(2)}}^*$ are close, it is easy to see that the optimizer of $E[W]_K$ will be close to that of the expected service time, $b^{(1)}$. Thus even for low velocity variances,

$$L_{\text{ES}}^* \approx \arg\min_{\{L:\rho<1\}} b^{(1)} = L_{b(1)}^*.$$

We see that this is true even for many general velocity profiles as will be shown in examples of Section 9.4.

9.3.8   Non-elastic calls: block and drop probabilities

As before the system can handle at maximum of K parallel calls. The call is picked up immediately (by the BS of the SC in which the call is originated) only if the macrocell is serving fewer than K users at the time of its arrival. If all the K servers are busy the user is dropped. When an active user reaches the boundary of a macrocell, its call is continued in the next macrocell only if the new macrocell has free servers. Each macrocell can thus be modeled by an $M/G/K/K$ queue. And its call block probability is given by the Erlang loss formula,

$$P_{\mathrm{B}}(L) = \frac{\rho(L)^K/K!}{\sum_{k=0}^{K}\rho(L)^k/k!},\tag{9.20}$$

where ρ denotes the stability factor as was defined in the previous section. It is interesting to note that $P_{\mathrm{B}}(L)$ and ρ are both differentiable with respect to L and further that if the derivative $d\rho/dl$ is zero at a L^* so is the derivative dP_{B}/dL. By taking the second derivative, we can in fact show that their minimizers are the same. Hence,

$$L^*_{P_B} = \arg\min_{\{L:\rho(L)<1\}} P_{\mathrm{B}}(L) = L^*_{\rho}.\tag{9.21}$$

Further, at fixed velocities, $L^*_{\rho} = L^*_{b^{(1)}}$, and so, we have the following theorem:

Theorem 9.5   *The minimizer, $L^*_{b^{(1)}}$ also minimizes the block probability, P_B, for fixed velocities. For any velocity profile, L^*_{ρ} also minimizes the block probability.*

Drop probability: Under the assumptions stated earlier, only a new call can reach the boundary and not a call that has already been handed over once. Further, an active call is dropped only when it reaches the macrocell boundary and the new macrocell is busy. By independence of the two events (status of the new macrocell prior to handover is independent of the call that is handed over), the drop probability is

$$P_D(L) = P_{\mathrm{ho}}P_{\mathrm{B}}(L).$$

One can design an optimal system, catering to NES calls, either by jointly minimizing both the block and drop probabilities or by minimizing one of the probabilities while placing a constraint on the other. Usually systems are designed with more stringent requirements on P_{Dr} than on P_{B}. We note from the above calculations that P_{Dr} is directly proportional to P_{B} and will be smaller than P_{B} by a factor P_{ho}. We make in the rest of this subsection, a commonly made assumption that, the location of the call arrivals is uniformly distributed, i.e., that $X \sim \mathcal{U}[-D, D]$. Under this

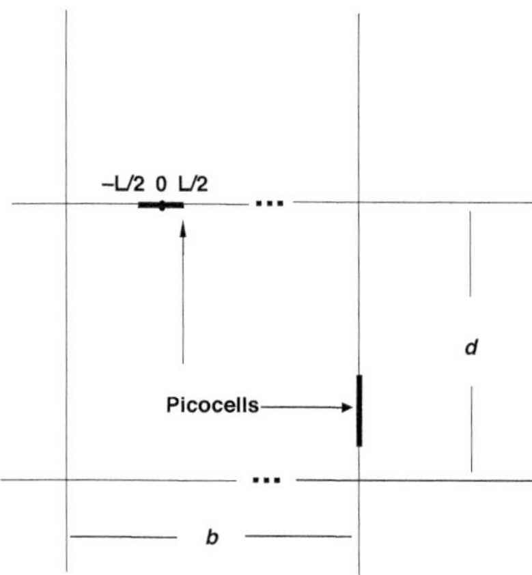

Figure 9.3 Two-dimensional network for rectangular-grid SCNs.

assumption:

$$P_{\mathrm{ho}} = P_n\left(T_\partial(X, V) < T_{\mathrm{c}}(V, S)\right) = \mathbb{E}_{n,S,V}\left[P_X(D - X < V T_{\mathrm{c}}(V, S))\right] \quad (9.22)$$

$$= \frac{\mathbb{E}_{n,S,V}\left[\min\left\{2D, \frac{VSL}{\eta(L) - V B_{\mathrm{h}}}\right\}\right]}{2D} \quad (9.23)$$

$$= \mathbb{E}_{n,V}\left[e^{-\mu 2D(\eta(L) - V B_{\mathrm{h}})/VL}\right] \quad (9.24)$$

$$+ \frac{\mathbb{E}\left[1 - e^{-\mu 2D(\eta(L) - V B_{\mathrm{h}})/VL}\left(1 + \frac{2D\mu(\eta(L) - V B_{\mathrm{h}})}{VL}\right)\right]}{2D\mu} \quad (9.25)$$

$$< \mathbb{E}_{n,V}\left[e^{-\mu 2D(\eta(L) - V B_{\mathrm{h}})/VL}\right] + \frac{1}{2D\mu}. \quad (9.26)$$

Thus P_{ho} decreases with $2D$, the macrocell size. Macrocells are large in dimension and hence P_{Dr} can be certain to be within the prescribed limits (the limit is a design parameter) by directly minimizing P_{B} itself. Thus we propose to choose cell size L to minimize P_{B} and hence equivalently ρ:

$$L_{\mathrm{NES}}^* := L_\rho^* = \arg\min_{\{L:\rho(L)<1\}} \rho(L) \approx L_{b^{(1)}}^*.$$

Remark 9.2 *Thus for both ES and NES applications one needs to minimize the first moment of the service time, $b^{(1)}$, to obtain the optimal cell size. This optimal cell size has been discussed in the previous section for fixed velocities and for velocity profiles with small variances. The general situation is studied in Section 9.4 via numerical examples.*

9.3.9   Mobility on a street grid

We assume that users are moving in a rectangular grid overlaying a large area $[-D, D] \times [-D, D]$ with grid size b, d as shown in Figure 9.3. This example is typical of urban

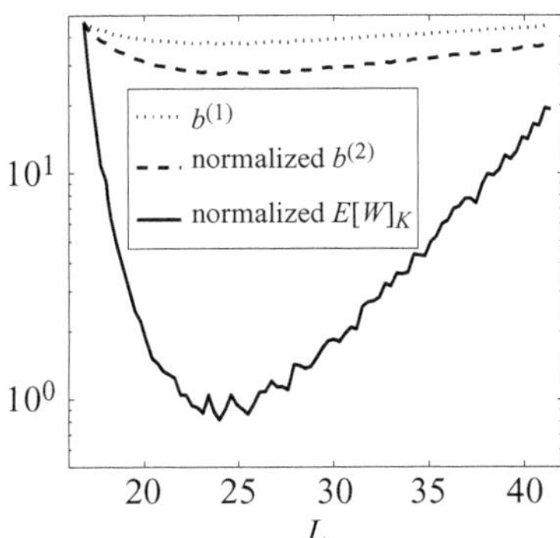

Figure 9.4 Moments of the service time and the expected waiting time (y-axis) versus L (x-axis). ©2010 IEEE. Reprinted, with permission, from [18].

areas where the streets are in a criss-cross manner (this is a well known model, see for example [10, 11]) and hence is an interesting example for study.

In this case, we assume that the location of arrival X is uniformly distributed on the lines, i.e., $X \sim \mathcal{U}[\mathcal{G}]$, where the grid

$$\mathcal{G} := \cup_{i=1}^{D/d}[-D, D] \times \left\{ id + \frac{d}{2} \right\} \ \cup \ \cup_{i=1}^{D/b} \left\{ ib + \frac{b}{2} \right\} \times [-D, D].$$

A one-dimensional vector V represents the speed of the vehicle, which is uniformly distributed, i.e., $V \sim \mathcal{U}[0, V_{\max}]$. Its direction depends upon the position of arrival X: it is horizontal if X is on a horizontal line and is vertical if on a vertical line. One can easily extend the analysis to include zig-zag paths. In either case we assume it is equiprobable in the two possible directions; toward left or right in case of a horizontal line and toward up or down in case of a vertical line.

Any SC is a line segment of a street and a BS is placed at the center of this cell (if we neglect the small number of SCs that might possibly span across two intersecting streets). The mobiles may change directions as they take a turn, but the rate they obtain with their BS once again follows a periodic pattern as explained in Section 9.3.1. Hence the time to reach the boundary T_{∂}, the time to serve S bytes B_S, and hence the service time T_D are just the same as those derived in the previous sections. Thus *the analysis and the results of all the previous sections are applicable to the grid structure. So in the grid structure, the two-dimensional analysis actually boils down to one-dimensional analysis itself.*

9.4 Mobility examples

In the numerical examples of this section, we consider uniformly distributed velocity profiles. The position of arrival X is also uniformly distributed in the macrocell $[-D, D]$. We use Monte Carlo simulations to estimate $b^{(1)}$, $b^{(2)}$ and use exhaustive search to find the optimizers. In Figure 9.4 we plot normalized values of $b^{(1)}$, $b^{(2)}$, and $E[W]_K$ versus

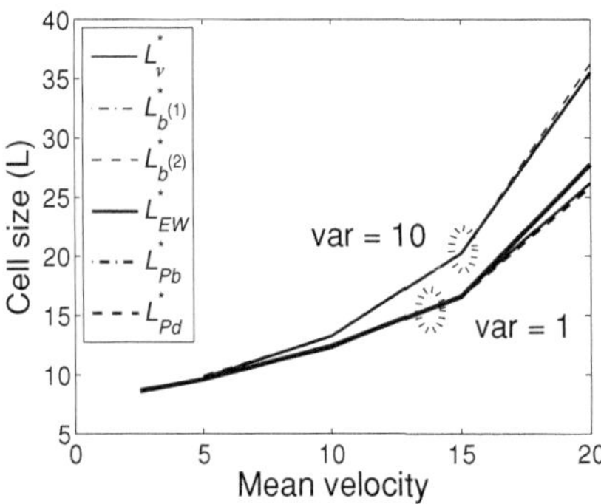

Figure 9.5 Optimal cell size (x-axis) versus mean velocity (y-axis) for different variances. ©2010 IEEE. Reprinted, with permission, from [18].

L. As discussed earlier we notice that the various performance measures decrease with cell size initially, reach an optimal value, and increase again from then on. In fact, all the performance measures have a unique minimum at the same L. We study more details of these minimizers in the following.

In Figure 9.5 we plot the optimal cell size (optimal with respect to the moments of the service time $b^{(1)}$, $b^{(2)}$, block and drop probabilities P_B, P_{Dr} of NES calls and the expected waiting time $\mathbb{E}[W]_K$ of ES calls) versus the mean velocity for two different values of variance. We set $d_0 = 5$, $\lambda = 0.1$, $B_h = 2$, $P = 1$, $\mu = 5$, $K = 20$ and consider a macrocell of size $D = 1000$. We also plot L_v^*, which is the maximizer of $v(\mathbb{E}_n[V], L)$. Note that L_v^* is a single curve in the figure, while the remaining five optimizers ($L_{b^{(1)}}^*$, $L_{b^{(2)}}^*$, $L_{P_B}^*$, $L_{P_{Dr}}^*$, and $L_{\mathbb{E}[W]}^*$) are plotted for two values of variance and hence there are two curves for each of these five optimizers. In fact, L_v^* is not visible separately as it completely coincides with the minimizer $L_{\mathbb{E}[W]}^*$, plotted for variance equal to 1. For small velocity variance (variance equal to 1), all the minimizers are close to L_v^*. For large velocity variance, we notice that all the minimizers (together) are away from L_v^* but, however, are close to each other for most cases. That is, the minimizers of expected waiting time are the same as that of block as well as drop probabilities and all of them equal $L_{b^{(1)}}^*$. This suggests that *even for velocity profiles with high variances, it is sufficient to optimize the average service time $b^{(1)}$ for both ES as well as NES calls and hence the optimal cell size again remains independent of the application. However, for high variance it is not sufficient to minimize $v(\mathbb{E}_n[V], L)$, rather one needs to minimize $b^{(1)}$ directly.*

In Figure 9.6 we further illustrate the same, by plotting the various optimizers now as a function of velocity variance. We set $\mathbb{E}[V] = 10$, $d_0 = 5$, $\lambda = 0.1$, $B_h = 2$, $P = 1$, $\mu = 5$, $K = 20$. We once again note that all the minimizers are close to each other for many cases. We also note that all the minimizers are close to L_v^* for low velocity variances. We further observe that the optimal cell size increases with increase in the variance also. Thus *the larger the velocities the system has to support, the larger are the optimal cell sizes.*

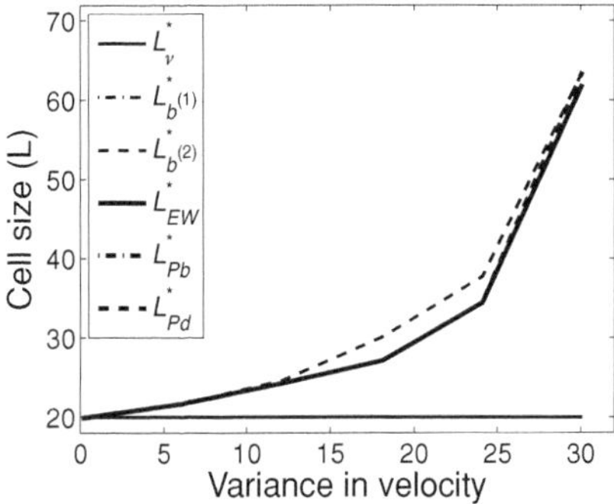

Figure 9.6 Optimal cell size (x-axis) versus variance of the velocity (y-axis). ©2010 IEEE. Reprinted, with permission, from [18].

We notice that, in both Figures 9.5 and 9.6 only the optimizer of the second moment $L_{b^{(2)}}^*$, is sometimes different from the rest of the minimizers. However, even when $L_{b^{(2)}}^*$ is different from $L_{b^{(1)}}^*$, the minimizer L_{EW}^* (minimizing the expected waiting time) is equal to $L_{b^{(1)}}^*$ and thus for both the types of traffic $L_{b^{(1)}}^*$ (minimizer of expected service time) gives the optimal cell size.

9.5   Conclusions

In this chapter we characterize the performance of SCNs in the presence of mobility. We modeled various traffic types between base stations and mobiles as different types of queues. Furthermore, we derived explicit expressions for expected waiting time, service time, and drop/block probabilities for the various queuing models considered for both fixed as well as random velocity of mobiles. We showed that there exists an optimal cell size for a given velocity profile, which minimizes the service time for elastic applications as well as the drop and block probabilities for non-elastic applications. We obtained (approximate) closed form expressions for this optimal cell size when the velocity variations of the mobiles is very small. Our findings show that if the call is long enough, the optimal cell size depends mainly on the velocity profile of the mobiles, its mean and variance; it is independent of the traffic type or duration of the calls. We show that for any fixed power of transmission, there exists a maximum velocity beyond which successful communication between the mobile and the system is not possible. This maximum possible velocity increases with the power of transmission. Further, for any given power, the optimal cell size increases when either the mean or the variance of the mobiles' velocity increases.

Appendix 9A Discussion on call drops at picocell boundaries and finite number of transmission rates

So far in our analysis, we have assumed that there are no call drops during handover when the mobile traverses across picocells. We hence called this as just BSS. In practice

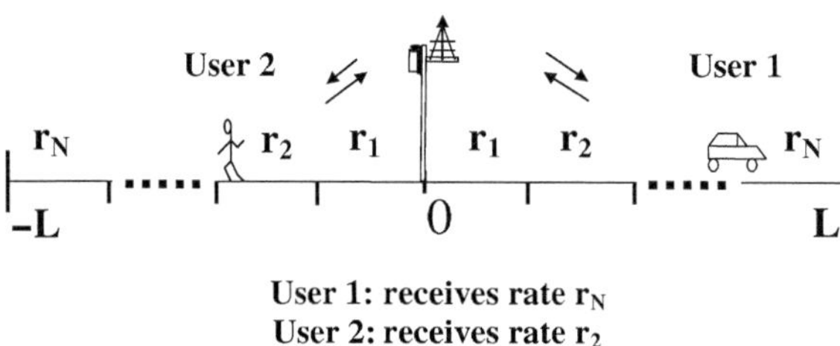

Figure 9.7 One-dimensional cell, rate partitioning, and user's movement.

even the handovers at picocell boundaries can fail, though with an extremely small probability.

In [18, Section 6], we obtained some initial results considering possible call drops at picocell boundaries for the special case of uniform arrivals, uniform velocity profile, and the network catering to NES calls. Every picocell handover, as before, needs B_h bytes to perform BSS and in addition has a (small) possibility of not being successful. We now call the BSS near a picocell boundary also as a handover. The picocell admits a call (handover or a new one) as before, K simultaneous calls at maximum. We discussed various call admission controls, both centralized and distributed, which give higher priority to handover calls. Using similar computations as in the previous section, explicit expressions for drop and busy probabilities and closed form expressions for optimal cell sizes (see [18]) were obtained.

In [18, Section 6], it was again considered that the communication rates can change continually based on the continual change in the distance between the user and the BSs. However, this is not a practically feasible procedure. Most of the practical systems can support only a finite number of transmission rates and a rate is allocated in the current time slot (from amongst a finite set), based on the estimated SNR of the operating channel. In a recent study in [21], the case with finite number of transmission rates is considered. In this we consider two sets of users: (i) high-speed users moving in one direction (in a line) and with fixed but random velocity as in the previous section; and (ii) randomly wandering users moving in a one-dimensional or two-dimensional cell, respectively (see Figures 9.7 and 9.8). Most of the details of the system model are as in the previous section except for the following:

Power per transmission: Let η represent the dimension of the cell, i.e., $\eta = 1$ for a one-dimensional and 2 for a two-dimensional cell. It might be a better idea to decide an optimal cell size, maintaining the total power budget in the system constant. That is, we will replace per power transmissions P (see for example, equation (9.4)) with a P_L, indicating it can depend upon L, the cell size. If the total power in the system has to remain constant[6] then $P_L = PL^\eta$. However, we will notice (in a short while) that it is not sufficient that P_L scales by η^+ with L for the case of a finite number of transmission rates.

---

[6] With η representing the dimension of the cell, the number of picocells, when each is of dimension L, is proportional to $L^{-\eta}$ and hence total power would be proportional to $P_L L^{-\eta}$. Thus to maintain the total power constant, $P_L = PL^\eta$ for some constant $P > 0$.

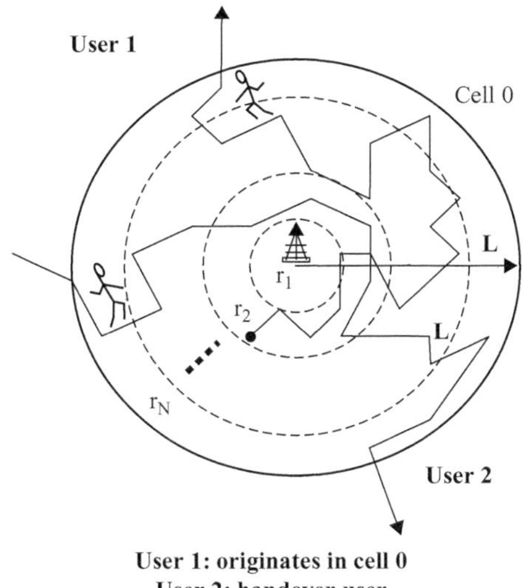

Figure 9.8 Two-dimensional cell, rate partitioning, and user's movement.

Rate regions: The cell is divided into $2N$ (or N for two-dimensional) disjoint segments (based on the distance from BS) such that the users in a segment are served with the same transmission rate. Let $\{\mathbb{A}_n\}_{n\in\mathbb{N}}$ represent these rate regions (see Figures 9.7 and 9.8), where with $|.|$ representing the norm or the absolute value, we have

$$\mathbb{A}_n := \begin{cases} \left[\frac{(n-1)L}{N}, \frac{nL}{N}\right] 1_{\{n>0\}} + \left[\frac{nL}{N}, \frac{(n+1)L}{N}\right] 1_{\{n<0\}}, & \text{if one-dimensional} \\ \left\{ \mathbf{x} \in \mathcal{R}^2 : \frac{(n-1)L}{N} \leq |\mathbf{x}| \leq \frac{nL}{N} \right\}, & \text{if two-dimensional} \end{cases} \tag{9.27}$$

$$\text{and } \mathbb{N} := \begin{cases} \{-N, \cdots, -1, 1, \cdots, N\}, & \text{if 1D} \\ \{1, \cdots, N\}, & \text{if 2D.} \end{cases} \tag{9.28}$$

A user in region n receives service at rate $r_{|n|}$. Let $\mathbb{R} := \{r_1, \ldots, r_N\}$ represent the ensemble of all possible transmission rates. Note that this set is arranged in decreasing order. For example in a two-dimensional (circular) cell of Figure 9.8, each annular ring is served with a common rate and these common rates decrease as the distance from the center (where the BS is located) increases. The rate at which the service is offered changes once the user switches from one region to another.

Embedded (rate) Markov chain: The users can be located anywhere in one of the rate regions $\{\mathbb{A}\}_n$. We represent the user location at time step k by Φ_k. When $\Phi_k = n$, it implies that the user is wandering in segment $\mathbb{A}_n$ and is receiving service at rate $r_{|n|}$ at time k. Let W_n represent the time for which the user remains in nth region, $\mathbb{A}_n$. This represents (for any k), the actual time for which the kth step lasts, given that $\Phi_k = n$. Note here, we are inherently assuming that the consequent times the user spends in the same rate region, are independent and identically distributed (i.i.d.). However, these times can depend upon the region in which the user is wandering. After wandering in a certain rate region n for time W_n the user either moves to region $n + 1$ with probability p_n or to region $n - 1$ with probability $1 - p_n$. *Note that $p_1 = 1$ always for two-dimensional.* That is, $\{p_n\}$ represent the transition probabilities of the embedded Markov chain $\{\Phi_k\}$.

All the quantities $\{p_n\}$, $\{W_n\}$, $\{r_n\}$ can depend upon the dimension L (which we are trying to optimize). Every arrival brings along with it the marks (Φ, S), where $\Phi \in \mathbb{N}$ is the position of arrival with distribution $\Pi := \{\pi_n\}$ and S the number of bytes to be transmitted, which is exponentially distributed, i.e., $S \sim \mu exp^{-\mu t} dt$ for some $\mu > 0$.

An example of $\mathbb{R}$: *One can choose the set of possible transmission rates, $\mathbb{R}$ and N based on the practical channel coding schemes that are going to be used in the network design. The analysis presented can be utilized to study a system with any given $\mathbb{R}$ and N.* However, in this paper, we consider a specific example. This specific $\mathbb{R}$ is obtained using low SNR approximation of the following theoretical (capacity) rate as in the previous section:

$$r(d) := P_L \left(1_{\{d \leq d_0\}} + r_0 |d|^{-\beta} 1_{\{d > d_0\}}\right) \text{ with } r_0 = d_0^{\beta},$$

where $r(d)$ is the rate at distance d, d_0 is a small lossless distance, while β is the propagation coefficient. We consider a specific system that supports transmission at the maximum possible rate for the entire region. For example in $\mathbb{A}_n$ the farthest user will be at distance $|n|L/N$ and hence maximal transmission rate, that can be allocated, equals

$$r_n = r(|n|L/N) = r_0 P_L N^\beta L^{-\beta} |n|^{-\beta}. \tag{9.29}$$

Alternatively, if the system under consideration can design modulation and/or channel coding schemes so as to achieve (almost) v percent of the theoretical rates where $v < 1$ is a fixed coefficient, then again the above rate structure is applicable (after absorbing v into r_0 of (9.29)).

We study such small cells again using queuing theoretic tools and obtain relevant performance metrics such as the expected service time, call busy and/or drop probability, etc. For the case of randomly wandering users we required extra tools to analyze, that of the conditional expectations and the theory of Markov chains [21]. We obtain the explicit expressions for the case in which the wandering times are exponentially distributed. We also derive *capacity per cell*, a notion that gives the maximum number of bytes that can be transferred while the user traverses in a cell divided by the cell size. Using this notion we establish the following important phenomenon, which arises with finite number of transmission rates. In the following we discuss only this important phenomenon in detail and the reader is referred to [21] for remaining details.

Capacity per cell: Capacity of a cell is the average number of the *maximum*[7] bytes (the maximum information) per cell dimension, that can be transmitted, while a user moves in the cell which can support N distinct rates. If the total power in the system has to remain constant then $P_L = PL^\eta$ (where η equals 1 for one-dimensional or 2 for two-dimensional). With P_L scaling as PL^η, we notice (see [21] for details) that C_{cell} decreases with L. This implies that the optimal cell size (optimizing the fundamental limit C_{cell}) is Nd_0, which is practically an infeasible cell dimension. In other words,

---

[7] By *maximum* we mean the theoretical maximum possible rate, given the rate partitioning. The rates given by (9.29) exactly represent this *maximum* rate when $v = 1$.

the total power budget has to be increased with L, to design cells with practical values of cell dimension. The necessary growth rate is obtained as below (see [21] for details):

Lemma 9.1 (β^+ **scaling**) *Capacity per cell increases with L only if the power per transmission scales with L according to*

$$P_L = PL^{\beta+\gamma} \text{ for some } \gamma > 0.$$

The above lemma only says that the fundamental capacity can improve monotonically with cell size only when you use $P_L = PL^{\beta+\gamma}$. Henceforth, we call this as β^+ *scaling*. This is the major difference in comparison with the continual rates case. For continual rates, the capacity C_{cell} equals $E[g(L)]L^{-\eta}$ where $g(L)$ is defined in (9.4) and as we notice from (9.4) this does not decrease with L once the power is scaled as P^η (i.e., when P in (9.4) is replaced with PL^η). In other words, for continual rates one just needs η^+ power scaling (in place of β^+ scaling as required with a finite number of transmission rates), which is obtained once we spend the same total average power in the system, independent of L the cell size. This is mainly because, with larger cells, the user spends more time in the same cell and hence more bytes can be transferred while moving in a cell, increasing $E[g(L)]$ with L, which finally saturates because of propagation losses even if the power per transmission is maintained constant. But once you have a rate region, one must allocate a rate that can be sustained by all the users of the same region and hence will be decided by the farthest user in that region (we right now consider only distance-based propagation losses). So, whenever the cell size increases, with N remaining the same, the rate regions are also expanded by a proportional factor. This expansion in the rate regions results in a smaller rate allocated to users. The reduction in the number of bytes transferred because of these reduced rates could be more significant than the extra bytes transferred because of the user moving in a larger cell. Further, the reduction in rates allocated is proportional to the extra distance each rate region is expanded, by a factor given by the path-loss coefficient β. One can counteract this effect, either by increasing N (the number of feasible rates) proportional to L or by increasing the power using β^+ scaling.

The above discussion is just about a fundamental limit that captures the maximum that can be derived out of the system, given the choice of transmission rates. But this fundamental limit does not consider the losses due to handover. However, one would be more interested in designing systems that are optimal with respect to measures such as drop/busy probability. In [21], we obtain closed-form expressions for the above-mentioned performance measures with β^+ power scaling and obtain (for some asymptotic cases) closed-form expressions for optimal cell size. In this work, we consider both high-speed unidirectional users and randomly wandering users, albeit with independent resources. We plan to study the case when the two sets of users share the same resources. We are also looking at power allocation based on user classes based on their average speeds. In all these case studies, we make one common observation, the optimal cell size increases with the increase in user speeds.

Appendix 9B Calculations related to macro queue

9B.1 Proof of Theorem 9.2

The communication time is finite with probability 1 if and only if

$$\mathbb{P}(B_{\mathrm{h}} > g(L)) = P_{n,V}(V B_{\mathrm{h}} > \eta(L)) = 0, \quad \text{where } \eta(L) := V g(L).$$

Note from (9.4) that η is only a function of P, L and hence this probability depends only on the velocity profile V, cell size L, and the transmit power P. Because of the path loss, for any fixed P, $\eta(L)$ increases as L increases and finally saturates (when $\beta > 1$). Thus for all L (when $\beta > 1$),

$$\eta(L) \leq \eta_\infty := \lim_{L \to \infty} \eta(L) < \infty.$$

When $\beta = 1$, $\eta_\infty = \infty$ and this proves the first statement of the theorem. As $\mathbb{P}(B_{\mathrm{h}} > g(L)) = \mathbb{P}(V B_{\mathrm{h}} > \eta(L))$, there exists a cell size L with $P(B_{\mathrm{h}} > g(L)) = 0$ if and only if

$$\mathbb{P}\left(V > \frac{\eta_\infty}{B_{\mathrm{h}}}\right) = 0. \tag{9.30}$$

Thus for a given power P, the system can handle all velocities that are strictly less than

$$V_{lim} := \frac{\eta_\infty}{B_{\mathrm{h}}} = \frac{2 P d_0 \beta}{\beta - 1}. \quad \square \tag{9.31}$$

9B.2 Moments of service time and its derivatives

The moments can be rewritten as,

$$b^{(k)} = \frac{1}{(1 + P_{\mathrm{ho}})} \int_0^\infty \int_{-D}^D \int_0^{V_{\max}} \left[T_D(x, v, s)^k + T_D(-D, v, s)^k P_{\mathrm{ho},v}\right]$$

$$\times f_{n,V}(v) \mathrm{d}v \ f_{n,X}(x) \mathrm{d}x \ \mu e^{-\mu s} \mathrm{d}s.$$

$$= \mathbb{E}_n \left[\frac{T_D(x, v, s)^k + T_D(-D, v, s)^k P_{\mathrm{ho},v}}{1 + P_{\mathrm{ho}}}\right]. \tag{9.32}$$

By bounded convergence theorem (BCT), all $b^{(k)}$ are continuously differentiable (c.d.) in L and the derivative is

$$\frac{\mathrm{d}b^{(k)}}{\mathrm{d}L} = \mathbb{E}_n \left[\frac{\mathrm{d}}{\mathrm{d}L}\left(\frac{T_D(x, v, s)^k + T_D(-D, v, s)^k P_{\mathrm{ho},v}}{1 + P_{\mathrm{ho}}}\right)\right].$$

Because,

1. T_D is almost surely c.d.
2. $P_{\mathrm{ho},v}$ is c.d. everywhere in L.
3. P_{ho} is c.d.
4. All the derivatives involved are uniformly bounded almost surely.

And hence by virtue of the mean value theorem, terms like

$$\frac{|T_D(X, V, S; L + \delta) - T_D(X, V, S; L)|}{\delta}, \quad \frac{|P_{\mathrm{ho},v}(L + \delta) - P_{\mathrm{ho},v}(L)|}{\delta}, \text{ etc.}$$

can be bounded uniformly by a constant.

9B.3 v has a unique maximizer

From equation (9.4), g and hence $\eta = vg$ are both concave in L on $(0, \infty)$ for every v. Thus from (9.11), for any fixed velocity v, v has a unique maxima,

$$L_v^*(v) := \arg\max_L v(v, L),$$

which satisfies $\partial v/\partial L = 0$ (as clearly $L_v^*(v) > 0$ for all v).

9B.4 Derivatives $\mathrm{d}b^{(k)}/\mathrm{d}L$ vanish only at $L_v^*(\bar{v})$ when $V \equiv \bar{v}$

Define $\Psi(L) := \mathbb{E}_n\left[T_D(X, V, S)^k + T_D(-D, V, S)^k P_{\mathrm{ho},V}\right]$. Then from (9.14),

$$b^{(k)} = \frac{\Psi(L)}{1 + P_{\mathrm{ho}}}. \tag{9.33}$$

From (9.7), T_{c} depends upon L only via the function v given by (9.11) and hence so is the service time $T_D(x, v, s) = \min\{T_{\mathrm{c}}(v, s), T_\partial(x, v)\}$ for all x, v, s. Similarly from (9.13) and (9.12), $P_{\mathrm{ho},v}$ and P_{ho} depend upon L only via the function v. Hence with

$$\Theta(v) := -\frac{\partial P_{\mathrm{ho},v}}{\partial v} = \frac{\mu}{v}\mathbb{E}_{n,X}\left[(D - X)e^{-\mu v(v,L)(D-X)/v}\right] \tag{9.34}$$

$$\frac{\mathrm{d}b^{(k)}}{\mathrm{d}L} = \frac{1}{1 + P_{\mathrm{ho}}}\mathbb{E}_n\left[\frac{\partial v(V, L)}{\partial L}\left(\frac{\partial T_D(X, V, S)^k}{\partial v} + P_{\mathrm{ho},V}\frac{\partial T_D(-D, V, S)^k}{\partial v} + \frac{\partial P_{\mathrm{ho},V}}{\partial v}T_D(-D, V, S)^k\right)\right]$$

$$\quad - \frac{1}{(1 + P_{\mathrm{ho}})^2}\Psi(L)\frac{\mathrm{d}P_{\mathrm{ho}}}{\mathrm{d}L}$$

$$= \frac{1}{1 + P_{\mathrm{ho}}}\mathbb{E}_n\left[\frac{\partial v(V, L)}{\partial L}\left(-kT_D(X, V, S)^{k-1}\frac{S1_{\{SV <(D-X)v(V,L)\}}}{v(V, L)^2}\right.\right.$$

$$\quad \left.\left. -kP_{\mathrm{ho},V}T_D(-D, V, S)^{k-1}\frac{S1_{\{SV <2Dv(V,L)\}}}{v(V, L)^2} - T_D(-D, V, S)^k\Theta(V)\right)\right]$$

$$\quad + \frac{1}{(1 + P_{\mathrm{ho}})^2}\Psi(L)\mathbb{E}_n\left[\frac{\partial v(V, L)}{\partial L}\Theta(V)\right]$$

$$= \mathbb{E}_n\left[\frac{\partial v(V, L)}{\partial L}\left(\frac{-k\frac{S^k 1_{\{SV <(D-X)v(V,L)\}}}{v(V,L)^{k+1}} - kP_{\mathrm{ho},V}\frac{S^k 1_{\{SV <2Dv(V,L)\}}}{v(V,L)^{k+1}} - T_D(-D, V, S)^k\Theta(V)}{1 + P_{\mathrm{ho}}}\right.\right.$$

$$\quad \left.\left. + \frac{\Psi(L)\Theta(V)}{(1 + P_{\mathrm{ho}})^2}\right)\right]. \tag{9.35}$$

Thus the derivatives will have the form

$$\frac{db^{(k)}}{dL} = \mathbb{E}_{n,V}\left[\frac{\partial v(V,L)}{\partial L}\mathbb{E}_{n,X,S}[\Gamma^{(k)}(X,V,S,v(V,L))]\right] \tag{9.36}$$

for some functions $\Gamma^{(k)}$. Thus for fixed velocities, i.e., when $V \equiv \bar{v}$

$$\frac{db^{(k)}}{dL} = \frac{\partial v(\bar{v},L)}{\partial L}\mathbb{E}_{n,X,S}[\Gamma^{(k)}(X,\bar{v},S,v(\bar{v},L))]. \tag{9.37}$$

Claim: The term $\mathbb{E}_{n,X,S}[\Gamma^{(k)}(X,\bar{v},S,v(\bar{v},L))]$ is strictly negative.
Proof of claim: For any velocity v, $T_D(X,v,S) \le T_D(-D,v,S)$ for all (X,S). Thus, for fixed velocities,

$$\Psi(L) \le (1 + P_{\mathrm{ho},\bar{v}})\mathbb{E}_{n,S}[T_D(-D,\bar{v},S)^k].$$

Further $P_{\mathrm{ho}} = \mathbb{E}_{n,V}[P_{\mathrm{ho},V}] = P_{\mathrm{ho},\bar{v}}$. Hence the sum of the last two inner terms of the equation (9.35),

$$\mathbb{E}_{n,X,V}\left[-\frac{T_D(-D,V,S)^k\Theta(V)}{1 + P_{\mathrm{ho}}} + \frac{\Psi(L)\Theta(V)}{(1 + P_{\mathrm{ho}})^2}\right] \le 0.$$

The remaining two inner terms of (9.35) are always negative and this proves the claim.   □

From (9.37), by the virtue of the claim, derivative $db^{(k)}/dL$ is zero only at a zero of $\partial v/\partial L$. But $\partial v/\partial L$ has a unique zero at the maximizer $L_v^*(\bar{v})$. Thus $L_v^*(\bar{v})$ is the only zero of all the service time moments.

Appendix 9C Calculations related to pico queue

9C.1 Pico handover speed distribution

In the following the event $\{\partial\ arrival\} = \{\partial\}$ means that the arrival in cell 0 was at the boundary (i.e., it was a handover arrival from cell -1), the event $\{int\ arrival\} = \{int\}$ means a new arrival in cell 0 and the event $\{ho\}$ implies that the cell 0 active user has reached the boundary before completing his service (i.e., the call has to be handed over to cell 1). Note that a handover from cell 0 occurs either due to an already handed over call or new call, whose service could not be completed before reaching the boundary of cell 0 and thus:

$$\mathbb{P}(\mathrm{ho}) = \frac{\lambda_L P_{\mathrm{ho}} + \lambda_{hL} P_{\mathrm{ho},\partial}}{\lambda_L + \lambda_{hL}}.$$

Note that $\mathbb{P}(\mathrm{ho}) \to 1$ as $L \to 0$ (as both P_{ho}, $P_{\mathrm{ho},\partial}$ converge to 1). With the above notations, the fixed point equation for the handover speed density can be obtained as:

$$f_{h,V}(v) = \mathbb{P}(V \in vdv \,|\, \text{ho } \textit{from cell } 0) \tag{9.38}$$

$$= \mathbb{P}(V \in vdv \ \textit{int arrival}\,|\,\text{ho}) + \mathbb{P}(V \in vdv \ \text{ho } \textit{arrival}\,|\,\text{ho}) \tag{9.39}$$

$$= \frac{\mathbb{P}(\text{ho } V \in vdv \ \textit{int}) + \mathbb{P}(V \in vdv \ \partial \ \text{ho})}{\mathbb{P}(\text{ho})} \tag{9.40}$$

$$= \frac{\mathbb{P}(\text{ho}\,|\,\textit{int } v)\mathbb{P}(V \in vdv\,|\,\textit{int})\mathbb{P}(\textit{int})}{\mathbb{P}(\text{ho})} \tag{9.41}$$

$$+ \frac{\mathbb{P}(\text{ho}\,|\,\partial \ v)\mathbb{P}(V \in vdv\,|\,\partial)\mathbb{P}(\partial)}{\mathbb{P}(\text{ho})} \tag{9.42}$$

$$= \frac{P_{\text{ho},v}\, f_{n,V}(v)}{\mathbb{P}(\text{ho})} \frac{\lambda_L}{\lambda_L + \lambda_{hL}} + \frac{P_{\text{ho},v,\partial}\, f_{h,V}(v)}{\mathbb{P}(\text{ho})} \frac{\lambda_{hL}}{\lambda_L + \lambda_{hL}}. \tag{9.43}$$

Solving (when $f_{n,V}(v) = 1/V_{\max}$ for all $v \le V_{\max}$) the handover speed density is,

$$f_{h,V}(v) = \frac{\frac{\lambda_L}{\mathbb{P}(\text{ho})(\lambda_L + \lambda_{hL})V_{\max}} P_{\text{ho},v}}{1 - \lambda_{hL} \frac{P_{\text{ho},v,\partial}}{\mathbb{P}(\text{ho})(\lambda_L + \lambda_{hL})}}.$$

Thus as L tends to zero ($P_{\text{ho},v}$, $P_{\text{ho},v,\partial}$, $\mathbb{P}(\text{ho})$, $P_{\text{ho}} \to 1$), the speed of a handover arrival will tend to a uniform distribution. This effect is seen sooner if the packet size S is larger.

9C.2 Pico stability factor

The stability factor ρ in a pico queue is given by:

$$\rho_{pico}(L) = \frac{(\lambda_L + \lambda_{hL})b^{(1)}}{K} = \frac{1}{K}\mathbb{E}_V[\lambda_L \mathbb{E}_X[\mathbb{E}_S[B_L(S, X, V)]]$$

$$+ \lambda_{hL}\mathbb{E}_S[B_L(S + B_{\text{h}}, 0, V)]] \tag{9.44}$$

$$\approx \frac{1}{K}\left(\lambda_L(L - \mathbb{E}[X]) + L\lambda_{hL}\right) E\left[\frac{1}{V}\right] = \frac{1}{K}\left(\lambda_L \frac{L}{2} + L\lambda_{hL}\right) E\left[\frac{1}{V}\right] \tag{9.45}$$

$$= c_0 L(\lambda_L + 2\lambda_{hL}) \overset{\text{as } P_{\text{ho}} \approx 1}{\approx} c_0 L \lambda_L \frac{1 - P_{\text{ho},\partial} + 2}{1 - P_{\text{ho},\partial}} \approx \frac{2c_0 \lambda_L L}{1 - P_{\text{ho},\partial}} \tag{9.46}$$

where $c_0 = 1/(2K)E[1/V]$ is a constant independent of L. Using the approximation in (equation (15) in [18]),

$$\rho_{pico}(L) = \frac{\lambda p L}{K D \mu v(\bar{v}_{\text{inv}}, L)}. \tag{9.47}$$

9C.3 Drop probability

The drop probability $P_{D,pico}$ considering possible pico drops can be calculated as below:

$$P_{D,pico} = \mathbb{P}\,(\text{Call ever dropped before completion}\,|\, \text{Call is picked up}) \tag{9.48}$$

$$= P_{\text{ho}}\left(P_{\text{ho},D}(1 - P_{\text{busy, pico}}) + P_{\text{busy, pico}}\right) + (1 - P_{\text{ho}})(0) \tag{9.49}$$

where $P_{\text{ho},D}$ is defined as the probability of call drop at any of the future instances of handovers, given that the current handover (first handover in the context of the above equation) is successful. Because of the memoryless nature of S, this probability does not depend upon the number of the handover. Probability $P_{\text{ho},D}$ can be calculated by first conditioning on the event that the call is completed in the current cell (call it as C) and then on the event that the call is not picked up in the next cell. Note that $P_{\text{ho}}(C^c) = P_{\text{ho},\partial}$ and $P_{\text{ho}}(S|C^c) = P_{\text{ho,fail}} = 1 - P_{\text{busy, pico}}$. Thus, by conditioning

$$
\begin{aligned}
P_{\text{ho},D} &= P_{\text{ho}}\left(\text{Call dropped} \cap C\right) + P_{\text{ho}}\left(\text{Call dropped} \cap C^c\right) \\
&= 0 + P_{\text{ho},\partial} P_{\text{ho}}\left(\text{Call dropped} \,|\, C^c\right) \\
&= P_{\text{ho},\partial}\left(P_{\text{ho}}\left(\text{Call dropped} \cap S^c | C^c\right) + P_{\text{ho}}\left(\text{Call dropped} \cap S | C^c\right)\right) \\
&= P_{\text{ho},\partial}\left(P_{\text{ho},D}(1 - P_{\text{busy, pico}}) + 1 P_{\text{busy, pico}}\right) \\
&\overset{Solving}{=} \frac{P_{\text{busy, pico}} P_{\text{ho},\partial}}{1 - P_{\text{ho},\partial}(1 - P_{\text{busy, pico}})} \quad \text{and hence,}
\end{aligned}
$$

$$
\begin{aligned}
P_{D,\text{pico}} &= \frac{P_{\text{ho}} P_{\text{busy, pico}}}{1 - P_{\text{ho},\partial}(1 - P_{\text{busy, pico}})} \approx \frac{P_{\text{busy, pico}}}{1 - P_{\text{ho},\partial} + P_{\text{busy, pico}}} \\
&= \frac{P_{\text{busy, pico}}}{\frac{\mu L \nu(\bar{v}_{\text{inv}}, L)}{\bar{v}_{\text{inv}}} + P_{\text{busy, pico}}} = \frac{P_{\text{busy, pico}}}{\frac{\mu(\eta(L) - \bar{v}_{\text{inv}} B_{\text{h}})}{\bar{v}_{\text{inv}}} + P_{\text{busy, pico}}}.
\end{aligned}
\tag{9.50}
$$

Acknowledgments

The work of the first author is sponsored by FCAM and Ganesh associate NRIA team. The work of the second author is carried out under the framework of INRIA – Alcatel Bell Labs joint research lab on Self Organized Networks, and ANR Ecocells project. The work of the third author is sponsored by ANR Ecocells, IFCAM, and Ganesh associated INRIA team.

References

[1] S. Sen, A. Arunachalam, K. Basu, and M. Wernik, "A QoS management framework for 3G wireless mobile network," in *Proc. IEEE Wireless Commun. Networking Conf. (WCNC)*, New Orleans, LA, Sep. 1999, pp. 1273–7.

[2] Alcatel-Lucent, "Beyond the base station router," Alcatel-Lucent, Tech. Rep., 2008.

[3] P. V. Orlik and S. S. Rappaport, "On the handoff arrival process in cellular communications," *Wirel. Netw.*, vol. 7, no. 2, Mar.–Apr. 2001.

[4] S. Dharmaraja, K. S. Trivedi, and D. Logothetis, "Performance analysis of cellular networks with generally distributed handoff interarrival times," *Elsevier Comput. Commun.*, vol. 26, no. 15, pp. 1747–56, Sep. 2003.

[5] X. Huang and R. F. Serfozo, "Spatial queueing processes," *Math. Oper. Res.*, vol. 24, no. 4, pp. 865–86, Nov. 1999.

[6] V. Kavitha and E. Altman, "Queueing in space: design of message ferry routes in sensor networks," in *Proc. Int. Teletraffic Congress (ITC)*, Paris, France, Sep. 2009, pp. 1–8.

[7] W. Saad, Z. Han, T. Basar, M. Debbah, and A. Hjorungnes, "A selfish approach to coalition formation among unmanned air vehicles in wireless networks," in *Proc. Int. Conf. Game Theory for Networks (GAMENETS)*, Istanbul, Turkey, May 2009, pp. 259–67.

[8] F. Baccelli, B. Blaszczyszyn, and M. Karray, "A spatial markov queuing process and its applications to wireless loss systems," INRIA, Tech. Rep., 2010.

[9] R. W. Wolff, *Stochastic Modeling and the Theory of Queues*. Prentice-Hall, 1989.

[10] J. G. Markoulidakis, G. L. Lyberopoulos, D. F. Tsirkas, and E. D. Sykas, "Mobility modeling in third-generation mobile telecommunications systems," *IEEE Pers. Commun.*, vol. 4, no. 4, pp. 41–56, Aug. 1997.

[11] J. B. Andersen, T. S. Rappaport, and S. Yoshida, "Propagation measurements and models for wireless communications channels," *IEEE Commun. Mag.*, vol. 33, no. 1, pp. 42–9, Jan. 1995.

[12] M. Grossglauser and D. N. C. Tse, "Mobility increases the capacity of ad hoc wireless networks," *IEEE/ACM Trans. Netw.*, vol. 10, no. 4, pp. 477–86, Aug. 2002.

[13] N. Bansal and Z. Liu, "Capacity, delay and mobility in wireless ad-hoc networks," in *Proc. IEEE Int. Conf. on Computer Commun. (INFOCOM)*, San Francisco, CA, Apr. 2003, pp. 1553–63.

[14] T. Bonald, S. Borst, N. Hegde, M. Jonckheere, and A. Proutiere, "Flow-level performance and capacity of wireless networks with user mobility," *Queueing Syst.: Theory Appl.*, vol. 63, no. 1–4, pp. 131–64, Dec. 2009.

[15] T. Bonald, S. Borst, and A. Proutiere, "How mobility impacts the flow-level performance of wireless data systems," in *Proc. IEEE Int. Conf. on Computer Commun. (INFOCOM)*, Hong Kong, Mar. 2004, pp. 1872–81.

[16] S. Borst, A. Proutiere, and N. Hegde, "Capacity of wireless data networks with intra- and inter-cell mobility," in *Proc. IEEE Int. Conf. on Computer Commun. (INFOCOM)*, Barcelona, Spain, Apr. 2006, pp. 1–12.

[17] S. Borst, N. Hegde, and A. Proutiere, "Mobility-driven scheduling in wireless networks," in *Proc. IEEE Int. Conf. on Computer Commun. (INFOCOM)*, Rio de Janeiro, Brazil, Apr. 2009, pp. 1260–8.

[18] V. Kavitha, S. Ramanath, and E. Altman, "Spatial queuing analysis for design and dimensioning of picocell networks with mobile users," *Elsevier Perform. Evaluation*, vol. 68, no. 8, pp. 710–27, Aug. 2011.

[19] S. Ramanath, E. Altman, V. Kumar, and M. Debbah, "Optimizing cell size in pico-cell networks," in *Proc. Int. Workshop on Resource Allocation in Wireless Networks (RAWNET)*, Seoul, South Korea, June 2009, pp. 1–9.

[20] Y. Takahashi, "An approximation formula for the mean waiting time of an M/G/c queue," *J. Operations Research Society of Japan*, pp. 150–63, 1977.

[21] V. Kavitha, S. Ramanath, and E. Altman, "Analysis of small cells with randomly wandering users," in *Proc. Ing. Conf. Mobile, Ad hoc and Wireless Networks (WiOpt)*, Paderborn, Germany, May 2012.

10 Cognitive radio resource management in autonomous femtocell networks

Kwang-Cheng Chen and Shao-Yu Lien

10.1 Introduction

Because of the use of common radio resources in small cell networks [1, 2], destructive interference may occur not only between the femtocell and the macrocell, but also among femtocells whose coverage areas are highly overlapped with each other (collocated femtocells), as shown in Figure 10.1. To avoid interference, one typical solution is to divide the entire available spectrum into several frequency bands, and then each femtocell and the macrocell utilize different frequency bands from each other [3]. This deployment is referred to as "dedicated channel" deployment in 3rd Generation Partnership Project (3GPP) [4]. However, the performance of this solution is limited by the assigned bandwidth, which makes it infeasible for dense femtocell deployments where each femtocell can only utilize a very limited fraction of the bandwidth. An alternative solution is to adopt spatial domain frequency reuse [5]; however, this is infeasible for user deployed femtocells without centralized and perfect planning. As a result, a practical solution turns out to be "co-channel" deployment, where all femtocells and the macrocell can utilize all the available spectrum. To mitigate interference in co-channel deployment, dynamic power adaptation in femtocells has been proposed for code division multiple access (CDMA) systems [6–11] to combat interference due to the near–far problem. Considering that orthogonal frequency division multiple access (OFDMA) has been adopted by 3GPP LTE-Advanced (LTE-A) and wireless interoperability for microwave access (WiMAX), new interference mitigation techniques are needed [12]. In OFDMA, the major cause of interference is that multiple networks occupy the same radio resources (subcarriers and orthogonal frequency division multiplexing (OFDM) symbols) simultaneously. To avoid this, a cross-tier centralized radio resource allocation scheme has been proposed in [13]. In this centralized scheme, however, the potentially high computational complexity and the large amount of information exchange create challenges to scalability, which cannot be applied to a scenario with a very large number of femtocells. As a result, femtocells should be able to "autonomously" mitigate interference to/from the macrocell as well as to/from other collocated femtocells. In addition, femtocells are required to provide quality of service (QoS) guarantees while fully utilizing the radio

Small Cell Networks: Deployment, PHY Techniques, and Resource Management, ed. Tony Q. S. Quek, Guillaume de la Roche, İsmail Güvenç, and Marios Kountouris. Published by Cambridge University Press. © Cambridge University Press 2013.

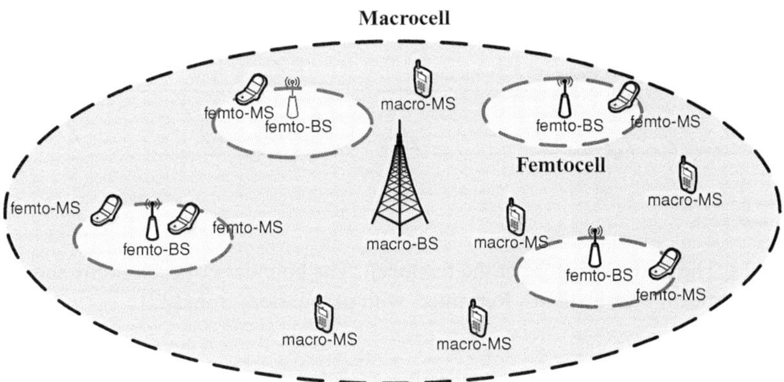

Figure 10.1 Network model of the considered scenario. ©2011 IEEE. Reprinted, with permission, from [22].

resources [14], which has been recognized as an essential capability for new wireless technologies [15, 16]. Consequently, an effective radio resource management scheme for femtocells is needed, which shall jointly resolve following three major challenges: (i) autonomous interference mitigation; (ii) QoS guarantee provisioning; and (iii) efficient utilization of radio resources.

To tackle the above challenges, *cognitive radio* technology, enabling a station to recognize and adapt to communication environments so as to reach the optimum overall network performance [17–20] is an effective solution.Without imposing additional complexity to current macrocell protocols and operations, *cognitive radio* is well suited for femtocell networks, treating the macrocell network as a *primary system* [21–23]. Under this framework, we propose that each femtocell performs periodic channel sensing to obtain the radio resource usage of the macrocell. By only utilizing radio resources sensed as *unoccupied* by the macrocell, interference to/from the macrocell can be mitigated.

To provide QoS guarantees while fully exploiting the radio resources in femtocells, we make use of the concept of effective capacity. Effective capacity, introduced by Wu and Negi [24], is a link-layer channel model specifying the maximum constant arrival rate that the system can support while satisfying given QoS requirement; therefore, effective capacity exactly serves our purpose. By analytically deriving effective capacities of the proposed interference mitigation schemes, we provide systematic procedures to control the sensing period and the radio resource allocation for femtocells. Simulation results show that the performance of the proposed solution outperforms that of the existing solutions, while effectively supporting smooth transmissions of real-time voice and video streams over the femtocell.

This chapter is organized as follows. A cognitive radio resource management (CRRM) scheme for femtocells to autonomously mitigate interference to/from the macrocells is presented in Section 10.2. The effective capacity of the CRRM is also derived. Using the derived effective capacity of the CRRM, systematic procedures to control the sensing period and the radio resource allocation for each femtocell are proposed in Section 10.3,

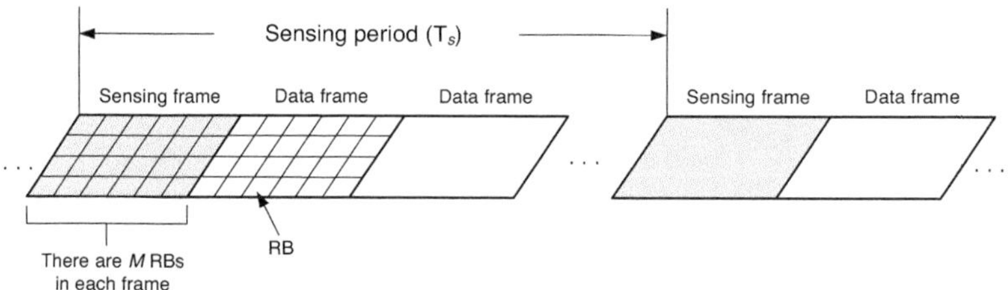

Figure 10.2 The frame structure of the femtocell. The boundary of each frame should align to that of the macrocell. ©2011 IEEE. Reprinted, with permission, from [22].

so as to provide statistical delay guarantees while achieving full radio resource utilization. Performance evaluation results are provided in Section 10.4 and this chapter is concluded in Section 10.5.

10.2   Cognitive radio resources management for femtocells

Consider the femtocell deployment scenario shown in Figure 10.1, where each femtocell is composed of a base station (femto-BS) and multiple mobile stations (femto-MSs) connected to the femto-BS. The macrocell is composed of a base station (macro-BS) and multiple mobile stations (macro-MSs) connected to the macro-BS. All femtocells and the macrocell adopt OFDMA and share all available spectrum (co-channel deployment).

Considering that all femtocells and the macrocell belong to the same wireless system (LTE-A or WiMAX), the frame structure of these two classes of networks are the same and boundaries of frames of femtocells should align to that of the macrocell [1–3]. To reduce the complexity of the radio resource allocation, state-of-the-art OFDMA-based systems typically allocate radio resources in the unit of resource block (RB) [1, 2]. It is considered that there are M RBs in a frame and that each RB is composed of F_{RB} successive subcarriers over T_{RB} OFDM symbols, as depicted in Figure 10.2. Some important notations are listed in Table 10.1.

10.2.1   Power control considerations

In our femtocell model, power control is used to combat channel fading. We choose to adopt *truncated channel inversion* [25] as the power control scheme for femtocells, by which the signal to interference plus noise ratio (SINR) on each RB can be maintained at a required value, so that all RBs in a frame can carry the same number of bits [25], except some RBs suffering deep fades, thus not being utilized by the system. As a result, there is no difference in selecting different RBs. This setting simplifies the problem of RB allocation, and as we can see in the following sections, the QoS requirements of a femto-MS can be fulfilled if the femto-MS can be allocated a sufficient number of RBs without suffering interference. In this chapter, we define an RB without interference if this RB is not simultaneously utilized by more than one network.

Table 10.1 Important notations used in this chapter.

Notation	Description
M	Number of RBs in a frame
F_{RB}, T_{RB}	Number of subcarriers and OFDM symbols in an RB
B, B'	The buffer length and a certain threshold of the buffer length
ζ	The probability that the buffer is not empty
θ	QoS exponent
$d_{\max}$	Delay bound
δ	A constant jointly determined by the arrival process and the service process
$E_C(\theta), E_C(\theta)$	Effective capacity and effective bandwidth
T_s	Sensing period
M_{macro}	Number of RBs occupied by the macrocell in a frame
ρ	Traffic load of the macrocell
η	RB allocation correlation probability among frame of the macrocell
M_η	Number of RBs with non-zero η in a frame
φ	Fraction of correlated RB allocation of the macrocell
n	Number of bits carried by an RB of the femtocell
l	Number of RBs utilized by a femtocell
$E_C^l(\theta)$	Effective capacity of the femtocell adopting the CRRM utilizing l RBs
ϖ^l	As defined in (10.11)

10.2.2 Statistical quality of service guarantees and effective capacity

Real-time services typically require bounded delays. In [24] it has been shown that due to the impact of time-varying fading channels, providing *deterministic* QoS guarantees over a Rayleigh fading channel is not possible. As a result, a practical solution turns out to provide *statistical* QoS guarantees; i.e., the probability that the delay violates a given requirement is bounded. For this purpose, *large deviation theory* [26, 27] shows that, for stationary arrival and service processes under sufficient conditions, the probability that the buffer length B exceeds a certain threshold B' decays exponentially fast as the threshold B' increases. That is,

$$\mathbb{P}\{B > B'\} \approx e^{-\theta B'}, \tag{10.1}$$

where θ is a positive constant called *QoS exponent*. When delay is the main QoS metric of interest, an expression similar to (10.1) is given by

$$\mathbb{P}\{Delay > d_{\max}\} \approx e^{-\theta \delta d_{\max}}, \tag{10.2}$$

where $d_{\max}$ is the delay bound and δ is jointly determined by the arrival process and the service process. From (10.2), it is can be observed that a small θ implies that the system can only support a *weak* QoS requirement, while a large θ means that a *strong* QoS requirement can be supported.

To provide statistical delay guarantees, the effective bandwidth and the effective capacity provide significant foundations. The effective bandwidth [26, 27], denoted by $E_B(\theta)$, specifies the *maximum constant service rate needed by the given arrival process subject to a given* θ. On the other hand, the effective capacity, denoted by $E_C(\theta)$, is the dual of the effective bandwidth and specifies the *maximum constant arrival rate that can be supported by the system subject to a given θ* [24]. If θ^* can be found as the solution of $E_B(\theta^*) = E_C(\theta^*)$, δ can be obtained as [28]

$$\delta = E_B(\theta^*) = E_C(\theta^*). \tag{10.3}$$

Consequently, the system can achieve the statistical delay guarantee

$$\mathbb{P}\{Delay > d_{\max}\} \approx e^{-\theta^* \delta d_{\max}}. \tag{10.4}$$

The effective capacity can be formally defined as [24]

$$E_C(\theta) \triangleq -\frac{\Lambda_c(-\theta)}{\theta} = -\lim_{t \to \infty} \frac{1}{\theta t} \log \left(\mathbb{E} \left[e^{-\theta \sum_{i=1}^{t} R[i]} \right] \right), \tag{10.5}$$

where $\mathbb{E}[\cdot]$ denotes the expected value, $\sum_{i=1}^{t} R[i]$ is the partial sum of the discrete-time stationary and ergodic service process $\{R[i], i = 1, 2, \ldots\}$ and

$$\Lambda_c(\theta) = \lim_{t \to \infty} \left(\frac{1}{t} \right) \log \left(\mathbb{E} \left[e^{\theta \sum_{i=1}^{t} R[i]} \right] \right) \tag{10.6}$$

is a convex function differentiable for all real θ. To achieve statistical delay guarantees, approaches to derive the effective bandwidth of real-time streams have been widely discussed (e.g., the effective bandwidth of the voice traffic can be obtained by the method proposed in [29]). However, the effective capacity depends on the specific system design. In this chapter, we derive the effective capacity of the proposed CRRM so as to control the system to achieve the required QoS requirements.

10.2.3　Interference mitigation between the femtocell and the macrocell

Since femtocells suffer from interference from the macrocell if femtocells utilize RBs occupied by the macrocell, each femto-BS should avoid allocating RBs occupied by the macrocell to its femto-MSs. As a result, the key idea of cross-tier interference mitigation in this chapter is that each femto-BS autonomously estimates the RB usage of the macrocell (i.e., which RB is occupied by the macrocell within a frame). Then, each femto-BS avoids allocating these occupied RBs to its femto-MSs. For this purpose, the CRRM is proposed to incorporate the channel sensing capability into the femto-BS as follows:

Cognitive radio resource management (CRRM)

1. The femto-BS periodically senses the channel to identify which RB is occupied by the macrocell. The sensing period is T_s frames and each channel sensing persists for one frame, as depicted in Figure 10.2. Therefore, for a femtocell, a frame is referred to as a "sensing frame" if the femto-BS performs channel sensing at that frame. On the other hand, a frame is referred to as a "data frame" if the femto-BS does not perform channel sensing at that frame. The femto-BS cannot perform data transmission and reception within the sensing frame.

2. The femto-BS senses the *received interference power* on each RB within the sensing frame.
   a. If the *received interference power* on an RB exceeds a certain threshold, the RB is identified as being occupied by the macrocell.
   b. Otherwise, the RB is unoccupied by the macrocell.
3. In subsequent data frames, the femto-BS only allocates unoccupied RBs sensed in the sensing frame to its femto-MSs.
4. The femto-BS also extracts the following parameters from channel sensing: (i) the traffic load of the macrocell; (ii) the RB allocation correlation probability of the macrocell; (iii) the fraction of correlated RB allocation of the macrocell, which are detailed below.

In the CRRM framework, the femto-BS considers that the macrocell RB usage in the data frame is the same as that in the latest sensing frame. If the macrocell changes its RB allocation very frequently among frames, there could be estimation errors on the RB usage of the macrocell in data frames. Therefore, a small T_s can decrease the estimation error and thus decrease the probability that the femto-BS allocates an occupied RB to the femto-MS in data frames. However, channel sensing is an overhead since the femtocell cannot perform data transmission and reception in a sensing frame. A small T_s implies that more radio resources (more sensing frames) are used for channel sensing. As a result, there is a tradeoff between T_s and interference, or in other words a tradeoff between the radio resource utilization efficiency and interference. A good tradeoff should provide statistical delay guarantees for femto-MSs while fully utilizing radio resources. To achieve this goal, each femto-BS should appropriately control T_s and the RB allocation. The concept of effective capacity facilitates the control of T_s and the RB allocation, and the CRRM effective capacity should be derived. For this purpose, the femto-BS needs more information about the characteristics of the RB allocation in the macrocell. That is, whether the macro-BS utilizes RBs in a high correlation manner or in a low correlation manner among frames, which depends on the time-varying channel conditions of each macro-MS, the connection time of each macro-MS, and the scheduling algorithm adopted by the macro-BS. To capture this characteristic, the femto-BS is proposed to extract the following parameters from the results of channel sensing. These parameters are also noted by 3GPP LTE-Advanced [30].

Definition 10.1 *Let M_{macro} be the number of RBs occupied by the macrocell in a frame, the traffic load of the macrocell, ρ, is defined as $\rho \triangleq M_{\text{macro}}/M$.*

Definition 10.2 *When an RB is occupied by the macrocell in a frame, the RB allocation correlation probability of the macrocell, η, is the probability that the macrocell will still occupy this RB in the subsequent frame.*

Definition 10.3 *Let M_η be the number of RBs with non-zero η in a frame, the fraction of correlated RB allocation of the macrocell, φ, is defined as $\varphi \triangleq M_\eta/\rho M$.*

The merit of adopting these three parameters is that in order to obtain these parameters, the femto-BS only needs to sense the *received interference power* on each RB, by which the femto-BS can identify the traffic load of the macrocell, ρ. By further using the time-domain correlation of the channel sensing results, η and φ can be obtained, which capture the correlation of the RB allocation among frames in the macrocell. When both η and φ are large, it implies that the macro-BS allocates RB to macro-MSs in a high correlation manner among frames. On the other hand, when both η and φ are small, it means that the macro-BS allocates RBs to macro-MSs in a low correlation manner. By obtaining these parameters in step 4 of the CRRM, we start the derivation of the effective capacity of the CRRM by introducing the following lemma.

Lemma 10.1 *Considering that all RBs carry the same number of bits (say n bits), the effective capacity of the femtocell utilizing one RB without suffering interference is given by*

$$E_C^1(\theta) = -\frac{1}{\theta} \log\left(e^{-n\theta}\right), \tag{10.7}$$

Proof. Based on (10.5), we have

$$E_C^1(\theta) = -\lim_{t \to \infty} \frac{1}{\theta t} \log\left(\mathbb{E}\left[e^{-\theta \sum_{i=1}^{t} R[i]}\right]\right). \tag{10.8}$$

For the block fading channel wherein the service process $\{R[i], i = 1, 2, \ldots\}$ is uncorrelated, (10.8) can be rewritten as

$$E_C^1(\theta) = -\frac{1}{\theta} \log\left(e^{-\theta R[i]}\right). \tag{10.9}$$

Since each RB carries n bits, $R[i] = n$ for all i. We therefore obtain (10.7) and complete the proof. $\qquad\qquad\square$

In (10.7), it is assumed for the moment that the RB utilized by the femtocell does not suffer from interference. To generalize (10.7), the effective capacity of the femtocell utilizing l RBs with potential interference due to estimation errors of the RB usage of the macrocell is given by the following theorem.

Theorem 10.1 *Considering that the CRRM is applied to the femto-BS, the average effective capacity per frame of the femtocell that utilizes l RBs in each data frame is given by*

$$E_C^l(\theta) = l\varpi^l E_C^1(l\varpi^l \theta), \tag{10.10}$$

where

$$\varpi^l = \frac{T_s - 1}{T_s} \cdot \left(1 - \frac{\sum_{g=0}^{\min(l,(1-\eta\varphi)\rho M)} g C_g^l C_{(1-\eta\varphi)\rho M - g}^{(1-\eta\varphi\rho)M - l}}{l \cdot C_{(1-\eta\varphi)\rho M}^{(1-\eta\varphi\rho)M}}\right) \tag{10.11}$$

and $C_b^a \triangleq \frac{a!}{b!(a-b)!}$.

Proof. Due to the potential estimation error on the RB usage of the macrocell in the data frame, we need to derive the probability that there are g RBs suffering from interference among l RBs utilized by the femtocell in the data frame. For this purpose, we can divide the RBs occupied by the macrocell into two classes: (i) there are $\eta\varphi\rho M$ RBs whose allocations are the same in each frame; (ii) there are $(1 - \eta\varphi)\rho M$ RBs whose allocations certainly change in each frame. Therefore, there are $C_{(1-\eta\varphi)\rho M}^{(1-\eta\varphi\rho)M}$ possibilities of RB arrangement of the macrocell in each frame. In a data frame, the femto-BS only allocates RBs sensed as unoccupied in the sensing frame to the femto-MS. Therefore, there are $C_g^l C_{(1-\eta\varphi)\rho M-g}^{(1-\eta\varphi\rho)M-l}$ possibilities that there are g RBs suffering interference among l RBs utilized by the femtocell. Therefore, the probability that there are g RBs suffering interference among l RBs utilized by the femtocell in the data frame is

$$\frac{C_g^l C_{(1-\eta\varphi)\rho M-g}^{(1-\eta\varphi\rho)M-l}}{C_{(1-\eta\varphi)\rho M}^{(1-\eta\varphi\rho)M}}. \tag{10.12}$$

Since the maximum value of g is $\min(l, (1 - \eta\varphi)\rho M)$, the expected value of g is

$$\sum_{g=0}^{\min(l,(1-\eta\varphi)\rho M)} g \frac{C_g^l C_{(1-\eta\varphi)\rho M-g}^{(1-\eta\varphi\rho)M-l}}{C_{(1-\eta\varphi)\rho M}^{(1-\eta\varphi\rho)M}}. \tag{10.13}$$

Therefore, among l RBs utilized by the femtocell, the expected number of RBs that do not suffer from interference in each data frame is

$$l - \sum_{g=0}^{\min(l,(1-\eta\varphi)\rho M)} g \frac{C_g^l C_{(1-\eta\varphi)\rho M-g}^{(1-\eta\varphi\rho)M-l}}{C_{(1-\eta\varphi)\rho M}^{(1-\eta\varphi\rho)M}}. \tag{10.14}$$

Further considering the overhead of sensing frames, we can obtain

$$l\varpi_C^l = \frac{T_s - 1}{T_s} \left(l - \sum_{g=0}^{\min(l,(1-\eta\varphi)\rho M)} g \frac{C_g^l C_{(1-\eta\varphi)\rho M-g}^{(1-\eta\varphi\rho)M-l}}{C_{(1-\eta\varphi)\rho M}^{(1-\eta\varphi\rho)M}} \right) \tag{10.15}$$

as the average number of RBs without interference in each frame given that the femtocell utilizes l RBs in each data frame. By applying the results in [31, 32], the effective capacity of the femtocell per data frame that the femtocell utilizes l RBs in each data frame without interference is given by

$$E_C^l(\theta) = lE_C^1(l\theta). \tag{10.16}$$

By substituting l in (10.16) by $l\varpi_C^l$ in (10.15), we can obtain (10.10) as the average effective capacity per frame of the femtocell that utilizes l RBs in each data frame. $\quad\square$

With the facilitation of Theorem 10.1, the femto-BS can control T_s and the RB allocation to achieve the required statistical delay guarantees for each femto-MS. This part will be detailed in the following section.

10.3   Control of the sensing period

In this section, systematic procedures are proposed to control the sensing period T_s and the RB allocation for the femtocell. Let $d_{\max}$ and ε be the delay bound and the acceptable delay bound violation probability, respectively. The procedures of the control of T_s and the RB allocation are defined as follows.

The control of T_s and the RB allocation:

1. The femto-BS calculates the effective bandwidth $E_B(\theta)$ of the real-time traffic.
2. To efficiently utilize the radio resources, T_s is initially set to a predetermined value.
3. The femto-BS first allocates $l = 1$ RB to the femto-MS in the data frame.
4. The femto-BS obtains the effective capacity by (10.10).
5. Find the solution of θ such that

$$E_B(\theta) = E_C^l(\theta) = \delta. \tag{10.17}$$

6. Derive the delay violation probability by

$$\mathbb{P}\{\text{Delay} > d_{\max}\} = e^{-\theta \delta d_{\max}}, \tag{10.18}$$

   a. If $e^{-\theta \delta d_{\max}} > \varepsilon$, l is determined by

$$\min_{1 \leq l \leq L} \{l\}, \ s.t. \ e^{-\theta \delta d_{\max}} \leq \varepsilon \tag{10.19}$$

   L is the maximum number of RBs that can be allocated to the femto-MS in a data frame.

   b. If (10.19) is not satisfied, decrease T_s by one if $T_s > 2$ and repeat step 4 to step 6 to find the appropriate l and T_s such that (10.19) can be satisfied.

From our performance evaluation (see next section), an appropriately large T_s improves the effective capacity. This observation facilitates to further reduce the computational complexity. By selecting an appropriate predetermined T_s in step 2 (e.g., 24 frames), the effective capacity can be sufficiently close to the optimum value. Therefore, step 6(b) can be avoided to reduce the complexity.

10.4   Performance evaluations

The performance of the proposed solution is evaluated by adopting the system parameters of 3GPP LTE-Advanced [1], as listed below. There are $M = 100$ RBs in each frame for the 20 MHz bandwidth, $F_{RB} = 12$ subcarriers and $T_{RB} = 7$ OFDM symbols. Quadrature phase shift keying (QPSK) modulation is used on each RB and the length of a frame is 1 ms.

Resource block allocation of the macrocell among frames is typically correlated and this cannot be controlled by femtocells. Therefore, we should evaluate the performance of the femtocell under different correlations levels of macrocell RB allocations. In this performance evaluation, we consider the macrocell allocating RBs in a high correlation manner ($\eta = \varphi = 0.8$) and in a low correlation manner ($\eta = \varphi = 0.3$). Note also the extreme case in which we select the macrocell allocating RBs in a low correlation manner. Therein, when the macrocell experiences a typically high traffic load ($\rho = 0.8$), there are only $\lceil 0.3 \times 0.3 \times 0.8 \times 100 \rceil = 8$ RBs that the macrocell will persistently occupy in each frame. In other words, the macrocell changes the occupancies of the remaining 72 RBs in each frame.

Without employing CRRM to estimate the macrocell RB usage, the interference mitigation scheme that is widely adopted by the state-of-the-art OFDMA systems, is

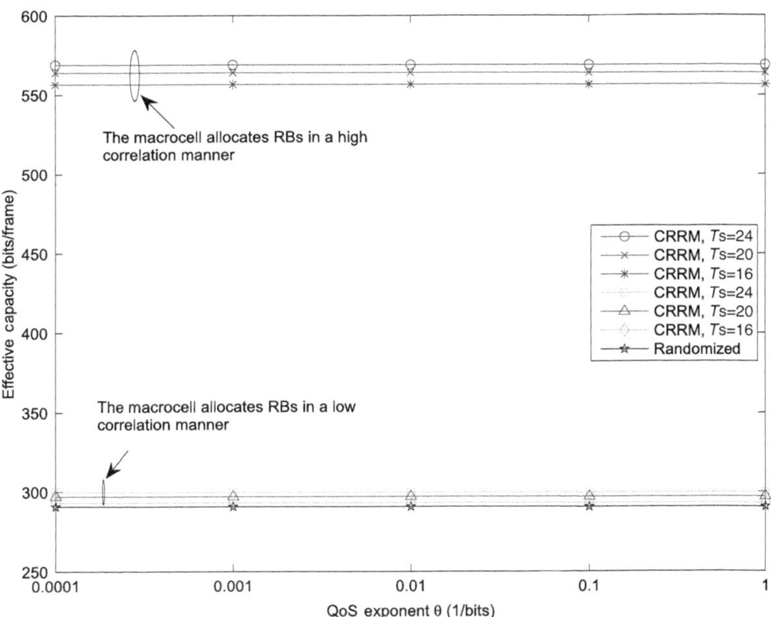

Figure 10.3 Effective capacities of the femtocell adopting the CRRM under different T_s and the randomized scheme ($l = 6$, $\rho = 0.8$). ©2011 IEEE. Reprinted, with permission, from [22].

to randomize the RB allocation in each frame. Such a randomized scheme is similar to the concept of the interleaved RB allocation for block fading channels [33]. In this performance evaluation, the randomized scheme is selected as a benchmark.

In the CRRM, channel sensing is accounted for overhead. Therefore, the first thing we need to investigate is whether it is effective to perform channel sensing to mitigate interference. It can be observed in Figure 10.3 that, when T_s is appropriately selected, the CRRM outperforms the randomized scheme when the macrocell allocates RBs in a high correlation manner. Even in the extreme case that the macrocell allocates RBs in a low correlation manner, the performance of the CRRM is around the same level as that of the randomized scheme. These results support the effectiveness of using channel sensing for cross-tier interference mitigation. Furthermore, we can observe that the performance of the CRRM can be enhanced when T_s increases, while this performance enhancement becomes marginal when T_s keeps increasing. Therefore, by selecting an appropriate T_s (e.g., 24 frames), the performance can be sufficiently close to the optimum value. However, a very large T_s is not suggested, since this will increase the estimation errors of ρ, η, and φ of the macrocell, thus reducing the performance gain of the CRRM.

Next, we investigate the performance of the femtocell utilizing different numbers of RBs in Figure 10.4. Since the femtocell using CRRM can identify the unoccupied RBs to utilize, the probability of interference can be decreased, especially when the number of RBs needed by the femtocell is large.

In Figure 10.5, the performance of the femtocell under different traffic loads of the macrocell is investigated. It can be observed that when the traffic load of the macrocell

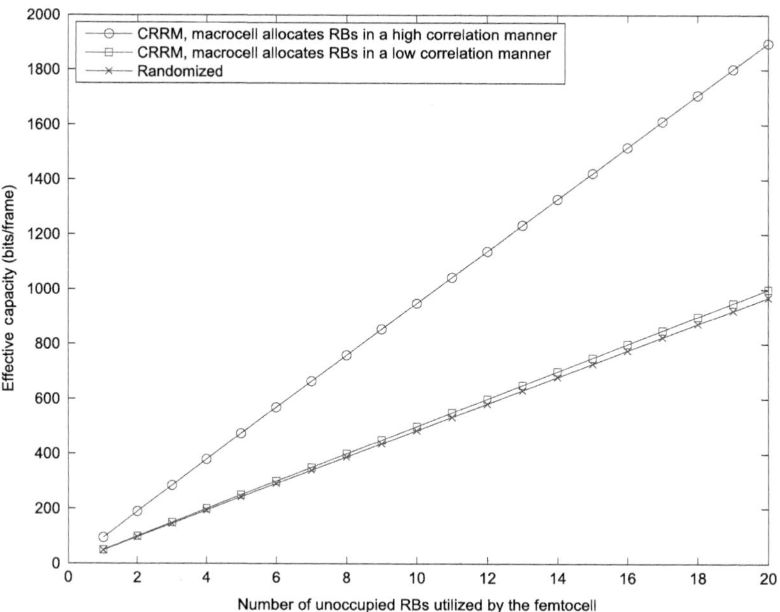

Figure 10.4 Effective capacities of the femtocell adopting the CRRM and the randomized scheme under different l ($T_s = 24$, $\rho = 0.8$, $\theta = 0.01$). ©2011 IEEE. Reprinted, with permission, from [22].

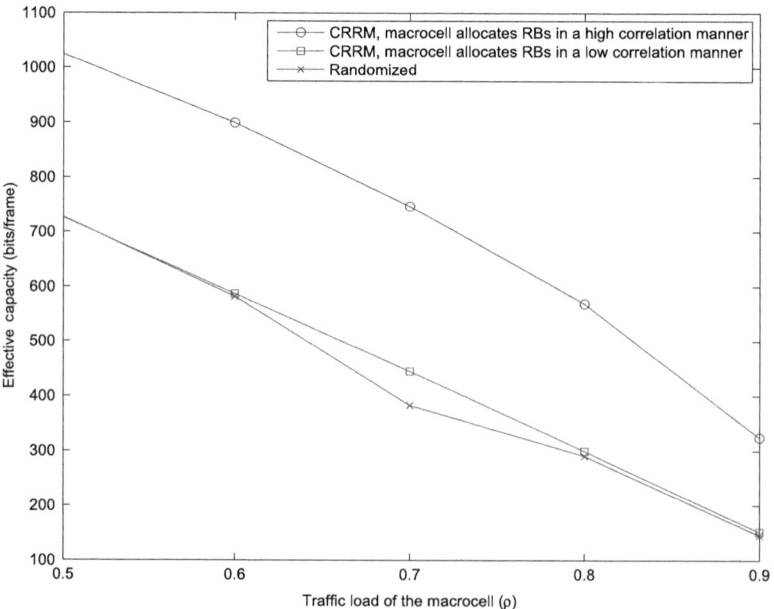

Figure 10.5 Effective capacities of the femtocell adopting the CRRM and the randomized scheme under different traffic load of the macrocell ρ ($T_s = 24$, $l = 6$, $\theta = 0.01$). ©2011 IEEE. Reprinted, with permission, from [22].

Table 10.2 QoS requirements and simulation results on the delay bound violation probability for the voice over IP (VoIP) and high-quality MPEG-4 video transmissions in the femtocell [22].

Traffic	Delay bound d_{max}	Required violation prob. ε	Violation prob. of CRRM ($\eta = \varphi = 0.3$)	Violation prob. of CRRM ($\eta = \varphi = 0.8$)	Violation prob. of Randomized ($\eta = \varphi = 0.3$)	Violation prob. of Randomized ($\eta = \varphi = 0.8$)
Star Wars	40 ms	0.02	0.0199	0.0066	0.0464	0.0331
VoIP	20 ms	0.02	0.0012	0.0003	0.0013	0.0008

increases, both the effective capacities of the femtocell with CRRM and of those using the randomized scheme decrease. The reason is that when the number of RBs occupied by the macrocell increases, the probability that the femtocell and the macrocell utilizing the same RB also increases. However, since the femtocell using the CRRM only utilizes the unoccupied RBs, the CRRM evidently outperforms the randomized scheme.

To evaluate the capability of providing the statistical delay guarantee, Table 10.2 shows the simulation results of the CRRM for real-time voice and video transmissions. For the voice traffic, a low-bit-rate VoIP stream is considered. The arrival process of the voice stream is the well known ON–OFF fluid model. The holding times in "ON" and "OFF" states are exponentially distributed with means 6.1 s and 8.5 s, respectively. The data rate of the "ON" state is 32 kbps. The delay bound is set to 20 ms and the delay bound violation probability is set to 0.02. For the video traffic, a high-quality MPEG-4 movie (*Star Wars*) is considered [34]. The delay bound of the video traffic is 40 ms and the delay bound violation probability is 0.02. By applying the systematic procedures proposed in Section 10.3, Table 10.2 shows the effectiveness of the CRRM on providing statistical delay guarantee, whereas video traffic cannot be supported by the randomized scheme.

10.5   Conclusions

In this chapter, we provided the following solutions for OFDMA femtocells: (i) the proposed CRRM for femtocells enables autonomous cross-tier and intra-tier interference mitigation, providing scalability for a dense deployment without any impact on the existing macrocell infrastructure; (ii) by analytically deriving the effective capacity of CRRM, the proposed systematic procedures for controlling the sensing period and the RB allocation provide statistical delay guarantees while achieving full radio resource utilization. Our performance evaluation results showed that the proposed CRRM significantly outperforms the randomized RB allocation scheme and enables smooth real-time voice and video transmissions over the femtocell. Yielding a limited computational complexity and imposing no impact on state-of-the-art femtocell architecture, the proposed solutions can be smoothly applied to 3GPP LTE-Advanced and WiMAX femtocells to serve urgent needs in the progress of standardization.

References

[1] 3GPP TS36.300 V11.2.0, "Evolved universal terrestrial radio access (E-UTRA) and evolved universal terrestrial radio access network (E-UTRAN)," in *(Release 11)*, July 2012.

[2] IEEE Std 802.16-2009, "IEEE standard for local and metropolitan area networks part 16: air interface for broadband wireless access systems," in *revision of IEEE Std 802.16-2004*, May 2009.

[3] V. Chandrasekhar and J. G. Andrews, "Spectrum allocation in two-tier networks," *IEEE Trans. Commun.*, vol. 57, no. 10, pp. 3059–68, Oct. 2009.

[4] 3GPP TR25.967 V10.0.0, "Home node B radio frequency (RF) requirements (FDD)," in *(Release 10)*, Apr. 2011.

[5] I. Guvenc, M.-R. Jeong, F. Watanabe, and H. Inamura, "A hybrid frequency assignment for femtocells and coverage area analysis for co-channel operation," *IEEE Commun. Lett.*, vol. 12, no. 12, pp. 880–2, Dec. 2008.

[6] V. Chandrasekhar, J. G. Andrews, T. Muharemovic, Z. Shen, and A. Gatherer, "Power control in two-tier femtocell networks," *IEEE Trans. Wireless Commun.*, vol. 8, no. 8, pp. 4316–28, Aug. 2009.

[7] X. Li, L. Qian, and D. Kataria, "Downlink power control in co-channel macrocell femtocell overlay," in *Proc. Conf. on Information Sciences and Systems (CISS)*, Mar. 2009, pp. 383–8.

[8] H.-S. Jo, J.-G. Yook, C. Mun, and J. Moon, "A self-organized uplink power control for cross-tier interference management in femtocell networks," in *Proc. IEEE Military Communication Conf. (MILCOM)*, Nov. 2008.

[9] N. Arulselvan, V. Ramachandran, and S. Kalyanasundaram, "Distributed power control mechanism for HSDPA femtocells," in *Proc. IEEE Vehicular Tech. Conf. (VTC)*, Apr. 2009.

[10] H.-S. Jo, C. Mu, J. Moon, and J.-G. Yook, "Interference mitigation using uplink power control for two-tier femtocell networks," *IEEE Trans. Wireless Commun.*, vol. 8, no. 10, pp. 4906–10, Oct. 2009.

[11] V. Chandrasekhar and J. G. Andrews, "Uplink capacity and interference avoidance for two-tier femtocell networks," *IEEE Trans. Wireless Commun.*, vol. 8, no. 7, pp. 3498–509, July 2009.

[12] D. López-Pérez, A. Valcarce, G. de la Roche, and J. Zhang, "OFDMA femtocells: a roadmap on interference avoidance," *IEEE Commun. Mag.*, vol. 47, no. 9, pp. 41–8, Sep. 2009.

[13] K. Sundaresan and S. Rangarajan, "Efficient resource management in OFDMA femtocells," in *ACM International Symposium on Mobile Ad Hoc Networking and Computing, New Orleans, USA*, May 2009, pp. 33–42.

[14] D. N. Knisely, T. Toshizawa, and F. Favichia, "Standardization of femtocells in 3GPP," *IEEE Commun. Mag.*, vol. 47, no. 9, pp. 68–75, Sep. 2009.

[15] R. Srinivasan, J. Zhuang, L. Jalloul, R. Novak, and J. Park. (2008) IEEE 802.16m evaluation methodology document (EMD).

[16] 3GPP TS36.814 V9.0.0, "Further advancements for E-UTRA physical layer aspects," in *(Release 9)*, Mar. 2010.

[17] K. C. Chen and R. Prasad, *Cognitive Radio Networks*. John Wiley & Sons, 2009.

[18] S.-Y. Lien, C.-C. Tseng, and K.-C. Chen, "Carrier sensing based multiple access protocols for cognitive radio networks," in *Proc. IEEE Int. Conf. on Commun. (ICC)*, May 2008.

[19] ——, "Novel rate-distance adaptation of multiple access protocols in cognitive radio," in *Proc. IEEE Int. Symp. Personal, Indoor, Mobile Radio Commun. (PIMRC)*, Sep. 2007.

[20] S.-Y. Lien, N. R. Prasad, K.-C. Chen, and C.-W. Su, "Providing statistical quality-of-service guarantees in cognitive radio networks with cooperation," in *Proc. Wireless ViTAE*, May 2009.

[21] S.-Y. Lien, C.-C. Tseng, K.-C. Chen, and C.-W. Su, "Cognitive radio resource management for QoS guarantees in autonomous femtocell networks," in *Proc. IEEE Int. Conf. on Commun. (ICC)*, May 2010.

[22] S.-Y. Lien, Y.-Y. Lin, and K.-C. Chen, "Cognitive and game-theoretical radio resource management for autonomous femtocells with QoS guarantees," *IEEE Trans. Wireless Commun.*, vol. 10, no. 7, pp. 2196–206, July 2011.

[23] S.-M. Cheng, S.-Y. Lien, F.-S. Chu, and K.-C. Chen, "On exploiting cognitive radio to mitigate interference in macro/femto heterogeneous networks," *IEEE Wireless Commun. Mag.*, vol. 18, no. 3, pp. 40–7, June 2011.

[24] D. Wu and R. Negi, "Effective capacity: a wireless link model for support of quality of service," *IEEE Trans. Wireless Commun.*, vol. 12, no. 4, pp. 630–43, July 2003.

[25] A. J. Goldsmith and S.-G. Chua, "Variable-rate variable-power MQAM for fading channels," *IEEE Trans. Commun.*, vol. 45, no. 10, pp. 1218–30, Oct. 1997.

[26] C.-S. Chang, "Stability, queue length, and delay of deterministic and stochastic queueing networks," *IEEE Trans. Autom. Control*, vol. 39, no. 5, pp. 913–31, May 1994.

[27] C. Courcoubetis and R. Weber, "Effective bandwidth for stationary sources," *Probab. Eng. Inform. Sci.*, vol. 9, no. 2, pp. 285–94, 1995.

[28] J. Tang and X. Zhang, "Cross-layer modeling for quality of service guarantees over wireless links," *IEEE Trans. Wireless Commun.*, vol. 6, no. 12, pp. 4505–12, Dec. 2007.

[29] C.-S. Chang, *Performance Guarantees in Communication Networks*. Springer, 2000.

[30] Institute for Information Industry and Coiler Corporation, "R4-093196: Interference mitigation for HeNB by channel measurements," in *3GPP TSG RAN WG4 Meeting 52*, 2009.

[31] D. Wu and R. Negi, "Utilizing multiuser diversity for efficient support of quality of service over a fading channel," *IEEE Trans. Veh. Technol.*, vol. 54, no. 3, pp. 1198–206, May 2005.

[32] J. Tang and X. Zhang, "Cross-layer-model based adaptive resource allocation for statistical QoS guarantees in mobile wireless networks," *IEEE Trans. Wireless Commun.*, vol. 7, no. 6, pp. 2318–28, June 2008.

[33] S.-W. Lei and V. K. N. Lau, "Performance analysis of adaptive interleaving for OFDM systems," *IEEE Trans. Veh. Technol.*, vol. 51, no. 3, pp. 435–44, May 2002.

[34] [Online]. Available: http://www.tkn.tu-berlin.de/research/trace/trace.html

11 Decentralized reinforcement learning techniques for interference management in heterogeneous networks

Mehdi Bennis, Dusit Niyato, and Tansu Alpcan

Game theory (GT) is a mathematical tool that analyzes interactions among decision makers. Game theory is seen as a natural paradigm to study and analyze wireless networks where players compete for the same resources. The importance of studying the coexistence between macrocells and femtocells from a game theoretical perspective is multi-fold. First, as illustrated in Figure 11.1, by modeling the dynamic spectrum sharing among network players (macrocell base stations (MBSs), femtocell base stations (FBSs), mobile user equipment (MUE), and home user equipment (HUE)) as games, the behaviors and actions of players can be analyzed in a formalized structure, by which the theoretical achievements in GT can be fully utilized. Second, GT equips us with different optimality criteria for various spectrum sharing problems, which are of key importance when it comes to analyzing the equilibrium of the game. Third, the application of GT enables us to derive efficient distributed algorithms for self-organized networks relying only on partial information. In order to achieve this, the theory of strategic reinforcement learning is of utmost importance by allowing players to choose their *optimal* strategies and gradually learn from their environment through trial and error procedures. A comprehensive source of game theoretic approaches and their application to wireless communications can be found in [1].

In this chapter,[1] we propose the use of reinforcement learning techniques which can be seen as self-organizing network (SON) enablers for smooth coexistence among both macrocell and femtocell tiers.[2] Strategic learning has been used in a number of problems ranging from robotics, security, and biology among others, in which self-organization of agents is seen as instrumental. Due to its success, strategic learning has also received significant interest in the wireless community. For instance, in [2] and [3], reinforcement learning algorithms based on Q-learning were investigated in the context of network selection for multiple radio access technology (multi-RAT) networks, and

---

[1] This work has been partially funded by the European Union under ICT project ICT-4-248523 BeFEMTO.
[2] See Chapter 16 for a more detailed discussion of SON for small cell networks.

Small Cell Networks: Deployment, PHY Techniques, and Resource Management, ed. Tony Q. S. Quek, Guillaume de la Roche, İsmail Güvenç, and Marios Kountouris. Published by Cambridge University Press. © Cambridge University Press 2013.

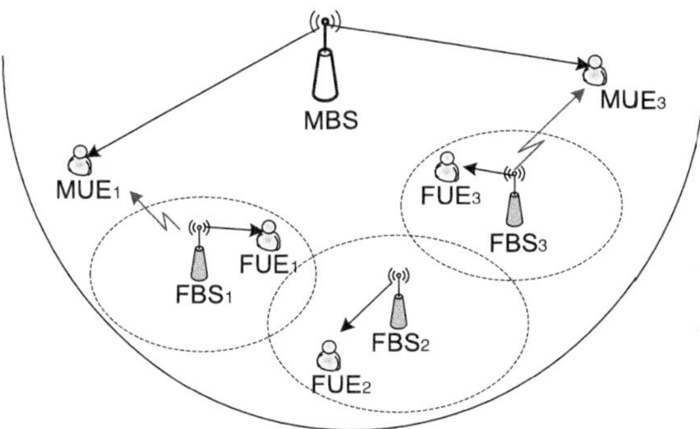

Figure 11.1 Network topology with one macrocell underlaid with three femtocell networks. MUE and HUE stand for macrocell/femtocell user equipment, respectively. MBS and FBS stand for macrocell and femtocell base station, respectively. ©2011 IEEE. Reprinted, with permission, from [10].

channel selection in multi-user cognitive radios, respectively. Recently, in [4] and [5] a reinforcement learning framework based on Q-learning was studied for interference mitigation among femtocells. In [6], macrocell and femtocell coexistence was modeled as a Stackelberg game where macrocell base stations are assumed to have enough information to predict the response of the femtocells for given macrocell power profile. In [7], the authors investigate the problem of interference management using a multi-armed bandit approach. Nevertheless, the existing studies often require information exchange among transmitters, which represents a non-affordable increment of signaling messages, and falls short of examining the non-trivial tradeoff between information exchange and performance improvement. Recently, a decentralized algorithm for cross-tier interference mitigation was investigated in [8], in which femtocells converge toward the Logit equilibrium (also known as ϵ-Nash equilibrium). In the case where femtocells do cooperate, a number of studies can be found such as [9], in which femtocell users are used to relay macrocell users' data in return for reward by the macrocell [9].

In this chapter, we analyze the strategic coexistence between macrocell and femtocell tiers using tools from machine learning and evolutionary game theory under different information exchange scenarios. Various decentralized algorithms pertaining to different deployment scenarios are devised aiming at mitigating interference toward the macrocell tier. First, when femtocells are able to exchange information, hereinafter referred to as networked femtocells, evolutionary game theory is used to devise optimal strategies adopted by femtocells to mitigate their interference toward the macrocell tier. Owing to the semi-centralized nature of this solution, femtocells can periodically review their strategies by imitating other femtocells in the population. Yet, another formulation is explored when femtocells maximize their payoffs given the empirical frequency of other femtocells' actions, referred to as fictitious play (FP). Similarly, due to the signaling overhead of FP, a low-complexity variant called smoothed FP is proposed [11]. On the other hand, when information exchange among femtocells is no longer possible (i.e., stand-alone femtocell scenario), each femtocell engages in an exploration/exploitation paradigm by gradually learning its strategies and interacting with its local environment through trial and error. This procedure is referred to as Q-learning.

The chapter is organized as follows: Section 11.1 introduces the system model. Section 11.2 revisits the basics and definitions from game theory and strategic learning, which will prove useful throughout the chapter. Section 11.3 delves into the networked femtocell deployment scenario with information exchange among femtocells, followed in Section 11.4 by stand-alone femtocell deployment where information exchange is no longer possible. Section 11.5 presents numerical results of the various strategic learning algorithms along with a discussion highlighting interesting tradeoffs facing small cells. Finally, conclusions are drawn in Section 11.6.

11.1 Background on game theory

In the following, we discuss some basic concepts and elements of game theory, which will be used throughout this chapter.

11.1.1 Strategic-form games

The cross-tier spectrum sharing game between the macrocells and femtocells can be modeled by the following static game in strategic form:

$$macrocell\ G = \left(\mathcal{K}, (\mathcal{P}_k)_{k \in \mathcal{K}}, (u_k)_{k \in \mathcal{K}} \right), \tag{11.1}$$

where $\mathcal{K}$ is the set of players (i.e., FBSs). An action of a given FBS $k \in \mathcal{K}$ is a power allocation scheme, i.e., an N-dimensional power allocation vector $p_k = (p_k^{(1)}, \ldots, p_k^{(N)}) \in \mathcal{P}_k$, where $\mathcal{P}_k$ is the set of all possible power allocation vectors that FBS k can use. Finally, $u_k(p_k, \mathbf{p}_{-k})$ represents the utility function (or payoff) of player $k \in \mathcal{K}$ where a value is assigned to every player for every possible outcome of the game. $\mathbf{p}_{-k}$ refers to the power allocation vector of all other players except player k.

11.1.2 Rationality versus bounded rationality

Rationality is one of the most common assumptions in game theory. It refers to the fact that at every time instant, a given player aims at optimizing its own utility given the strategies taken by all other players in the game. On the other hand, with bounded rationality, players have limited knowledge of their environment and thus engage in a learning process, in which they first learn their environment by exploring a number of strategies, after which players go back to their rational behavior. In this chapter, we also use elements of bounded rationality such as smoothed best response learning dynamics.

11.1.3 Non-cooperative and cooperative games

In a non-cooperative game, each player is selfish and unconcerned about all other players' performances. In particular, each player chooses its strategy to optimize its own performance metric under the assumption that all players are rational and adopt the same selfish behavior. Thus, each player's utility function is defined in terms of local target quality of service (e.g., individual transmission rate, outage probability, minimum transmit power, etc.). In contrast, in a cooperative approach each player aims at maximizing a common benefit for the set of players assuming that all other players have adopted the same cooperative behavior. A common benefit could be interpreted for instance as the sum of individual benefits (social welfare problem).

Often, if the performance metric (utility function) is well chosen, non-cooperative games might be played by each player using only local information (e.g., channel gains and power constraints regarding only a given player). However, cooperative games often require information regarding all the players' local information. Hence, cooperative games are often used either when there exists a central controller (e.g., a base station or a gateway) that has complete information about all players or when communications among all the players is possible (e.g., common signaling is available and affordable) as well as in some particular wireless communication settings such as broadcast systems.

11.1.4   The concept of equilibrium

Our interest is to find a strategy profile $sp = (sp_1, \ldots, sp_K) \in \mathcal{P}$ such that no player is interested in changing or deviating from its own strategy. Once the network configuration sp^* is reached, any unilateral deviation of a given player decreases its own utility. A network configuration sp^* is known as a Nash equilibrium (NE) [12].[3]

The NE is an important concept in the field of game theory wherein an NE corresponds to a profile of strategies $p^* = (p_1^*, \ldots, p_K^*)$ for which each player's strategy $p_1^* \in \mathcal{P}$ is an NE if it satisfies:

$$\forall k \in \mathcal{K} \quad \text{and} \quad \forall p_k \in \mathcal{P}_K, \quad u_k(p_k^*, p_{-k}^*) \geq u_k(p_k, p_{-k}^*). \tag{11.2}$$

Said otherwise, an NE is a set of strategies such that no player has incentives to unilaterally change its strategies. That is, at the NE, any unilateral deviation from the strategy profile p_k of player k, $\forall k \in \mathcal{K}$ will not increase its utility function. Hence, at the NE there does not exist any motivation for a player to deviate from the NE strategy profile. As players are selfish and decide by themselves their strategy, one question arises: does an NE lead to an efficient game outcome? The NE is generally *inefficient* as compared to the case when players do cooperate or when correlation exists among players' strategies. However, it is a lower bound in terms of performance when players are non-cooperative. Furthermore, the existence of the NE is a difficult problem to prove and there may be an infinite number of NEs. Therefore, the selection of the best NE is a difficult task in game theory. However, there exist many criteria to choose which equilibrium to select, such as the Pareto optimality defined below.

---

[3] It is worth noting that the Nash equilibrium is only one of the possible encountered equilibria in the large literature. Other interesting equilibria are correlated, coarse-correlated equilibria and their variants, whose application in small cells can be found in [13].

Definition 11.1 (Pareto optimality) *In the game macrocell $G = (\mathcal{K}, (\mathcal{P}_k)_{k \in \mathcal{K}},$ $(u_k)_{k \in \mathcal{K}})$, let $\mathbf{s}p = (p_1, \ldots, p_K)$ and $\mathbf{s}p' = (p'_1, \ldots, p'_K)$ be two different strategy profiles in $\mathcal{P}$. Then if,*

$$\forall k \in \mathcal{K} \quad u_k(p_k, sp_{-k}) \geq u_k(p'_k, sp'_{-k}) \tag{11.3}$$

with strict inequality for at least one player, the strategy profile $\mathbf{s}p$ is Pareto-superior to the strategy profile $\mathbf{s}p'$. If there exists no strategy that is Pareto superior to p, then p is Pareto optimal, which allows us to assess the gap between non-cooperative and cooperative approaches.

Very often, the NE strategy profile is not Pareto optimal and thus the loss of performance observed in a non-cooperative game due to the lack of cooperation is a common optimality measure called the price of anarchy (PoA).

Definition 11.2 (Price of anarchy) *Let the game macrocell $G = (\mathcal{K}, (\mathcal{P}_k)_{k \in \mathcal{K}},$ $(u_k)_{k \in \mathcal{K}})$ be a non-cooperative game and let $\mathbf{s}p^* = (p_1^*, \ldots, p_K^*)$ be the NE strategy profile that minimizes (if several NEs) $\sum_{k=1}^{K} u_k(p_k^*, sp_{-k})$. Then the ratio:*

$$PoA = \frac{\max_{p \in \mathcal{P}} \sum_{k=1}^{K} u_k(p_k, sp_{-k})}{\min_{p^*} \sum_{k=1}^{K} u_k(p_k^*, sp_{-k}^*)} \tag{11.4}$$

is the PoA of the game macrocell G.

Definition 11.3 (Best response correspondence)   In a non-cooperative game described by the 3-tuple *macrocell $G = \left(\mathcal{K}, (\mathcal{P}_k)_{\forall k \in \mathcal{K}}, (u_k)_{\forall k \in \mathcal{K}}\right)$*, the relation $\mathrm{BR}_k : \mathcal{P}_{-k} \to \mathcal{P}_k$ such that

$$\mathrm{BR}_k(sp_{-k}) = \arg\max_{sq_k \in \mathcal{P}_k} u_k(sq_k, sp_{-k}), \tag{11.5}$$

is defined as the best response correspondence of player $k \in \mathcal{K}$, given the actions $\mathbf{s}p_{-k}$ adopted by all the other players.

Definition 11.4 (Best response dynamics) Let the action profile $sp(t) = (sp_1(t), \ldots, sp_K(t))$ be the result of a best response dynamics at time t. Then, for all $k \in \mathcal{K}$, and for all $t \in \mathbb{N}$ and $t > 0$, the vector $sp_k(t)$ can be obtained as follows:

- In the sequential best response dynamics (round-robin order):

$$sp_k(t) \in \mathrm{BR}_k(sp_1(t), \ldots, sp_{k-1}(t), sp_{k+1}(t-1), \ldots, sp_K(t-1)), \tag{11.6}$$

- In the simultaneous best response dynamics:

$$sp_k(t) \in \mathrm{BR}_k(sp_{-k}(t-1)), \tag{11.7}$$

where $sp(0)$ can be any vector $sp \in \mathcal{P}$.

11.2   System model

Let us assume that there exists $M = 1$ MBS operating over a set $\mathcal{N} = \{1, \ldots, N\}$ of N frequency bands. Let $\Gamma_0 = (\Gamma_0^{(1)}, \ldots, \Gamma_0^{(N)})$ denote the minimum time-average signal to interference plus noise ratio (SINR) offered by the MBS to its MUE over its corresponding spectrum band. Consider now a set $\mathcal{K} = \{1, \ldots, K\}$ of K femtocells underlaying the macrocell. Each femtocell can use any of the available frequency bands to serve its corresponding HUE as long as it does not induce a lower time-average SINR than the minimum required by the MUE.

Let t be a discrete time index and let MUE denote the scheduled macro-user connected to its serving MBS. Designate MBS's transmit power on a given subcarrier to be $p_0^{(n)}$. Let $|h_{0,0}^{(n)}|^2$ denote the channel gain between the MBS and its associated MUE in subcarrier $n \in \mathcal{N}$. Likewise, $|h_{i,j}^{(n)}|^2$ denotes the channel gain between transmitter i and receiver j on subcarrier n. Moreover, let σ_n^2 be the variance of additive white Gaussian noise at MUE, which is assumed to be constant over all subcarriers for simplicity. The transmit power of FBS $k \in \mathcal{K}$ on subcarrier n is denoted by $p_k^{(n)}$.

The downlink SINR at the MUE is

$$\gamma_0^{(n)} = \frac{|h_{0,0}^{(n)}|^2 p_0^{(n)}}{\sigma^2 + \underbrace{\sum_{k \in \mathcal{K}} |h_{k,0}^{(n)}|^2 p_k^{(n)}}_{\text{femtocells}}}. \tag{11.8}$$

Likewise, the SINR at the HUE served by FBS $k \in \mathcal{K}$ is

$$\gamma_k^{(n)} = \frac{|h_{k,k}^{(n)}|^2 p_k^{(n)}}{\sigma^2 + \underbrace{|h_{0,k}^{(n)}|^2 p_0^{(n)}}_{\text{macrocell}} + \underbrace{\sum_{j \in \mathcal{K} \setminus \{k\}} |h_{j,k}^{(n)}|^2 p_j^{(n)}}_{\text{femtocells}}}. \tag{11.9}$$

11.3   Macrocell–femtocell coexistence with information exchange among femtocells

In this section, the focus is on cross-tier interference mitigation algorithms with information exchange among femtocells. We start with the evolutionary game with replication dynamics, followed by the replication by imitation case.

11.3.1   Evolutionary game: replication dynamics

This interference mitigation mechanism carried out by femtocells is based on the concept of evolutionary game theory (EGT) [14], where each FBS chooses its strategy against other FBSs within the same network (referred to as population). Femtocells observe the behavior of other interfering femtocells, learn from these observations, and thus make the best decision based on their instantaneous payoff, as well as the *average* payoff of all other femtocells. Let us denote by $\mathcal{G}^{(ev)} = \left(\mathcal{K}, \{\mathcal{P}_k\}_{k\in\mathcal{K}}, \{u_k\}_{k\in\mathcal{K}}\right)$ the EGT model. Here, $\mathcal{K}$ represents the set of deployed femtocells in the network. For all $k \in \mathcal{K}$, the set of actions of FBS k is the set of power allocation vectors, $\mathcal{P}_k = \{q_k^{(l,n)} : l \in \{0, \ldots, L_k\},$ and $n \in \mathcal{N}\}$, where $\mathcal{L}_k = \{1, \ldots, L_k\}$ and $L_k \in \mathbb{N}$ is the number of discrete power levels of FBS k, i.e., $\frac{p_{k,\max}}{L_k}, \ldots, p_{k,\max}$ and $l \in \mathcal{L}_k$ is the transmit power level. The power allocation vector when FBS k transmits over subcarrier n with power level l is

$$q_k^{(l,n)} = \frac{l}{L} p_{k,\max}, \tag{11.10}$$

and, hence, FBS k has $N_k' = (L_k + 1)N$ possible power allocation vectors $q_k^{(0,0)}, q_k^{(1,1)}, \ldots, q_k^{(L_k,N)}$. Finally, $u_k : \mathcal{P} \to \mathbb{R}$ is the performance metric of femtocell $k \in \mathcal{K}$ and the *average* payoff of the entire femtocell population at time t is defined as $\bar{u}(t) = \frac{\sum_{k\in\mathcal{K}} u_k(t)}{K}$. The objective of every femtocell is to maximize its own achievable transmission rate $u_k(p_k, sp_{-k})$.

In this scenario, an entity that is referred to as Home NodeB (HNB)-gateway, collects the payoffs for all femtocells and calculates the average payoff of the entire femtocell network. The payoff $u_k(t)$ of FBS k is then compared with the average payoffs $\bar{u}(t)$ and in the case when it is less than the average payoff of the femtocell network, a random strategy from the transmission set is chosen (exploration phase) and the whole process is repeated again. Let $\zeta_a^{(l,n)}(t) = \sum_{k=1}^{K} \mathbf{1}_{\{p_k(t)=q_k^{(l,n)}=a\}}$ represent the total number of femtocells choosing strategy $q_k^{(l,n)}$, and $x_a(t) = \frac{\zeta_a^{(l,n)}(t)}{\sum_a \zeta_a^{(l,n)}(t)}$ is the percentage of femtocells using strategy $a \in \mathcal{P}$. Hence the replication dynamic equation can be defined as follows, $\forall (l, n) \in (\mathcal{L} \times \mathcal{N})$:

$$\dot{x}_a(t) = x_a(t)\big(u_a(t) - \bar{u}(t)\big), \tag{11.11}$$

where $u_a(t)$ is the payoff at time t of femtocell k when using action a, and $\bar{u}(t)$ is the corresponding average payoff of the entire femtocell population over all strategies $a \in \mathcal{P}$ where $\bar{u}(t) = \sum_{a\in\mathcal{P}} x_a u_a(t)$. It is worth noting that a controller is required to compute $\bar{u}(t)$. Finally, it is worth noting that the rationale behind (11.11) is that femtocells gather into distinct clusters or populations adopting strategy a. The pseudo-code of the replication dynamics learning algorithm is given in Algorithm 11.1.

11.3.2　Evolutionary game: replication by imitation

In the case where femtocells adopt the replication by dynamic learning approach, a semi-centralized controller is needed which is responsible for computing the average payoffs of the femtocell population, based on which femtocells gather themselves. In this replication by imitation game theoretic learning approach, each player sticks to a pure strategy for some time period, and every now and then reviews its strategy, sometimes resulting in a change of strategy. First, a specification is defined for the time rate at which femtocells review their strategy depending on the current performance and other aspects of the current population state. Let $r_k^i(x)$ denote the average review rate of FBS $k \in \mathcal{K}$ when using strategy $i \in \mathcal{P}_k$. Second, the probability that FBS k in population x

---

Algorithm 11.1 Replication dynamics

Require: Number of iterations, Number of random femto configurations
  repeat
    repeat
      Calculate utility of all the femtocells, U_k
      Calculate players with payoff better than average payoff, U_{avg}, of the population
      for all Players $k \in \mathcal{K}$ **do**
        if $U_k < U_{\text{avg}}$ AND rand $< (U_{\text{avg}} - U_k)/U_{\text{avg}}$ **then**
          Imitate a random better off player
        end if
      end for
    until Required iterations
  until Eventual convergence
  Calculate ensemble average of utility

---

will change its strategy from i to j is denoted by $pr_i^j(x)$, which depends on the current performance of these strategies and other aspects of the current population state. Let $\mathbf{s}pr_i(x) = \left(pr_i^{(1)}(x), \ldots, pr_i^{(N_k)}(x)\right)$ denote the probability distribution vector over the set of $\mathcal{P}_k$ of pure strategies; $pr_i^i(x)$ means that player k does not change her strategy i. Assuming that all players' Poisson processes are statistically independent, the aggregate of reviewing times in population i is itself a Poisson process with arrival rate $x_i r_i(x)$, with x_i being the femtocell population using strategy i. Assuming that strategy switches are statistically independent random variables across players, the arrival rate of the aggregate Poisson process of switches from strategy i to strategy j is $x_i r_i(x) pr_i^j(x)$. Finally, the dynamics of the replication by imitation are given as follows:

$$\dot{x}_i = \sum_{j \in \mathcal{P}_k}(1 - x_i) r_i(x) pr_j^i(x) - \sum_{j \in \mathcal{P}_k} x_i pr_i^j . r_i(x), \tag{11.12}$$

where the first term in the right hand side of (11.12) regards the share of the femtocell population that imitates another femtocell and as a result changes its strategy from j to i. The second term is the remaining population, which changes its strategy i to other strategies. Note that in (11.12), there is no need to know the average payoff of the whole population as in the replication dynamics. The pseudo-code of the replication by imitation learning algorithm is given in Algorithm 11.2.

11.3.3   Fictitious play with complete information

The FP formulation [11] can be defined by $\mathcal{G}^{(fp)} = \left(\mathcal{K}, \{\mathcal{P}_k\}_{k \in \mathcal{K}}, \{u_k\}_{k \in \mathcal{K}}\right)$. Let us first assume that femtocells have *complete* and *perfect* information, i.e., they know the structure of the game and *observe* at each time t the power allocation vector taken by all other femtocells.[4] Each FBS $k \in \mathcal{K}$ assumes that all of its counterparts play

---

[4] In other words, femtocells do know the transmission strategy of competing femtocells.

Algorithm 11.2 Replication by imitation

Require: Number of iterations, Number of random femto configurations, intensity of review $\sigma > 0$
  repeat
    repeat
      Calculate utility of all the femtocells, U_k
      Calculate population distribution over the action set
      for all Players $k \in \mathcal{K}$ **do**
        if rand $< 1 - \exp(-\sigma U_k)/\mathrm{mean}(\exp(-\sigma U_k))$ **then**
          Select a random player based upon the population distribution over the action set
          Imitate the selected player
        end if
      end for
    until Required iterations
  until Eventual convergence
  Calculate ensemble average of utility

---

independent and stationary (*time-invariant*) mixed strategies π_j, $\forall j \in \mathcal{K} \setminus \{k\}$, where $\pi_j = (\pi_{j,q_j^{(1,1)}}, \ldots, \pi_{j,q_j^{(L_j,N)}})$ and

$$\pi_{j,q_j^{(l,n)}} = P\left(p_j(t) = q_j^{(l,n)}\right). \tag{11.13}$$

Under these conditions, femtocell k is able to build an empirical probability distribution over each set $\mathcal{P}_j$, for all $j \in \mathcal{K} \setminus \{k\}$. Let

$$f_{k,p_k(t)} = \frac{1}{t} \sum_{s=1}^{t} \mathbf{1}_{\left\{p_k(s)=q_k^{(l,n)}\right\}}, \tag{11.14}$$

be the empirical probability with which players $j \in \mathcal{K} \setminus \{k\}$ observe that player k plays action $q_k^{(l,n)} \in \mathcal{P}_k$. Hence, $\forall k \in \mathcal{K}$ and $\forall p_k \in \mathcal{P}_k$, the following recursive expression holds:

$$f_{k,p_k}(t+1) = f_{k,p_k}(t) + \frac{1}{t+1}\left(\mathbf{1}_{\left\{p_k(t)=q_k^{(l,n)}\right\}} - f_{k,p_k}(t)\right). \tag{11.15}$$

Let $\bar{f}_{k,p_{-k}}(t) = \prod_{j \neq k} f_{j,p_j}(t)$ be the probability with which player k observes the action profile $p_{-k} \in \mathcal{P}_{-k}$ at time t, for all $k \in \mathcal{K}$. Let the $|\mathcal{P}_{-k}|$ dimensional vector $\mathbf{s}f_k(t) = (\bar{f}_{k,p_{-k}})_{\forall p_{-k} \in \mathcal{P}_{-k}}$ be the empirical probability distribution over the set $\mathcal{P}_{-k}$ observed by player k. In what follows, the vector $\mathbf{s}f_k(t)$ represents the beliefs of player k over the strategies of all its corresponding counterparts. Hence, at each time t and based on its own beliefs $\mathbf{s}f_k(t)$, each FBS k chooses its action $p_k(t) = q_k^{(l,n)}$, i.e.,

$$(l,n) \in \arg \max_{(l,n) \in (\mathcal{L} \times \mathcal{N})} \bar{u}_k\left(p_k(t), \mathbf{s}f_k(t)\right), \tag{11.16}$$

where for all

$$k \in \mathcal{K}, \bar{u}_k(\pi) = \mathbb{E}_\pi \left[u_k(p_k, \mathsf{s}p_{-k}) \right]. \tag{11.17}$$

From (11.16), it can be implied that in the FP game formulation, players behave *myopically* in which they build their beliefs of strategies used by all the other players in the network.[5] Moreover, at each time t, players choose the action that maximizes their instantaneous expected utility as per the best response dynamics given in (11.6). Note that this fictitious formulation requires full information in which every femtocell perfectly observes the actions of the other femtocells and that the network is time-invariant. The pseudo-code of the fictitious play learning algorithm is given in Algorithm 11.3.

Algorithm 11.3 Fictitious play with complete information

Require: Number of iterations, Number of random femto configurations
  repeat
    repeat
      Calculate utility of all the femtocells, U_k
      for all Players $k \in \mathcal{K}$ **do**
        Count the relative frequency with which a femtocell selects the an action
      end for
      for all Players $k \in \mathcal{K}$ **do**
        Calculate the strategy that maximizes U_k given the strategy of other players
        Use that strategy in the next iteration
      end for
    until Required iterations
  until Eventual convergence
  Calculate ensemble average of utility

11.3.4 Fictitious play with incomplete information

Unlike the *classical* fictitious play, it turns out that in many practical settings femtocells observe imperfectly the actions of their counterparts. Before investigating modeling these information limitations, we first define a *smoothed* version of fictitious play, where a specific (in this case *entropy*) term is added to the utility function of players in order to *randomize* their actions. Such randomization is motivated by the fact that players (i.e., femtocells) seek to maximize their *long-term* utility function whereby the transmission probability distribution probability that strikes a balance between the paradigm of exploration and exploitation is given by the Boltzmann distribution, also known as soft-max function. This is formalized as follows.

Let us redefine $p_k := [p_{k1} \dots p_{kN_k}]^T$ as the probability distribution on the action set of a player k with cardinality N_k, such that $0 \le p_{kj} \le 1 \; \forall j$ and $\sum_j p_{kj} = 1$. A smoothed

---

[5] By myopic it is meant that femtocells are interested in maximizing their instantaneous transmission rates. In contrast, when femtocells seek to explore their environment it is referred to as foresightedness.

utility function U_k is given by

$$U_k(p_k, s f_k) := \bar{u}_k(p_k, s f_k) + \tau_k H(p_k), \tag{11.18}$$

where the entropy function, H, is used as a smoothing factor and is defined as

$$H(p_k) := -\sum_{j=1}^{N_k} p_{kj} \log(p_{kj}) = -p_k^T \log(p_k),$$

and τ_i represents how much player k wants to randomize its own actions [11]. Note that this term is independent of the actions of other players.

The best response of a player based on its strategy beliefs used by all the other players in the network, is defined as

$$B R_k(s f_k) = \arg\max_{p_k} U_k(p_k, s f_k),$$

which leads to a soft-max solution if $\bar{u}_k$ is formulated as a K-player matrix game:

$$B R_k(s f_k) = \sigma \left(\frac{A \otimes s f_k}{\tau_i} \right),$$

where A is the K-dimensional game array (or rank-K tensor) and the vector-valued soft-max function σ is defined as

$$\sigma_j(x) = \frac{e^{x_j}}{\sum_{j=1}^{N_k} e^{x_j}}, \quad j = 1, \ldots, N_k. \tag{11.19}$$

The range of the soft-max function is in the interior of the probability simplex and $\sigma_j(x) > 0 \, \forall j$.

Fictitious play with limited information is an active research topic [15] in which decentralized variants of FP are currently under investigation aiming at relaxing the need for *full information*. One interesting approach is when players need to simultaneously estimate their utility function and optimize their transmission probability (power level and frequency). This approach significantly relaxes the need to know the actions taken by the other femtocells, where a simple feedback is necessary from the HUE to its serving FBS. Some early works in this direction can be found in [8] and [13] where femtocells estimate their long-term objectives by tuning their transmission strategies across different subcarriers.

11.4   Macrocell–femtocell coexistence with no information exchange

Unlike Section 11.3, in this section information exchange among femtocells is no longer possible for certain scenarios. In the following, we investigate two online learning procedures adopted by femtocells to mitigate interference toward the macrocell tier, namely the reinforcement learning procedure followed by its cooperative variant.

11.4.1　Reinforcement learning (Q-learning)

One of the decentralized reinforcement learning approaches available in the literature is Q-learning. The Q-learning model consists of a set of states S and actions A aiming at finding a policy that maximizes the observed rewards over the interaction time of the players (i.e., femtocells). Every FBS $k \in \mathcal{K}$ explores its environment, observes its current state s, and takes a subsequent action a, according to a decision policy $\pi : s \rightarrow a$. With their ability to learn, the knowledge about other players' strategies is not needed. Instead, a Q-function maintains the knowledge about other players in the network, based on which decisions can be made accordingly. Let us denote by $\mathcal{G}^{(Q)} = (\mathcal{K}, \{\mathcal{P}_k\}_{k \in \mathcal{K}}, \{u_k\}_{k \in \mathcal{K}})$ the Q-learning game. Here, the players are the FBSs $k \in \mathcal{K}$, $\mathbf{s}s_k(t) = (s_k^{(1)}(t), \ldots, s_k^{(N)}(t))$ is the composite state of FBS k at time t in subcarrier $n \in \mathcal{N}$. Let $s_k^{(n)}(t) \in \{0, 1\}$ indicate whether the SINR requirements of MUEs on their allocated resource blocks is violated. The transmission strategy (or action) of every FBS is denoted as $a_k^{(n)} = \frac{l}{L} p_{k,\max}$. Finally, $\mathbf{u}_k(t) = (u_k^{(1)}(t), \ldots, u_k^{(N)}(t))$ is the utility function or payoff vector of FBS k at time-instant t. The *expected* discounted reward over an infinite horizon is given by

$$V^\pi(s) = \mathbb{E}\left[\gamma^t \times r(s_t, \pi^*(s_t))|s_0 = s\right], \tag{11.20}$$

where $0 \leq \gamma \leq 1$ is a discount factor and r is the player's reward at time t. Equation (11.20) can be rewritten as follows:

$$V^\pi(s) = R(s, \pi^*(s)) + \gamma \sum_{v \in S} P_{s,v}(\pi(s)) V^\pi(v), \tag{11.21}$$

where $R(s, \pi^*(s)) = \mathbb{E}\{r(s, \pi(s))\}$ is the mean value of reward $r(s, \pi(s))$, and $P_{s,v}$ is the transition probability from state s to v. Moreover, the optimal policy π^* satisfies the optimality criterion:

$$V^*(s) = V^{\pi^*}(s) = \max_{a \in \mathcal{A}} \left(R(s, a) + \gamma \sum_{v \in \mathcal{S}} P_{s,v}(a) V^*(v) \right). \tag{11.22}$$

It is generally difficult to explicitly calculate the reward $R(s, a)$ and transition probability $P_{s,v}(a)$. However, through Q-learning, the knowledge of these values can be gradually learned and reinforced with time. For a given policy π, define a Q-value as follows:

$$Q^*(s, a) = R(s, a) + \gamma \sum_{v \in S} P_{s,v}(a) V^\pi(v), \tag{11.23}$$

which is the expected discounted reward when executing action a at state s and then following policy π thereafter.

Here, we use the Q-learning algorithm to iteratively approximate the state-action value function $Q(s, a)$. The agent keeps trying all actions in all states with non-zero probability and must sometimes explore by choosing at each step a random action with probability $\epsilon \in (0, 1)$, and the greedy action with probability $(1 - \epsilon)$. This is referred to as ϵ-greedy exploration [16].

The Q-learning process aims at finding $Q(\mathbf{s}, \mathbf{a})$ in a recursive manner where the update equation is given as:

$$Q_t(\mathbf{s}, \mathbf{a}) = (1 - \alpha)Q_{t-1}(\mathbf{s}, \mathbf{a}) + \alpha\left[r_t(\mathbf{s}, \mathbf{a}) + \gamma V_{t-1}(v_t)\right], \qquad (11.24)$$

where $V_{t-1}(v_t) = \max_{b \neq a} Q_{t-1}(v, b)$ and α is the learning rate. The pseudo-code of the Q-learning algorithm is given in Algorithm 11.4.

Algorithm 11.4 Q-learning

Require: Number of iterations, Number of random femto configurations
Require: Exploration probability ϵ, discount factor γ, $\alpha = 1$
  repeat
    repeat
      for all Players $k \in \mathcal{K}$ **do**
        if rand $< \epsilon$ **then**
          Choose a random strategy to explore
        else
          Calculate strategy that maximizes Q_k
        end if
        Update Q_k
      end for
      Calculate utility of all the femtocells
      $\alpha = t/t + 1$
    until Required iterations
  until Eventual convergence
  Calculate ensemble average of utility

11.4.2   Cooperative Q-learning

Unlike the classical case where femtocells build their Q-tables in a non-cooperative manner, femtocells have the possibility to adopt another scheme coined *learning-by-expert*, where cooperation between femtocells takes place based on temporal difference of their utility function [17]:

$$d_k(t) = u\big(p_k(t), p_{-k}(t)\big) - u\big(p_k(t-1), p_{-k}(t-1)\big). \qquad (11.25)$$

As a consequence, if a femtocell i finds a femtocell j to have similar difference such that

$$|d_i(t) - d_j(t)| \leq \zeta, \qquad (11.26)$$

then it considers FBS j as an expert to which an appropriate weight is assigned. As a result, a given femtocell modifies accordingly its Q-table by learning from a small group of femtocells dubbed as *experts*. Note that this information exchange among femtocells takes place periodically. It is worth noting that this cooperative variant of Q-learning requires an exchange of information among femtocells, which may be impaired by errors and delays over the backhaul. While this variant of Q-learning yields higher performance, there is no proof of convergence and it requires a sizeable amount of information to be exchanged. The pseudo-code of the cooperative Q-learning algorithm is given in Algorithm 11.5.

Algorithm 11.5 Cooperative Q-learning

Require: Number of iterations, Number of random femto configurations, Cooperation interval
Require: Exploration probability ϵ, discount factor γ
Require: Impressionability β, $\alpha = 1$
  repeat
    repeat
      if Time to cooperate **then**
        for all Players $k \in \mathcal{K}$ **do**
          Assign expertness weight to each user, $w(i, j)$
          Update Q_k
        end for
      end if
      for all Players $k \in \mathcal{K}$ **do**
        if rand $< \epsilon$ **then**
          Choose a random strategy to explore
        else
          Calculate strategy that maximizes Q_k
        end if
        Update Q_k
      end for
      Calculate utility of all the femtocells
      $\alpha = t/t + 1$
    until Required iterations
  until Eventual convergence
  Find ensemble average of utility

11.5 Simulation results

In order to substantiate our theoretical findings and evaluate the performance of the reinforcement learning interference mitigation algorithms, we consider the case where one macrocell with radius $R_m = 500\,\text{m}$ is underlaid with K femtocells each of radius $R_f = 20\,\text{m}$, transmitting over $N = 8$ subcarriers. We assume that femtocells have $L = 3$ transmit power levels and service only one FUE at a time.[6] The minimum SINR of the MUEs is given by $\Gamma_0 = (\Gamma_0^{(1)}, \ldots, \Gamma_0^{(N)})$. It is assumed that $\Gamma_0^{(1)} = \cdots = \Gamma_0^{(N)} = 3\,\text{dB}$. The transmission power of the MBS is $43\,\text{dBm}$, whereas the fBS adjusts its power

---

[6] Arbitrary number of power levels L can be considered without loss of generality.

through the various learning schemes to a value of maximum 10 dBm. The channel is in line with those of 3GPP composed of path-loss fading and log-normal shadowing with standard deviation of 8 and 4 dBm for outdoor and indoor communications [18]. For learning through replication by imitation, we set the intensity of review (average review rate) to be $r = 0.01$, for all players. For Q-learning and cooperative Q-learning approaches, we set the discount factor $\gamma = 0.95$, exploration probability $\epsilon = 0.1$, and impressionability $\beta = 0.3$. For cooperative Q-learning, cooperation was set to occur after every 4 intervals of time.

The considered path-loss modelcell from macrocell base station (BS) to macrocell and HUEs is

$$PL(d) = 28 + 36\log_{10}(d) + W, \tag{11.27}$$

the considered path-loss model from femtocell BS to macrocell and HUEs is

$$PL(d) = 7 + 56\log_{10}(d) + W, \tag{11.28}$$

and, the considered path-loss model from femtocell BS to its own HUEs is

$$PL(d) = 37 + 20\log_{10}(d) + W, \tag{11.29}$$

where W represents the wall penetration loss assumed to be 15 dB. All path-loss values are in dB.

The interference mitigation metric considered herein is

$$u(p_k, p_{-k}) = \sum_{n=1}^{N} \log_2\left(1 + \gamma_k^{(n)}\right) \cdot \mathbf{1}_{\left\{\gamma_0^{(n)} > \Gamma_0^{(n)}\right\}}. \tag{11.30}$$

This metric is different from zero only if the macrocell satisfies the minimum SINR level required for its own communications. Hence, as long as the macrocell sees its QoS requirement satisfied, femtocells obtain a positive reward. This models certain altruism from the behavior of the FBSs, which sacrifice their performance to guarantee the QoS of the macrocell system. It is worth noting that other utility functions can be considered. Finally, a uniform power allocation is assumed for the macrocell transmission, which corresponds to the full load scenario (i.e., MUEs use all resource blocks).

11.5.1 Convergence behavior of the reinforcement learning based interference management algorithms

In Figure 11.2, we plot the average femtocell sum-rate for $K = 50$ FBSs underlaying one macrocell over $N = 8$ subcarriers, highlighting the convergence behavior for different learning algorithms. From this plot, we notice that replication dynamics, fictitious play, Q-learning, and cooperative Q-learning schemes eventually converge to some steady state, whereas the replication by imitation converges to zero. This is attributed to the fact

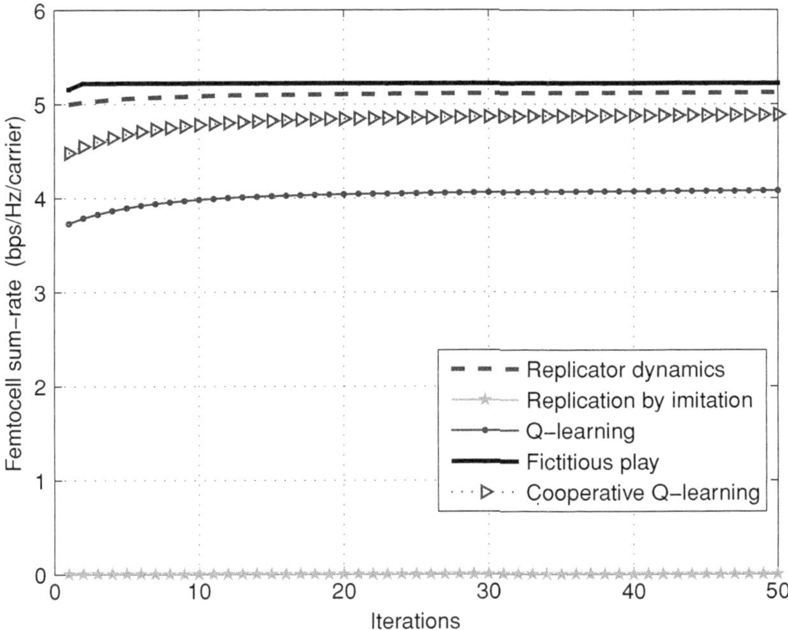

Figure 11.2 Convergence of various learning algorithms and their impact on the average femtocell sum-rate for $K = 50$ femtocells and $N = 8$ subcarriers. ©2011 IEEE. Reprinted, with permission, from [10].

that the *reviewing* femtocell imitates a random femtocell with the most *used* transmission strategy, and not necessarily the femtocell with better payoff (*expert*). Thus the reviewing femtocell has a higher probability of choosing the more "popular" strategy, instead of a performance enhancing strategy. As a result, this leads femtocells to congregate around few strategies, causing higher interference to the macrocell and yielding forced femtocell shutdowns.

When information can be exchanged among femtocells, better performance in terms of average spectral efficiency is obtained. This highlights the tradeoff of better performance with the cost of information exchange. Notably, the cooperative Q-learning approach outperforms its classical counterpart, whereas the fictitious play and replication dynamics schemes outperform the cooperative Q-learning approach. Nevertheless, this comes at the expense of having a gateway or an aggregator for exchanging information among femtocells. In addition, it is worth noting that while yielding lower average performance (compared to the other approaches), the classical Q-learning is totally decentralized and relies mainly on a HUE-to-FBS feedback. Nevertheless, other possibilities exist where a smarter initialization of the Q-tables yield faster convergence. In addition, in terms of convergence, it is important to note that there is no proof of optimality for both Q-learning approaches; however, there are sufficient conditions for both fictitious play and replication dynamics, and only for a number of cases. For more information, the interested reader is referred to [1].

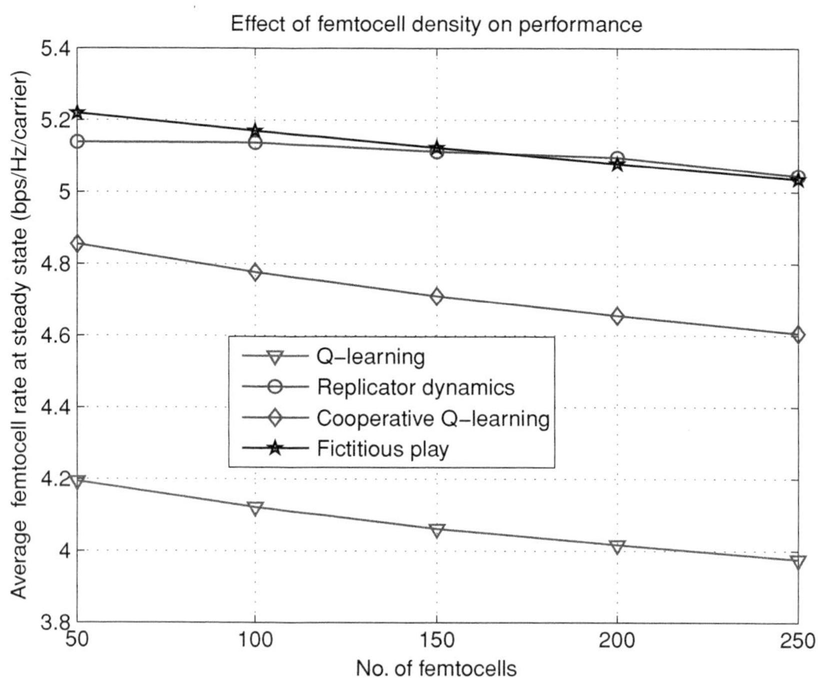

Figure 11.3 Effect of femtocell density on the average femtocell sum-rate, for different learning algorithms and different femtocell densities. ©2011 IEEE. Reprinted, with permission, from [10].

11.5.2   Effect of femtocell density on the performance of learning algorithms

Figure 11.3 plots the impact of the femtocell density on the average femtocell sum-rate for different learning algorithms for $K = \{50, 100, 150, 200, 250\}$ femtocells. As expected, a general decrease in performance is witnessed as the network densifies. Nonetheless, we see that the rate of decrease in performance is not the same for all reinforcement learning algorithms. In particular, for $K = 250$ femtocells, the average sum-rate is around 5 bps/Hz using replication dynamics and fictitious play interference management techniques, whereas approximately 4 bps/Hz is obtained via Q-learning and 4.6 bps/Hz for the cooperative Q-learning, respectively. Figure 11.4 plots the cumulative distribution function for the average femtocell rate for the various reinforcement learning algorithms. It is seen that the fictitious play learning scheme outperforms the Q-learning, cooperative Q-learning, and replication by dynamics schemes, for all percentile ranges.

Figure 11.5 plots the cumulative distribution function (CDF) of the average femtocell sum-rate for both dedicated and shared bands. It can be seen that the co-channel deployment outperforms the dedicated one for both Q-learning and its cooperative variant. Besides, the dedicated case for both fictitious play and replication dynamics slightly outperforms the co-channel deployment case. This is attributed to the fact that with a larger action space, femtocells can fine tune their transmissions so as to maximize their transmission rates.

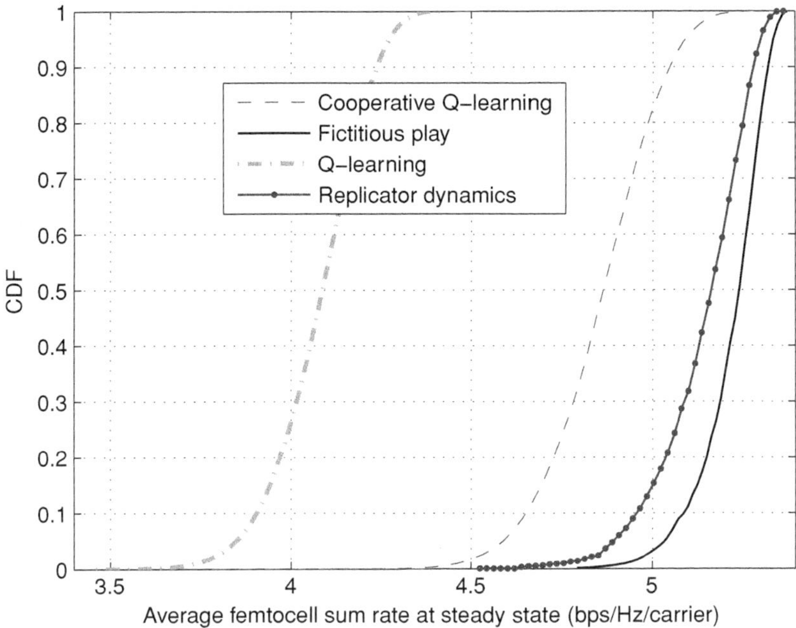

Figure 11.4 Cumulative distribution function (CDF) of the average femtocell spectral efficiency for different reinforcement learning algorithms. ©2011 IEEE. Reprinted, with permission, from [10].

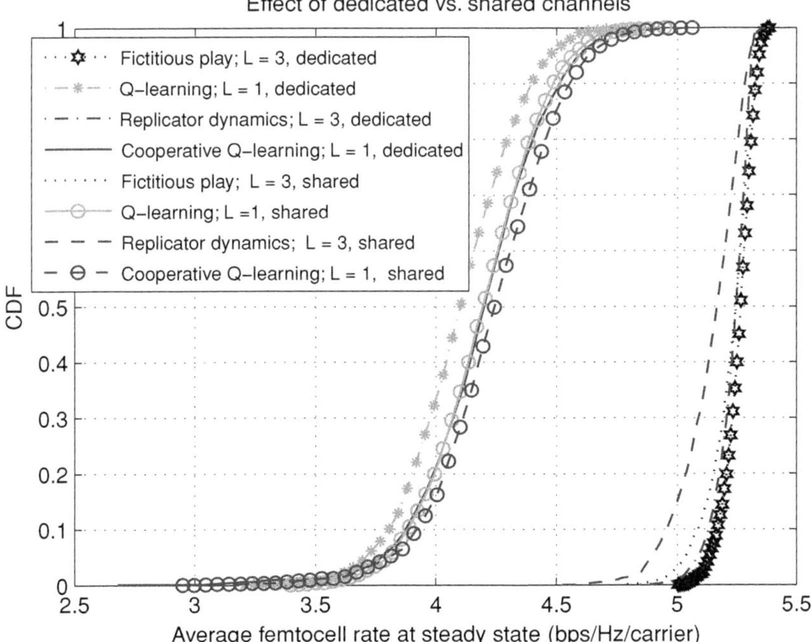

Figure 11.5 Cumulative distribution function (CDF) of the average femtocell spectral efficiency of $K = 50$ femtocells at steady state. ©2011 IEEE. Reprinted, with permission, from [10].

11.5.3   General remarks and inherent tradeoffs facing small cells

The purpose of this subsection is to provide additional insights about the performance of the various reinforcement learning based interference management algorithms, pertaining to different information assumption available at the femtocells. The femtocell average spectral efficiency relies on whether femtocells need to exchange information or not. Therefore every femtocell needs to strike a balance on whether it should exchange information with its potential interferers and possibly improving its performance or eventually learning on its own, albeit lower spectral efficiency. In detail, as per their formulation and information assumptions, the classical fictitious play formulation requires a significant amount of information about all femtocells in the network. On the other hand, the Q-learning based reinforcement learning algorithm is fully decentralized whereby each FBS k needs only to observes its corresponding achieved utility function at each iteration through the feedback of its respective receiver. Finally, the fictitious play requires that the network should be static during the whole learning process, nonetheless that can be problematic for high-speed scenarios. Robust learning schemes that do not rely on synchronization are essential for an efficient operation of femtocells. Having said this, fast and robust reinforcement learning algorithms are needed that bridge the gap between fully decentralized algorithms with slower convergence versus semi-centralized approaches with faster convergence. In addition, a *gamut* of context aware learning enablers should be considered such as the location of interferers and the impact of heterogeneous backhauls for exchanging information among femtocells. Finally, the decentralized interference management tools investigated in this chapter can be readily applied in other avenues and problem formulations such as coordinated multi-point transmissions (CoMPs), multi-radio access technology, and multi-layers.

11.6   Conclusions

In this chapter, the problem of co-tier interference management was studied from a game theoretic learning perspective. Due to their decentralized nature, femtocells have a variety of tools that allow them to smoothly coexist with the legacy macrocell network, which is oblivious to their presence. Two types of learning schemes were investigated, namely with and without information exchange among femtocells. It was shown that while the performance of the macrocell network is satisfied, femtocells have the tradeoff of either being self-interested relying only on their own learning capabilities, or eventually cooperating with other femtocells. While the former has less signaling at the price of longer convergence, the latter exhibits a faster convergence *albeit* requiring information exchange. It is worth noting that there are still significant challenging problems that need to be tackled, such as joint access and backhaul. Finally, driven by the increasing interest of CoMP techniques in both the downlink and uplink, decentralized approaches for cooperative and coordinated interference management for dense small cells are needed.

References

[1] S. Lasaulce and H. Tembine, *Game Theory and Learning for Wireless Networks: Fundamentals and Applications*, 1st edn. Academic Press, 2011.

[2] D. Niyato and E. Hossain, "Dynamics of network selection in heterogeneous wireless networks: an evolutionary game approach," *IEEE Trans. Veh. Technol.*, vol. 58, no. 4, pp. 2008–17, May 2009.

[3] H. Li, "Multi-agent Q-learning of channel selection in multi-user cognitive radio systems: a two by two case," in *IEEE Int. Conf. on Systems, Man and Cybernetics (SMC)*, Oct. 2009, pp. 1893–8.

[4] M. Bennis and D. Niyato, "A Q-learning based approach to interference avoidance in self-organized femtocell networks," in *Proc. IEEE Global Telecommun. Conf. (GLOBECOM) Workshops*, Dec. 2010, pp. 706–10.

[5] M. Galindo-Serrano, L. Giupponi, and M. Majoral, "On implementation requirements and performances of Q-learning for self-organized femtocells," in *Proc. IEEE Global Telecommun. Conf. (GLOBECOM) Workshops*, Dec. 2011, pp. 706–10.

[6] S. Guruacharya, D. Niyato, E. Hossain, and D. I. Kim, "Hierarchical competition in femtocell-based cellular networks," in *Proc. IEEE Global Telecommun. Conf. (GLOBECOM)*, Dec. 2010, pp. 1–5.

[7] A. Feki and V. Capdevielle, "Autonomous resource allocation for dense LTE networks: a multi armed bandit formulation," in *Proc. IEEE Int. Symp. Personal, Indoor, Mobile Radio Commun. (PIMRC)*, Dec. 2010, pp. 706–10.

[8] M. Bennis and S. M. Perlaza, "Decentralized cross-tier interference mitigation in cognitive femtocell networks," in *Proc. IEEE Int. Conf. on Commun. (ICC)*, June 2011, pp. 1–5.

[9] F. Pantisano, M. Bennis, W. Saad, and M. Debbah, "Spectrum leasing as an incentive towards uplink macrocell and femtocell cooperation," *IEEE J. Sel. Areas Commun. (JSAC)*, vol. 30, no. 3, pp. 617–30, Apr. 2012.

[10] S. Bennis, M. Gurucharya and D. Niyato, "Distributed learning strategies for interference mitigation in femtocell networks," in *Proc. IEEE Global Telecommun. Conf. (GLOBECOM)*, June 2011, pp. 1–5.

[11] J. Shamma and G. Arslan, "Dynamic fictitious play, dynamic gradient play, and distributed convergence to Nash equilibria," *IEEE Trans. Autom. Control*, vol. 50, no. 3, pp. 312–27, Mar. 2005.

[12] J. F. Nash, "Equilibrium points in n-person games," *P. Nat. Acad. Sci. USA*, vol. 36, no. 1, pp. 48–9, 1950.

[13] M. Bennis, S. Perlaza, and M. Debbah, "Learning coarse-correlated equilibria in two-tier networks," in *Proc. IEEE Int. Conf. on Commun. (ICC)*, June 2012, pp. 1–5.

[14] J. W. Weibull, *Evolutionary Game Theory*. The MIT Press, 1997.

[15] T. Alpcan and T. Basar, *Network Security: A Decision and Game Theoretic Approach*, 1st edn. Cambridge: Cambridge University Press, 2011.

[16] S. Hart and A. Mas-Colell, "A simple adaptive procedure leading to correlated equilibrium," *Econometrica*, vol. 68, no. 5, pp. 1127–50, Sep. 2000.

[17] A. Nedic and A. Ozdaglar, "Distributed subgradient methods for multi-agent optimization," *IEEE Trans. Autom. Control*, vol. 54, no. 1, pp. 48–61, Jan. 2009.

[18] "3rd Generation Partnership Project; Technical Specification Group Radio Access Networks; 3G Home NodeB Study Item Technical Report (Release 8)," 3GPP, 3GPP TR 25.820, Mar. 2008.

12 Resource allocation optimization in heterogeneous wireless networks

Chee Wei Tan

12.1 Introduction

Wireless cellular networks are designed to provide network coverage over large areas and support many users. Most recently, studies in 3GPP LTE-advanced have looked at the deployment of heterogeneous wireless networks to improve system performance as well as to enhance network coverage, especially in-building coverage [1–6]. Heterogeneous wireless networks use a mix of higher tier macrocells to extend network reach and lower tier small cells to enhance performance within the same frequency band [1–6]. These smaller cells offload the traffic from the macrocells and connect the traffic to the cellular core network via broadband access networks. However, as user-installed small cells (femtocells) are often deployed in an ad hoc manner, this gives rise to the problem of interference between cells. For example, a macrocell with a femtocell or a femtocell with another femtocell. New resource allocation techniques are required to ensure that the users control their power to mitigate performance loss due to interference. To enhance decentralized deployment, users also need to adapt their power with minimal signaling overhead [1, 2]. For example, users in femtocell can use digital subscriber line (DSL) or cable modem to exchange messages through the cellular core network to adjust their transmit powers to reduce the interference caused to the macrocell users.

Resource allocation is fundamental to coordinate the sharing of spectrum resource between the macrocell and the femtocell, and then coordinate the sharing of spectrum resource between the femtocells. This is especially important because sharing must be done fairly in a collaborative and distributed manner. Maintaining a balanced operation in the macrocell–femtocell heterogeneous wireless network, however, becomes highly problematic since interference rises rapidly with increasing ad hoc femtocell deployment. Without appropriate resource coordination, a network can become unstable or operate in a highly inefficient and unfair manner [7, 8]. This fairness can depend on factors such as the channel variation, power allocation, interference level, etc. A reasonable fairness scheme is an egalitarian approach. Resource allocation techniques have to allocate the power of different users in such a way as to compensate for the negative effects of interference on users with the worst-case performance in the system.

Small Cell Networks: Deployment, PHY Techniques, and Resource Management, ed. Tony Q. S. Quek, Guillaume de la Roche, İsmail Güvenç, and Marios Kountouris. Published by Cambridge University Press. © Cambridge University Press 2013.

Power control generates the optimal set of signal to interference plus noise ratio (SINR) to control important performance metrics in the wireless environment such as the outage probability [9, 10]. A link outage is declared when the received SINR falls below a given threshold that is often computed from the quality of service (QoS) requirement. The statistics of the SINR, and consequently that of the outage probability, are also dependent on other network parameters, including imperfections due to the statistical channel fading, the additive background noise, user mobility, and the dynamics of power control updates. Channel fading can be modeled by a Rayleigh, a Rician, or a Nakagami distribution depending on the environment under consideration [11, 12]. In this chapter, Rayleigh fading, which is relevant to the in-building coverage model and heavily built-up urban environments, will be studied in detail.

Many recent works have recognized that power control is an appropriate mechanism to tackle wireless resource allocation problems [9, 10]. In [13], the authors propose tracking Rayleigh-fading fluctuation to reduce outage in a macrocell network. In [14], algorithms that adapt SINR requirements for utility maximization are studied for a macrocell–femtocell network. The authors in [15] study subcarrier and power control to minimize outage in an OFDMA cellular network. The authors in [16] study the worst outage probability problem and the total power minimization problem in interference-limited Rayleigh-fading networks. The authors in [17] propose an iterative algorithm for the total power minimization problem. The authors in [18] use geometric programming to tackle a suite of non-convex power control problems that include outage constraints. These optimization problems are often difficult to solve due to the non-trivial non-linear functions of the performance metrics such as the outage probability, which are functions of SINR, which is in turn a non-linear (and neither convex nor concave) function of powers.

In this chapter, several resource allocation optimization problems using power control will be studied. These problem formulations take into account typical performance metrics in the literature and will be studied using an approach that leverages a non-linear Perron-Frobenius theory in [19–21] and non-negative matrix theory [22–24]. This approach can provide analytical solutions to the problems and a systematic way to design algorithms that adapt the transmission powers of all the users. In particular, these resource allocation problems that guarantee egalitarian fairness to all the users can be viewed as non-linear eigenvalue problems, whose optimal value and optimal solution are associated with the spectral radius (also the largest eigenvalue) and its right eigenvector of particular non-negative matrices induced by non-linear positive mappings, respectively.

Besides optimality, it is also imperative to design heterogeneous wireless networks that are capable of *adapting* their behavior to be both spectral and energy efficient [1, 2]. The approach using non-linear Perron–Frobenius theory can also be used to design a dynamic power control algorithm that allows each femtocell user to adapt its outage probability specification to minimize the total energy consumption in the system and simultaneously guarantee a min-max fairness in terms of worst outage probability to all the femtocell users. The advantage is that users can adapt their performance quality in a dynamic environment without the need of a centralized admission control mechanism (which is often not practically feasible due to the ad hoc architecture).

Overall, the highlights of this chapter are summarized in the following:

1. The worst-outage probability problem, subject to a total power constraint (for the downlink scenario) or individual power constraints (for the uplink scenario) is studied and the optimal solution is given analytically. A power control algorithm that converges geometrically fast to the optimal solution is then proposed. As a by-product, it solves the open problem of convergence for a previously proposed algorithm in [16] for the interference-limited special case.
2. A tight relationship between the worst-outage probability problem and its certainty-equivalent margin counterpart (the max-min weighted SINR problem) is established. This is utilized to find useful bounds and the convergence rate of the power control algorithm. A by-product of the analysis solves the open problem of convergence for a previously proposed algorithm to a max-min weighted SINR problem in [5].
3. The total power minimization problem with both outage specification and individual power constraints is studied. The feasibility condition of this problem is characterized analytically using the results established for the worst-outage probability problem. This feasibility can be hard to assess in practice, and so feasibility bounds that are based on the problem parameters, e.g., SINR thresholds, the outage thresholds, are derived for practical purposes.
4. Based on the established feasibility conditions, a dynamic power control algorithm is proposed for the graceful handling of infeasibility in a femtocell–macrocell network. When there is infeasibility in the system, the algorithm optimizes the overall energy consumption based on an adaptive outage probability specification that guarantees the worst-outage probability to all the users in both the femtocell and macrocell.
5. The numerical simulations demonstrate the value of deploying closed-access femtocells in a macrocell using the dynamic algorithm in terms of the total energy savings. When there is infeasibility in the system, the percentage of macrocell users meeting their outage probability specification is higher in comparison to a baseline non-adaptive algorithm.

This chapter is organized as follows. The system model is introduced in Section 12.2. In Section 12.3, the problem of minimizing the worst-outage probability is studied, and its analytical solution and a geometrically fast convergent algorithm are given. In Section 12.4, the feasibility condition of a total power minimization problem is studied, and a dynamic power control algorithm that adapts the outage probability specification to minimize the total energy is proposed. Further discussions are given in Section 12.5. In Section 12.6, the numerical performance of the power control algorithms is illustrated. The conclusion is then given in Section 12.7. Figure 12.1 gives an overview of the connection between the various optimization problems and topics studied in this chapter.

The following notation is used. Boldface uppercase letters denote matrices, boldface lowercase letters denote column vectors, italics denote scalars, and $\mathbf{u} \geq \mathbf{v}$ denotes componentwise inequality between vectors $\mathbf{u}$ and $\mathbf{v}$. Let $(\mathbf{By})_l$ denote the l-th element of $\mathbf{By}$. Let $\mathbf{x}/\mathbf{y}$ denote the vector $[x_1/y_1, \ldots, x_L/y_L]^T$. Also, $\mathbf{B} \geq \mathbf{F}$ if $B_{ij} \geq F_{ij}$ for all i, j. The Perron–Frobenius eigenvalue of a non-negative matrix $\mathbf{F}$ is denoted as $\rho(\mathbf{F})$, and the Perron (right) and left eigenvector of $\mathbf{F}$ associated with $\rho(\mathbf{F})$ are denoted by

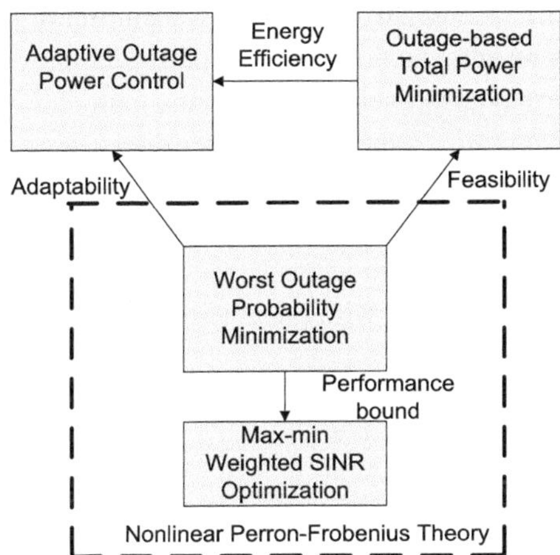

Figure 12.1 An overview that connects the different optimization problems in this chapter. The two problems contained in the dashed box are solved by the non-linear Perron–Frobenius theory. Central to this chapter is the worst-outage probability problem. It is linked to the max-min weighted SINR problem to deduce performance bounds. It can also be used to assess the feasibility issue of a total power minimization problem with outage constraints and to design an adaptive outage based power control algorithm. ©2011 IEEE. Reprinted, with permission, from [6].

$\mathbf{x}(\mathbf{F}) \geq \mathbf{0}$ and $\mathbf{y}(\mathbf{F}) \geq \mathbf{0}$ (or, simply $\mathbf{x}$ and $\mathbf{y}$, when the context is clear) respectively. Recall that the Perron–Frobenius eigenvalue of $\mathbf{F}$ is the eigenvalue with the largest absolute value. Assume that $\mathbf{F}$ is an irreducible non-negative matrix. Then $\rho(\mathbf{F})$ is simple and positive, and $\mathbf{x}(\mathbf{F}), \mathbf{y}(\mathbf{F}) > \mathbf{0}$ [25]. The superscript $(\cdot)^T$ denotes transpose. Let $\mathbf{e}_l$ denote the l-th unit coordinate vector and $\mathbf{I}$ as the identity matrix. For any vector $\tilde{\boldsymbol{\gamma}} = (\tilde{\gamma}_1, \ldots, \tilde{\gamma}_L)^T \in \mathbb{R}^L$, let $e^{\tilde{\boldsymbol{\gamma}}} = (e^{\tilde{\gamma}_1}, \ldots, e^{\tilde{\gamma}_L})^T$.

12.2   System model

Consider a multi-user communication system with L users (logical transmitter/receiver pairs) sharing a common frequency. Each user employs a single-user decoder, i.e., treating interference as additive Gaussian noise, and has perfect channel state information at the receiver. Our system with L users can be modeled by a Gaussian interference channel having the baseband signal model $y_l = h_{ll}x_l + \sum_{j \neq l} h_{lj}x_j + z_l$, where $y_l \in \mathbb{C}^{1 \times 1}$ is the received signal of the l-th user, $h_{lj} \in \mathbb{C}^{1 \times 1}$ is the channel coefficient between the transmitter of the j-th user and the receiver of the l-th user, $x \in \mathbb{C}^{N \times 1}$ is the transmitted (information carrying) signal vector, and z_l's are the independent and identically distributed (i.i.d.) additive complex Gaussian noise coefficient with variance $n_l/2$ on each of its real and imaginary components (thus n_l being the noise power at the l-th receiver). At each transmitter, the signal is constrained by an average power constraint, i.e., $\mathbb{E}[|x_l|^2] = p_l$, which is assumed to be upper bounded by $\bar{p}$ for all l. The vector $(p_1, \ldots, p_L)^T$ is the transmit power vector, which is the optimization variable of interest.

Under Rayleigh fading, the power received from the j-th transmitter at l-th receiver is given by $G_{lj} R_{lj} p_j$ where G_{lj} represents the non-negative path gain between the j-th transmitter and the l-th receiver (it may also encompass antenna gain and coding gain) that is often modeled as proportional to $d_{lj}^{-\gamma}$, where d_{lj} denotes distance, γ is the power fall-off factor, and R_{lj} models Rayleigh fading and is independent and exponentially distributed with unit mean. The distribution of the received power from the j-th transmitter at the l-th receiver is then exponential with mean value $E\left[G_{lj} R_{lj} p_j\right] = G_{lj} p_j$.

Next, let us define the non-negative matrix $\mathbf{F}$ with the entries:

$$F_{lj} = \begin{cases} 0, & \text{if } l = j \\ \frac{G_{lj}}{G_{ll}}, & \text{if } l \neq j \end{cases} \tag{12.1}$$

and

$$\mathbf{v} = \left(\frac{n_1}{G_{11}}, \frac{n_2}{G_{22}}, \dots, \frac{n_L}{G_{LL}} \right)^T. \tag{12.2}$$

Moreover, $\mathbf{F}$ is assumed to be irreducible,[1] i.e., each user has at least an interferer.

The SINR at the receiver (e.g., a linear matched filter) output of the l-th user can be given by [16]:

$$\mathsf{SINR}_l(\mathbf{p}) = \frac{R_{ll} p_l}{\sum_{j \neq l} F_{lj} R_{lj} p_j + v_l}. \tag{12.3}$$

Now, (12.3) is a random variable that depends on Rayleigh fading. The transmission from the l-th transmitter to its receiver is successful if $\mathsf{SINR}_l(\mathbf{p}) \geq \beta_l$ (zero-outage), where β_l is a given threshold for reliable communication. If there is channel fading in the system, a fading-induced outage event occurs at the l-th receiver when $\mathsf{SINR}_l(\mathbf{p}) < \beta_l$. Let us denote this outage probability by $\mathbb{P}(\mathsf{SINR}_l(\mathbf{p}) < \beta_l)$.

Power constraint is a key design parameter in a wireless network [11]. It reflects the available resource budget in the system. For example, in a code division multiple access (CDMA) cellular network, it is common to have a power constraint $\mathcal{P}$ being either a total power constraint for CDMA cellular downlink or individual power constraints for CDMA cellular uplink, i.e.,

$$\mathcal{P} = \{\mathbf{p} \mid \mathbf{p} \geq \mathbf{0}, \, \mathbf{1}^T \mathbf{p} \leq \bar{P}\} \quad \text{or} \quad \mathcal{P} = \{\mathbf{p} \mid \mathbf{p} \geq \mathbf{0}, \, p_l \leq \bar{p} \; \forall l\}. \tag{12.4}$$

Besides the ad hoc deployment feature of femtocells, the presence of new power constraints, which differ from typical CDMA cellular networks, is also one of the key features of femtocells. These power constraints are imposed either due to regulatory policy for health considerations or to limit excessive interference to macrocell users. For example, a basic premise in femtocells is that the following two requirements are satisfied [5]:

---

[1] A non-negative matrix $\mathbf{F}$ is said to be irreducible if there exists a positive integer m such that the matrix $\mathbf{F}^m$ has all entries positive.

1. A femtocell user receives adequate levels of transmission quality within the femtocell.
2. The femtocell users do not cause unacceptable levels of interference to the macrocell users in the larger macrocell.

In order to satisfy (2) above, one possible way is to impose power constraints on the femtocell users so as to limit the amount of interference caused to the macrocell users. Let us illustrate an example of such a power constraint for the case of a single macrocell user with multiple femtocells given in [5]. Assume that there is no channel fading. Let us denote the macrocell user by the index 0, and the femtocell users have indices $1, \ldots, L$. The macrocell user transmits with a fixed power P_0, where $P_0 \geq \gamma_0 v_0$. This means that the macrocell user can satisfy his or her desired SINR threshold γ_0 even when there is no interference from the femtocells. In the presence of interference from the femtocells, the SINR of this macrocell user has to satisfy $\frac{P_0}{\sum_{j=1}^{L} F_{0j} p_j + v_0} \geq \gamma_0$, which can be rewritten as a single power constraint to yield

$$\mathcal{P} = \left\{ \mathbf{p} \mid \mathbf{p} \geq \mathbf{0}, \ \sum_{j=1}^{L} F_{0j} p_j \leq (P_0/\gamma_0 - v_0) \right\} \tag{12.5}$$

that must be satisfied by the transmit powers of all the femtocell users. In general, a power constraint of the form $\mathbf{a}^T \mathbf{p} \leq 1$ for some positive constant vector $\mathbf{a}$ can be formulated for interference management in a macrocell–femtocell network.

In order to satisfy (1) above, some form of metrics that measure the performance have to be associated with each link in the network to assess the adequate level of transmission quality. For example, in [5, 21, 26] the authors consider maximizing the minimum adequate level of SINR among all the users in the network. In [6], the metric used for performance analysis is the worst-outage probability. These performance metrics will be studied in the following.

12.3 Worst-outage probability minimization

The problem of minimizing the worst-outage probability can be formulated as

$$\begin{aligned} \text{minimize} \ \max_{l} \ & \mathbb{P}(\mathsf{SINR}_l(\mathbf{p}) < \beta_l) \\ \text{subject to} \ & \mathbf{p} \in \mathcal{P}, \\ \text{variables:} \ & \mathbf{p}. \end{aligned} \tag{12.6}$$

Let us denote the optimal worst-outage probability, i.e., the optimal value of (12.6), by $O^\star$.

Assuming independent Rayleigh fading at all signals, the outage probability of the l-th user can be given analytically by [16]:[2]

$$\mathbb{P}(\mathsf{SINR}_l(\mathbf{p}) < \beta_l) = 1 - e^{\frac{-v_l \beta_l}{p_l}} \prod_{j} \left(1 + \frac{\beta_l F_{lj} p_j}{p_l} \right)^{-1}. \tag{12.7}$$

---

[2] A closed form expression was first derived in [27], but another equivalent form derived in [16] is used here.

In the following, let us denote $\phi_l(\mathbf{p}) = 1 - e^{-v_l \beta_l / p_l} \prod_j \left(1 + \frac{\beta_l F_{lj} p_j}{p_l}\right)^{-1}$. Observe that the probability of successful transmission, i.e., the complement of (12.7), is simply the product of two factors, namely, $e^{-v_l \beta_l / p_l}$ and $\prod_j \left(1 + \frac{\beta_l F_{lj} p_j}{p_l}\right)^{-1}$, which are the probability of successful transmission in a noise-limited Rayleigh fading channel and the probability of successful transmission in an interference-limited Rayleigh-fading channel, respectively.

Using (12.7), (12.6) simplifies to a deterministic problem:

$$\text{minimize} \max_l \ \phi_l(\mathbf{p}) = 1 - e^{\frac{-v_l \beta_l}{p_l}} \prod_j \left(1 + \frac{\beta_l F_{lj} p_j}{p_l}\right)^{-1} \tag{12.8}$$

$$\text{subject to } \mathbf{p} \in \mathcal{P}.$$

Note that (12.8) is always feasible and its optimal solution is strictly positive. Previous work in the literature, e.g., [16], only considered (12.8) for the interference-limited case, i.e., $\mathbf{v} = 0$ and *without any power constraint*. In this special case, [16] showed that (12.8) can be reformulated as a geometric program (GP), and be solved efficiently using the interior point method [28].

In the following, a reformulation of (12.8) as a convex optimization problem will be given (not a convex GP formulation in general, but reduces to one in the interference-limited special case). By exploiting the non-linear Perron–Frobenius theory, a fast algorithm (no configuration whatsoever and orders of magnitude faster than standard convex optimization algorithms, e.g., interior-point method)[3] is given to solve (12.8) optimally. As a by-product, it resolves the open problem on the convergence of a previously proposed heuristic algorithm in [16]. Furthermore, the optimal value and solution of (12.8) are derived analytically in terms of the Perron–Frobenius eigenvalue and its eigenvector of a specially constructed non-negative matrix, respectively.

Let us next introduce the auxiliary variable τ and write (12.8) in the epigraph form (augmenting the constraint set with an additional L constraints):

$$\text{minimize } \tau$$
$$\text{subject to } 1 - e^{\frac{-v_l \beta_l}{p_l}} \prod_j \left(1 + \frac{\beta_l F_{lj} p_j}{p_l}\right)^{-1} \leq \tau \ \forall l, \tag{12.9}$$
$$\mathbf{p} \in \mathcal{P},$$
$$\text{variables: } \mathbf{p}, \ \tau.$$

By letting $\alpha = -\log(1 - \tau)$ and rewriting the L augmented constraints in (12.9), (12.9) is equivalent to the following problem:

$$\text{minimize } \alpha$$
$$\text{subject to } v_l \beta_l / p_l + \sum_j \log\left(1 + \frac{\beta_l F_{lj} p_j}{p_l}\right) \leq \alpha \ \forall l, \tag{12.10}$$
$$\mathbf{p} \in \mathcal{P},$$
$$\text{variables: } \mathbf{p}, \ \alpha.$$

---

[3] Some key computational considerations are the extremely fast signal processing requirement at the transceiver chip and the decentralized environment in a heterogeneous wireless network.

Let us call the first L constraints of (12.10) the *outage constraints*, and denote the optimal solution to (12.10) by $(\mathbf{p}^\star, \alpha^\star)$. Note that $\mathbf{p}^\star$ is also the optimal solution to (12.8). Hence, this results in the optimal worst-outage probability $O^\star = \phi_l(\mathbf{p}^\star)$ and also $O^\star = 1 - e^{-\alpha^\star}$.

Now, (12.10) is non-convex in $(\mathbf{p}, \alpha)$. However, by making a logarithmic change of variable in $\mathbf{p}$, i.e., $\tilde{p}_l = \log p_l$ for all l, (12.10) can be converted into the following convex optimization problem in $(\tilde{\mathbf{p}}, \alpha)$:[4]

$$
\begin{aligned}
&\text{minimize } \alpha \\
&\text{subject to } v_l \beta_l e^{-\tilde{p}_l} + \sum_j \log \left(1 + \beta_l F_{lj} e^{\tilde{p}_j - \tilde{p}_l}\right) \le \alpha \ \ \forall\, l, \\
&\qquad\qquad e^{\tilde{\mathbf{p}}} \in \mathcal{P}, \\
&\text{variables: } \tilde{\mathbf{p}}, \ \alpha.
\end{aligned}
\tag{12.11}
$$

Though solving the non-convex problem (12.10) is equivalent to solving the convex problem (12.11), a non-linear Perron–Frobenius theory-based approach is used to solve (12.10) optimally. Using non-negative matrix theory, an analytical relationship is established to connect (12.10) to the Lagrange duality of (12.11) (cf. Lemma 12.2 later).

Lemma 12.1 *At optimality of (12.10), the outage constraints in (12.10) are tight:*

$$
v_l \beta_l / p_l^\star + \sum_j \log \left(1 + \frac{\beta_l F_{lj} p_j^\star}{p_l^\star}\right) = \alpha^\star \ \ \forall\, l.
\tag{12.12}
$$

Furthermore, if $\mathcal{P} = \{\mathbf{p} \mid \mathbf{1}^T \mathbf{p} \le \bar{P}\}$, then $\mathbf{1}^T \mathbf{p}^\star = \bar{P}$, and if $\mathcal{P} = \{\mathbf{p} \mid p_l \le \bar{p} \ \forall\, l\}$, then $p_i^\star = \bar{p}$ for some i.

Proof. First, we note that it has been pointed out in [16] that all the outage constraints are tight for the interference-limited case, i.e., $\mathbf{v} = \mathbf{0}$. Let us prove the first part of Lemma 12.1 for the general case here. Clearly, the function on the left hand side of the l-th outage constraint in (12.10) is monotone increasing in p_j, $j \ne l$, and monotone decreasing in p_l. Suppose the l-th constraint is not tight at optimality, i.e., $v_l \beta_l / p_l^\star + \sum_j \log \left(1 + \frac{\beta_l F_{lj} p_j^\star}{p_l^\star}\right) < \alpha^\star$. Then, choose a feasible power $p_l < p_l^\star$ such that the evaluated value of $v_l \beta_l / p_l + \sum_j \log \left(1 + \frac{\beta_l F_{lj} p_j^\star}{p_l}\right)$ is still less than $\alpha^\star$. Now, $v_j \beta_j / p_j^\star + \sum_{k \ne l} \log \left(1 + \frac{\beta_j F_{jk} p_k^\star}{p_j^\star}\right) + \log \left(1 + \frac{\beta_j F_{jl} p_l}{p_j^\star}\right)$ for all $j \ne l$. This implies that the value of α can be further decreased, i.e., $\alpha < \alpha^\star$, which contradicts the assumption. Hence, the l-th constraint must be tight at optimality for all l.

Let us next prove the second part for $\mathcal{P} = \{\mathbf{p} \mid p_l \le \bar{p} \ \forall\, l\}$. Suppose $p_l^\star < \bar{p}$ at optimality for all l. Let a positive scalar $a = \min_l \bar{p}/p_l^\star > 1$, and choose a feasible power $\mathbf{p} = a\mathbf{p}^\star$, which evaluates the outage constraints as $v_l \beta_l / p_l + \sum_j \log \left(1 + \frac{\beta_l F_{lj} p_j}{p_l}\right) = v_l \beta_l / a p_l^\star + \sum_j \log \left(1 + \frac{\beta_l F_{lj} p_j^\star}{p_l^\star}\right) < v_l \beta_l / p_l^\star + \sum_j \log \left(1 + \frac{\beta_l F_{lj} p_j^\star}{p_l^\star}\right) = \alpha^\star$ for all l. This implies that α can be further decreased, i.e., $\alpha < \alpha^\star$, which contradicts the

---

4 Note that (12.10) cannot be rewritten as a standard GP formulation, as has been done in [16]. Nevertheless, after a logarithmic change of variables, a convex form can still be obtained as shown here.

assumption. Hence, $p_i^\star = \bar{p}$ for some i. A similar proof can be given when $\mathcal{P} = \{\mathbf{p} \mid \mathbf{1}^T \mathbf{p} \le \bar{P}\}$ and is omitted. $\qquad\qquad\qquad\qquad\qquad\qquad\qquad\qquad\qquad\square$

Remark 12.1 *An analytical solution to (12.12) and thus the optimal solution of (12.8) is given in Section 12.3.3 (see Table 12.1 later).*

By exploiting a connection between the non-linear Perron–Frobenius theory in [19, 20] and the algebraic structure of (12.10), the following algorithm (with geometric convergence rate and no configuration whatsoever) is proposed to compute the optimal solution of (12.10). Let k index discrete time slots.

Algorithm 12.1 Worst-outage probability minimization

---

1. Update power $\mathbf{p}(k+1)$:

$$p_l(k+1) = -\log\left(1 - \phi_l(\mathbf{p}(k))\right) p_l(k) \quad \forall\, l. \tag{12.13}$$

2. Normalize $\mathbf{p}(k+1)$:

$$\mathbf{p}(k+1) \leftarrow \frac{\mathbf{p}(k+1) \cdot \bar{P}}{\mathbf{1}^T \mathbf{p}(k+1)} \quad \text{if } \mathcal{P} = \{\mathbf{p} \mid \mathbf{1}^T \mathbf{p} \le \bar{P}\}. \tag{12.14}$$

$$\mathbf{p}(k+1) \leftarrow \frac{\mathbf{p}(k+1) \cdot \bar{p}}{\max_j p_j(k+1)} \quad \text{if } \mathcal{P} = \{\mathbf{p} \mid p_l \le \bar{p} \;\forall\, l\}. \tag{12.15}$$

Theorem 12.1 *Starting from any initial point $\mathbf{p}(0)$, $\mathbf{p}(k)$ in Algorithm 12.1 converges geometrically fast to the optimal solution of (12.8).*[5]

Proof. Let us write the left-hand side of the l-th outage constraint in (12.10) as $\frac{f_l(\mathbf{p})}{p_l} \le \alpha$, where

$$f_l(\mathbf{p}) = v_l \beta_l + \sum_j p_l \log\left(1 + \frac{\beta_l F_{lj} p_j}{p_l}\right). \tag{12.16}$$

In the following, it will be shown that $f_l(\mathbf{p})$ is a positive *concave self-mapping* on the standard cone $K = \mathbb{R}_+^L$. The definition of a concave self-mapping is given in [20] as follows.

Definition 12.1 (Concave Self-mapping [20]) *A mapping $T : K \to K$ is concave if*

$$T(a\mathbf{x} + (1-a)\mathbf{y}) \ge aT\mathbf{x} + (1-a)T\mathbf{y} \;\; \forall\, \mathbf{x}, \mathbf{y} \in K, \;\; a \in [0,1],$$

and monotone if $\mathbf{0} \le \mathbf{x} \le \mathbf{y}$ implies $\mathbf{0} \le T\mathbf{x} \le T\mathbf{y}$.

Let $\|\cdot\|$ be a norm on $\mathbb{R}^L$ that is monotone, i.e., $\|\mathbf{x}\| \le \|\mathbf{y}\|$. A concave self-mapping of K is monotone on K and continuous on the interior of K with respect to $\|\cdot\|$ [20].

---

[5] Let $\|\cdot\|$ be an arbitrary vector norm. A sequence $\{\mathbf{p}(k)\}$ is said to converge geometrically fast to a fixed point $\mathbf{p}'$ if and only if $\|\mathbf{p}(k) - \mathbf{p}'\|$ converges to zero geometrically fast, i.e., there exists constants $A \ge 0$ and $\eta \in [0, 1)$ such that $\|\mathbf{p}(k) - \mathbf{p}'\| \le A\eta^k$ for all k [29].

Let us first show that $T = f_l(\mathbf{p})$ is a cone mapping with respect to the interior of K. Taking the derivative of $f_l(\mathbf{p})$ with respect to p_l, the j-th entry of the first derivative $\nabla f_l(\mathbf{p})$ is given by:

$$(\nabla f_l(\mathbf{p}))_j = \begin{cases} \displaystyle\sum_k \left(\log\left(1 + \frac{\beta_l F_{lk} p_k}{p_l}\right) - \frac{\beta_l F_{lk} p_k}{p_l + \beta_l F_{lk} p_k} \right), & \text{if } j = l \\[2ex] \displaystyle\frac{\beta_l F_{lj} p_l}{p_l + \beta_l F_{lj} p_j}, & \text{if } j \neq l. \end{cases}$$

Since $z/(1 + z) \leq \log(1 + z)$ for $z \geq 0$, $(\nabla f_l(\mathbf{p}))_l \geq 0$. Thus, $(\nabla f_l(\mathbf{p}))_j \geq 0$ for all j, i.e., $f_l(\mathbf{p})$ increases monotonically in $\mathbf{p}$. Now, the following result [30] is stated.

Theorem 12.2 (Proposition 3.2 in [30]) *Let $\mathcal{K}$ be the set of cone mappings with respect to the interior of the positive standard cone. Suppose $T : K \to K$ is differentiable and the following inequalities hold for the component mappings: $T_l : K \to \mathbb{R}_+$ for all l:* $\sum_j | \frac{\partial T_l}{\partial p_j}(\mathbf{p}) | \leq T_l \mathbf{p}$ *on* K. *Then* $T \in \mathcal{K}$.

Now, we have

$$\sum_j \left| \frac{\partial T_l}{\partial p_j}(\mathbf{p}) \right| = \sum_j p_j (\nabla f_l(\mathbf{p}))_j$$

$$= \sum_k \left(p_l \log\left(1 + \frac{\beta_l F_{lk} p_k}{p_l}\right) - \frac{\beta_l F_{lk} p_k p_l}{p_l + \beta_l F_{lk} p_k} \right) + \sum_{j \neq l} \frac{\beta_l F_{lj} p_l p_j}{p_l + \beta_l F_{lj} p_j}$$

$$= \sum_k p_l \log\left(1 + \frac{\beta_l F_{lk} p_k}{p_l}\right) \leq f_l(\mathbf{p}). \tag{12.17}$$

Hence, by Theorem 12.2, $f_l(\mathbf{p})$ is a strictly positive and monotone cone mapping on K.

Next, let us show that $T = f_l(\mathbf{p})$ is a concave self-mapping. Taking the second derivative, the Hessian $\nabla^2 f_l(\mathbf{p})$ is obtained with entries given by:

$$(\nabla^2 f_l(\mathbf{p}))_{jk} = \begin{cases} -\dfrac{(\beta_l F_{lj})^2 p_j}{(p_l + \beta_l F_{lj} p_j)^2}, & \text{if } j = k, k \neq l \\[2ex] \dfrac{(\beta_l F_{lj})^2 p_j}{(p_l + \beta_l F_{lj} p_j)^2}, & \text{if } j \neq k, \text{ either } k \text{ or } j = l \\[2ex] -\displaystyle\sum_m \dfrac{(\beta_l F_{lm})^2 p_m^2 / p_l}{(p_l + \beta_l F_{lm} p_m)^2}, & \text{if } j = k = l \\[2ex] 0, & \text{otherwise.} \end{cases}$$

Now, the Hessian $\nabla^2 f_l(\mathbf{p})$ is indeed negative definite: for all real vectors $\mathbf{z}$, we have

$$\mathbf{z}^T \nabla^2 f_l(\mathbf{p}) \mathbf{z} = -\frac{1}{p_l} \sum_k \frac{(\beta_l F_{lk})^2 (p_l z_k - p_k z_l)^2}{(p_l + \beta_l F_{lk} p_k)^2} < 0.$$

Another proof is to observe that $t \log(1 + x/t)$ is strictly concave in (x, t) for strictly positive t, as it is the perspective function of the strictly concave function $\log(1 + x)$ [28]. Hence, $f_l(\mathbf{p})$ is a sum of strictly concave perspective function, and therefore $f(\mathbf{p})$ is strictly concave in $\mathbf{p}$.

Note that any concave self-mapping of K is in $\mathcal{K}$ and it is monotone and continuous [30]. Indeed, $f_l(\mathbf{p})$ is monotone increasing in $\mathbf{p}$ as has been shown earlier.

Let us first state the following key theorem in [20].[6] It also plays a key role in later results for other related optimization problems.

Theorem 12.3 (Krause's theorem [20]) *Let $\| \cdot \|$ be a monotone norm on $\mathbb{R}^L$. For a concave mapping $f : \mathbb{R}_+^L \to \mathbb{R}_+^L$ with $f(\mathbf{z}) > 0$ for $\mathbf{z} \geq 0$, the following statements hold. The conditional eigenvalue problem $f(\mathbf{z}) = \lambda\mathbf{z}$, $\lambda \in \mathbb{R}$, $\mathbf{z} \geq 0$, $\|\mathbf{z}\| = 1$ has a unique solution $(\lambda^*, \mathbf{z}^*)$, where $\lambda^* > 0$, $\mathbf{z}^* > 0$. Furthermore, $\lim_{k \to \infty} \tilde{f}(\mathbf{z}(k))$ converges geometrically fast to $\mathbf{z}^*$, where $\tilde{f}(\mathbf{z}) = f(\mathbf{z})/\|f(\mathbf{z})\|$.*

The total and individual power constraints in (12.4) are the monotone norm $\|\mathbf{p}\|_1 = \bar{P}$ and $\|\mathbf{p}\|_\infty = \bar{p}$ respectively. By Theorem 12.3, the convergence of the iteration

$$\mathbf{p}(k+1) = \frac{f(\mathbf{p}(k))}{\|f(\mathbf{p}(k))\|}$$

to the *unique* fixed point $\mathbf{p} = f(\mathbf{p})/\|f(\mathbf{p})\|$ is geometrically fast, regardless of the initial point. $\square$

Remark 12.2 *To compute $\phi_l(\mathbf{p}(k))$ in (12.13), the l-th user measures separately the received interfering power $\{F_{lj}p_j(k)\}$, $j \neq l$. The normalization at Step 2 can be made distributed using gossip algorithms to compute either $\max_l p_l(k+1)$ or $\mathbf{1}^T\mathbf{p}(k+1)$ [31].*

In the following, let us first derive useful bounds to $O^\star$ given in terms of the problem parameters of (12.8). Next, (12.8) is solved analytically in the interference-limited special case without any power constraint, and the analysis is then extended to the general case with power constraints.

12.3.1   Worst-outage probability bounds

Useful lower and upper bounds for the worst-outage probability $O^\star$ are developed using another optimization problem, which is a non-convex problem that can also be solved by the non-linear Perron–Frobenius theory [21]. This is the maximization of the minimum weighted SINR problem:[7]

$$\text{maximize} \quad \min_l \ \frac{\text{SINR}_l(\mathbf{p})}{\beta_l} \tag{12.18}$$
$$\text{subject to} \ \mathbf{p} \in \mathcal{P}, \ \mathbf{p} \geq 0.$$

Now, the optimal value and solution of (12.18) can be obtained analytically despite the fact that (12.18) is non-convex [21, 26]. Let us summarize the result as follows.

---

[6] Notably, a special case of Theorem 12.3 is $f(\mathbf{p}) = \mathbf{F}\mathbf{p}$ where $\mathbf{F}$ is an irreducible non-negative matrix. This reduces to the classical linear Perron–Frobenius theory [19, 20] in linear algebra. Then, $\lambda^* = \rho(\mathbf{F})$ and $\mathbf{z}^*$ is the right eigenvector of $\rho(\mathbf{F})$ in Theorem 12.3 for this special case. The iteration $\mathbf{z}(k+1) = \mathbf{F}\mathbf{z}(k)/\|\mathbf{F}\mathbf{z}(k)\|$ is simply the classical power method in linear algebra.

[7] This is also known as the certainty equivalent margin (CEM) problem in [16]. The CEM problem replaces the statistical variation in the desired signal and the interference of (12.3) by their mean values.

Theorem 12.4 *Suppose $\mathcal{P} = \{\mathbf{p} \mid p_l \leq \bar{p} \ \forall l\}$. Then, an optimal solution of (12.18) is such that the weighted SINR for all the users are equal. This weighted SINR is given by*

$$\frac{1}{\rho(diag(\boldsymbol{\beta})(\mathbf{F} + (1/\bar{p})\mathbf{ve}_i^T))},\tag{12.19}$$

where

$$i = \arg\max_l \rho(diag(\boldsymbol{\beta})(\mathbf{F} + (1/\bar{p})\mathbf{ve}_i^T)).\tag{12.20}$$

Further, all links i that achieve the minimum in (12.20) transmit at maximum power $\bar{p}$ and the rest do not. Further, the optimal $\mathbf{p}$, denoted by $\mathbf{p}^*$, is $t\mathbf{x}(diag(\boldsymbol{\beta})(\mathbf{F} + (1/\bar{p})\mathbf{ve}_i^T))$ for a constant $t = \bar{p}/x_i$.

Suppose $\mathcal{P} = \{\mathbf{p} \mid \mathbf{1}^T\mathbf{p} \leq \bar{P}\}$. Then, the optimal value and solution of (12.18) is given by

$$\frac{1}{\rho(diag(\boldsymbol{\beta})(\mathbf{F} + (1/\bar{P})\mathbf{v1}^T))}\tag{12.21}$$

and

$$(\bar{P}/\mathbf{1}^T\mathbf{x}(diag(\boldsymbol{\beta})(\mathbf{F} + (1/\bar{P})\mathbf{v1}^T)))\ \mathbf{x}(diag(\boldsymbol{\beta})(\mathbf{F} + (1/\bar{P})\mathbf{v1}^T))\tag{12.22}$$

respectively.

In addition, the optimal solution of (12.18) can be computed by the following algorithm:

Algorithm 12.2 Max-min weighted SINR

---

1. Update power $\mathbf{p}(k + 1)$:

$$p_l(k + 1) = \frac{\beta_l}{\text{SINR}_l(\mathbf{p}(k))} p_l(k), \quad \forall\, l.\tag{12.23}$$

2. Normalize $\mathbf{p}(k + 1)$:

$$\mathbf{p}(k + 1) \leftarrow \frac{\mathbf{p}(k + 1) \cdot \bar{P}}{\mathbf{1}^T\mathbf{p}(k + 1)} \quad \text{if } \mathcal{P} = \{\mathbf{p} \mid \mathbf{1}^T\mathbf{p} \leq \bar{P}\}.$$

$$\mathbf{p}(k + 1) \leftarrow \frac{\mathbf{p}(k + 1) \cdot \bar{p}}{\max_j p_j(k + 1)} \quad \text{if } \mathcal{P} = \{\mathbf{p} \mid p_l \leq \bar{p} \ \forall l\}.$$

Remark 12.3 *Theorem 12.4 can be proved by the application of Theorem 12.3 to a concave self-mapping: $f_l(\mathbf{p}) = (\mathbf{Fp})_l + v_l$ for all l.*

Remark 12.4 *Interestingly, (12.23) in Algorithm 12.2 is simply the distributed power control (DPC) algorithm in [32], where the l-th user has a virtual SINR threshold of β_l in the downlink transmission. The normalization at Step 2 of Algorithm 12.2 can be computed centrally at the base station. In cases when it cannot be expected to know the total number of users in the system, the normalization can be made distributed using* *gossip algorithms to compute $\mathbf{1}^T\mathbf{p}(k+1)$ at each user [31]. Furthermore, Algorithm 12.2 converges geometrically fast.*

Remark 12.5 *When $\mathcal{P}$ in (12.18) is given by (12.5), Theorem 12.4 can be suitably modified to provide an analytical closed form solution to the max-min SINR problem of the macrocell–femtocell case in [5]. In particular, (12.5), written in its general form as $\mathbf{a}^T\mathbf{p} \leq 1$ for some positive $\mathbf{a}$, can be associated with a weighted ℓ-1 monotone norm, which permits the use of Theorem 12.3 (also see Table 12.1). Therefore, the iteration given by $\mathbf{p}(k+1) = (\mathbf{Fp}(k)+\mathbf{v})/\mathbf{a}^T(\mathbf{Fp}(k)+\mathbf{v})$ converges geometrically fast to the optimal solution of the macrocell–femtocell problem in [5], thereby resolving the open problem in [5].*

Let us now come back to provide analytical performance bounds to (12.10). We have the following bounds using the analytical optimal value of the max-min weighted SINR problem (also in terms of the constant problem parameters of (12.10)).

Corollary 12.1 *Let i be given by (12.20). If $\mathcal{P} = \{\mathbf{p} \mid p_l \leq \bar{p} \ \forall l\}$, the worst-outage probability $O^\star$ satisfies*

$$\frac{\rho(diag(\boldsymbol{\beta})(\mathbf{F}+(1/\bar{p})\mathbf{ve}_i^T))}{1+\rho(diag(\boldsymbol{\beta})(\mathbf{F}+(1/\bar{p})\mathbf{ve}_i^T))} \leq O^\star = 1 - e^{-\alpha^\star} \leq 1 - e^{-\rho(diag(\boldsymbol{\beta})(\mathbf{F}+(1/\bar{p})\mathbf{ve}_i^T))}, \quad (12.24)$$

and $\alpha^\star$ is the optimal value to (12.10).

Proof. Using the inequalities $1 + \sum_{l=1}^{L} z_l \leq \prod_{l=1}^{L}(1 + z_l) \leq e^{\sum_{l=1}^{L} z_l}$ for non-negative $\mathbf{z}$ (cf. [16]), a lower and upper bound on $\alpha^\star$ can be given by $1/(1 + \text{CEM}) \leq \alpha^\star \leq 1 - e^{-1/\text{CEM}}$, where CEM is the optimal value of (12.18) and is given by $1/\rho(diag(\boldsymbol{\beta})(\mathbf{F} + (1/\bar{p})\mathbf{ve}_i^T))$, where i is given by (12.20) [21]. The bounds on $O^\star = 1 - e^{-\alpha^\star}$ can thus be obtained, hence proving Corollary 12.1. $\qquad\square$

Remark 12.6 *Note that the lower bound in Corollary 12.1 is not necessarily the tightest, but the bounds in Corollary 12.1 illustrate that the max-min weighted SINR problem, i.e., the spectral information of a concave self-mapping $T\mathbf{p} = diag(\boldsymbol{\beta})(\mathbf{F} + \mathbf{v})\mathbf{p}$ (cf. [21]) can provide useful quick bounds to the worst-outage probability. Corollary 12.1 reduces to a result in [16] in the interference-limited case. Results similar to Corollary 12.1 can also be obtained for the case when $\mathcal{P} = \{\mathbf{p} \mid \mathbf{1}^T\mathbf{p} \leq \bar{P}\}$.*

Example 12.1 Figure 12.2 plots the worst-outage probability and the bounds for a system with 20 femtocell users using parameters in [1], where each user has a common SINR threshold β. Observe that the bounds are concave in β, and the worst-outage probability is very close to its upper bound. In fact, our numerical observations indicate that the optimal power $\mathbf{p}^\star$ is also close in value to the optimal solution of (12.18), i.e., $\mathbf{x}(diag(\boldsymbol{\beta})(\mathbf{F} + (1/\bar{p})\mathbf{ve}_i^T))$, where i is given by (12.20) (cf. [21]). Further, for small β, the bounds are very close, which suggests that in low-power femtocell networks, the max-min weighted SINR solution can give a sufficiently good approximation to the worst-outage probability.

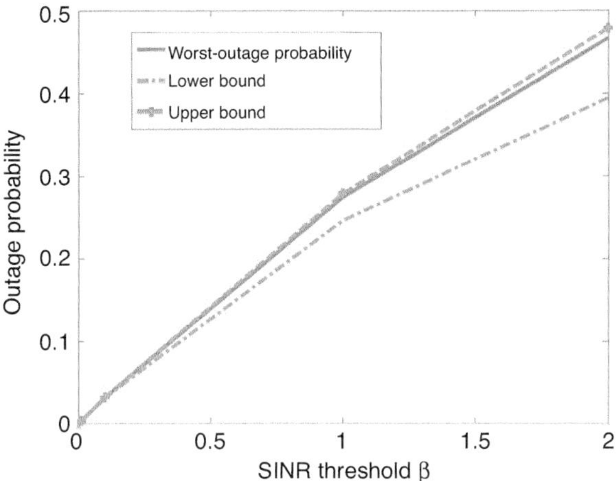

Figure 12.2 Outage probability versus the SINR threshold β for a system with 20 femtocell users, where each user has a common SINR threshold β. ©2011 IEEE. Reprinted, with permission, from [6].

12.3.2 Interference-limited case

Let us now turn to solve (12.8) analytically for the interference-limited case, i.e., $\mathbf{v} = \mathbf{0}$. In this case, (12.16) is in addition a primitive positive homogeneous function of degree. 1. Next, let us define the non-negative matrix $\mathbf{B}$ with the entries (that are functions of $\mathbf{p}$):

$$B_{lj} = \begin{cases} 0, & \text{if } l = j \\ \frac{p_l}{\beta_l p_j} \log\left(1 + \frac{\beta_l F_{lj} p_j}{p_l}\right), & \text{if } l \neq j. \end{cases} \tag{12.25}$$

Note that $\mathbf{B}(\mathbf{p})$ is irreducible whenever $\mathbf{F}$ is. Using (12.25), the optimal value $\alpha^\star$ and optimal solution $\mathbf{p}^\star$ of (12.10) in the interference-limited special case can be written as

$$\alpha^\star = \rho(\text{diag}(\boldsymbol{\beta})\mathbf{B}(\mathbf{p}^\star)) \tag{12.26}$$

$$\text{and } \mathbf{p}^\star = \mathbf{x}(\text{diag}(\boldsymbol{\beta})\mathbf{B}(\mathbf{p}^\star)) \text{ (up to a scaling constant)} \tag{12.27}$$

respectively. Thus, $\mathbf{p}^\star$ is a fixed point of

$$\mathbf{p} = \frac{1}{\rho(\text{diag}(\boldsymbol{\beta})\mathbf{B}(\mathbf{p}))}\text{diag}(\boldsymbol{\beta})\mathbf{B}(\mathbf{p})\mathbf{p}. \tag{12.28}$$

Further, it is interesting to note the following result of $\mathbf{p}^\star$.

Corollary 12.2 *In the interference-limited case, the optimal power of (12.10), $\mathbf{p}^\star$, satisfies*

$$\mathbf{p}^\star = arg\max_{\mathbf{p}>0} \rho(diag(\boldsymbol{\beta})\mathbf{B}(\mathbf{p})). \tag{12.29}$$

Proof. Using (12.25), let us rewrite the outage constraints in (12.10) in matrix form as

$$\text{diag}(\boldsymbol{\beta})\mathbf{B}(\mathbf{p})\mathbf{p} \leq \alpha\mathbf{p}. \tag{12.30}$$

Next, the following result from [25] is stated.

Theorem 12.5 (Theorem 1.6, [25] (subinvariance theorem)) *Let $\mathbf{A}$ be an irreducible non-negative matrix, s a positive number, and $\mathbf{z} \geq \mathbf{0}$, a vector satisfying $\mathbf{A}\mathbf{z} \leq s\mathbf{z}$. Then, (i) $\mathbf{z} > \mathbf{0}$; (ii) $s \geq \rho(\mathbf{A})$. Moreover, $s = \rho(\mathbf{A})$ if and only if $\mathbf{A}\mathbf{z} = s\mathbf{z}$.*

Applying Theorem 12.5 to (12.30) (let $\mathbf{A} = \mathrm{diag}(\boldsymbol{\beta})\mathbf{B}(\mathbf{p})$, $\mathbf{z}$ be a feasible $\mathbf{p}$ and $s = \alpha$), we have $\rho(\mathrm{diag}(\boldsymbol{\beta})\mathbf{B}(\mathbf{p})) \leq \alpha$ for any feasible $\mathbf{p}$ and α. But $\alpha^\star = \rho(\mathrm{diag}(\boldsymbol{\beta})\mathbf{B}(\mathbf{p}^\star))$. Hence, $\rho(\mathrm{diag}(\boldsymbol{\beta})\mathbf{B}(\mathbf{p})) \leq \rho(\mathrm{diag}(\boldsymbol{\beta})\mathbf{B}(\mathbf{p}^\star))$. This proves Corollary 12.2.   □

The following method was first proposed in [16] to compute the optimal solution for the interference-limited case without any power constraint:

$$\mathbf{p}(k+1) = \frac{1}{\rho(\mathrm{diag}(\boldsymbol{\beta})\mathbf{B}(\mathbf{p}(k)))}\mathrm{diag}(\boldsymbol{\beta})\mathbf{B}(\mathbf{p}(k))\mathbf{p}(k). \tag{12.31}$$

The authors in [16] observed that this iteration converges numerically to an acceptable accuracy with a fixed initial vector $\mathbf{p}(0) = \mathbf{x}(\mathrm{diag}(\boldsymbol{\beta})\mathbf{F})$ (optimal solution of (12.18) without power constraints and noise power). The issues of convergence and existence of a fixed point were, however, left open in [16].

Our Algorithm 12.1 in fact reduces to an update similar to (12.31) when $\mathbf{v} = \mathbf{0}$, and computes a solution that is equal to that of (12.31) up to a scaling constant. The scaling factor in (12.31) tends to $\rho(\mathrm{diag}(\boldsymbol{\beta})\mathbf{B}(\mathbf{p}^\star))$ with increasing k. From Theorem 12.1, this means that (12.31) converges from *any initial point* to the fixed point in (12.28) geometrically fast, thus resolving the open problem of convergence in [16].

12.3.3   Duality by Lagrange and Perron–Frobenius

The optimal value $\alpha^\star$ and optimal solution $\mathbf{p}^\star$ of (12.10) can be derived analytically from the spectral information of a specially constructed rank-one perturbation of $\mathrm{diag}(\boldsymbol{\beta})\mathbf{B}$, where $\mathbf{B}$ is given in (12.25). The following result is obtained based on the non-linear Perron–Frobenius theory in [19] and the Friedland–Karlin inequality in [22].

Lemma 12.2 *The optimal solution $(\mathbf{p}^\star, \alpha^\star)$ of (12.10) satisfies*

$$\log \alpha^\star = \log \rho(\mathrm{diag}(\boldsymbol{\beta})(\mathbf{B}(\mathbf{p}^\star) + \mathbf{v}\mathbf{c}_*^T)) = \max_{\|\mathbf{c}\|_D = 1} \log \rho(\mathrm{diag}(\boldsymbol{\beta})(\mathbf{B}(\mathbf{p}^\star) + \mathbf{v}\mathbf{c}^T))$$

$$= \max_{\lambda \geq 0, \mathbf{1}^T\lambda = 1} \ \min_{\mathbf{p} \in \mathcal{P}} \sum_l \lambda_l \log \frac{\beta_l(\mathbf{B}(\mathbf{p})\mathbf{p} + \mathbf{v})_l}{p_l} \tag{12.32}$$

$$= \min_{\mathbf{p} \in \mathcal{P}} \ \max_{\lambda \geq 0, \mathbf{1}^T\lambda = 1} \sum_l \lambda_l \log \frac{\beta_l(\mathbf{B}(\mathbf{p})\mathbf{p} + \mathbf{v})_l}{p_l}, \tag{12.33}$$

where the optimal $\mathbf{p}$ in (12.32) and (12.33) are both given by $\mathbf{x}(\mathrm{diag}(\boldsymbol{\beta})(\mathbf{B}(\mathbf{p}^\star) + \mathbf{v}\mathbf{c}_*^T))$ (which is equal to $\mathbf{p}^\star$ up to a scaling constant), and the optimal λ in (12.32) and (12.33) are both given by the Schur product of $\mathbf{x}(\mathrm{diag}(\boldsymbol{\beta})(\mathbf{B}(\mathbf{p}^\star) + \mathbf{b}\mathbf{c}_*^T))$ and $\mathbf{y}(\mathrm{diag}(\boldsymbol{\beta})(\mathbf{B}(\mathbf{p}^\star) + \mathbf{b}\mathbf{c}_*^T))$).

Furthermore, $\mathbf{p} = \mathbf{x}(\mathrm{diag}(\boldsymbol{\beta})(\mathbf{B}(\mathbf{p}^\star) + \mathbf{v}\mathbf{c}_*^T))$ is the dual of $\mathbf{c}_*$ with respect to $\|\cdot\|_D$.[8]

---

[8] A pair $(\mathbf{x}, \mathbf{y})$ of vectors of $\mathbb{R}^L$ is said to be a dual pair with respect to $\|\cdot\|$ if $\|\mathbf{y}\|_D\|\mathbf{x}\| = \mathbf{y}^T\mathbf{x} = 1$.

Proof. The proof outline of Lemma 12.2 is to first consider the Lagrange duality of (12.11) and then apply the non-negative matrix theory result in [19, 22]. Let us first state the following lemma that extends a result in [26]:

Lemma 12.3 *Let $\mathbf{A}$ be an irreducible non-negative matrix, $\mathbf{b}$ a non-negative vector and $\|\cdot\|$ a norm on $\mathbb{R}^L$ with a corresponding dual norm $\|\cdot\|_D$. Then,*

$$\log \rho(\mathbf{A} + \mathbf{b}\mathbf{c}_*^T) = \max_{\|\mathbf{c}\|_D=1} \log \rho(\mathbf{A} + \mathbf{b}\mathbf{c}^T)$$

$$= \max_{\lambda \geq 0, \mathbf{1}^T\lambda=1} \min_{\|\mathbf{p}\|=1} \sum_l \lambda_l \log \frac{(\mathbf{A}\mathbf{p} + \mathbf{b})_l}{p_l} \tag{12.34}$$

$$= \min_{\|\mathbf{p}\|=1} \max_{\lambda \geq 0, \mathbf{1}^T\lambda=1} \sum_l \lambda_l \log \frac{(\mathbf{A}\mathbf{p} + \mathbf{b})_l}{p_l}, \tag{12.35}$$

where the optimal $\mathbf{p}$ in (12.34) and (12.35) are both given by $\mathbf{x}(\mathbf{A} + \mathbf{b}\mathbf{c}_*^T)$, and the optimal λ in (12.34) and (12.35) are both given by $\mathbf{x}(\mathbf{A} + \mathbf{b}\mathbf{c}_*^T) \circ \mathbf{y}(\mathbf{A} + \mathbf{b}\mathbf{c}_*^T)$.
Furthermore, $\mathbf{p} = \mathbf{x}(\mathbf{A} + \mathbf{b}\mathbf{c}_*^T)$ is the dual of $\mathbf{c}_*$ with respect to $\|\cdot\|_D$.

First, let us express the outage constraints in (12.11) using the matrix $\mathbf{B}$ and rewrite (12.11) as the following equivalent problem (let $\tilde{\alpha} = \log \alpha$):

$$\begin{aligned}
&\text{minimize } \tilde{\alpha} \\
&\text{subject to } \log \left(\frac{\beta_l(v_l + \mathbf{B}(e^{\tilde{\mathbf{p}}})e^{\tilde{\mathbf{p}}})_l}{e^{\tilde{p}_l}} \right) \leq \tilde{\alpha} \; \forall \, l, \\
&\qquad e^{\tilde{\mathbf{p}}} \in \mathcal{P}, \\
&\text{variables: } \tilde{\mathbf{p}}, \; \tilde{\alpha}.
\end{aligned} \tag{12.36}$$

Next, by augmenting only the outage constraints, the partial Lagrangian function of (12.36) is given by

$$L(\tilde{\alpha}, \tilde{\mathbf{p}}, \lambda) = \left(1 - \sum_{l=1}^L \lambda_l\right) \tilde{\alpha} + \sum_{l=1}^L \lambda_l \log \left(\frac{\beta_l(v_l + \mathbf{B}(e^{\tilde{\mathbf{p}}})e^{\tilde{\mathbf{p}}})_l}{e^{\tilde{p}_l}} \right). \tag{12.37}$$

Now, the Lagrange dual function of (12.36) is finite only if $\sum_{l=1}^L \lambda_l = 1$ for all feasible λ. Hence, the Lagrange dual function is given by

$$\min_{\tilde{\alpha}, \tilde{\mathbf{p}} \in \mathcal{P}} L(\tilde{\alpha}, \tilde{\mathbf{p}}, \lambda) = \sum_{l=1}^L \lambda_l \log \left(\frac{\beta_l(v_l + \mathbf{B}(e^{\tilde{\mathbf{p}}^\star})e^{\tilde{\mathbf{p}}^\star})_l}{e^{\tilde{p}_l^\star}} \right), \tag{12.38}$$

where $\tilde{\mathbf{p}}^\star$ is the optimal solution to (12.36).

Now, we are ready to apply Lemma 12.3 to (12.38). In particular, $\mathbf{c}_*$ can be computed explicitly depending on the choice of $\mathcal{P}$. For the case when $\mathcal{P} = \{\mathbf{p}|\mathbf{1}^T\mathbf{p} = \bar{P}\}$, let $\mathbf{A} = \text{diag}(\beta)\mathbf{B}(\mathbf{p}^\star)$, $\mathbf{b} = \text{diag}(\beta)\mathbf{v}$ and $\mathbf{c}_* = (1/\bar{P})\mathbf{1}$ in Lemma 12.3. For the case when $\mathcal{P} = \{\mathbf{p}|p_l \leq \bar{p}\,\forall\,l\}$, let $\mathbf{A} = \text{diag}(\beta)\mathbf{B}(\mathbf{p}^\star)$, $\mathbf{b} = \text{diag}(\beta)\mathbf{v}$ and $\mathbf{c}_* = (1/\bar{p})\mathbf{e}_i$, where $i = \arg\max_l \rho(\text{diag}(\beta)(\mathbf{B}(\mathbf{p}^\star) + \mathbf{v}\mathbf{e}_l^T))$ in Lemma 12.3. $\qquad \square$

Note that the optimal dual variable in (12.11) is equal to the optimal λ in (12.32) and (12.33). Using Lemma 12.2, the optimal value and solution of (12.10) can now be given analytically when $\mathcal{P} = \{\mathbf{p} \mid p_l \leq \bar{p} \; \forall l\}$.

Remark 12.7 *Lemma 12.2 has several implications. In [21], Lemma 12.2 can be used to provide an alternative algorithmic approach, different from that in Theorem 12.4, to solve (12.18). In [26, 33, 34], Lemma 12.2 also enables a characterization of the jointly optimal beamformer and power in a system model with multiple antenna.*

Corollary 12.3 *The optimal value and solution of (12.10) is given respectively by*

$$\alpha^\star = \rho\left(diag(\boldsymbol{\beta})(\mathbf{B}(\mathbf{p}^\star) + (1/\bar{p})\mathbf{v}\mathbf{e}_i^T)\right) \tag{12.39}$$

and

$$\mathbf{p}^\star = \mathbf{x}\left(diag(\boldsymbol{\beta})(\mathbf{B}(\mathbf{p}^\star) + (1/\bar{p})\mathbf{v}\mathbf{e}_i^T)\right), \tag{12.40}$$

where $i = \arg\max_l \rho\left(diag(\boldsymbol{\beta})(\mathbf{B}(\mathbf{p}^\star) + (1/\bar{p})\mathbf{v}\mathbf{e}_i^T)\right).$ $\tag{12.41}$

Furthermore, $p_i^\star = \bar{p}$ *for the* i *(not necessarily unique) in (12.41).*

Remark 12.8 *In general, the optimal index* i *in (12.41) differs from (12.20). Our simulations show that both are empirically the same when the max-min weighted SINR solution is close to* $\mathbf{p}^\star$*, especially so in the low-power regime (cf. Figure 12.2). Unlike (12.41), (12.20) can be computed a priori from the problem parameters.*

Table 12.1 summarizes the connection between the non-linear Perron–Frobenius spectrum (of the respectively different concave self-mappings) and the optimal value and solution of the optimization problems under the certainty-equivalent margin model and the Rayleigh-fading model subject to the different power constraints (individual and total power constraints).

12.4   Total power minimization and adaptive outage power control

In this section, the total power minimization problem subject to both outage specification and individual power constraints is first studied, and its feasibility conditions are then addressed using the results in the previous sections. An adaptive algorithm is then proposed to minimize the total power consumption and simultaneously to guarantee a min-max fairness in terms of worst-outage probability.

The problem of minimizing the total power subject to given outage specification under Rayleigh fading and individual power constraints can be formulated as

$$\begin{aligned}
&\text{minimize } \mathbf{1}^T\mathbf{p} \\
&\text{subject to } 1 - e^{\frac{-v_l \beta_l}{p_l}} \prod_j \left(1 + \frac{\beta_l F_{lj} p_j}{p_l}\right)^{-1} \leq \bar{O}_l \; \forall \, l, \\
&\qquad\qquad \mathbf{p} \in \{\mathbf{p} \mid p_l \leq \bar{p} \; \forall l\}, \; \mathbf{p} \geq \mathbf{0}, \\
&\text{variables: } \mathbf{p},
\end{aligned} \tag{12.42}$$

Table 12.1 Non-linear Perron–Frobenius characterization of the weighted max-min SINR problem and the worst-outage probability problem. The first and second rows tabulate the results for the worst-outage probability problem with individual and total power constraints, respectively. The third and fourth rows tabulate the results for the max-min weighted SINR problem with individual and total power constraints, respectively. The fifth row tabulates the results for the max-min SINR problem with a weighted power constraint for femtocell users in a single macrocell.

Concave self-mapping $(T\mathbf{p})_l$	Perron eigenvalue $\alpha^\star$	Perron eigenvector $\mathbf{p}^\star$	Remark
$v_l\beta_l + \sum_j p_l \log\left(1 + \frac{\beta_l F_{lj} p_j}{p_l}\right)$ $i = \arg\max_l \rho\left(\mathrm{diag}(\boldsymbol{\beta})(\mathbf{B}(\mathbf{p}^\star) + (1/\bar{p})\mathbf{ve}_i^T)\right)$	$\rho\left(\mathrm{diag}(\boldsymbol{\beta})(\mathbf{B}(\mathbf{p}^\star) + (1/\bar{p})\mathbf{ve}_i^T)\right)$	$\mathbf{x}\left(\mathrm{diag}(\boldsymbol{\beta})(\mathbf{B}(\mathbf{p}^\star) + (1/\bar{p})\mathbf{ve}_i^T)\right)$	[6], Corollary 12.3 herein
$v_l\beta_l + \sum_j p_l \log\left(1 + \frac{\beta_l F_{lj} p_j}{p_l}\right)$	$\rho\left(\mathrm{diag}(\boldsymbol{\beta})(\mathbf{B}(\mathbf{p}^\star) + (1/\bar{P})\mathbf{v1}^T)\right)$	$\mathbf{x}\left(\mathrm{diag}(\boldsymbol{\beta})(\mathbf{B}(\mathbf{p}^\star) + (1/\bar{P})\mathbf{v1}^T)\right)$	[6]
$(\mathrm{diag}(\boldsymbol{\beta})(\mathbf{Fp} + (1/\bar{p})\mathbf{v}))_l,$ $i = \arg\max_l \rho\left(\mathrm{diag}(\boldsymbol{\beta})(\mathbf{F} + (1/\bar{p})\mathbf{ve}_i^T)\right)$	$\rho\left(\mathrm{diag}(\boldsymbol{\beta})(\mathbf{F} + (1/\bar{p})\mathbf{ve}_i^T)\right)$	$\mathbf{x}\left(\mathrm{diag}(\boldsymbol{\beta})(\mathbf{F} + (1/\bar{p})\mathbf{ve}_i^T)\right)$	[21], Theorem 12.4 herein
$(\mathrm{diag}(\boldsymbol{\beta})(\mathbf{Fp} + (1/\bar{P})\mathbf{v}))_l$	$\rho\left(\mathrm{diag}(\boldsymbol{\beta})(\mathbf{F} + (1/\bar{P})\mathbf{v1}^T)\right)$	$\mathbf{x}\left(\mathrm{diag}(\boldsymbol{\beta})(\mathbf{F} + (1/\bar{P})\mathbf{v1}^T)\right)$	[26]
$(\mathbf{Fp} + \mathbf{v})_l$	$\rho\left(\mathbf{F} + \mathbf{va}^T\right)$	$\mathbf{x}\left(\mathbf{F} + \mathbf{va}^T\right)$	[5], Remark 12.5 herein

where $0 < \bar{O}_l < 1$ is a given outage probability bound for the l-th user. Depending on the given parameters $\bar{O}_l$ for all l, (12.42) may or may not be feasible. This is unlike the worst-outage probability problem in (12.6), which is always feasible.

Next, using (12.16), (12.42) can be rewritten as

$$
\begin{aligned}
&\text{minimize } \mathbf{1}^T \mathbf{p} \\
&\text{subject to } \frac{f_l(\mathbf{p})}{p_l} \leq \alpha_l \ \ \forall \ l, \\
&\mathbf{p} \in \{\mathbf{p} \mid p_l \leq \bar{p} \ \forall l\}, \ \mathbf{p} \geq \mathbf{0},
\end{aligned}
\tag{12.43}
$$

where $\alpha_l = -\log(1 - \bar{O}_l)$ for all l.

If (12.42) is feasible, it can be shown that all the L outage constraints in (12.42) are tight at optimality [17]. The feasibility condition of (12.42) will now be deduced from (12.43) in the following result.

Lemma 12.4 *There is a unique and finite optimal $\mathbf{p}$ in (12.42) if and only if*

$$
\max_{\mathbf{p} \in \{\mathbf{p} \mid p_l \leq \bar{p} \ \forall l\}} \rho(diag(\boldsymbol{\beta}/\boldsymbol{\alpha})\mathbf{B}(\mathbf{p})) < 1.
\tag{12.44}
$$

Furthermore, $\rho(diag(\boldsymbol{\beta}/\boldsymbol{\alpha})\mathbf{B}(\mathbf{p})) < 1$ if $\alpha^\star < \min_l \alpha_l$, i.e. $1 - e^{-\alpha^\star} < \bar{O}_l$ for all l.

Proof. To show the feasibility condition, let us examine the condition under which there is a fixed point to

$$
f_l(\mathbf{p}) = \beta_l((\mathbf{B}(\mathbf{p})\mathbf{p})_l + v_l) = \alpha_l p_l \quad \text{for all} \ \ l,
$$

which can be rewritten in matrix form as

$$
(\mathbf{I} - diag(\boldsymbol{\beta}/\boldsymbol{\alpha})\mathbf{B}(\mathbf{p}))\mathbf{p} = diag(\boldsymbol{\beta}/\boldsymbol{\alpha})\mathbf{v}.
\tag{12.45}
$$

We first state the following result from [25].

Theorem 12.6 *A necessary and sufficient condition for a solution $\mathbf{z} \geq \mathbf{0}, \mathbf{z} \neq \mathbf{0}$ to the equations $(\mathbf{I} - \mathbf{A})\mathbf{z} = \mathbf{c}$ to exist for any $\mathbf{c} \geq \mathbf{0}, \mathbf{c} \neq \mathbf{0}$ is that $\rho(\mathbf{A}) < 1$. In this case there is only one solution $\mathbf{z}$, which is strictly positive and given by $\mathbf{z} = (\mathbf{I} - \mathbf{A})^{-1}\mathbf{c}$.*

Since $diag(\boldsymbol{\beta}/\boldsymbol{\alpha})\mathbf{B}$ is an irreducible non-negative matrix, it follows from Theorem 12.6 (letting $\mathbf{A} = diag(\boldsymbol{\beta}/\boldsymbol{\alpha})\mathbf{B}$, $\mathbf{c} = diag(\boldsymbol{\beta}/\boldsymbol{\alpha})\mathbf{v}$) that $\mathbf{p}$ in (12.45) is unique and strictly positive if and only if $\rho(diag(\boldsymbol{\beta}/\boldsymbol{\alpha})\mathbf{B}(\mathbf{p})) < 1$ for all $\mathbf{p} \in \{\mathbf{p} \mid p_l \leq \bar{p} \ \forall l\}$. This is equivalent to stating

$$
\max_{\mathbf{p} \in \{\mathbf{p} \mid p_l \leq \bar{p} \ \forall l\}} \rho(diag(\boldsymbol{\beta}/\boldsymbol{\alpha})\mathbf{B}(\mathbf{p})) < 1,
$$

thus proving the first part.

To show the second part, let us note that $\rho(diag(\boldsymbol{\beta}/\boldsymbol{\alpha})\mathbf{B}(\mathbf{p})) \leq (1/\min_l \alpha_l) \rho(diag(\boldsymbol{\beta})\mathbf{B}(\mathbf{p})) \leq (1/\min_l \alpha_l)\alpha^\star$. Thus, a sufficient condition that $(1/\min_l \alpha_l)\alpha^\star < 1$ implies that $\rho(diag(\boldsymbol{\beta}/\boldsymbol{\alpha})\mathbf{B}(\mathbf{p})) < 1$. $\qquad\square$

12.4.1   Feasibility bounds

From (12.44) in Lemma 12.4, it can be seen that verifying the feasibility of (12.42) requires solving a Perron–Frobenius eigenvalue maximization problem (a non-convex problem). Let us next provide useful (tight) bounds on this non-convex optimal value $\rho(\mathrm{diag}(\boldsymbol{\beta}/\boldsymbol{\alpha})\mathbf{B}(\mathbf{p}))$ in (12.44) that exploit the optimal value and solution of the worst-outage probability problem in Section 12.3.

Theorem 12.7 *Let $\alpha^\star$ and $\mathbf{p}^\star$ be given in (12.26) and (12.27), respectively. Then, it is shown that*

$$\prod_l (\alpha_l)^{-x_l(\mathbf{B}(\mathbf{p}^\star))y_l(\mathbf{B}(\mathbf{p}^\star))}\alpha^\star \le \max_{\mathbf{p}\in\{\mathbf{p}\mid p_l\le\bar{p}\ \forall l\}} \rho(diag(\boldsymbol{\beta}/\boldsymbol{\alpha})\mathbf{B}(\mathbf{p})) \le \max_l(\alpha^\star/\alpha_l). \tag{12.46}$$

Further, equality is achieved in the lower and upper bounds when α_l's are equal for all l.

Proof. In the following, let us first state the Friedland–Karlin inequalities that were initiated in [22] and later extended in [23].

Theorem 12.8 *Let $\mathbf{A}\in\mathbb{R}_+^{L\times L}$ be an irreducible non-negative matrix. Assume that $\mathbf{x}(\mathbf{A})=(x_1(\mathbf{A}),\dots,x_L(\mathbf{A}))^T$, $\mathbf{y}(\mathbf{A})=(y_1(\mathbf{A}),\dots,y_L(\mathbf{A}))^T>0$ are left and right Perron–Frobenius eigenvectors of $\mathbf{A}$, normalized such that $\mathbf{x}(\mathbf{A})\circ\mathbf{y}(\mathbf{A})$ is a probability vector. Suppose $\boldsymbol{\gamma}$ is a non-negative vector. Then,*

$$\rho(\mathbf{A})\prod_{l=1}^{L}\gamma_l^{(\mathbf{x}(\mathbf{A})\circ\mathbf{y}(\mathbf{A}))_l} \le \rho(diag(\gamma)\mathbf{A}). \tag{12.47}$$

If γ is a positive vector then equality holds if and only if all γ_l are equal. Furthermore, for any positive vector $\mathbf{z}=(z_1,\dots,z_L)^T$, the following inequality holds:

$$\rho(\mathbf{A})\le\prod_{l=1}^{L}\left(\frac{(\mathbf{A}\mathbf{z})_l}{z_l}\right)^{(\mathbf{x}(\mathbf{A})\circ\mathbf{y}(\mathbf{A}))_l}. \tag{12.48}$$

If $\mathbf{A}$ is an irreducible non-negative matrix with positive diagonal elements, then equality holds in (12.48) if and only if $\mathbf{z}=t\mathbf{x}(\mathbf{A})$ for some positive t.

By applying the Friedland–Karlin inequalities, the function $\rho(\mathrm{diag}(\boldsymbol{\beta}/\boldsymbol{\alpha})\mathbf{B}(\mathbf{p}))$ can be bounded by

$$\prod_l(\alpha_l)^{-x_l(\mathbf{B})y_l(\mathbf{B})}\rho(\mathrm{diag}(\boldsymbol{\beta})\mathbf{B}) \le \rho(\mathrm{diag}(\boldsymbol{\beta}/\boldsymbol{\alpha})\mathbf{B}) \le \max_l(1/\alpha_l)\rho(\mathrm{diag}(\boldsymbol{\beta})\mathbf{B}) \tag{12.49}$$

for any feasible $\mathbf{p}\in\{\mathbf{p}\mid p_l\le\bar{p}\ \forall l\}$. Now, from the lower bound in (12.49), we have

$$\prod_l(\alpha_l)^{-x_l(\mathbf{B}(\mathbf{p}^\star))y_l(\mathbf{B}(\mathbf{p}^\star))}\rho(\mathrm{diag}(\boldsymbol{\beta})\mathbf{B}(\mathbf{p}^\star))$$

$$\le \max_{\mathbf{p}\in\{\mathbf{p}\mid p_l\le\bar{p}\ \forall l\}}\left\{\prod_l(\alpha_l)^{-x_l(\mathbf{B}(\mathbf{p}))y_l(\mathbf{B}(\mathbf{p}))}\rho(\mathrm{diag}(\boldsymbol{\beta})\mathbf{B}(\mathbf{p}))\right\} \tag{12.50}$$

$$\le \max_{\mathbf{p}\in\{\mathbf{p}\mid p_l\le\bar{p}\ \forall l\}}\rho(\mathrm{diag}(\boldsymbol{\beta}/\boldsymbol{\alpha})\mathbf{B}(\mathbf{p})),$$

where $\mathbf{p}^{\star} \in \{\mathbf{p} \mid p_l \leq \bar{p} \; \forall l\}$ is given by (12.27). Using $\alpha^{\star}$ given in (12.26), (12.46) is thus established. The condition under which equalities in (12.49) are achieved follows from the application of the Friedland–Karlin inequalities in [22]. $\qquad\square$

These bounds can be easily computed in the two-user case.[9]

Example 12.2 In the two-user case, it can be shown that

$$\frac{\alpha^{\star}}{\sqrt{\alpha_1 \alpha_2}} \leq \max_{\mathbf{p} \in \{\mathbf{p} \mid p_l \leq \bar{p} \; \forall l\}} \rho(\mathrm{diag}(\boldsymbol{\beta}/\boldsymbol{\alpha})\mathbf{B}) \leq \max \left\{ \frac{\alpha^{\star}}{\alpha_1}, \frac{\alpha^{\star}}{\alpha_2} \right\}.$$

Combining Theorem 12.7 with Corollary 12.1, simplified bounds in terms of the max-min weighted SINR solution (and more directly in terms of the problem parameters) can be obtained:

$$\prod_l (\alpha_l)^{-x_l(\mathbf{B}(\mathbf{p}^{\star}))y_l(\mathbf{B}(\mathbf{p}^{\star}))} \log(1 + \rho(\mathrm{diag}(\boldsymbol{\beta})(\mathbf{F} + (1/\bar{p})\mathbf{ve}_i^T)))$$
$$\leq \max_{\mathbf{p} \in \{\mathbf{p} \mid p_l \leq \bar{p} \; \forall l\}} \rho(\mathrm{diag}(\boldsymbol{\beta}/\boldsymbol{\alpha})\mathbf{B})$$
$$\leq \max_l (1/\alpha_l) \rho(\mathrm{diag}(\boldsymbol{\beta})(\mathbf{F} + (1/\bar{p})\mathbf{ve}_i^T)),$$

where i is given by (12.20) and the fact that $\max_l \rho(\mathrm{diag}(\boldsymbol{\beta})(\mathbf{B}(\mathbf{p}^{\star}) + (1/\bar{p})\mathbf{ve}_i^T)) \leq \max_l \rho(\mathrm{diag}(\boldsymbol{\beta})(\mathbf{F} + (1/\bar{p})\mathbf{ve}_i^T))$ has been used in the last inequality.

Let us now state the following algorithm proposed in [17].

Algorithm 12.3 Total power minimization

$$p_l(k+1) = \min\{-\log(1 - \phi_l(\mathbf{p}(k)))\, p_l(k)/\alpha_l, \; \bar{p}\} \; \forall \, l. \tag{12.51}$$

Let us now establish the necessary and sufficient condition under which Algorithm 12.3 converges. This condition is also necessary and sufficient for (12.42) to be feasible.

Corollary 12.4 *Starting from any initial point* $\mathbf{p}(0)$, $\mathbf{p}(k)$ *in Algorithm 12.3 converges geometrically fast to the optimal solution of (12.42) if and only if* $\rho(\mathrm{diag}(\boldsymbol{\beta}/\boldsymbol{\alpha})\mathbf{B}(\mathbf{p})) < 1$ *for all* $\mathbf{p} \in \{\mathbf{p} \mid p_l \leq \bar{p} \; \forall l\}$.

Proof. The necessary and sufficient condition under which (12.42) is feasible is given in Lemma 12.4. If (12.42) is feasible, the convergence proof for Algorithm 12.3 can be found in [17]. Hence, Corollary 12.4 is proved. $\qquad\square$

---

[9] The Schur product of the Perron and left eigenvectors of a zero-diagonal 2×2 positive matrix equals $[1/2, 1/2]$, simplifying the computation in (12.46).

12.4.2 Adaptive outage-based power control

Heterogeneous wireless networks have to be adaptive in order to be spectral and energy efficient. When the system is infeasible, it is important that the resource requirements are adapted appropriately. In this section, an adaptive outage-based power control (AOPC) algorithm is proposed for the minimization of the total energy in the network.

Algorithm 12.4 Adaptive outage-based power control (AOPC)

1. Update the auxiliary variable $\mathbf{z}(k+1)$:

$$z_l(k+1) = -\log\left(1 - \phi_l(\mathbf{z}(k))\right) z_l(k) \quad \forall\, l. \tag{12.52}$$

2. Normalize $\mathbf{z}(k+1)$:

$$\mathbf{z}(k+1) \leftarrow \mathbf{z}(k+1) \cdot \bar{p} / \max_j z_j(k+1). \tag{12.53}$$

3. Update the transmit power $\mathbf{p}(k+1)$:

$$p_l(k+1) = \min\left\{ \frac{-\log\left(1 - \phi_l(\mathbf{p}(k))\right) p_l(k)}{\max\{\alpha_l, -\log\left(1 - \phi_l(\mathbf{z}(k))\right)\}}, \ \bar{p} \right\} \forall\, l. \tag{12.54}$$

Corollary 12.5 *Starting from any initial point $\mathbf{z}(0)$ and $\mathbf{p}(0)$, $\mathbf{p}(k)$ in Algorithm 12.4 converges geometrically fast to the optimal solution of (12.43), where the right hand side of the outage constraints in (12.43) are replaced by $\max\{\alpha_l, \rho(diag(\boldsymbol{\beta})(\mathbf{B}(\mathbf{p}^\star) + (1/\bar{p})\mathbf{ve}_i^T))\}$, where $\mathbf{p}^\star$ and i are given by (12.40) and (12.41), respectively, for all l.*

Proof. Theorem 12.1 proves the convergence of $\mathbf{z}(k)$ in Step 1 and 2 of Algorithm 12.4, and also $\lim_{k\to\infty} -\log(1 - \phi_l(\mathbf{z}(k))) \to \rho(diag(\boldsymbol{\beta})(\mathbf{B}(\mathbf{p}^\star) + (1/\bar{p})\mathbf{ve}_i^T))$. From Corollary 12.4, $\mathbf{p}(k)$ converges to a point $\mathbf{p}'$ that satisfies $-\log(1 - \phi_l(\mathbf{p}')) = \alpha' = \max\{\alpha, \rho(diag(\boldsymbol{\beta})(\mathbf{B}(\mathbf{p}^\star) + (1/\bar{p})\mathbf{ve}_i^T))\}$, which always satisfies $\rho(diag(\boldsymbol{\beta}/\alpha')\mathbf{B}(\mathbf{p})) < 1$. This proves Corollary 12.5. $\qquad\square$

Remark 12.9 *If $\alpha_l < \rho(diag(\boldsymbol{\beta})(\mathbf{B}(\mathbf{p}^\star) + (1/\bar{p})\mathbf{ve}_i^T))$ for all l, then $\lim_{k\to\infty} \mathbf{z}(k) = \lim_{k\to\infty} \mathbf{p}(k) = \mathbf{p}^\star$ in Algorithm 12.4.*

12.5 Further discussions

It is important to understand whether the non-linear Perron–Frobenius theory can be used to solve (12.6) for other practical system models in wireless networks. In this chapter, (12.6) has been solved for the Rayleigh fading case. It is interesting to investigate its applications to other fading channel models such as the Rice, Weibull, and Nakagami distributions, which are especially important to communications engineers [11, 12]. Finding a suitable concave self-mapping for these different fading models can be challenging, and it may be advantageous to approximate the exact distribution expressions, if it is too complicated, in order to gain simplicity and insight into the analysis at the expense of a limited validity of the approximation. It is also interesting to find out how this theory-driven approach can lead to other adaptive algorithms that are applicable to a wireless heterogeneous network. It is possible that these adaptive algorithms can be combined with decentralized admission control network protocols to overcome the barrier of infeasibility in a system.

Better performance can be obtained by a joint optimization of multiple input multiple output (MIMO) transmit and receive beamforming with power controlled single input multiple output (SIMO) system, the non-linear Perron–Frobenius theory can be used to provide an analytical characterization to the optimal beamformer and power control. This approach also leads to additional insights on network duality for joint beamformer and power control optimization. Using this network duality, the authors in [26, 33, 34] propose an algorithm that combines the iterative algorithm in Theorem 12.3 and a linear minimum mean square error (LMMSE) beamformer update to alternately optimize the power and the beamformers in a distributed manner. In [35], a network duality for cognitive radio networks has also been developed to design distributed algorithms. In [36], techniques developed from random matrix theory and non-linear Perron–Frobenius theory are employed to derive optimal solution for the max-min weighted SINR problem in a large system where both the number of users and antennas are asymptotically large.

Techniques introduced in this chapter can also be extended to cross-layer resource allocation problems. The authors in [37] study the weighted max-min fairness flow rate allocation problem in a multi-user wireless network. They show that the joint flow rate and power control problem can be decoupled into two separate problems on fairness – one at the network layer and one at the physical link layer – where the latter problem assumes the case where all sources in the network layer are sending at a common rate. At the network layer, the non-linear Perron–Frobenius theory characterizes how network features such as the topology of the wireless network and bandwidth adaptation shape the end-to-end flow rate allocation. At the physical link layer, the non-linear Perron–Frobenius theory characterizes how a large class of link rate functions that includes the well-known Shannon capacity function and the CDMA rate function shapes the sharing of the spectrum resources. Putting them together, these two separate characterizations then in turn provide a holistic solution to the original joint cross-layer resource allocation problem.

12.6   Numerical evaluation

In this section, the performance of Algorithms 12.1, 12.3, and AOPC are evaluated. Figure 12.3 illustrates the basic macrocell–femtocell interference network model used in [13] for our simulations. A femtocell user is a link between the home access terminal and access point (box in Figure 12.3), and interference comes from the macrocell base station/access terminals and other femtocell users. Let us consider a single macrocell base station with 50 access terminals that consist of both macrocell and femtocell users. We assume an unplanned deployment of closed-access femtocells (with one user in each femtocell) by the same cellular operator and the femtocells are connected via DSL or cable modem to the cellular core network. Each macrocell user and each femtocell user communicates with its respective base station over independent Rayleigh fading

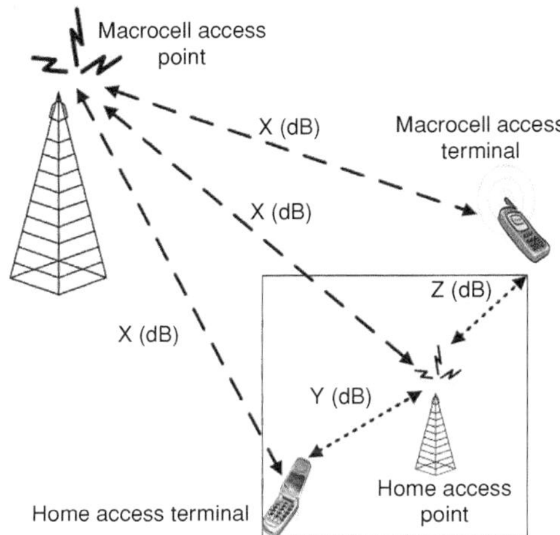

Figure 12.3 A macrocell–femtocell model. The parameters X dB, Y dB, and Z dB denote path gain between macrocell access point and home access point, between home access point and home access terminal, and between home access point and macrocell access terminal, respectively (see Table 2 of [13] for values of X, Y, Z). ©2011 IEEE. Reprinted, with permission, from [6].

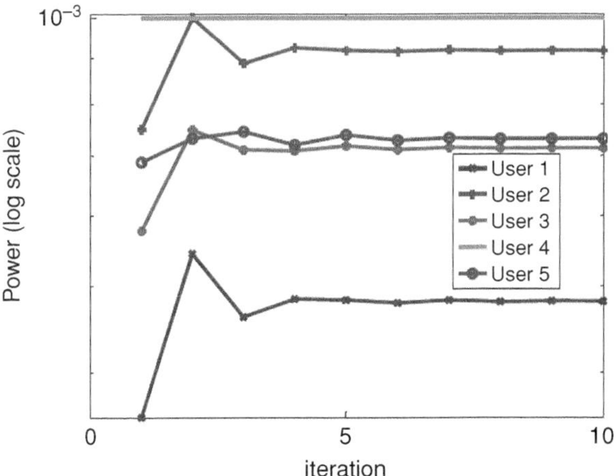

Figure 12.4 Experiment 1. Convergence in power (in watts and illustrated above in logarithmic scale) for five femtocell users using Algorithm 12.1.

channels. Hence, all the users experience inter-femto interference and also interference between the macrocells and the femtocells. We use the dense-urban propagation parameters in Table 2 of [13].

12.6.1   Experiment 1 (convergence of Algorithm 12.1)

Figure 12.4 shows the convergence of Algorithm 12.1 for five femtocell users. Simulations show that convergence happens in less than ten iterations even for thousands of users and a large power range, e.g., 125 mW to 2 W (maximum output of UMTS/3G Power Class 4 to Class 1 mobile phone, respectively). From Theorem 12.3, Algorithm 12.1 can be viewed as a *non-linear* power method in linear algebra. It is well known that the convergence rate of the power method is determined by the ratio of the second dominant eigenvalue to the Perron–Frobenius eigenvalue [25]. The method converges slowly if this ratio is close to 1. Now, this ratio for $\mathbf{B}(\mathbf{p}^\star)$ (cf. fourth and fifth row of Table 12.1) determines only the *local convergence rate* of Algorithm 12.1 near the fixed point $\mathbf{p}^\star$. Since the max-min weighted SINR solution is numerically observed to give a good approximation to $\mathbf{p}^\star$ (especially in the regime of low power and small β), this ratio is conceivably close to those computed using the max-min weighted SINR solution (cf. second and third row of Table 12.1), which on the other hand determines a *global convergence rate* for the max-min weighted SINR solution. This ratio can be empirically determined to be in the range of 0.2 to 0.4 using the parameters in [1]. This thus determines numerically the overall convergence rate of Algorithm 12.1.

12.6.2   Experiment 2 (comparison between Algorithms 12.3 and AOPC)

Next, numerical examples to compare the total power consumption using Algorithms 12.3 and AOPC are provided. Figure 12.5 shows the total power evolution in a network with initially 100 and 200 femtocell users. Then, 10 and 50 percent of the users are removed at time slots 30 and 70, respectively. On each graph, the difference in total power and the fraction of users that satisfy their outage probability threshold α are recorded. As illustrated, Algorithm 12.3 can produce power savings of 50% or more in all cases at the expense of a smaller number of users meeting α. On the other hand, problem infeasibility causes users who run Algorithm 12.3 and are unable to achieve α to transmit at $\bar{p}$. By enforcing a worst-outage probability fairness across all users, Algorithm AOPC computes powers that are typically smaller than $\bar{p}$, thus leading to a smaller total power.

12.6.3   Experiment 3 (performance comparison between macrocell and femtocell users)

In this experiment, the performance of Algorithms 12.3 and AOPC are evaluated in a macrocell network with 15 randomly located femtocell users, as illustrated in Figure 12.6. The cell radius is 1.4 km. Each macrocell user and each femtocell user communicates with its respective base station over independent Rayleigh fading channels with a path loss of $L = 128.1 + 37.6\log_{10}(d)$dB and $L = 98.5 + 20\log_{10}(d)$dB, where d is the distance in kilometers, respectively. Each user has a maximum power constraint of 500 mW (27 dBm), and each user is served one independent data stream from its base station. The noise power spectral density is set to -162 dBm/Hz. The outage probability of 2% and SINR thresholds of 5 dB are assigned to all the macrocell and femtocell users.

Let us analyze the performance of three scenarios, where Scenario 1: when all the femtocell base stations are switched off, i.e., all the access terminals are macrocell users

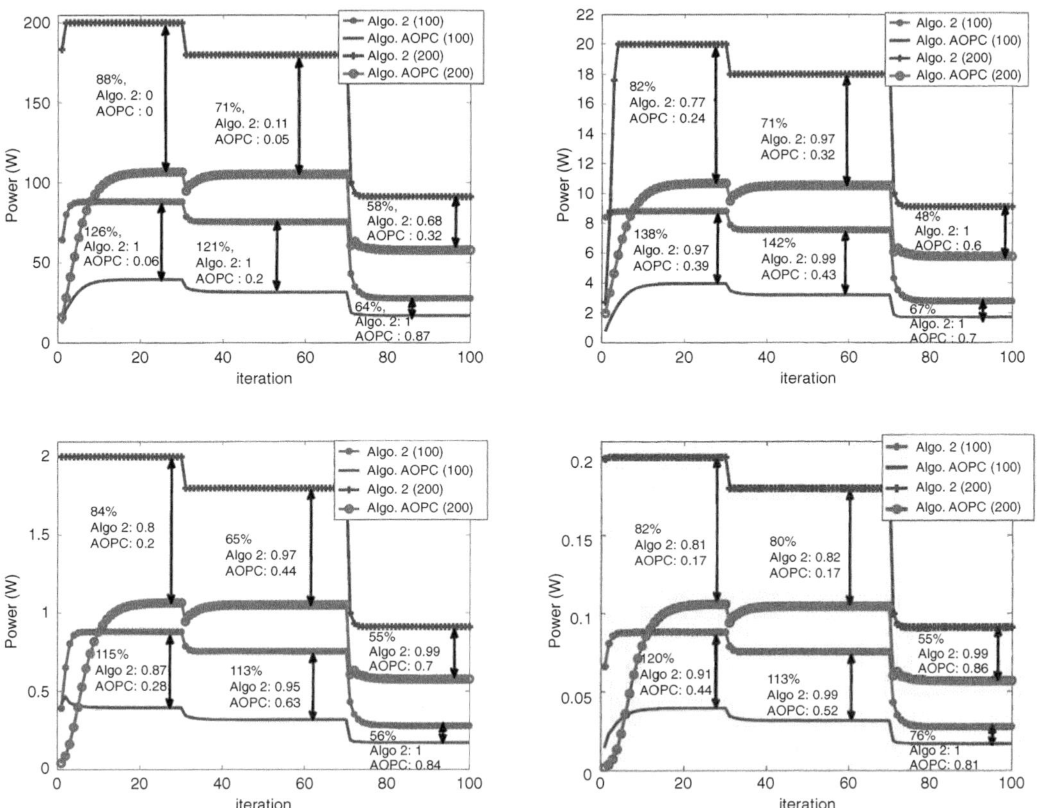

Figure 12.5 Experiment 2. The figures plot the evolution of the total power consumption for two networks with 100 and 200 femtocell users as $\bar{p}$ is varied. A time slot on the x-axis of each graph refers to a power update iteration. Ten and fifty percent of the users are removed at time slot 30 and 70, respectively. The top left graph, top right graph, bottom left graph, and bottom right graph show the case for $\bar{p} = 1\ W$, $\bar{p} = 0.1\ W$, $\bar{p} = 10\ \text{mW}$, and $\bar{p} = 1\ \text{mW}$, respectively. Power savings between Algorithms 12.3 and AOPC, and the fraction of the number of users meeting the given threshold α are in text form.

and use Algorithm 12.3; Scenario 2: when all the femtocell base stations are switched on and all the users use Algorithm 12.3; and Scenario 3: all macrocell users use Algorithm 12.3 and all femtocell users use Algorithm AOPC. The normalization of $\mathbf{z}(k)$ at Step 2 of Algorithm AOPC can be performed by the femtocell users in a distributed manner assuming that the exchange of messages can be performed by each femtocell base stations over the cellular core network. The three scenarios are run over 5,000 numerical simulations with 15 femtocell users randomly placed in the cell at each simulation run. Let us compare the average percentage of macrocell users that meet the given outage constraints among the three scenarios in Figure 12.7. Figure 12.8 compares the total power consumption for the three scenarios. By comparing Scenarios 1 and 2 as shown in Figures 12.7 and 12.8, the deployment of the femtocell improves the percentage of users meeting the given outage constraints, but the total power savings may not be appreciable. By

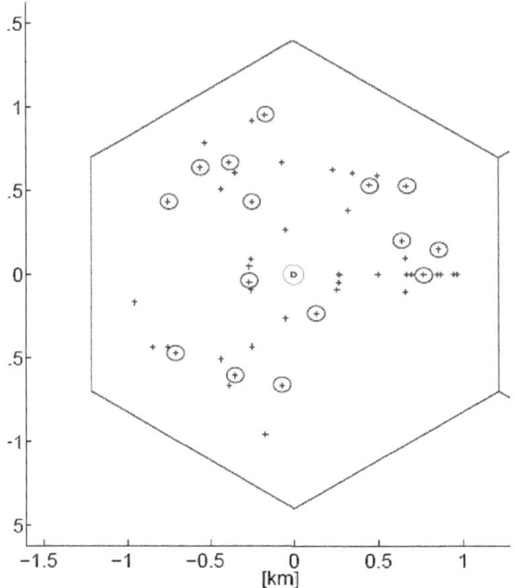

Figure 12.6 A macrocell with 15 randomly located femtocell users in the closed access mode. The macrocell base station is circled in red in the middle. The femtocell users are the dots circled in black. The other dots are the macrocell users.

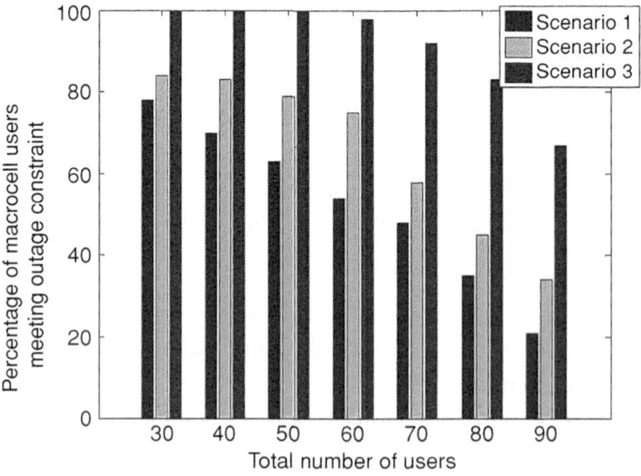

Figure 12.7 Performance of the three scenarios comparing the average percentage of macrocell users meeting the outage constraints over 5,000 simulations.

comparing the performance of Scenarios 2 and 3, it can be observed that when we allow femtocell users to have a min-max outage probability fairness, the number of macrocell users meeting their outage constraint and the power savings obtained can increase by as much as 50%.

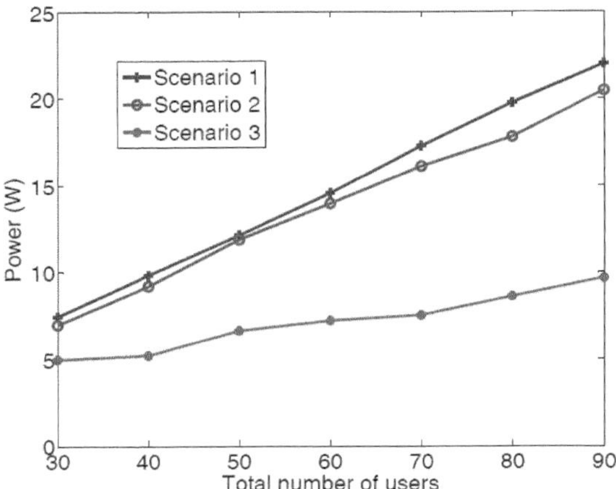

Figure 12.8 Performance of the three scenarios comparing the average total power consumption of all the users over 5,000 simulations.

12.7 Conclusion

In this chapter, we have studied a class of practical resource allocation problems in heterogeneous wireless networks using tools from the non-linear Perron–Frobenius theory and non-negative matrix theory. These resource allocation optimization problems, though seemingly non-convex, can be solved by viewing them as non-linear eigenvalue problems. The optimal value and solution can be characterized by the spectral property of matrices induced by non-linear positive mappings. This approach also provides a systematic way to derive distributed fast algorithms often free of parameter tuning and to evaluate the fairness of resource allocation in wireless networks. In particular, we have seen how the worst-outage probability problem with power constraints in a multi-user Rayleigh-faded network can be solved analytically. A tight relationship between this worst-outage probability problem and its certainty-equivalent margin counterpart, also known as the max-min weighted SINR problem, has been established and used to provide performance bounds. Similarly, the max-min weighted SINR problem can be solved analytically. The proposed algorithms that solve the worst-outage probability problem and the max-min weighted SINR problem are computationally fast, free of parameter tuning, and distributed. The illustrated techniques are general enough to handle constraints that are associated with interference management and resource budget. We have also studied a total power minimization problem with given outage specification constraints, and have established its feasibility condition by connecting the worst-outage probability problem to it. A dynamic algorithm has been proposed to adapt the outage probability specification to minimize the total energy in a heterogeneous wireless network. This permits a graceful handling of outage infeasibility in a distributed manner, and also guarantees a min-max fairness in terms of the worst-outage probability. Numerical studies have shown that the proposed algorithms are effective to manage interference, regulate the total power expenditure, and to ensure fairness in the system.

Acknowledgments

The author gratefully acknowledges the collaborations and interactions on this topic with many colleagues, including Desmond W. H. Cai, Mung Chiang, Shmuel Friedland, Yichao Huang, K. R. Krishnan, Steven H. Low, John C. S. Lui, Tony Q. S. Quek, Bhaskar D. Rao, R. Srikant, and Ao Tang. The research reported here was supported in part by grants from the Research Grants Council of Hong Kong, Project No. RGC CityU 125212, Qualcomm Inc., and the Science, Technology and Innovation Commission of Shenzen Municipality, Project on Green Communications in Small-cell Mobile Networks.

References

[1] M. Yavuz, F. Meshkati, S. Nanda, A. Pokhariyal, N. Johnson, B. Raghothaman, and A. Richardson, "Interference management and performance analysis of UMTS/HSPA+ femtocells," *IEEE Commun. Mag.*, vol. 47, no. 9, pp. 102–9, Sep. 2009.

[2] R. Madan, J. Borran, A. Sampath, N. Bhushan, A. Khandekar, and T. Ji, "Cell association and interference coordination in heterogeneous LTE-A cellular networks," *IEEE J. Sel. Areas Commun. (JSAC)*, vol. 28, no. 9, pp. 1479–89, Dec. 2010.

[3] S. Rangan and R. Madan, "Belief propagation methods for intercell interference coordination in femtocell networks," *IEEE J. Sel. Areas Commun. (JSAC)*, vol. 30, no. 3, pp. 631–40, Apr. 2012.

[4] W. C. Cheung, T. Q. S. Quek, and M. Kountouris, "Throughput optimization, spectrum allocation, and access control in two-tier femtocell networks," *IEEE J. Sel. Areas Commun. (JSAC)*, vol. 30, no. 3, pp. 561–74, Apr. 2012.

[5] K. R. Krishnan and H. Luss, "Power selection for maximizing SINR in femtocells for specified SINR in macrocell," *Proc. IEEE Wireless Commun. Networking Conf. (WCNC)*, Mar. 2011.

[6] C. W. Tan, "Optimal power control in Rayleigh-fading heterogeneous networks," *Proc. IEEE Int. Conf. on Computer Commun. (INFOCOM)*, Apr. 2011.

[7] S. Shakkottai and R. Srikant, "Network optimization and control," *Found. Trends Network.*, vol. 2, no. 3, pp. 271–379, 2007.

[8] X. Lin, N. B. Shroff, and R. Srikant, "A tutorial on cross-layer optimization in wireless networks," *IEEE J. Sel. Areas Commun. (JSAC)*, vol. 24, no. 8, pp. 1452–63, Aug. 2006.

[9] M. Chiang, P. Hande, T. Lan, and C. W. Tan, "Power control in wireless cellular networks," *Found. Trends Network.*, vol. 2, no. 4, pp. 381–533, 2008.

[10] M. Chiang, S. H. Low, A. R. Calderbank, and J. C. Doyle, "Layering as optimization decomposition: A mathematical theory of network architectures," *IEEE Proceedings*, vol. 95, no. 1, pp. 255–312, Jan. 2007.

[11] D. N. C. Tse and P. Viswanath, *Fundamentals of Wireless Communication*, 1st edn. Cambridge: Cambridge University Press, 2005.

[12] H. Kobayashi, B. L. Mark, and W. Turin, *Probability, Random Processes, and Statistical Analysis: Applications to Communications, Signal Processing, Queueing Theory, and Mathematical Finance*. Cambridge: Cambridge University Press, 2012.

[13] S. Jagannathan, M. Zawodniok, and Q. Shang, "Distributed power control of cellular networks in the presence of Rayleigh fading channel," *Proc. IEEE Int. Conf. on Computer Commun. (INFOCOM)*, Mar. 2004.

[14] V. Chandrasekhar and J. G. Andrews, "Uplink capacity and interference avoidance for two-tier femtocell networks," *IEEE Trans. Wireless Commun.*, vol. 8, no. 7, pp. 3498–509, July 2009.

[15] S. V. Hanly, L. H. Andrew, and T. Thanabalasingham, "Dynamic allocation of subcarriers and transmit powers in an OFDMA cellular network," *IEEE Trans. Inf. Theory*, vol. 55, no. 12, pp. 5445–62, 2009.

[16] S. Kandukuri and S. Boyd, "Optimal power control in interference-limited fading wireless channels with outage-probability specifications," *IEEE Trans. Wireless Commun.*, vol. 1, no. 1, pp. 46–55, Jan. 2002.

[17] J. Papandriopoulos, J. Evans, and S. Dey, "Optimal power control for Rayleigh-faded multiuser systems with outage constraints," *IEEE Trans. Wireless Commun.*, vol. 4, no. 6, pp. 2705–15, Nov. 2005.

[18] M. Chiang, C. W. Tan, D. P. Palomar, D. O'Neill, and D. Julian, "Power control by geometric programming," *IEEE Trans. Wireless Commun.*, vol. 6, no. 7, pp. 2640–51, July 2007.

[19] V. D. Blondel, L. Ninove, and P. V. Dooren, "An affine eigenvalue problem on the nonnegative orthant," *Linear Algebra and its Applications*, vol. 404, pp. 69–84, 2005.

[20] U. Krause, "Concave Perron-Frobenius theory and applications," *Nonlinear Anal.*, vol. 47, no. 2001, pp. 1457–66, 2001.

[21] C. W. Tan, M. Chiang, and R. Srikant, "Fast algorithms and performance bounds for sum rate maximization in wireless networks," *Proc. IEEE Int. Conf. on Computer Commun. (INFOCOM)*, Apr. 2009.

[22] S. Friedland and S. Karlin, "Some inequalities for the spectral radius of non-negative matrices and applications," *Duke Math. J.*, vol. 42, no. 3, pp. 459–90, Sep. 1975.

[23] C. W. Tan, S. Friedland, and S. H. Low, "Nonnegative matrix inequalities and their application to nonconvex power control optimization," *SIAM J. Matrix Anal. A.*, vol. 32, no. 3, pp. 1030–55, 2011.

[24] C. W. Tan, "Spectrum management in multiuser cognitive wireless networks: optimality and algorithm," *IEEE J. Sel. Areas Commun. (JSAC)*, vol. 29, no. 2, pp. 421–30, Feb. 2011.

[25] E. Seneta, *Non-negative Matrices and Markov Chains*, 2nd edn. Springer, 2006.

[26] C. W. Tan, M. Chiang, and R. Srikant, "Maximizing sum rate and minimizing MSE on multiuser downlink: optimality, fast algorithms and equivalence via max-min SINR," *IEEE Trans. Signal Process.*, vol. 59, no. 12, pp. 6127–43, Dec. 2011.

[27] Y. D. Yao and A. Sheikh, "Investigations into cochannel interference in microcellular mobile radio systems," *IEEE Trans. Veh. Technol.*, vol. 41, no. 2, pp. 114–23, May 1992.

[28] S. Boyd and L. Vanderberghe, *Convex Optimization*. Cambridge: Cambridge University Press, 2004.

[29] D. P. Bertsekas and J. N. Tsitsiklis, *Parallel and Distributed Computation: Numerical Methods*. Englewood Cliffs, NJ: Prentice Hall, 1989.

[30] U. Krause, "A local–global stability principle for discrete systems and difference equations," *Proc. 6th Int. Conf. on Difference Equations (2001)*, vol. CRC Press, pp. 167–80, 2004.

[31] S. Boyd, A. Ghosh, B. Prabhakar, and D. Shah, "Randomized gossip algorithms," *IEEE Trans. Inf. Theory*, vol. 52, no. 6, pp. 2508–30, June 2006.

[32] G. J. Foschini and Z. Miljanic, "A simple distributed autonomous power control algorithm and its convergence," *IEEE Trans. Veh. Technol.*, vol. 42, no. 4, pp. 641–6, Nov. 1993.

[33] D. W. H. Cai, T. Q. S. Quek, and C. W. Tan, "A unified analysis of max–min weighted SINR for MIMO downlink system," *IEEE Trans. Signal Process.*, vol. 59, no. 8, pp. 3850–62, Aug. 2011.

[34] ——, "Coordinated max–min SIR optimization in multicell downlink – duality and algorithm," *Proc. IEEE Int. Conf. on Commun. (ICC)*, June 2011.

[35] L. Zheng and C. W. Tan, Cognitive Radio Network Duality and Algorithms for Utility Maximization, *IEEE Journal on Selected Areas in Communications*, vol. 31, no. 3, Mar. 2013.

[36] Y. Huang, C. W. Tan, and B. D. Rao, "Asymptotic analysis of max–min weighted SINR in multiuser downlink," *Proc. Conf. on Information Sciences and Systems (CISS)*, Mar. 2012.

[37] D. W. H. Cai, C. W. Tan, and S. H. Low, "Optimal max–min fairness rate control in wireless networks: Perron–Frobenius characterization and algorithms," *Proc. IEEE Int. Conf. on Computer Commun. (INFOCOM)*, Mar. 2012.

13 New strategies for femto–macro cellular interference control

Sundeep Rangan

13.1 Femto–macro interference coordination problem

Femtocells are small wireless access points that are typically installed in a subscriber's premises, but operate in a cellular provider's licensed spectrum. Since femtocells can be manufactured at a very low cost, require minimal network maintenance by the operator, and can leverage the subscriber's backhaul, femtocells offer the possibility of expanding cellular capacity at a fraction of the cost of traditional macrocellular deployments. With the surge in demand for wireless data services, femtocells have thus attracted considerable recent attention, both in cellular standards bodies such as the 3rd Generation Partnership Project (3GPP) [1–3] and academic research [4, 5].

However, one of the key technical challenges in deploying femtocells is the interference between the underlay of small femtocells and the overlay of comparatively large macrocells – an issue raised in virtually every survey on femtocells [5–8]. While interference is a fundamental challenge in any cellular system, the so-called *cross-tier* interference in femtocell networks has two particularly challenging aspects:

1. *Strong and varied interference:* Due to *closed access* or *restricted association*, mobile terminals (or user equipment (UE) in 3GPP terminology) may not be able to connect to a given femtocell even when it provides the closest serving base station [9]. Such restrictions can result in strong interference both from the macrocell UE transmitter onto the femtocell uplink and from the femtocell downlink onto the macrocell UE receiver. In addition, since the femtocell access points are often deployed in an essentially ad hoc manner, interference conditions are much more varied than traditional planned macrocellular networks.

2. *Need for distributed and adaptive control:* Femtocells are installed by the subscriber and thus need to be largely plug-and-play and self-organizing in operation. Moreover, femtocells are generally not connected directly into the operator's core network like a traditional macrocell, but instead may go through the subscriber's private Internet service provider (ISP). Therefore usage of backhaul signaling for interference coordination (such as the X2 interface in long term evolution (LTE)) is often more limited. In addition, over-the-air signaling may also be difficult due to the power

Small Cell Networks: Deployment, PHY Techniques, and Resource Management, ed. Tony Q. S. Quek, Guillaume de la Roche, İsmail Güvenç, and Marios Kountouris. Published by Cambridge University Press. © Cambridge University Press 2013.

disparities between the femtocells and macrocells and the large number of femto-cells per macrocell. These characteristics necessitate that interference coordination be performed in a much more distributed and autonomous nature than traditional macrocellular networks.

Recognizing these challenges, standards bodies have expended considerable effort in studying femtocell interference, including work in the Femto Forum [10] and 3GPP [1, 3]. Advanced methods for femtocell interference coordination have also been a major driving factor in the 3GPP LTE-Advanced standardization effort [11, 12]. Particularly for 3G code division multiple access (CDMA) femtocells, the dominant method for interference coordination is for the femtocells and macrocells to operate in a co-channel manner and regulate the interference through power control [13–15]. This strategy derives from the basic paradigm of CDMA macrocellular networks where the network is operated in universal frequency reuse (all transmitters spread their signals across the entire band), interference is treated as Gaussian noise, and resource allocation is coordinated via power and rate control [16–19].

However, while universal frequency reuse with power control has been immensely successful in macrocellular networks, this chapter argues that the two trends in interference in heterogeneous networks – stronger and varied interference combined with the need for autonomous and distributed control – motivate a reconsideration of universal frequency reuse and the power control framework. Our argument is essentially information theoretic: considering a simple Gaussian interference channel [20], we know that universal frequency reuse and treating interference as noise is optimal only when the interference is weak. This situation, however, corresponds well with macrocellular networks, where mobiles can generally connect to the strongest cell and there is rarely any single large interferer.

However, in heterogeneous networks with closed access, interference is much stronger. In this case, results from the Gaussian interference channel suggest that in the case of strong interference, orthogonalization or more sophisticated methods based on interference cancellation can have significant gains.

To understand the possibility of these gains, we consider techniques based on subband scheduling – a prominent feature of 4G orthogonal frequency division multiplexing (OFDM) systems such as the LTE standard. Subband scheduling essentially allows combined use of both traditional reuse 1 and inter-cell frequency orthogonalization in the same network. While subband scheduling and the related concept of fractional frequency reuse (FFR) have been widely studied in the context of macrocellular systems [21–24], we show in this chapter that these techniques can offer significantly improved performance for femto–macro mixed networks. Moreover, we review a power allocation strategy in [25] that shows how subband partitions in the macrocellular network can be maximally exploited by short-range femtocell links in a distributed manner with minimal coordination with the macrocell. Thus, these strategies may offer a tractable and readily implementable method for commercial heterogeneous networks.

We also briefly consider more sophisticated techniques based on interference cancellation that have also been considered in the macrocellular context [26–28]. Specifically, we consider the case where the femtocell receiver (the femto base station in the uplink and femto mobile in the downlink) does not treat the macrocellular interference as noise, but instead jointly decodes the interference with the desired signal from the femtocell transmitter. We argue that, when such capabilities exist, femto links have improved opportunities to transmit. Our simulations, however, only show modest gains thus suggesting that interference cancellation may not be warranted.

13.2 Frequency reuse revisited

13.2.1 An historical perspective

A key design parameter in the way that cellular systems manage interference is the level to which frequencies are shared or "reused" in neighboring cells. The subband partitioning methods that will be discussed in this chapter can be seen as hybrid techniques that combine two basic strategies: frequency orthogonalization and universal frequency reuse 1. To understand why such hybrid schemes may be useful in emerging femtocellular networks, it is instructive to begin by briefly reviewing the basics of orthogonalization and reuse 1, the historical roles each have played in cellular systems to date, and compare the relative merits of each of the two methods.

When a cellular system employs inter-cell *frequency orthogonalization*, it means that neighboring cells operate on disjoint frequencies, either on different carriers, or in the case of OFDM systems such as LTE, on different frequency resource blocks. Orthogonalization eliminates nearby inter-cell interference but reduces the bandwidth available to each cell. The level of orthogonalization in a cellular system is usually characterized by its *reuse factor* – a reuse factor of K indicating that each frequency is used only once per K cells. As a result, each cell can be orthogonal to up to $K - 1$ of its neighbors. Early cellular systems such as global system for mobile communication (GSM) were typically deployed with a frequency reuse factor of 4, allowing a high level of orthogonalization between neighboring cells [29].

One of the key differentiating features of CDMA cellular systems that emerged in the 1990s was the advocacy of *universal frequency reuse*, or *frequency reuse 1*, where every link spreads its signal across the entire bandwidth. All transmissions interfere with one another, and therefore the rate per unit bandwidth is typically lower than orthogonal systems. However, since each link has access to the entire bandwidth, it was realized that the overall cell capacity could be higher than orthogonal systems and this fact was a major motivation for the transition to CDMA [30].

However, proper operation of CDMA systems requires power control to ensure that no one transmission overwhelms another (the well known near–far effect [31]), and rate control to ensure that the transmission rate is correctly adapted to the interference level. Fortunately, power and rate control have been extensively studied in macrocellular systems [16–18], and there are now a large number of power and rate control algorithms with simple distributed implementations, with provable performance and convergence guarantees, and excellent performance. See, for example, the monograph [19].

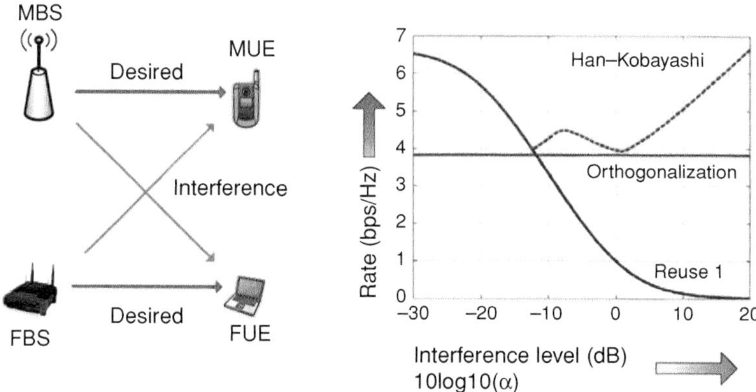

Figure 13.1 Lessons from the Gaussian interference channel. The choice between frequency reuse 1 and orthogonalization depends on the strength of interference. Multi-tier networks with closed access will experience varied mixes of strong and weak interference, thus necessitating adaptive strategies. Moreover, the possibility of very strong interference suggests possible gains from more sophisticated techniques such as Han–Kobayashi rate splitting and interference cancellation.

To apply the frequency reuse 1 concept to femtocells, the femtocells and macrocells would transmit in the same channel and control the cross-tier interference via power and rate control. Due to its success in macrocells, power control with frequency reuse 1 is perhaps the most widely studied method for cross-tier interference control, particularly in CDMA femtocells. For example, [14, 28] study adaptive attenuation and power control methods for universal mobile telecommunication system (UMTS) femtocells, while [13, 15] study more sophisticated power control methods based on distributed utility maximization. These power control methods have the benefit that they can be largely implemented in existing cellular systems, and based on simulations such as the Femto Forum study [10] appear to work well in a range of circumstances. Moreover, frequency reuse 1 eliminates the need for frequency planning, which is difficult in femtocellular networks that are deployed in an ad hoc manner. In addition, power control admits precise analyses. For example, the studies [32–35] provide analytic expressions for the outage capacity under power control with various cell selection methods.

13.2.2   Lessons from the Gaussian interference channel

However, despite the benefits of frequency reuse 1 and power control in macrocellular systems, reuse 1 may not be optimal for femtocells. Orthogonalization and other strategies may have significant benefits if used selectively and intelligently. To understand this point, it is instructive to consider a simple two link femto–macro interference scenario shown in Figure 13.1. The figure shows simultaneous downlink transmissions on two links: one link from a macrocell base station (MBS) to a mobile user equipment (MUE), and a second link from a femtocell base station (FBS) to femtocell user equipment (FUE).

One could compare orthogonalization and reuse 1 in a range of different parameters. However, for illustration and simplicity, suppose that the FUE and MUE have identical thermal noise N and are positioned so that they have the same desired signal power P. Thus, in the absence of interference, each link would achieve a signal to noise ratio (SNR) of P/N. As a measure of the cross-tier interference suppose that each link creates an interference power of αP on the receiver of the other link, where α is a measure of the cross-link gain. This scenario is then identical to the classic Gaussian interference channel in information theory [20].

The right panel of Figure 13.1 compares the theoretical rates as a function of the cross-tier interference α for various strategies. The curve labeled "reuse 1," represents the case when both the femto and macro base stations spread their energy across the entire bandwidth and the femto and macro receivers treat the interference from the other base station as Gaussian noise. The signal to interference plus noise ratio (SINR) in this case will be $P/(N + \alpha P)$, and the curves plot the rates assuming the transmitters adjust their rate to the Shannon capacity for this SINR, namely $\log_2(1 + P/(N + \alpha P))$. The figure plots this rate with $P/N = 20\,\text{dB}$. We can see that reuse 1 with treating interference as noise does well when the interference is weak, in which case it can actually be shown to be optimal [36–38].

This property explains the benefits of frequency reuse 1 in traditional CDMA macrocellular systems. In these systems, mobiles generally connect to the strongest serving cells, and experience interference from the aggregate of weak interference from a large numbers of users. From Figure 13.1, it is this regime that frequency reuse 1 with simple power control does well.

However, as discussed above, a key attribute of cross-tier interference in femtocellular networks, particularly with closed access, is the presence of much stronger interference. Figure 13.1 shows that in strong interference, standard reuse 1 performs poorly and other strategies offer much better rates. For example, the curve labeled "orthogonalization" shows the rate when the femtocell and macrocell split the bandwidth and transmit on half the bandwidth each. In this case, there is no interference, so each link sees an SINR of P/N in its bandwidth, irrespective of the cross-tier gain α. Therefore, the Shannon capacity per unit of total bandwidth is simply $(1/2)\log_2(1 + P/N)$. When the interference α is strong, this simple orthogonalization strategy can provide considerably higher rates than reuse 1.

Figure 13.1 also shows a more complicated Han–Kobayashi (HK) method [39] based on rate splitting and interference cancellation. This strategy is significantly more complicated to implement than orthogonalization or reuse 1 and treating interference as noise. However, the method can achieve even greater rates with strong interference. We will return to this method later in Section 13.4.

Of course, the cross-tier interference in femtocellular networks is neither uniformly strong nor weak. Interference levels in femtocellular networks may span a large range with possibilities of much stronger interference than in traditional macrocellular systems. This property motivates the need for *dynamic* and *adaptive* methods that can appropriately select the level of orthogonalization and reuse depending on the interference conditions of individual mobiles.

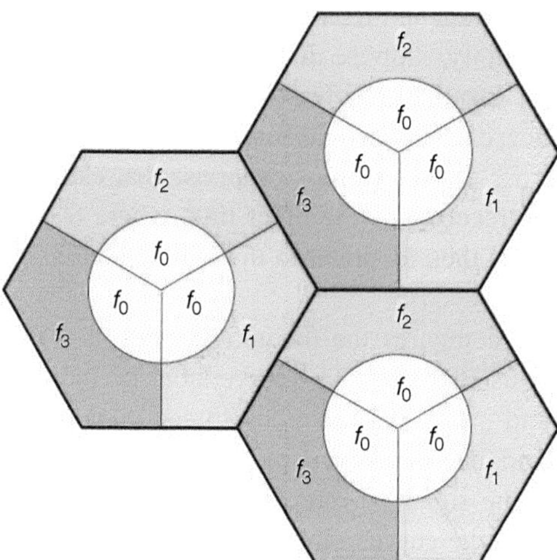

Figure 13.2 Typical subband partitioning of a macrocellular network. Each cell site occupies one hexagon with the cell site partitioned into three 120 degree sectors called cells. The frequency subband f_0 is used in all cells and serves mobiles close to the cell. The other subbands f_1 to f_3 are used in only one cell per cell site and serve mobiles close to the cell edge.

13.3   Interference control with subband partitioning

13.3.1   Subband partitioning in the macrocell

The motivating idea of subband partitioning is to capture the benefits of both reuse 1 and orthogonalization. The spectrum in both the uplink and downlink is divided into disjoint frequency intervals called *subbands*, with different types of reuse patterns used in different subbands. Base stations can then schedule each mobile in a subband with an appropriate reuse pattern corresponding to its interference. Mobiles close to the cell experiencing weak interference can be scheduled on reuse 1 subbands, while mobiles at the cell edge under strong interference can be scheduled on subbands with higher reuse factors. Although subband partitioning can, in principle, be applied in any multi-carrier system, the partitioning is particularly easy to implement in orthogonal frequency division multiple access (OFDMA) systems such as LTE where the frequency is naturally divided into frequency resource blocks [40].

A common type of partitioning for a macrocellular network with three-way sectorized cell sites is shown in Figure 13.2. Following the 3GPP terminology, we call each sector in the cell site a *cell*. With this sectorization, the figure shows a partitioning of the uplink (UL) and downlink (DL) bands into four subbands whose center frequencies are denoted f_0 to f_3. With some abuse of notation, we will use f_k to denote the subband as well as its center frequency.

One of the four subbands, f_0, is reused in every cell, and thus serves as a reuse 1 subband. This subband can be used to serve mobiles in weak interference. Since these mobiles are generally close to the cell, the area served by the f_0 subband is drawn in Figure 13.2 as an inner circle of the cell site. Each of the remaining subbands – f_1 to

f_3 – is used in only one cell per cell site, and therefore has reuse 3 pattern. The reuse 3 subbands are arranged so that neighboring cells do not use the same subbands, so the interference in these subbands is low. These subbands are used to serve the mobiles at the edge of the cell where the mobile would experience high interference if served in the reuse 1 subband.

Although we will not explore the concept in this chapter, subband partition can be further improved through a related concept called *fractional frequency reuse* (FFR). Looking at Figure 13.2, we see that two of the subbands go unused in each cell. However, the cell could transmit a small power in each of these cells causing minimal interference onto mobiles in neighboring cells being served in that subband. With the small power, the cell can then serve mobiles that are very close to the cell, thereby gaining extra bandwidth with minimal cost. Although for simplicity in this chapter we will just consider the simple subband partitioning with on–off power allocations, many of the concepts can be extended to the FFR case as well.

Early descriptions of subband partitioning and FFR can be found in [21, 22] with further analyses for distributed algorithms in [23, 24]. Those and other studies show that subband partitioning can generally provide a significant improvement in the cell edge rate with only a minimal loss in the average cell capacity. The partitioning thus creates a more uniform throughput experience through the cell, which is particularly valuable for data users.

13.3.2   Subband partitioning and femtocell links

While subband partitioning can often improve the rate distribution in the macrocellular network, it was observed in [41] that subband partitioning can also create improved opportunities for short-range femtocell transmissions with minimal interference into the macrocell network.

To see this property, consider, for example, the transmissions of two neighboring macrocells in Figure 13.3. Macrocell 1 serves two macro UEs, MUE1 and MUE2, while macrocell 2 serves macro UEs MUE3 and MUE4. MUE1 and MUE3 are close to their serving cells and thus experience little interference from the other cell. These mobiles can be served in the reuse 1 subband, f_0, where the full bandwidth of the subband can be reused in both cells. In contrast, MUE2 and MUE4 are at the cell edge. If served in the same subband, the interference would be high. For these mobiles, the rate may be higher by serving them on orthogonal bandwidths, f_1 and f_2, which eliminates the interference from the other cell.

Now suppose a short-range link, such as link between a femto base station and femto UE is added to the network. Figure 13.3 shows three possible locations for these short-range links, labeled SR1 to 3. For now, ignore the short range links SR4 to 7, which we will discuss in Section 13.4.

First consider a short-range link, such as SR1, at the macrocell edge (i.e., between the two macrocells). A short-range link such as SR1 can operate in the reuse 1 subband, f_0, in either the UL or DL. For the UL, the short-range link is far from either macrocell, so the transmissions should have minimal interference at the macrocell receivers. Also,

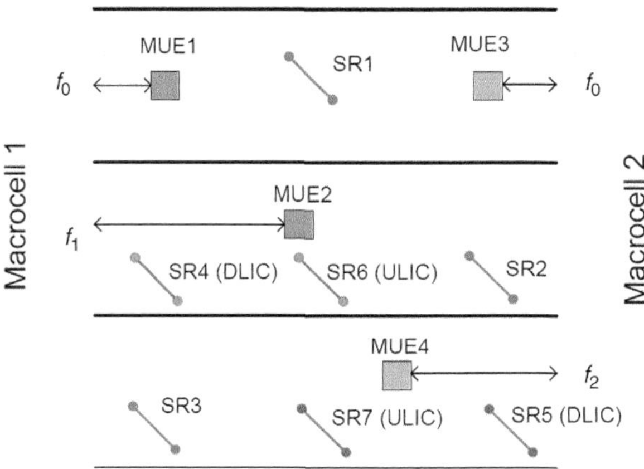

Figure 13.3 Typical locations of macro UEs (MUE1–4) in each subband, along with possible locations for short-range links (SR1–7). SR1–3 can operate in either the UL or DL and do not require interference cancellation (IC). SR4 and 5 would operate in the DL with IC of the DL signal from the macrocell base stations. SR6 and 7 operate in the UL with IC of the macro UE transmissions.

since the UEs in f_0 are typically close to their cell, they will also be far from SR1. Therefore, transmissions on SR1 in the DL f_0 will also have minimal interference at the macro UEs such as MUE1 and MUE3.

Now consider the cases, such as SR2 or SR3, when the short-range links are close to one of the macrocells. In these cases, the short-range links can transmit in one of the subbands f_1 or f_2 not being used by the macrocell that it is close to. In this way, the subband partitioning provides, wherever the short-range link is located, frequency locations on which it can transmit where the interference will be minimal.

13.3.3   Optimized power control with subbands

To capture the benefits of subband partitioning in the macrocellular network, the femtocell transmitters must correctly distribute their power across the subbands to minimize their effect on the macrocell while providing good throughput to the desired femtocell receiver. Ideally, the power allocations should be done in a distributed manner with minimal coordination with the macrocellular network.

In the method in [41], each femtocell transmitter (the femto base station (BS) in the DL and femto UE in the UL) is statically assigned to a single subband based on the DL SINR. In this chapter, we will review a slightly more sophisticated technique from [25] where each femtocell transmitter can simultaneously use all subbands, with a power level in each subband based on the precise interference conditions. This method enables a larger fraction of the bandwidth to be used and to select the powers in the subbands based on the actual interference conditions.

To describe the proposed method mathematically, suppose that there are K total subbands. For example, in the subband partitioning of the previous section, there would be $K = 4$ subbands in the UL or DL. We will assume there are N short-range links denoted SR_j, $j = 1, \ldots, N$. We will use the term short-range links, as opposed to femto links, since the method does not require that short-range links using the same technology as the macro. Also, the macrocellular system could be time division duplex (TDD) or frequency division duplex (FDD). In the case of TDD, the short-range links will use the UL and DL time periods aligned with the macro.

We will assume that in subband k, there are $M(k)$ macro receivers, which we will denote by MRX_{ik}, $i = 1, \ldots, M(k)$. In an UL subband, the macro receivers are the macrocell base stations using that subband, and in the DL, the macro receivers are the UEs using that subband.

Now let G_{ijk} be the path gain from the short-range transmitter $SRTX_j$ to the macro receiver MRX_{ik} in subband k. We will let p_{jk} be the transmit power of $SRTX_j$ in subband k, so that the total interference at MRX_{ik} in subband k from all short-range links is

$$q_{ik} = \sum_{j=1}^{N} G_{ijk} p_{jk}. \tag{13.1}$$

To implement the femto–macro interference power control, we will assume that each short-range transmitter, $SRTX_j$, can learn the path gains G_{ijk} to all the macro receivers in all the K subbands. To perform this estimation in the UL subbands, the macrocell base stations would need to intermittently transmit some reference signal in the UL. Of course, this would require that the regular UL macrocellular transmissions are suspended in some quiet period and thus requires some overhead. Similarly, in the DL, the path-losses to the macro UEs could be estimated by having a quiet period in the DL where the macro UEs suspend their regular reception and transmit reference signals that the short-range transmitters can hear. Alternatively, in a TDD system, a short-range transmitter could estimate a UL path gain to the base stations from a DL reference such as the pilot signals, and an DL path gain to the macro UEs from the UEs UL power control signals. These reference signals can also be used in an FDD system, but with some error due to fast fading.

Now let $\mathbf{p}_j$ be the vector of powers used by $SRTX_j$ across the K subbands,

$$\mathbf{p}_j := [p_{j1} \cdots p_{jK}]^T. \tag{13.2}$$

The power control problem is for each $SRTX_j$ to select the power vector $\mathbf{p}_j$ to control the interference levels q_{ik} in (13.1), while maximizing the throughput. This is a distributed control problem since all the short-range transmitters have some effect on the interference levels.

Under the assumptions we will make below, this power control problem is somewhat trivial and can be solved by a large number of distributed methods. See, for example, the survey [19]. As one possible solution, we present a *load-spillage* method similar to [42]. Each macro receiver MRX_{ik} broadcasts a *load factor*, denoted s_{ik}, on subband k. The load factor is simply some positive scalar. Each short-range transmitter $SRTXj$ then computes a per subband *spillage* given by

$$r_{jk} = \sum_{i=1}^{M} G_{ijk} s_{jk}. \tag{13.3}$$

The SRTX j can then select any set of powers p_{jk} subject to

$$\sum_{k=1}^{K} r_{jk} p_{jk} \leq \lambda_j, \tag{13.4}$$

where λ_j is some predetermined positive scalar, which we will call the *power constraint bound*.

The load-spillage rule (13.4) can be understood qualitatively as follows: the rule (13.4) permits the short-range links to transmit a high power p_{jk} precisely when the corresponding spillage r_{jk} is small. Now, the spillage (13.3) r_{jk} is high when there is some macro receiver MRX_{ik} that is either close to the short-range transmitter (so that G_{ijk} is large) or the macro receiver broadcasts a high load factor s_{ik}. Thus, the spillage rule (13.4) will cause the short-range transmitters to allocate power away from subbands where there are closeby macro receivers or macro receivers indicating high load.

We can, in fact, mathematically show that under certain simplifying assumptions, the load spillage rule can be optimal. Specifically suppose that, associated with each short-range link j, there is some "utility" $U_j(\mathbf{p}_j)$ of the powers $\mathbf{p}_j$. This utility could be, for example, the rate or some function of the rate. The implicit assumption is that the utility does not depend on the power levels in other short-range links. That is, the interference at the short-range receivers is dominated by the interference from the macro transmissions.

Given such a utility function and power constraint (13.4), a natural strategy for each short-range transmitter SRTX_j is to select the power vector $\mathbf{p}_j$ to maximize the utility:

$$\max_{\mathbf{p}} U_j(\mathbf{p}_j) \tag{13.5}$$

subject to (13.4). In the case when $U_j(\mathbf{p}_j)$ is the total rate across the K subbands, and the rate is given by the Shannon capacity, this optimization reduces to a standard waterfilling procedure [20].

However, whatever the utility function $U_j(\mathbf{p}_j)$, we can show the following result. Say a selection of power vectors $\mathbf{p} = \{\mathbf{p}_j, j = 1, \ldots, N\}$ is *Pareto optimal* if any other selection of power vectors that results in a lower interference q_{ik} on some subband k for some macro receiver MRX_{ik}, or higher utility U_j for some short-range link SR_j, must also result in a higher interference for some other macro receiver or lower utility in some other link.

Theorem 13.1 *Suppose that the utility functions $U_j(\mathbf{p}_j)$ are strictly increasing in each component p_{jk}. Then,*

(a) *If the load factors s_{ik} and power constraint bounds λ_j are positive and each short-range link j selects the power vector according to (13.5), then the resulting power vectors are Pareto optimal.*

(b) *If, in addition, the utility functions are concave and twice differentiable and $G_{ijk} > 0$ for all i, j, and k, then the converse is true. That is, if $\mathbf{p}$ is any set of power vectors that is Pareto optimal, then there exists non-negative s_{ik} and λ_j such that each $\mathbf{p}_j$ satisfies (13.5).*

Proof. See Appendix 13A. $\qquad\qquad\square$

The consequence of the theorem is that the load factors s_{ik} and λ_j provide a parameterization of all the Pareto optimal power vectors. Thus, the power vectors obtained for any s_{ik} and λ_j are guaranteed to be Pareto optimal. Moreover, by appropriately selecting the parameters, any optimal point can be achieved.

However, it is important to recognize some caveats. Most importantly, as mentioned above, the implicit assumption in the utility function is that the power selection in one short-range link does not affect other short-range links. This assumption is valid when the interference at the short-range links is dominated by either thermal noise or interference from the macro transmitters. When there is significant interference between short-range links, the problem is significantly more complex. In fact, when $K > 1$, the problem is similar to the parallel channel interference power control problem that arises in digital subscriber line (DSL) cross-talk [43]. This problem is generally non-convex and NP-hard. The load spillage method can be extended to provide a suboptimal solution to the problem. For example, following the lines of [44], the short-range receivers could themselves broadcast load factors and have their interference counted in the spillage term (13.3). This procedure would at best converge to some local maxima.

Also, while the theorem states that, in theory, any Pareto optimal power vector is achievable, it provides no mechanism for selecting the appropriate load factors s_{ik} or bounds λ_j. In the simulations below, the macro receivers will select the load factors s_{ik} simply based on measuring the noise rise and adjusting the load factors to keep the noise rise at a given target. However, selection of the bounds λ_j or optimal selection of the load factors s_{ik} would generally require some femto–macro communication, which we are not considering.

There are indeed significant possibilities for more sophisticated interference coordination here. However, we will not consider such methods in this chapter, as our main objective here is to show that even simple power control methods that use minimal coordination can provide good performance.

13.4   Interference cancellation and joint detection

13.4.1   Beyond reuse 1 and orthogonalization

The above subband partitioning scheme creates two types of subbands: reuse 1 subbands for femto links with weak interference with the macro, and reuse 3 subbands for links with strong interference. Femto transmitters optimally allocate their power to maximally combine the subbands depending on their actual interference level. In the context of the discussion in Section 13.2, the strategy can thus be seen as a way to combine the two basic interference techniques: reuse 1 and orthogonalization.

However, as we alluded to in the discussion of Figure 13.1, reuse 1 and orthogonalization are not the only two methods for interference control. Under strong interference, further gains may be possible with *interference cancellation* [20, 45].

To understand IC, observe that in the reuse 1 method described so far, the interference at the femtocell receiver is treated as Gaussian noise. However, the interference is not

Gaussian and actually comes from some macrocell transmitter. When the interfering signal from the macrocell transmitter is sufficiently strong, the femtocell receiver can thus attempt to decode the signal prior to or jointly with the decoding of the desired signal.

For a simple two-link system, a well-known technique to exploit this type of IC is the HK method [39]. Each link splits its transmissions into a private component, decodable at the intended receiver, and a public component, which is decodable and cancelable at both receivers. The rate with the HK method is shown in Figure 13.1, where we see that increasing the interference actually increases the achievable rate and exceeds the performance from orthogonalization. The promise of such IC techniques is alluring: interference can be entirely canceled under suitable conditions.

Unfortunately, IC comes with significantly greater computational requirements at the femtocell receiver. The receiver must not only decode its desired signal from the femto transmitter, but also decode the interfering signal from the macro transmitter. Moreover, to obtain the full benefits of IC, the femto receiver should ideally decode both signals jointly, not successively. This joint detection is generally implemented with iterative decoding that adds further computational requirements. Nevertheless, improved computational resources have recently made IC implementable in practical receiver circuits [26–28] and it is thus worth considering if the strong interference in femtocell networks warrants the additional computational complexity.

13.4.2   Interference cancellation with subband partitioning

In the context of the subband partitioning method, the capabilities to perform IC can be seen as a way to enable a greater number of locations where femtocells transmission are possible. To see this, suppose that the femtocell or other short-range receiver has the computational ability to perform IC on the macrocellular signals. Now return again to Figure 13.3 and consider first the short-range links SR4 and SR5. SR4, being close to macrocell 1, will receive the DL signal on f_1 strongly. Also, the signal on that subband will typically be at low rate since it is intended for an edge of cell mobile, MUE2. Thus, with a high probability, the receiver on SR4 will be able to jointly decode and cancel the interfering signal from macrocell 1 while receiving the signal from its desired transmitter. Similarly, the receiver SR5 can perform IC with high likelihood on the DL signal from macrocell 2 on f_2.

Now for the UL, consider short-range links SR6 and SR7. These links are close to the macro UEs, MUE2 and MUE3. Since MUE2 and MUE3 are far from their serving macrocell, they will likely transmit at a low rate, and thus can be canceled by the nearby short-range receivers.

In this way, we see that when the short-range link receivers are capable of IC, with high likelihood, the interference cancellation may be able to remove interference in a number of the subbands. Moreover, IC can be performed entirely opportunistically in that the macro does not need to change its scheduler policy at all, or even be aware of the short-range links. The regular operation of the macro under the subband partitioning will naturally open up possibilities where IC can be used.

Table 13.1 Parameters for simulation of the femto–macro interference power control algorithm.

Parameter	Value
Macro topology	24 hexagonal cell sites with wraparound and 3 cells per site.
Femto layout	Uniformly distributed at density of 10 per macrocell
Number of macro UEs	10 per macrocell
Number of femto UEs	1 per femtocell
Macrocell radius	500 m
Femto link distance	20 m
Macro antenna pattern	$A(\theta) = -\min\left\{ 12\left(\frac{\theta}{\theta_{3dB}}\right)^2, A_m \right\}$, $\theta_{3dB} = 70$ degrees and $A_m = 20$ dB
Femto antenna pattern	Omni
Lognormal shadowing (macro BS involved)	8.1 dB, 50% inter-site correlation; 100% intra-site correlation
Lognormal shadowing (macro BS not involved)	4 dB, no correlation

13.5 Numerical simulation

The method was evaluated under a simulation model similar to the Femto Forum interference management study in [10] with parameters shown in Table 13.1. The macro network is composed of a 24 hexagonal cell sites arranged in a 4×6 grid with wraparound to eliminate edge effects. Each site has three cells with the antenna pattern $A(\theta)$ given in Table 13.1. The bandwidth is partitioned into reuse 1 and reuse 3 subbands as described in Section 13.3, with a fraction $p_1 = 0.5$ of the bandwidth allocated to the reuse 1 subband.

Macro UEs are distributed uniformly with an average of ten macro UEs per macrocell. We assume that the macro UEs connect to the strongest macrocell. In each cell, the fraction p_1 of the macro UEs with the highest SINR are assigned to the reuse 1 subband, and the remaining fraction of $1 - p_1$ macro UEs are assigned to the reuse 3 subband. We assume that the macro network uses OFDMA and the macro UEs within each subband are assigned equal, orthogonal fractions of the subband bandwidth. In the DL, we assume each macro BS transmits equal power density to all the macro UEs. In the UL, we assume power control at each cell sets the macro UEs served in that cell to have equal receive power density. Moreover, we assume that the receive power densities are adjusted so that all cells experience an interference to thermal (IoT) of 10 dB. In both the UL and DL, the rate is then selected based on the received SINR. We assume that the links achieve a rate equal to 3 dB below Shannon capacity, with a maximum spectral efficiency of 5 bps/Hz.

The underlay of femtocell BSs are randomly placed with a uniform distribution at a density of ten femto BSs per macrocell. Each femto BS serves one femto UE at a random location 20 m from the femtocell BS. Note that the femtocell links are considerably smaller than the macrocell radius of 500 m.

Table 13.2 Path-loss models based on [10]. L_{ow} is the outer wall loss and R is the distance in meters.

Link	Path-loss (dB)
MBS ↔ MUE	$15.3 + 37.6\log_{10}(R)$
MBS ↔ FUE or FBS	$15.3 + 37.6\log_{10}(R) + L_{ow}, L_{ow} = 10$
FBS ↔ FUE (serving link)	$38.46 + 20\log_{10}(R) + 0.7R$
FBS or FUE ↔ FUE or FBS in different femtocell	$15.3 + 37.6\log_{10}(R) + L_{ow}, L_{ow} = 20$
FBS or FUE ↔ MUE	$15.3 + 37.6\log_{10}(R) + L_{ow}, L_{ow} = 10$

The path-loss models shown in Table 13.2 are also based on [10]. The wall loss values are the lower values in [10] under the assumption that all the macro UEs are outside, each femtocell (both the BS and UE) is inside a house, with different femtocells in different houses. We have chosen the lower wall loss values, since higher values would result in a more favorable case for the proposed algorithm as the separation between the femto and macro would be greater.

Under these assumptions, we generated a random "drop" of the femto and macro network elements and then ran two simulations of the proposed femto–macro interference control algorithm in Section 13.3.3: one simulation for the DL and a second for the UL. In each simulation, the load factors in the load-spillage algorithm were set in an iterative manner as follows.

In the DL, the "victims" of the femto interference are the macro UEs. Each macro UE begins by broadcasting some initial load factors across the subbands. The femto BS transmitters compute the corresponding spillage terms and then adjust their power across the subbands to maximize the rate to the femto UE. The macro UEs then compute the resulting interference and compare the interference to a target level. In this simulation, the target level was set to the interference that would degrade the rate of macro UE by no more than 10%. The load factors are then increased or decreased depending on whether the measured interference is above or below the target. The algorithm was run for 50 iterations at which point we observed that more than 99% of the macro UE experienced an interference level below 0.1 dB of the target.

Similarly, in the UL, the victims of the femto interference are the macro BS receivers. In this case, the load factors were adjusted to bound the interference rise from the femtos to 0.5 dB. This noise rise corresponds to an approximately 10% loss in rate in the power-limited regime.

The distribution of the DL and UL rates of the femto links after the final iteration of the load-spillage algorithm are shown in Figure 13.4. The curves labeled "subband" are the cumulative distribution functions (CDFs) without IC, while "subband + JD" includes IC via joint detection of the macro interfering signal and desired femto signal. As a link-layer model for joint detection, we assumed the rate achievable via joint detection matched the Shannon rate with a 3 dB loss in both links and maximum spectral efficiency of 5 bps/Hz.

The curves show that the load-spillage algorithm is able to deliver a large rate to a substantial fraction of the femto links. For example, in the DL, the femto links achieve a median spectral efficiency of approximately 2.6 bps/Hz without IC. Moreover, the loss in

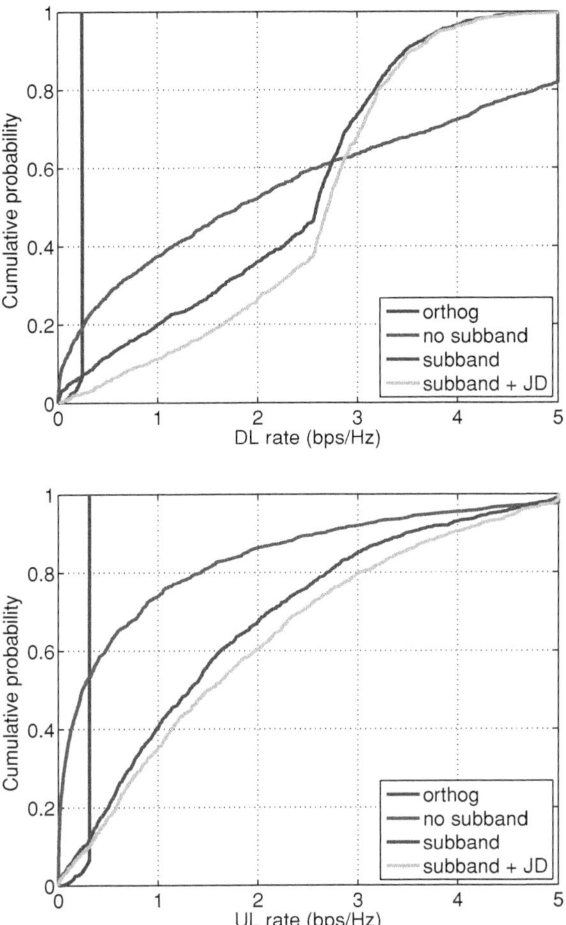

Figure 13.4 Distribution of rates available on short-range links using subband scheduling with interference cancellation ("subband"), and subband scheduling with interference cancellation via joint detection of the desired femto signal and interfering macro signal ("subband + JD"). Also shown are the rate distributions with simple orthogonalization ("orthog") and reuse 1 without subband partitioning ("no subband"). Top panel shows the DL and bottom is the UL. See text for simulation assumptions.

macro rate due to femto interference is small. Although the target interference level was set to a maximum 10% reduction in rate in the macro, many macro UEs experience an interference well below the target. Although not shown in the figure, the actual decrease in the macro rate observed in the simulation was 4.9%.

Adding IC via joint detection can improve the rate further, especially for low rate mobiles. However, some caveats are in order: most importantly, the gains with IC are modest. Also, our simulation assumed joint detection. In simulations of successive interference cancellation (SIC) (results are not shown), SIC alone produced only minimal gains. This is to be expected since SIC usually only has benefits for very strong interference. Since joint detection is computationally much harder than SIC, the practical value of IC may be limited.

Moving to the UL results in Figure 13.4, we see that the femto links were able to achieve median spectral efficiencies of approximately 1.4 bps/Hz without IC and 1.5 bps/Hz with IC. The loss in the macro rate was somewhat higher at 6.2%, but still small. The gains with IC in this case are minimal. This may be a result of the fact that we assumed a rate fair scheduler per cell in the uplink so no macro links come at very low rate. Other macro scheduling policies may show further improvements for IC, but this needs further study.

We conclude that subband scheduling in the macro combined with the load-spillage power control in the femto can enable a large rate for many femto links with minimal effect in the macro. In addition, modest gains are possible in the DL with IC, especially at low rates.

Figure 13.4 also shows a comparison of the proposed method to simple orthogonalization. In orthogonalization, the femtos and macros operate on different frequencies. Now, the proposed method resulted in a 4.9% drop in the average macro DL spectral efficiency and a 6.2% drop in the macro UL. To compare the proposed method to orthogonalization, we can subtract that fraction of the bandwidth from the macro and run the femtos with no macro interference in that bandwidth. The resulting CDF of the rates are shown in Figure 13.4 on the curve labeled "orthog." With orthogonalization from the macro, many of the femto links get the maximum rate of 5 bps/Hz times the bandwidth fraction. However, that rate is much smaller for the overwhelming majority of links.

Figure 13.4 also shows the rate distribution with reuse 1 and power control without subband partitioning. This can be simulated with the above framework, but with $K = 1$ subband. In the UL, we see that the proposed method offers a significantly greater rate. In the DL, the median rate is approximately the same, although the proposed method is more fair, in that the rate variation is less.

13.6   Conclusions

We have argued that femto–macro interference coordination differs from interference coordination in traditional macrocellular networks in two key ways: the presence of strong interference and a greater need for distributed control. An information theoretic argument based on the Gaussian interference channel suggests that, due to strong interference, traditional interference control methods that have been successful in CDMA macrocellular deployments may perform poorly in the femtocell context. It is therefore worthwhile to re-examine the basic paradigm of CDMA interference control, namely universal frequency reuse, treating interference as noise, and managing that interference via simple power and rate control.

In this vein, we have considered subband partitioning methods that can capture the benefits of both universal frequency reuse and orthogonal systems. While it is widely known that subband partitioning can be beneficial in the macrocell network, we have argued here that subband partitioning can also offer improved opportunities for femtocells or other short-range links to transmit with minimal disruption to the macro-cell receivers. To realize these opportunities, the femto transmitters must estimate the path losses to the macro receivers and then perform some optimal power allocation across the subbands based on the path-loss estimates. The estimation and allocation can be performed in an entirely distributed manner, with minimal coordination with the macrocell receivers. The only requirement is that the macro receivers broadcast some signal so that their presence and loading is known.

In the simulation scenario considered here, the proposed method offers significantly greater rates than simple orthogonalization between the femto and macros. The proposed method also appears to outperform reuse 1 without subband partitioning in the UL. In the DL, the proposed method has a similar median rate, but shows less rate variations. Further gains are also possible with interference cancellation in the DL, but this requires joint detection and not SIC alone. The gains in the UL for IC (even with joint detection) are small.

Further simulations are, however, needed. Most importantly, the current simulations have assumed perfect knowledge of the path losses and full buffer traffic without any dynamics. We have also not explicitly modeled the overhead for the signals to estimate the path losses.

Appendix 13A Proof of Theorem 13.1

To prove (a), fix the load factors $s_{ik} > 0$ and power constraint bounds $\lambda_j > 0$ and suppose that the powers $\mathbf{p}_j$ satisfy (13.5). To show that the set of $\mathbf{p}_j$'s is Pareto optimal, suppose that there is an alternative set of power vectors $\mathbf{p}_j^1$ with $U(\mathbf{p}_j^1) \geq U(\mathbf{p}_j)$ for all j with strict inequality for at least one j. Let q_{ik}^1 be the corresponding interference levels as in (13.1). We must show that $q_{ik}^1 > q_{ik}$ for some i and k.

First observe that since $U_j(\mathbf{p}_j)$ is strictly increasing in each p_{jk}, $U_j(\mathbf{p}_j^1) \geq U(\mathbf{p}_j)$ and $\mathbf{p}_j$ is the maxima of (13.5) subject to (13.4), we must have that for all j,

$$\sum_{k=1}^{K} p_{jk}^1 r_{jk} \geq \lambda_j + \epsilon_j, \quad \sum_{k=1}^{K} p_{jk} r_{jk} \leq \lambda_j$$

where $\epsilon_j \geq 0$ for all j with $\epsilon_j > 0$ for at least one j. Therefore,

$$\sum_{k=1}^{K} p_{jk}^1 r_{jk} \geq \sum_{k=1}^{K} p_{jk} r_{jk} + \epsilon_j$$

$$\overset{(a)}{\Leftrightarrow} \sum_{k=1}^{K} \sum_{i=1}^{M(k)} (p_{jk}^1 - p_{jk}) s_{ik} G_{ijk} \geq \epsilon_j$$

$$\overset{(b)}{\Rightarrow} \sum_{j=1}^{N} \sum_{k=1}^{K} \sum_{i=1}^{M(k)} (p_{jk}^1 - p_{jk}) s_{ik} G_{ijk} > 0$$

$$\overset{(c)}{\Rightarrow} \sum_{k=1}^{K} \sum_{i=1}^{M(k)} s_{ik} (q_{ik}^1 - q_{ik}) > 0$$

where (a) follows from (13.3); (b) follows from taking the sum over j and using the fact that $\epsilon_j \geq 0$ with strict inequality for at least one j; and (c) follows from the definition of q_{ik} in (13.1). Since $s_{ik} > 0$, we must have that $q_{ik}^1 > q_{ik}$ for at least one (i, k). This proves part (a) of Theorem 13.1.

To prove the converse part (b), we need the following standard result in linear algebra.

Lemma 13.1 *Suppose $\mathbf{A} \in \mathbb{R}^{m \times n}$ is a matrix such that for any $\mathbf{v} \in \mathbb{R}^n$, there exists some component $i = 1, \ldots, m$ such that $(\mathbf{A}\mathbf{v})_i \leq 0$. Then, there exist a $\mathbf{w} \in \mathbb{R}^m$ with $\mathbf{w}^T \mathbf{A} = 0$, $w_i \geq 0$ for all i and $\sum_i w_i = 1$.*

Proof. Let $\mathbf{W}$ be the simplex,

$$\mathbf{W} := \left\{ \mathbf{w} \in \mathbb{R}^m \ : \ w_i \geq 0, \ \sum_i w_i = 1 \right\}.$$

The hypothesis of the lemma implies that for every $\mathbf{v} \in \mathbb{R}^n$, there exists a $\mathbf{w} \in \mathbf{W}$ such that $\mathbf{w}^T \mathbf{A} \mathbf{v} \leq 0$. Therefore,

$$\max_{\mathbf{v} \in \mathbb{R}} \min_{\mathbf{w} \in \mathbf{W}} \mathbf{w}^T \mathbf{A} \mathbf{v} = 0.$$

Since this optimization is convex in $\mathbf{w}$ and concave in $\mathbf{v}$, there is a Nash equilibrium [46], so we can interchange the min and max to obtain

$$\min_{\mathbf{w} \in \mathbf{W}} \max_{\mathbf{v} \in \mathbb{R}} \mathbf{w}^T \mathbf{A} \mathbf{v} = 0.$$

But this can occur only if $\mathbf{w}^T \mathbf{A} = 0$. $\square$

We can use this result to prove the converse in part (b) as follows. Suppose $\mathbf{p}_j$, $j = 1, \ldots, N$ is a set of Pareto optimal powers. Since they are Pareto optimal, any small change in the powers must result in an increase in one of the interference levels q_{ik} or decrease in the utilities U_j. That is, for any non-zero set of Δp_{jk}'s, there must either be some i and k such that

$$\Delta q_{ik} := \sum_j G_{ijk} \Delta p_{jk} \geq 0, \tag{13.6}$$

or j such that

$$\Delta U_j := \frac{\partial U_j(\mathbf{p}_j)}{\partial p_{jk}} \Delta p_{jk} \leq 0. \tag{13.7}$$

Define the matrix $\mathbf{A}$ by

$$\mathbf{A} = \begin{bmatrix} -DU(\mathbf{p}) \\ \mathbf{G} \end{bmatrix},$$

where the columns are indexed by the pairs (j, k), $DU(\mathbf{p})$ is the matrix with the components $DU(\mathbf{p})_{j,jk} = \partial U_j(\mathbf{p}_j)/\partial p_{jk}$ and $\mathbf{G}$ is the matrix of gains $G_{i,jk}$. Now, (13.6) and (13.7) show that for every $\Delta\mathbf{p}$, $(\mathbf{A}\Delta\mathbf{p})_\ell \leq 0$ for some index ℓ. Hence, by Lemma 13.1, there must exist a $\mathbf{w} \geq 0$ with unit norm such that $\mathbf{w}^T \mathbf{A} = 0$. Partitioning

$$\mathbf{w} = [\mathbf{t}^T \ \mathbf{s}^T]^T$$

conformably with $\mathbf{A}$ we see that there exists a set of non-negative t_j and s_{ik} such that

$$t_j \frac{\partial U_j(\mathbf{p}_j)}{\partial p_{jk}} - \sum_{i=1}^{M(k)} s_{ik} G_{ijk} = 0. \tag{13.8}$$

Now $t_j \geq 0$ for all j and $s_{ik} \geq 0$ for all i and k, with at least one of the parameters being strictly greatly than zero. Since $G_{ijk} > 0$ and $\partial U_j / \partial p_{jk} > 0$ for all i, j, and k, it can be verified that (13.8) implies that that $t_j > 0$ for all j.

Now, using (13.3), (13.8) can be rewritten as

$$\frac{\partial L_j(\mathbf{p}_j)}{\partial p_{jk}} = 0, \tag{13.9}$$

where L_j is the Lagrangian

$$L_j(\mathbf{p}_j) := t_j U_j(\mathbf{p}_j) - \sum_{k=1}^{K} r_{jk} p_{jk}.$$

Let $\lambda_j = \sum_k r_{jk} p_{jk}$. Since $t_j > 0$, (13.9) shows that $\mathbf{p}_j$ is a critical point of the optimization (13.5) subject to (13.4). Since $U_j(\mathbf{p}_j)$ is concave, any critical point is the maxima. Thus, we have found s_{ik} and λ_j such that the set of power vectors $\mathbf{p}_j$ satisfy (13.5).

References

[1] 3GPP, "UTRAN architecture for 3G Home Node B (HNB); Stage 2," *TS 25.467 (release 9)*, 2010.

[2] ——, "Service requirements for Home NodeBs (UMTS) and eNodeBs (LTE)," *TS 22.220 (release 9)*, 2010.

[3] ——, "3G Home Node B Study Item Technical Report," *TR 25.820 (Release 9)*, 2010.

[4] S.-P. Yeh, S. Talwar, S.-C. Lee, and H. Kim, "WiMAX femtocells: a perspective on network architecture, capacity, and coverage," *IEEE Commun. Mag.*, vol. 46, no. 10, pp. 58–65, Oct. 2008.

[5] V. Chandrasekhar, J. G. Andrews, and A. Gatherer, "Femtocell networks: a survey," *IEEE Commun. Mag.*, vol. 46, no. 9, pp. 59–67, Sep. 2008.

[6] H. Claussen, "Co-channel operation of macro- and femtocells in a hierarchical cell structure," *Int. J. Wirel. Inf. Netw.*, vol. 15, no. 3–4, pp. 137–47, Dec. 2008.

[7] D. López-Pérez, A. Valcarce, G. de la Roche, and J. Zhang, "OFDMA femtocells: a roadmap on interference avoidance," *IEEE Commun. Mag.*, vol. 47, no. 9, pp. 41–8, June 2009.

[8] J. Zhang and G. de la Roche, *Femtocells: Technologies and Deployment*. John Wiley and Sons, Ltd, Jan. 2010.

[9] H. S. Jo, P. Xia, and J. G. Andrews, "Downlink femtocell networks: open or closed?" *Proc. IEEE Int. Conf. on Commun. (ICC)*, June 2011.

[10] Femto Forum, "Interference management in OFDMA femtocells," Whitepaper available at www.femtoforum.org, Mar. 2010.

[11] 3GPP, "New Work Item Proposal: Enhanced ICIC for non-CA based deployments of heterogeneous networks for LTE," RP-100372, 2010.

[12] A. Ghosh, J. G. Andrews, N. Mangalvedhe, R. Ratasuk, B. Mondal, M. Cudak, E. Visotsky, T. A. Thomas, P. Xia, H. S. Jo, H. S. Dhillon, and T. D. Novlan, "Heterogeneous cellular networks: from theory to practice," *IEEE Commun. Mag.*, vol. 50, no. 2, pp. 54–64, June 2012.

[13] V. Chandrasekhar, J. G. Andrews, T. Muharemovic, Z. Shen, and A. Gatherer, "Power control in two-tier femtocell networks," *IEEE Trans. Wireless Commun.*, vol. 8, no. 8, pp. 4316–28, Aug. 2009.

[14] M. Yavuz, F. Meshkati, S. Nanda, A. Pokhariyal, N. Johnson, B. Raghothaman, and A. Richardson, "Interference management and performance analysis of UMTS/HSPA+ femtocells," *IEEE Commun. Mag.*, vol. 47, no. 9, pp. 102–9, Sep. 2009.

[15] H.-S. Jo, C. Mun, J. Moon, and J.-G. Yook, "Interference mitigation using uplink power control for two-tier femtocell networks," *IEEE Trans. Wireless Commun.*, vol. 8, no. 10, pp. 4906–10, Oct. 2009.

[16] H. Holma and A. Toskala, *HSDPA / HSUPA for UMTS*. New York: John Wiley & Sons, 2006.

[17] E. Amaldi, A. Capone, and F. Malucelli, "Planning UMTS base station location: optimization models with power control and algorithms," *IEEE Trans. Wireless Commun.*, vol. 2, no. 5, pp. 939–52, Sep. 2003.

[18] A. Sampath, P. Sarath Kumar, and J. Holtzman, "Power control and resource management for a multimedia CDMA wireless system," in *Proc. IEEE Int. Symp. Personal, Indoor, Mobile Radio Commun. (PIMRC)*, July 1995, pp. 194–202.

[19] M. Chiang, P. Hande, T. Lan, and C. W. Tan, "Power control in wireless cellular networks," *Found. Trends Network.*, vol. 2, no. 4, July 2008.

[20] T. Cover and J. A. Thomas, *Elements of Information Theory*, 2nd edn. John Wiley & Sons, Inc., 2006.

[21] Huawei, "Soft frequency reuse scheme for UTRAN LTE," 3GPP R1-050507, May 2005.

[22] H. Lei, L. Zhang, X. Zhang, and D. Yang, "A novel multi-cell OFDMA system structure using fractional frequency reuse," in *Proc. IEEE Int. Symp. Personal, Indoor, Mobile Radio Commun. (PIMRC)*, Athens, Greece, Sep. 2007.

[23] S. Han, J. Park, T.-J. Lee, H.G.Ahn, and K. Jang, "A new frequency partitioning and allocation of subcarriers for fractional frequency reuse in mobile communication systems," *IEICE Trans. Comm.*, no. 8, p. 2748–51, Aug. 2008.

[24] A. Stolyar and H. Viswanathan, "Self-organizing dynamic fractional frequency reuse in OFDMA systems," in *Proc. IEEE INFOCOM*, Phoenix, AZ, Apr. 2009, pp. 691–98.

[25] S. Rangan, "Femto–macro cellular interference control with subband scheduling and interference cancelation," in *Proc. IEEE Global Telecommun. Conf. (GLOBECOM) Workshops*, Dec. 2010.

[26] J. Andrews, "Interference cancellation for cellular systems: a contemporary overview," *IEEE Wirel. Commun.*, vol. 12, no. 2, pp. 19–29, Apr. 2005.

[27] G. Boudreau, J. Panicker, N. Guo, R. Chang, N. Wang, and S. Vrzic, "Interference coordination and cancellation for 4G networks," *IEEE Commun. Mag.*, vol. 47, no. 4, pp. 74–81, Apr. 2009.

[28] Z. Shi and M. C. Reed, "Iterative maximal ratio combining channel estimation for multiuser detection on a time frequency selective wireless CDMA channel," in *Proc. IEEE Wireless Commun. Networking Conf. (WCNC)*, Hong Kong, Mar. 2007.

[29] T. S. Rappaport, *Wireless Communications: Principles and Practice*, 2nd edn. Prentice Hall, Dec. 2001.

[30] A. J. Viterbi, *CDMA: Principles of Spread Spectrum Communication*. Upper Saddle River, NJ: Prentice Hall, 1995.

[31] R. Lupas and S. Verdú, "Near–far resistance of multiuser detectors in asynchronous channels," *IEEE Trans. Commun.*, vol. 38, no. 4, pp. 496–508, Apr. 1990.

[32] S. Kishore, L. J. Greenstein, H. V. Poor, and S. C. Schwartz, "Capacity in a CDMA macrocell with a hotspot microcell: exact and approximate analyses," in *Proc. IEEE Vehicular Tech. Conf. (VTC)*, Atlantic City, NJ, Oct. 2001, pp. 1172–6.

[33] ——, "Soft handoff and uplink capacity in a two-tier CDMA system," *IEEE Trans. Wireless Commun.*, vol. 4, no. 4, pp. 1297–301, July 2005.

[34] V. Chandrasekhar and J. G. Andrews, "Uplink capacity and interference avoidance for two-tier femtocell networks," *IEEE Trans. Wireless Commun.*, vol. 8, no. 7, pp. 3498–509, July 2009.

[35] ——, "Spectrum allocation in two-tier networks," *IEEE Trans. Commun.*, vol. 57, no. 10, pp. 3059–68, Oct. 2009.

[36] S. Annapureddy and V. Veeravalli, "Sum capacity of the Gaussian interference channel in the low interference regime," in *Proc. Inform. Theory and Appl. Workshop (ITA)*, San Diego, CA, Feb. 2008.

[37] X. Shang, G. Kramer, and B. Chen, "A new outer bound and noisy-interference sum-rate capacity for the Gaussian interference channels," *IEEE Trans. Inform. Theory*, vol. 55, no. 2, pp. 689–99, Feb. 2009.

[38] A. S. Motahari and A. K. Khandani, "Capacity bounds for the Gaussian interference channel," *IEEE Trans. Inform. Theory*, vol. 55, no. 2, pp. 620–43, Feb. 2009.

[39] T. Han and K. Kobayashi, "A new achievable rate region for the interference channel," *IEEE Trans. Inform. Theory*, vol. 27, no. 1, pp. 49–60, Jan. 1981.

[40] E. Dahlman, S. Parkvall, J. Sköld, and P. Beming, *3G Evolution: HSPA and LTE for Mobile Broadband*, 2nd edn. Academic Press, Oct. 2008.

[41] C. Y. Oh, M. Y. Chung, H. Choo, and T.-J. Lee, "A novel frequency planning for femtocells in OFDMA-based cellular networks using fractional frequency reuse," in *Computational Science and Its Applications ICCSA 2010*, Fukuoka, Japan, Mar. 2010, pp. 96–106.

[42] P. Hande, S. Rangan, M. Chiang, and X. Wu, "Distributed uplink power control for optimal SIR assignment in cellular data networks," *IEEE/ACM Trans. Network.*, vol. 16, no. 6, pp. 1420–33, Dec. 2008.

[43] W. Yu, G. Ginis, and J. Cioffi, "Distributed multiuser power control for digital subscriber lines," *IEEE J. Sel. Areas Commun. (JSAC)*, vol. 20, no. 5, p. 1105–15, June 2002.

[44] R. Cendrillon, J. Huang, M. Chiang, and M. Moonen, "Autonomous spectrum balancing for digital subscriber lines," *IEEE Trans. Signal Process.*, vol. 55, no. 8, pp. 4241–57, Aug. 2007.

[45] R. Ahlswede, "Multi-way communication channels," in *Proc. IEEE Int. Symp. Inform. Theory (ISIT)*, Armenian S.S.R., Sep. 1971, pp. 23–52.

[46] S. P. Boyd, L. El Ghaoui, E. Feron, and V. Balakrishnan, *Linear Matrix Inequalities in System and Control Theory*. Philadelphia, PA: SIAM, 1994.

14 Femtocell interference control in standardization

Zubin Bharucha and Gunther Auer

User-deployed femtocells, each exclusively serving a set of registered users and sharing the same frequency spectrum as the overlay macrocells, are already defined in 3rd Generation Partnership Project (3GPP) specifications. Such a co-channel and random deployment of femtocells can cause heavy downlink (DL) control and data channel interference especially to mobile user equipment (MUE) in the vicinity of one or more femtocells and not belonging to their closed subscriber groups (CSGs). This chapter is dedicated to addressing this issue, termed as *inter-cell interference coordination (ICIC)*, paying particular attention to the control channel. In systems that employ full frequency reuse, such as long term evolution (LTE), the issue of inter-cell interference (ICI) is a very serious one and can severely compromise cell-edge performance. The situation is further exacerbated in systems with femtocells randomly distributed within the underlying macrocellular network. In such a system, ICI is not only experienced by MUEs at the edge of macrocells, but can also be experienced by those MUEs in the vicinity of one or more femtocells, whose CSGs they are not members of. While scheduling strategies do not come under the purview of LTE standardization, LTE does provide standardized signaling methods so that an appropriate signaling strategy may be employed to avoid excessive ICI for the data channels. However, these signaling methods are developed to be exchanged between macro base stations (BSs) over the X2 interface. It is expected that LTE femtocells will not have access to such an interface. Furthermore, unlike the data channel, the various control channels cannot be conveniently relocated in order to avoid interference. Therefore, either new signaling techniques are required to be defined for future LTE releases or methods of combating femto-to-macro control channel interference that do not rely on signaling should be investigated. In this chapter, we dwell on the latter. This is done because by not relying on new signaling techniques, backward compatibility can be ensured such that legacy LTE user equipment (UE) is still able to function in future LTE releases.

This chapter is organized as follows. Section 14.1 motivates the need for ICIC, especially in a heterogeneous network. Section 14.2 introduces ICIC in LTE and the existing signaling methods. Since this chapter deals with the LTE control channel, the various control channels are introduced and the methods by which they are mapped to

Small Cell Networks: Deployment, PHY Techniques, and Resource Management, ed. Tony Q. S. Quek, Guillaume de la Roche, İsmail Güvenç, and Marios Kountouris. Published by Cambridge University Press. © Cambridge University Press 2013.

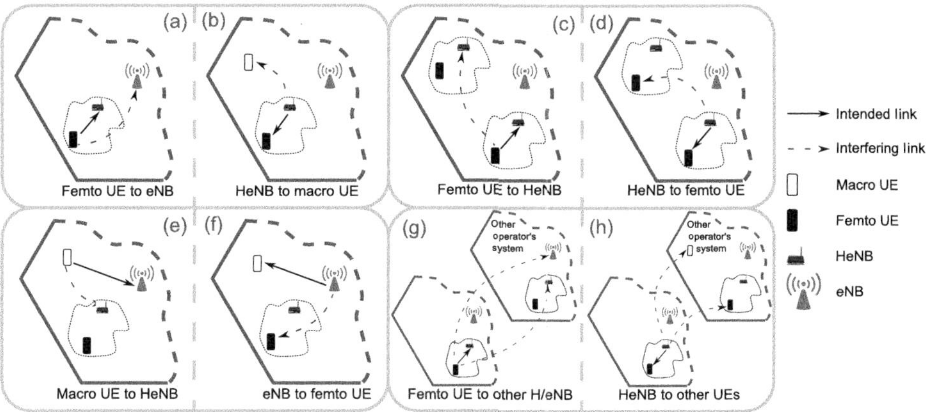

Figure 14.1 The various interference scenarios that arise through femtocell deployment.

resources in the frequency domain are described in Sections 14.3 and 14.4. Methods for femto-to-macro ICIC are detailed in Section 14.5. Section 14.6 contains the description of the system setup for simulations and the performance evaluation for state-of-the-art ICIC methods is captured in Section 14.7. Section 14.8 discusses the particularly poor performance of one of the control channels and proposes a solution. Finally, Section 14.9 contains the conclusions.

14.1 Motivation

Femtocells promise substantial gains in system spectral efficiency due to an enhanced reuse of radio resources [1–4]. They may be employed in a co-channel fashion such that the bandwidth used by the macrocell is reused by the femtocell. Femtocells maintain an exclusive list of UEs that they serve. This is known as the CSG configuration such that only authorized UEs are members of the CSG of that femtocell. Macro UEs are typically distributed randomly in a cellular network. Understandably, this leads to some of them being located in the vicinity of active femtocells. Unchecked network operation could wreak havoc on the performance of such macro UEs in the downlink and on the femtocells in the uplink. Figure 14.1 shows the various interference scenarios that can arise through uncoordinated femtocell deployment. The scenarios are paired according to the source of interference. Figures 14.1(a) and (b) depict interference originating from the femto layer and affecting the macro layer. Such interference scenarios are henceforth termed as femto-to-macro interference. Figure 14.1(a) shows uplink interference originating from the femto UE and affecting the macro NodeB (NB) located nearby. On the other hand, Figure 14.1(b) shows the DL interference originating from the Home NodeB (HNB), being experienced by a nearby macro UE that is not a member of the CSG of that HNB. Scenarios (c) and (d) depict inter-femtocell interference such that one femtocell interferes with another one nearby. Such interference scenarios are classified as femto-to-femto interference. Macro-to-femto interference scenarios are depicted in Figures 14.1(e) and (f). Finally, inter-operator interference scenarios are depicted in Figures 14.1(g) and (h), where the femtocell of one operator inflicts interference on the (macro and femto) network of another operator.

In a hierarchical (or heterogeneous) network where low-power nodes are embedded within a network served primarily by high-power nodes, priority is usually assumed to lie with the high-power nodes. In other words, every attempt must be made to ensure that the operation of the high-power nodes is not significantly disrupted by the introduction of the low-power nodes. Clearly, the macro network, consisting of high-power NBs, holds priority over the embedded low-power femtocells. Therefore, in this light, the scenarios shown in Figures 14.1(a) and (b) are more critical than the scenarios shown in (c) to (f), where the femto layer is the victim. In particular, since this chapter deals with DL femto-to-macro interference, the focus of attention is on the interference scenario portrayed in Figure 14.1(b). The rest of the depicted interference scenarios falls out of the scope of this chapter.

Much research has recently been devoted to the issue of interference in co-channel femto-to-macro DL interference [5–9]. These studies have consistently shown that a coverage-hole exists when co-channel closed-access femtocells are deployed in a macro-cell overlay network. Many of these studies propose that open access (OA) femtocells help alleviate this problem. However, in recent 3GPP discussions, it has been decided that femtocells solely operate in the CSG mode. The open-access variety of low-power nodes is known in LTE as picocells. These do not fall within the reach of this chapter. Any UE that is not a member of the CSG must attempt to maintain communication with its serving macro NB despite being in close proximity to the HNB. A rigorous study must therefore be conducted to assess the impact of femtocell deployment on co-channel macrocell UEs. Furthermore, a recent 3GPP study item [10] urges the investigation of interference management techniques that are backward compatible with LTE UEs for co-channel deployments of femtocells.

This chapter addresses an issue that has so far not been sufficiently discussed, i.e., the performance of the DL control channel for macro UEs in the presence of femtocell deployment. Femtocell research at this point is mature enough to allow for a detailed and rigorous performance assessment of the various elements that comprise the air interface. Emphasis is given to the control channel performance for macro UEs that are trapped in the coverage of one or more femtocells, i.e., it is assumed that femtocell operation must not hamper the operation of MUEs. The control channel is especially important because if it cannot be correctly decoded, the ensuing data is unintelligible. Based on this, certain contemporary LTE backward-compliant interference management techniques are introduced and assessed. It is shown that they help improve the performance of some but not all control channels for the vulnerable MUEs. A novel (and also backward compatible) technique, which improves the performance of the most vital control channel, is then introduced.

14.2　General introduction to LTE-A eICIC

Long term evolution systems are characterized by full frequency reuse. In other words, every LTE cell utilizes all the available frequency resources. This configuration, while maximizing the use of the available resources, is also responsible for injecting additional interference into the system. Clearly, with all cells using all resources simultaneously, the system performance is limited by the so-called ICI, where the interference experienced on a particular resource by a user in one cell arises from a neighboring cell also transmitting on the same resource. Long term evolution provides mechanisms for notifying neighboring cells of the possibility of high interference. However, the scheduling strategies used to avoid or mitigate ICI for data transmissions fall outside the scope of standardization. Thus, operators of LTE networks have a significant amount of freedom to implement their own interference avoidance techniques when it comes to resource allocation for the data channel.

Simple capacity calculations considering the effects of ICI show that when the interfering signal is comparable in magnitude to the desired signal, for a signal to noise ratio (SNR) of 0 dB, the capacity loss experienced by a user is around 40% [11]. A user at the edge of two cells (known henceforth as a cell-edge user) is likely to experience desired and interfering signals at similar magnitudes. Such a user will clearly experience an unacceptable loss of performance. This situation, of course, is true for a non-hierarchical or homogeneous network where only one type of node exists. However, in a heterogeneous network, the location of users experiencing equal interfering and desired signals is no longer limited to the edge of the macrocell. Obviously, in this case, a macro user may find itself very close to a low-power node to which it is not allowed access. In this case, again, the user might experience an interfering signal having power similar to or even greater than that of the desired signal. This not only motivates the need for some form of ICIC, but also shows how complicated it can be for heterogeneous networks.

Another simple assessment, as detailed in [12], shows that for a simplified two-cell scenario, with each cell containing one user, the optimal transmit power that maximizes the achievable capacity varies depending on the location of the users. If both users are located centrally within their respective cells, the capacity is maximized when both NBs transmit with the maximum possible power. On the other hand, when both users are located at the edges of their respective cells, capacity is maximized if only one of the NBs transmits, while the other one interrupts its transmission. Generalizing this, for such a setup with two cells, the research goes on to show that the optimum transmission power configuration in order to maximize capacity follows a binary law, i.e., either both NBs transmit with the maximum power simultaneously or only one of them transmits with maximum power while the other one remains silent. Naturally, the situation does not remain so simple as the network scenario becomes more complex by adding more cells. No doubt, finding the optimum power allocation for a heterogeneous network becomes even more convoluted.

The literature is replete with techniques developed to deal with ICI in homogeneous networks. One well-known technique is known as *partial/fractional frequency reuse* (FFR) [13]. Here, each cell in the network is divided into two regions – an inner and an outer region. In the inner region, ICI does not play a role significant enough to adversely affect the performance of UEs. Therefore, all cells employ full frequency reuse in the inner region. In contrast, in the outer region, since ICI is the cause for significant performance degradation, the scheduling is done such that resource allocations for cell-edge users are orthogonalized. Adopting this procedure for heterogeneous networks is not

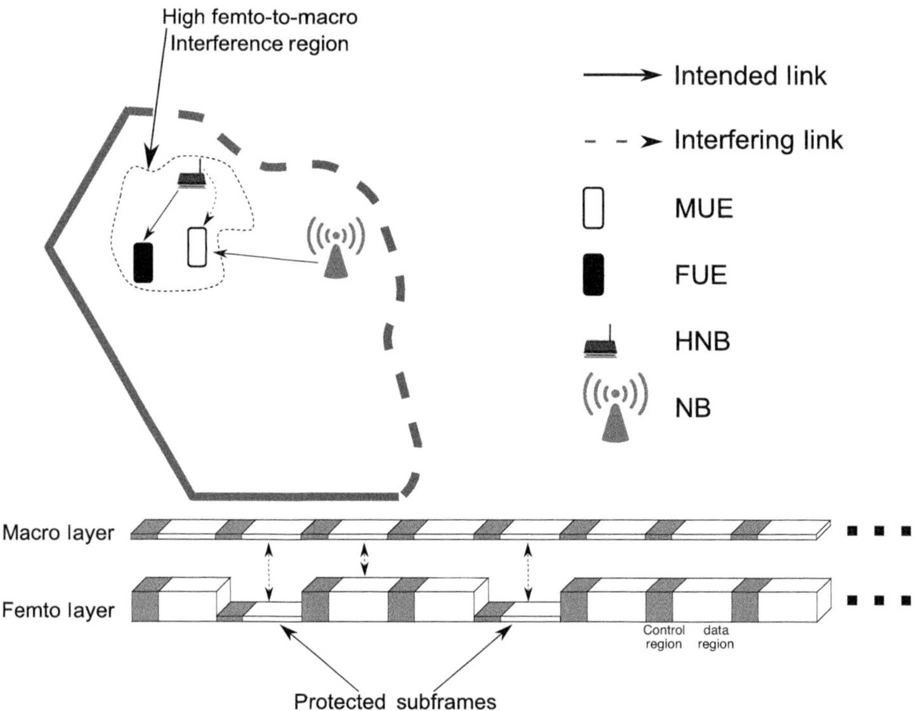

Figure 14.2 Femto-to-macro interference mitigation via power restrictions on the femto layer.

straightforward since femtocells are distributed randomly in the network. Furthermore, as is shown in Sections 14.3 and 14.5, such partitioning approaches are simply not possible for LTE control channels. Since the control channels span the entire system bandwidth, the possibilities for ICIC with respect to control channels are severely limited.

LTE-advanced (LTE-A) allows for carrier aggregation (CA), which is the scalable expansion of effective bandwidth delivered to a UE through concurrent utilization of radio resources across multiple carriers [14]. The femto-to-macro interference issue can be approached using CA such that the femto layer is forbidden from using certain carriers in order to leave them free of femto interference. However, this results in an inefficient usage of resources. Carrier aggregation will only be made available in LTE-A and beyond. Since this chapter deals exclusively with LTE, it is assumed that both the macro and femto layers use all available resources without selectively choosing carriers. Therefore, using CA to avoid femto-to-macro interference violates backward-compatibility and is beyond the scope of this chapter.

One possible method to alleviate DL femto-to-macro control channel interference in the case when both femtocells and macrocells are deployed in a co-channel fashion is to restrict the power of the femtocell transmissions in some *protected* subframes [15]. This restriction applies to the control as well as data regions of the subframe, thereby reducing interference in the entire subframe. This is depicted in Figure 14.2. It is expected that the macrocell will schedule MUEs experiencing high downlink interference from nearby femtocells on these subframes. In order to support such a technique, the pattern of the protected subframes needs to be signaled between the victim and interfering nodes. Such signaling may be possible between the macro and pico layers. However, this is not the case if femtocells are involved since it is expected that there is no interface between femtocells and macrocells. With this in mind, Figure 14.2 shows a subframe protection pattern synchronized with the eight-subframe timing of the LTE uplink (UL) hybrid automatic repeat request (HARQ) pattern. This is done so that the impact on the UL scheduling is minimized even if no UL scheduling grants can be transmitted in the protected subframes. It must be mentioned that this technique exhibits some obvious disadvantages:

- It is clear that the interference experienced by MUEs in the close vicinity of one or more femtocells will experience a higher interference on the non-protected subframes than the protected ones. Joint channel state information (CSI) measurements performed on both types of subframes will therefore result in an inaccurate channel estimate. In order to counter this, the UE needs to be configured with different CSI measurement subsets. However, this is not possible with legacy UEs.
- Obviously, decreasing the transmit power of the femtocell reduces its range. Therefore, the femtocell sacrifices the reliability of its own control channels and achievable capacity in the data region on protected subframes.
- Due to the lack of an interface between the femto and macro layers, the configuration of protected subframes must be static. This could significantly affect the efficiency of the femtocells.

Finally, for the sake of completeness, it must be mentioned that LTE provides standardized frequency-domain signaling solutions so that they may be used to implement an effective scheduling strategy for DL data transmission. In the DL, a bitmap known as the relative narrowband transmit power (RNTP) indicator is exchanged between NBs over the X2 interface [11]. Each bit in the RNTP indicator corresponds to one resource block (RB). A value of 0 indicates to the neighboring NBs if the informing NB plans to keep the transmit power on that RB below a predefined upper limit. A typical reaction for an NB receiving an indication that a neighboring cell will transmit on a certain RB with high power would be to schedule that RB to cell-center users [16]. In 3GPP discussions, no consensus has been reached on the interface connection between NBs and HNBs. In this light, it may not be possible for the HNBs to issue RNTP indicators to NBs for data channel interference reduction. In any case, these messages pertain to the data channel only and therefore cannot be used to manage control channel interference. Furthermore, the limitations in information exchange possibilities between the femtocell and the overlay macrocell pose severe challenges to the design of effective control channel interference mitigation techniques.

14.3 General introduction to downlink LTE-A control channels

The DL of an LTE orthogonal frequency division multiple access (OFDMA)-frequency division duplex (FDD) system is considered where the system bandwidth, W, is divided

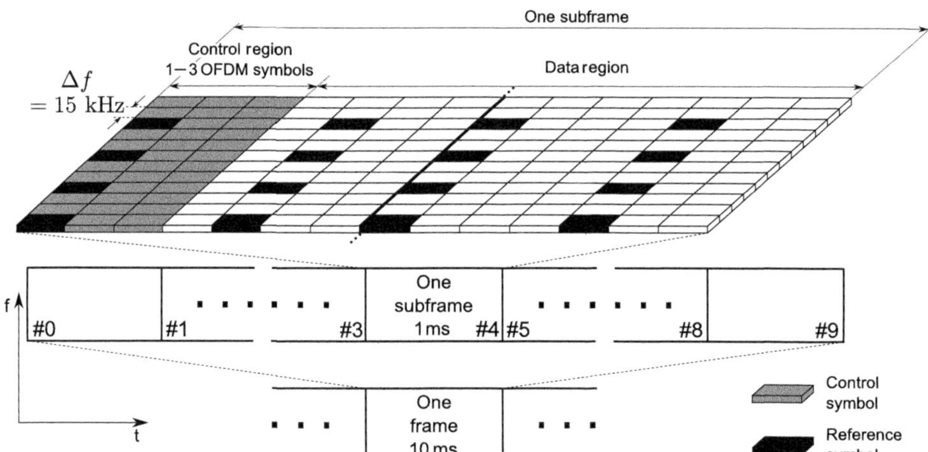

Figure 14.3 Frame structure for LTE systems operating in the FDD mode. Each 1 ms subframe consists of two slots with each slot containing several REs of 15 kHz bandwidth each. Ten subframes together make one frame.

into N_{RB} RBs, each of bandwidth W_{RB} so that $W = N_{RB} W_{RB}$. The RB represents the basic OFDMA time-frequency unit [11]. Each RB contains N_{RE} resource elements (REs). One slot in the time-domain consists of N_{OFDM} orthogonal frequency division multiplexing (OFDM) symbols. Two slots in the time-domain, i.e., $2N_{OFDM}$ symbols and N_{RB} RBs in the frequency-domain make one OFDM subframe as shown in Figure 14.3. The subframe consists of a control and a data region. Interspersed within these are common reference symbols (CRSs), which are used to estimate the channel. Figure 14.3 shows the CRS distribution for the two transmit antennas case. The density of CRS symbols doubles for the four transmit antenna case. The control region can span between one and three OFDM symbols and always occurs at the start of every subframe. Since the control region is pertinent to this chapter, it will be described in further detail in the following.

The DL control region consists of three physical channels. The physical control format indicator channel (PCFICH) carries the control format indicator, indicating the number of OFDM symbols (1, 2, or 3) used for the transmission of DL control channel information in each subframe. The physical hybrid-ARQ indicator channel (PHICH) carries the UL HARQ ACK/NACK information indicating whether the NB has correctly received an UL transmission. Four REs constitute a resource element group (REG). Nine REGs make a control channel element (CCE). The PCFICH always occupies four REGs and the number of REGs that the PHICH may occupy depends on the system bandwidth and a system-wide parameter, N_g. For further details, the reader may refer to [17]. The third physical channel is known as the physical downlink control channel (PDCCH) and it carries the DL control information, which includes transmission resource assignments and other control information for a UE or groups of UEs. The PDCCH belonging to any UE occupies any of {1, 2, 4, 8} CCEs depending on the prevailing channel conditions between the UE and the NB. The PCFICH always occurs exclusively on the first OFDM symbol of the control channel. The PHICH may occupy all available OFDM symbols, but this is not a mandatory requirement, depending on the type of PHICH being used. The PDCCH is distributed over all the available OFDM symbols of the control channel. It is important to note that in the context of the subframe, the control channel is crucial. If the PCFICH or PDCCH are incorrectly decoded, the subsequent data region is lost. This is the premise of the research documented in this chapter and is particularly important in a scenario with co-channel heterogeneous network deployment due to detrimental femto-to-macro interference.

The (four) redundant repetitions of the PCFICH and PHICH (three) are equally distributed in the frequency domain in order to exploit frequency selective fading [11]. The locations of these two physical channels are subject to a cyclic shift dependent on the physical cell identity (PCI) in order to reduce collisions of these channels between two neighboring cells. The mapping of PCFICH and PHICH to REs is detailed in Section 14.4 and for more intricate details, the reader may refer to [17].

In order to prevent the UE from attempting to blindly search the entire control channel in the frequency domain for its associated PDCCH, in LTE, each UE has a limited set of CCE locations where its PDCCH can be located. Each UE is assigned a cell radio network temporary identifier (C-RNTI), which is an identifier, unique within the cell and allocated by the NB. This C-RNTI is used to determine the possible location, known as the *search space*, of the UE's PDCCH. The CCEs available for use by PDCCHs are sequentially numbered and interleaved [18]. This is again done in order to exploit the channel's frequency selectivity. For details on PDCCH interleaving, the reader may refer to [18]. The PDCCH assigned to a UE occupies consecutively numbered CCEs, which, due to the interleaving, are spread in the frequency domain. Finally, the interleaving also depends on the subframe index within the frame being considered so as to further randomize PDCCH collisions from neighboring cells.

14.4 Control channel to resource element mapping

14.4.1 Common reference symbols and cell IDs

In LTE, the CRSs are regularly arranged on the two-dimensional lattice structure [11] (see Figure 14.3). Due to the maximum speed to be supported in LTE (500 km/h), the CRSs need to be repeated twice per slot, as shown in the figure. For this chapter, since we only deal with the control channel, only the CRSs occurring in the first two OFDM symbols matter. In the frequency domain, for a given OFDM symbol, there is one CRS every six REs. However, these are staggered so that in each time slot, there is one CRS every three REs.

The CRS carries one of the possible 504 different PCIs, $N_{\mathrm{ID}}^{\mathrm{cell}}$ [19]. This is the identity of an LTE cell and the H/NB is free to choose the PCI from a list of possible values. In LTE, there are 504 unique PCIs, grouped into 168 groups of three identities each [11]. In order to minimize inter-cell CRS interference, a cell-specific frequency shift, determined as $N_{\mathrm{ID}}^{\mathrm{cell}}$ mod 6, is also applied to the CRS sequence.

14.4.2   The physical control format indicator channel

The PCFICH carries the control format indicator (CFI), indicating the number of OFDM symbols (1, 2, or 3) used for the transmission of DL control channel information in each subframe. The CFI is made robust by adding redundancy to the codeword. Whereas two bits are sufficient to signal the three possible values of the CFI, it is actually 32 bits long. These 32 bits are mapped to 16 resource elements (REs) using quadrature phase shift keying (QPSK) modulation such that each RE carries two bits. Frequency diversity is achieved by distributing the 16 REs in the frequency domain in four groups of four REs (or one REG) each. The PCFICH always occurs on the first OFDM symbol. The reason for this is simple: the UE should be able to determine the size of the control region before being able to decode the other control information and, of course, the data transmitted to it. The cyclically rotated location of the PCFICH in each cell is dependent on the PCI of that cell so that PCFICH–PCFICH interference from neighboring cells is kept to a minimum. In addition, a cell-specific scrambling sequence is applied to the CFI codewords so that the UE can preferentially decode the PCFICH from its associated NB.

Using LTE nomenclature, a REG represents four consecutive REs free of CRS. Let $z(i) = z(0) \ldots z(3)$ represent the symbol quadruplet i. Each of these quadruplets is mapped to one REG [19]. The quadruplet $z(i)$ is mapped to the REG numbered k, where k is determined as

$$k = \bar{k} + \lfloor i \cdot N_{\mathrm{RB}}/2 \rfloor \, N_{\mathrm{RE}}/2, \tag{14.1}$$

where

$$\bar{k} = (N_{\mathrm{RE}}/2) \cdot \left(N_{\mathrm{ID}}^{\mathrm{cell}} \bmod 2N_{\mathrm{RB}} \right) \quad \text{and} \tag{14.2}$$

$$i = 1 \ldots 4. \tag{14.3}$$

Therefore, (14.1) and (14.2) ensure that the PCFICH is evenly distributed in the frequency domain and that it is cyclically rotated depending on the PCI of the cell.

14.4.3   The physical hybrid-ARQ indicator channel

The PHICH carries the UL HARQ ACK/NACK information indicating whether the NB has correctly received an UL transmission from the UE. The PHICH to RE mapping is somewhat similar to that of the PCFICH with two exceptions: it consists of three redundant repetitions in the frequency domain instead of four and it can occur on any combination of the available OFDM symbols, i.e., it is not restricted to just the first OFDM symbol. If the PHICH is smeared over two or three OFDM symbols, it is known as the *extended PHICH* configuration. Note that the extended PHICH occupies two OFDM symbols only in the multimedia broadcast single frequency network (MBSFN) subframe configuration (see Section 14.5.3). Details on PHICH to RE mapping can be found in [19]. The PHICH undergoes the same frequency shift as does the PCFICH. Therefore, its location is also entirely dependent on the PCI of the NB in question.

Figure 14.4 An example of the loading in part of the control region of a subframe.

14.4.4 The physical downlink control channel

Nine REGs make a CCE as mentioned before. The PDCCH carries the DL control information, which includes transmission resource assignments and other control information for a UE or groups of UEs. The PDCCH belonging to any UE occupies any of {1, 2, 4, 8} CCEs depending on the prevailing channel conditions between the UE and the NB. The REGs still free of PCFICH and PHICH are numbered consecutively and arranged in a matrix with a fixed number of columns and the appropriate number of rows. A known inter-column permutation pattern is applied to this matrix and the resulting matrix is reshaped into a vector, thus generating the interleaving pattern. This is comprehensively explained in [18, 20]. This interleaving pattern, like the CRS distribution and PCFICH, undergoes a cell-specific frequency shift so that the PDCCH locations in neighboring cells are pseudo-randomized.

Once the interleaving pattern is determined, the UE-specific PDCCH needs to be assigned. Each UE has a dedicated search space where it looks for its own PDCCH. This is done in order to reduce the number of *blind decoding* attempts needed to be performed by the UE [11]. The CCEs available in the system are numbered such that CCE 1 is composed of REGs 1 through 9 and so on. The candidate starting CCE indices in the search space of UE u in subframe k are calculated as

$$L^u \left\{ (Y_k + m) \mod \lfloor N_{\text{CCE},k}/L \rfloor \right\} + i, \tag{14.4}$$

where $N_{\text{CCE},k}$ is the number of CCEs available in subframe k, L^u is the aggregation level assigned to UE u, $m = 0, \cdots, M^{(L^u)}$ is the number of PDCCH candidates depending on the aggregation level L^u, and $i = 0, \ldots, L^u - 1$. Further details on the above procedure can be found in [21]. The variable Y_k is calculated as

$$Y_k = (A \cdot Y_{k-1}) \mod D, \tag{14.5}$$

where $Y_{-1} = n^u_{\text{RNTI}}$ (n^u_{RNTI} is the identifier of UE u; more details in [22]), $A = 39827$, $D = 65537$, and $k = n_s/2$ (n_s is the slot number within the LTE radio frame). Therefore, it is seen that changing the PCI of an NB changes the interleaving pattern used for PDCCH distribution and thus also changes the location of REGs carrying the PDCCH of UE u.

Figure 14.4 shows an example of the occupancy of the three control channels in a fragment of the control region of an LTE subframe. In this example, the control region is three OFDM symbols long and the CRS distribution for the two antenna case is shown. As is seen, the PDCCH is distributed across all available OFDM symbols. One repetition each of the PHICH (occupying 24 REs) and the PCFICH (occupying 4 REs) is shown.

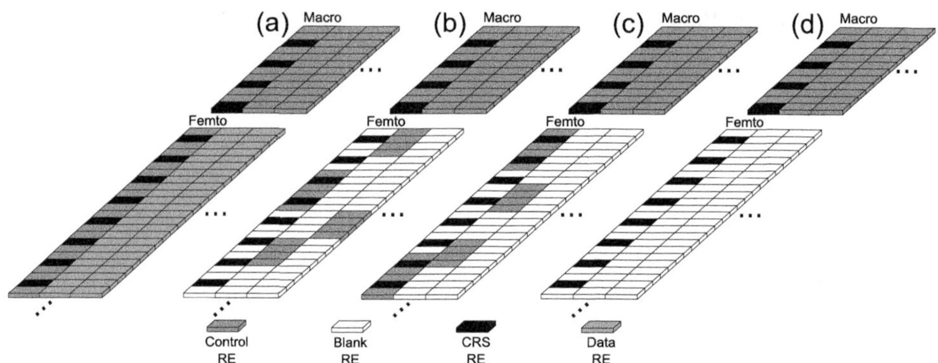

Figure 14.5 State-of-the-art femto-to-macro interference avoidance techniques. The system with no coordination is shown in (a); control channel sparseness is depicted in (b); the MBSFN configuration (with the control region spanning two OFDM symbols) is shown in (c); and the ABS configuration is shown in (d).

14.5   State-of-the-art methods for femto-to-macro control channel interference mitigation

14.5.1   No coordination

Typically, the HNB is expected to serve a very small set of femtocell user equipment (FUE) (3GPP assumes one FUE per femtocell). Furthermore, since the transmission distances between the serving NB and the MUE within the femtocell are much shorter than those of the overlay macrocell, channel conditions between transmitter and receiver are very good. Due to this, the aggregation levels (see Section 14.4.4) assigned to FUEs are usually low. Clearly, worse channel conditions lead to the usage of higher aggregation levels. As a result of the low number of physical control channels to be transmitted, unchecked LTE operation would then prescribe a control region of size one, i.e., one OFDM symbol in the femto layer, with the rest of the subframe utilized for data transmission in order to maximize the throughput delivered. The interference management techniques described in the following subsections exploit this knowledge in different ways. The uncoordinated subframe configuration is depicted in Figure 14.5(a). This figure shows the subframe configuration of the macro and femto layers. Since the macro NB generally serves many users, the macro subframe is shown as being fully occupied and always of size three OFDM symbols. Since the macro subframe is always fully occupied, just one RB is depicted here. For the femto layer, on the other hand, in order to precisely portray the various interference mitigation techniques detailed in the following, two RBs are illustrated.

14.5.2   Sparse control region

In Figure 14.5(b), the HNB is forced to use all three OFDM symbols for the control channel (even though this may not be needed due to the low number of FUEs served), thus making the control channel sparser [23, 24]. Making the femto control channel sparse has the advantage that the probability of collision on the PDCCH, PHICH, and PCFICH belonging to the trapped MUE is reduced. Furthermore, the added advantage of this technique is that the HNB may continue to transmit data in the rest of the subframe. The disadvantage of this method is that the data region of the trapped macro UE undergoes interference from the HNB. However, techniques such as resource partitioning to counter this situation have been proposed [9]. Another advantage is that the sparse control channel possibly reduces the interference among femtocells. However, the study of this does not fall within the scope of this chapter.

14.5.3 Multimedia broadcast single frequency network configuration

From its very first release, LTE has provided support for the MBSFN subframe configuration. Originally, this subframe configuration was intended to provide multimedia broadcast multicast services (MBMS) services whereby the same content is transmitted to many users located in the *MBMS service area* [15]. Each cell participating in this transmission configures a point-to-multipoint radio resource such that all UEs subscribed to the MBMS service simultaneously receive the same signal from all the cells. However, recently it has been observed that MBSFN configured subframes can be used in other scenarios too – for example in relaying or, in this context, for interference management in heterogeneous networks.

Multimedia broadcast single frequency network subframes consist of a control region of size one or (maximum) two OFDM symbols. The data region of the subframe is known as the *MBSFN region* and the contents of this region depend on the exact purpose of the MBSFN transmission. In the context of interference management, the data region can be left empty (as is done in this chapter). Such a configuration is depicted in Figure 14.5(c). The reason for having an occupied control region is to transmit UL control information to the UEs. Since system-specific information cannot be over-ridden or disturbed, MBSFN configured subframes cannot reside in subframes 0, 4, 5, and 9 (for the FDD case).

14.5.4 Almost blank subframe configuration

The almost blank subframe (ABS) configuration is depicted in Figure 14.5(d) [25]. In such a configuration, the subframe only carries CRS and is blank otherwise. This technique reduces the interference on all three control channels for the trapped macro UE. However, interference from CRSs in the femto layer still exists and data cannot be transmitted to femto UEs in the ABS configuration. Table 14.1 shows the differences between the various schemes in terms of control region size, presence of data, and control information and the PHICH configuration used.

14.5.5 Power control

Since favorable channel conditions exist within femtocells, power control can be utilized to reduce the control channel interference caused to trapped macro UEs. A simple power

Table 14.1 Details of ICIC schemes.

	Control channel size (OFDM symbols)	Presence of data	Presence of control channels	PHICH configuration
Uncoordinated	1	yes	yes	normal
Sparse	3	no	yes	normal
MBSFN	2	no	yes	normal
ABS	1	no	no	N/A

control scheme, which adjusts the HNB transmit power based on the reference signal received power, can be implemented [26] such that the HNB transmit power, P_{HNB}, is adjusted as

$$P_{\mathrm{HNB}} = \max\left(\min(x, P_{\max}), P_{\min}\right) \quad [\mathrm{dBm}], \qquad (14.6)$$

where $x = \alpha\left(\overline{R_{\mathrm{CRS}}} + 10\log_{10}\left(N_{\mathrm{RB}}^{\mathrm{DL}} N_{\mathrm{SC}}^{\mathrm{RB}}\right)\right) + \beta$. Here, $P_{\max}$ and $P_{\min}$ are the maximum and minimum allowable HNB transmit powers, respectively, $\overline{R_{\mathrm{CRS}}}$ is the mean measured reference signal received power (RSRP) from the strongest co-channel macrocell (in dBm), $N_{\mathrm{RB}}^{\mathrm{DL}}$ is the number of DL RBs, $N_{\mathrm{SC}}^{\mathrm{RB}}$ is the number of subcarriers per RB, α is a linear scalar that allows altering the slope of the power control mapping curve, and β is a parameter (in dB) that is used to alter the exact range of $\overline{R_{\mathrm{CRS}}}$ covered by the dynamic range of power control. Again, this technique is backward compatible since the power control method does not rely on any UE measurements. It is assumed that the HNB is equipped with a DL receiver in order to determine $\overline{R_{\mathrm{CRS}}}$. The tradeoff of using this technique is a reduced data capacity due to power control. However, past studies have shown that very high data capacities are possible within the femtocell. Therefore, despite power control, the femto UEs are expected to experience high data capacities.

14.6   System-level simulator setup

The simulation area comprises a one-tier tessellated hexagonal cell layout. Each hexagon represents a cell sector, thus implying that the NBs are situated at the junction of three hexagonal sectors. Statistics are taken from the central sector only. However, users are distributed in all sectors in order to simulate the wrap-around effect in terms of interference. For each sector, the azimuth antenna pattern, $A(\theta)$, is described as in [27].

According to the 3GPP simulation requirements described in [28], the dual stripe model is used in this chapter to simulate the femtocell distribution. This setup models a dense-urban HNB deployment, in which each block represents two multi-floor stripes of apartments, separated by a street. Furthermore, blocks are not allowed to overlap. An active HNB may exist in an apartment with probability p_{active}. Every apartment that contains an active HNB also contains exactly one associated femto UE. These are dropped randomly and uniformly within the apartment with a specified minimum separation from the HNB. In addition to this, as per the user distribution described

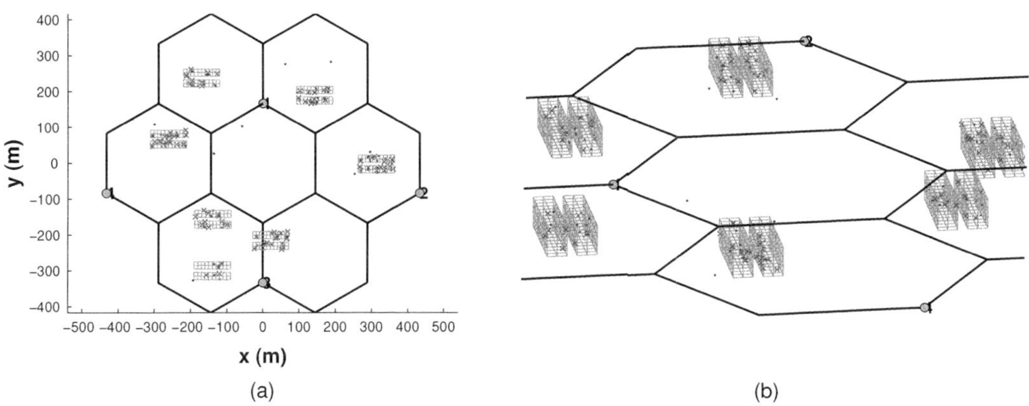

Figure 14.6 User distribution in a cellular system containing underlay femtocells. (a) A system containing one dual stripe and ten MUEs per macrocell sector. 80% of all MUEs lie indoors and every apartment has a 10% probability of containing an active HNB. (b) Each building of the dual stripe has several floors. The indoor MUEs may be located in any apartment on any floor. Each active HNB is associated with one FUE both of which are located randomly in the same apartment.

in [28], every MUE has a certain probability of lying within one of the apartments since the MUEs are distributed randomly in the entire network. Based on this probability, a MUE is randomly located either outdoors or indoors. Since the method of access is strictly closed, indoor MUEs are strictly served by the NB situated outdoors. In such a situation, these vulnerable MUEs suffer severely from DL interference originating from the nearby HNBs. Figure 14.6 shows the distribution of UEs and femtocells in a heterogeneous system.

Three path-loss models are used depending on the type of link [28], i.e., whether the link is purely outdoor, outdoor-to-indoor, or purely indoor. Fast fading channels are simulated using the delay profiles for the urban microcell (UMi) and indoor hotspot (InH) models provided in [29]. Due to the presence of multiple receive antennas, maximum ratio combining (MRC) is made use of. The gain from MRC is approximated by simulating two individual, uncorrelated receive streams and adding the achieved signal to interference plus noise ratio (SINR) on each of them [30].

It is assumed that the control region on the macro layer is always three OFDM symbols long. This is a reasonable assumption because the macro NB typically serves a large number of users. This, however, is not the case for the femto layer, where the size of the control region depends on the interference management scheme being used (see Section 14.5).

For each HNB, the three control channels are mapped to REs as explained in Section 14.4. The RE locations belonging to the PCFICH and PHICH are calculated based on the PCI assigned to the HNB in question. The candidate CCEs belonging to the search space of each UE are calculated as explained in Section 14.4.4. The macro UEs associated with each NB are sorted according to the SINRs they experience on the DL reference signals. Based on this sorting, the UEs are assigned the appropriate aggregation level such that each aggregation level is assigned to a quarter of the UEs belonging to that NB. In the femto layer, each FUE is assigned an aggregation level of 2.

As mentioned before, both macrocells and femtocells utilize the entire system bandwidth W. This means that all REs allocated to a macro UE are susceptible to interference from a nearby HNB because both layers utilize resources with a reuse factor of 1. The useful received signal power observed by UE_u (where u is the user index, regardless of whether it is an MUE or an FUE) on OFDM symbol $t \in \{1, 2, 3\}$ on RE n is given by

$$Y_{n,t}^u = G_{n,t}^{v,u} P_x, \tag{14.7}$$

where $G_{n,t}^{v,u}$ is the channel gain between UE_u and its serving HNB or NB, v, on symbol t, and the transmit power, P_x, represents either the NB or HNB transmit power depending on which entity the UE is served by. The aggregate interference $I_{n,t}^u$ seen by UE_u is composed of NB and HNB interference

$$I_{n,t}^u = \sum_{i \in \mathcal{M}_{\mathrm{int}}} G_{n,t}^{i,u} P_{\mathrm{m}} + \sum_{j \in \mathcal{F}_{\mathrm{int}}} G_{n,t}^{j,u} P_{\mathrm{f}}, \tag{14.8}$$

where $G_n^{y,u}$ accounts for the channel gain between interferer y and UE_u on symbol t. The set of instantaneous NB and HNB interferers on symbol t and RE n is denoted by $\mathcal{M}_{\mathrm{int}}$ and $\mathcal{F}_{\mathrm{int}}$, respectively. Note that the sets $\mathcal{M}_{\mathrm{int}}$ and $\mathcal{F}_{\mathrm{int}}$ change across OFDM symbols and subframes due to the PDCCH interleaving and cyclically shifted PCFICH and PHICH locations. The SINR observed on RE n and symbol t at UE_u therefore amounts to

$$\gamma_{n,t}^u = \frac{G_{n,t}^{v,u} P_x}{I_{n,t}^u + \eta}, \tag{14.9}$$

where η accounts for thermal noise per RE. In order to calculate the *effective* SINR across all the allocated REs on any of the control channels belonging to UE_u, a mapping to the capacity-domain is first made and this is then re-translated into the SINR domain as explained in [31]. Therefore, the effective SINR for UE_u on the control channel y (y representing either the PDCCH or PCFICH) is calculated as

$$\gamma_y^u = F^{-1} \left(\sum_{p \in \mathcal{N}_{\mathrm{RE},y}^u} F\left(\gamma_p^u\right) \right), \tag{14.10}$$

where $\mathcal{N}_{\mathrm{RE},y}^u$ is the set of REs allocated to the control channel y of UE_u. Each element $p \in \mathcal{N}_{\mathrm{RE},y}^u$ is an ordered set $p_1 = (n_1, t_1)$ containing an RE index in the frequency domain and a time instant, indicating the position of the RE in question. The capacity $F\left(\gamma_p^u\right)$ is calculated using the attenuated and truncated Shannon bound, as described in [9]. According to this bound, capacity saturates beyond a certain SINR or becomes zero below a certain SINR in order to avoid unrealistically high or low values.

The simulation parameters used are shown in Table 14.2.

Table 14.2 Simulation parameters.

Parameter	Value
Average dual stripes per macrocell sector	1
Average macro UEs per macrocell sector	10
Inter-site distance	500 m
Individual apartment dimensions	10 m × 10 m
Number of floors per building	3
HNB activation probability, p_{active}	10%
Probability of macro UEs lying indoors	80%
Number of REs per RB, N_{RE}	12
Total number of available RBs, N_{RB}^{DL}	50
Thermal noise density	−174 dBm/Hz
NB transmit power per sector	46 dBm
HNB transmit power, P_{max}	20 dBm
NB antenna gain	14 dBi
Number of antenna ports	2
Sectors per NB	3
Min. distance between macro UE and NB	35 m
Min. distance between femto UE and HNB	20 cm
Number of HNB/NB Rx antennas	2 Rx
Number of macro/femto UE Rx antennas	2 Rx
Wall penetration loss	20 dB
Power control slope factor, α	1
Power control dynamic range factor, β	70 dB

14.7 Performance comparison of inter-cell interference coordination methods

This section showcases the comparison of the performance of the various state-of-the-art interference management schemes described in Section 14.5. To put the results into perspective, the performances of the different schemes are compared against the *ideal* case, termed henceforth as the benchmark. In the benchmark system, the HNBs do not transmit, i.e., they are turned off. This case is ideal because it collapses the network into a homogeneous system such that there is no interference from the femto layer. The aim of the interference management techniques detailed in Section 14.5 is to bring the performance of the trapped MUEs as close as possible to that of the untrapped MUEs. To this end, pitting the performance of the various ICIC techniques against that of the benchmark system is a reasonable method of assessing how closely they are able to approach the performance of the ideal case.

Due to the different number of REs allocated to the different control channels, the tolerable effective SINRs (for a particular value of block error rate (BLER)) for the different control channels vary. When dealing with control channels, it is important to note that the transmission is considered to be successful if the achieved effective SINR is above the tolerable threshold. For the PCFICH and PHICH, the tolerable effective SINRs (for a BLER value of 1%) are approximately −6 dB [32] and −3 dB [33], respectively. For the PDCCH, the tolerable SINR depends on the aggregation level being considered.

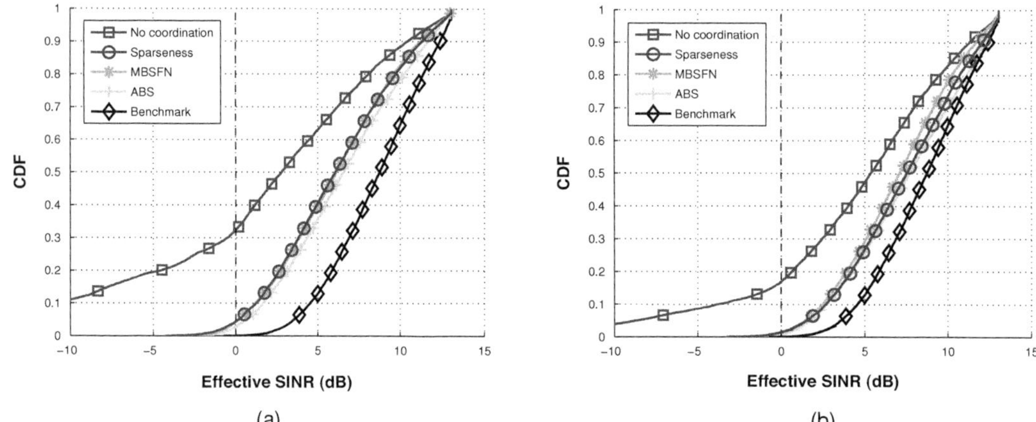

Figure 14.7 Performance of the PDCCH for all MUEs in the system (a) without power control; and (b) with power control.

Clearly, the higher the aggregation level, the greater the number of REs used, leading to a more robust PDCCH. For an aggregation level of 8, the tolerable effective SINR (for a BLER value of 1%) is approximately -4 dB [32]. For the lowest aggregation level of 1, the tolerable effective SINR increases to approximately 4 dB. In the following figures, these tolerable effective SINR thresholds are marked by a vertical line. In the case of PDCCH, this threshold is given a value of 0 dB as an average.

Figure 14.7 shows the cumulative distribution function (CDF) of the effective SINR achieved on the PDCCH for all MUEs in the system. The curves without and with power control are depicted in Figures 14.7(a) and 14.7(b), respectively. Concentrating on Figure 14.7(a), it is clear to see that the system without any interference coordination between the femto and macro layers exhibits very poor performance in comparison to the benchmark system which contains no femto interference. Using the 0 dB effective SINR reference (marked by line), in an uncoordinated system, it is seen that over 30% of all MUEs do not achieve this SINR target. The systems employing some sort of interference management technique are seen to significantly improve the PDCCH performance for the MUEs. Again, for the 0 dB reference effective SINR, the percentage of MUEs not achieving this target drastically diminishes to less than 5%. Seen another way, at the 50th percentile, without coordination, there is a deviation of approximately 6 dB in PDCCH SINR from the ideal case. With interference management, this difference improves to approximately 3 dB. However, the impact to the femto layer must not be dismissed. As mentioned in Section 14.5, the MBSFN and ABS configurations are such that data is not transmitted in that subframe. Clearly this results in a decrease of data capacity on the femto layer. For example, suppose a system is set up such that the MBSFN configuration is not used in subframes 0, 4, 5 and 9 but in all the other subframes as described in Section 14.5. This would result in a loss of 60% of femto data capacity. Similarly, a system configured to use ABSs would result in a data loss on the femto layer commensurate with the number of ABSs configured in the LTE frame. Fortunately, it has been shown in several previous studies [1, 34] that femtocells are capable of achieving very high data capacities. Therefore, even a reduction in half the capacity still leaves the femtocell achieving a significant capacity improvement in comparison to the macrocell. Nevertheless, it is still desirable to have an interference management technique that does not result in a significant capacity loss for the femto layer. To this end, the sparse control region technique exhibits attractive properties. Figure 14.7(a) shows that this technique results in a very similar (albeit very slightly degraded) PDCCH performance to the MBSFN and ABS schemes. However, it has the added advantage that data can still be transmitted on the femto layer. Referring again to Figure 14.5, we see that the sparse control region technique forces the HNB to always use a control region that is three OFDM symbols long. Therefore, it is expected that two OFDM symbols, which would otherwise be used for transmitting data, are sacrificed for control region interference mitigation. In comparison to the 60% capacity loss experienced by systems using the MBSFN technique, losing 2 out of 14 subframes using the sparse technique results in a considerably lower 15% loss in data capacity. Figure 14.7(b), which depicts the performance of the system with power control on the femto layer enabled (as described in Section 14.5.5), shows the same trends as the system with power control. However, it is seen that the difference between the uncoordinated system and the benchmark now diminishes to approximately 5 dB and that between the interference-managed systems and the benchmark reduces to approximately 2 dB. Of course, by changing the power control parameters α and β the degree of change from the full transmission power case can be effectively controlled at the cost of a degradation in femtocell PDCCH performance. Though those results are not shown, it has been observed that the loss of control channel performance in the femto layer through power control is very much within tolerable limits. Finally, it is seen that all the interference mitigation techniques achieve similar performance with the sparse technique showing a slightly degraded performance compared to the other two schemes. The reason for this is because the PDCCH is distributed in both the frequency and time domains. As is seen in Figure 14.7(a), the best performance is obtained by using the ABS configuration because other than CRS, the control region in such subframes is empty. The performance degrades somewhat with the MBSFN subframe configuration because, in this case, the control region is also occupied with the control channels. However, it performs slightly better than the sparse scheme because in the case of MBSFN, the third OFDM symbol is blank.

Figure 14.8 shows the CDF of the effective SINR achieved on the PCFICH for all MUEs in the system. Figure 14.8(a) depicts the PCFICH performance for all MUEs in the system without the use of DL power control on the femto layer. With the robust effective SINR reference of -6 dB in this case, approximately 15% of all MUEs in the uncoordinated system are unable to achieve the target, which is clearly beyond acceptable tolerance. Out of the interference mitigation schemes, the ABS configuration achieves the best performance. The reason for this is obvious: the ABS configuration on the femto layer results in the least populated subframe and therefore the chance of collision with the PCFICH of the trapped MUE is minimized. Unlike the PDCCH, which is distributed on all available OFDM symbols, the PCFICH strictly occurs only

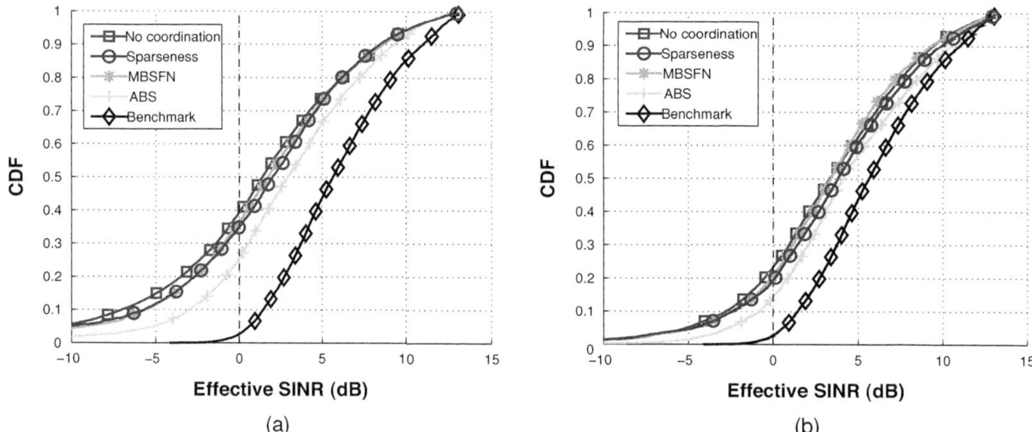

Figure 14.8 Performance of the PCFICH for all MUEs in the system (a) without power control; and (b) with power control.

in the first OFDM symbol as described in Section 14.4.2. As is explained before in Section 14.4, the first OFDM symbol also contains the other two control channels in addition to the PCFICH. This means that the first OFDM symbol in the control region of any NB is the most populated of all the OFDM symbols comprising the control region. Furthermore, the restricted location of the PCFICH (to the first OFDM symbol only) means that there is very limited flexibility in protecting this particular control channel. This fact is corroborated by the results, where it is observed that the systems with interference management perform nearly as badly as the uncoordinated system, in stark contrast to the case of the PDCCH seen in Figure 14.7. Finally, Figure 14.8(a) reveals that the systems with active femtocells lag the benchmark system by approximately 2.5 dB at the 50th percentile. From Figure 14.8(b), it can be seen that with femtocell power control, the gap between the benchmark system and the systems with active femtocells is reduced to approximately 1 dB at the 50th percentile. However, it is seen that approximately 5% of the MUEs are still unable to attain the -6 dB reference SINR target. As mentioned before, inability to decode the PCFICH will surely result in the loss of the entire subframe.

Figure 14.9, which shows the CDF of effective SINR on the PHICH, exhibits very similar trends to those seen in Figure 14.8. This is because when the extended PHICH configuration is not used, the PHICH distribution is similar to that of the PCFICH with the difference that the PHICH consists of three repetitions instead of four. Furthermore, since the PCFICH employs a very low code rate of 1/16, it has a lower tolerable effective SINR. In the case of the PHICH the tolerable SINR is -3 dB. With this reference, it is seen that approximately 20% of the UEs are unable to achieve the SINR. However, as in Figure 14.8, introducing interference mitigation techniques does not result in a significant improvement. Again, the ABS technique results in the greatest improvement, at the loss of data capacity. Power control brings the curves closer to the benchmark.

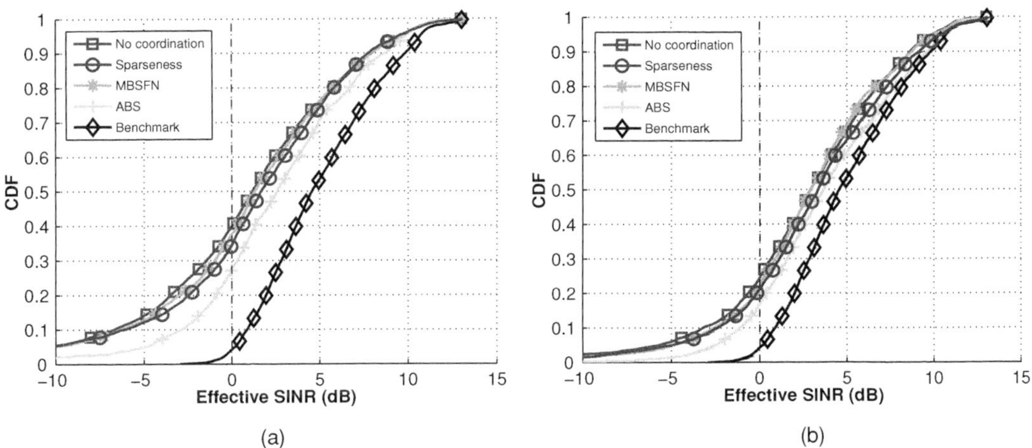

Figure 14.9 Performance of the PHICH for all MUEs in the system (a) without power control; and (b) with power control.

14.8   PCFICH performance improvement

In Section 14.5, it is seen that the interference management technique depicted in Figure 14.5(c) reduces interference on the control channel via the loss of all the information in the subframe (ABS). This causes a very significant loss in femto data capacity. Section 14.4 has shown that the position of all three control channels relies on the PCI of the NB. Currently in LTE, an HNB may arbitrarily have any one of the 504 possible PCIs. This means that there is a likelihood that the DL control channels of a HNB may collide detrimentally with the DL control channels of a UE trapped within the coverage of that HNB. In [26], it has been proposed that the HNB opportunistically switches its PCI between a default PCI and a CSG PCI in such a way that if a non-CSG UE is in the vicinity of the HNB, the CSG PCI is used so that the UE has access to the HNB. This method requires a large amount of signaling so that the HNB is aware of a non-CSG UE close to it. A very simple and efficient means of mitigating control channel interference without the need for any additional signaling is introduced here. The HNB must simply detect the PCI of the strongest neighboring macro NB. Long term evolution HNBs already have the ability to detect the strongest neighboring NB for power control purposes [26]. Identifying the PCI of the NB is straightforward as it can be done by decoding the NB's broadcast channel using the initial synchronization procedure [11]. Thus, if the HNB manipulates its PCI at startup, it can minimize the interference it causes from its CRS, PCFICH, PDCCH, and PHICH to the PCFICH of the trapped macro UE for as long as the macro NB maintains its PCI. This scheme offers several advantages:

- In order for this procedure to work, the HNB requires knowledge of the PCI belonging to the closest macro NB. This information does not need to be signaled because the HNB can simply infer it from the primary synchronization signal (PSS) of the macro NB.

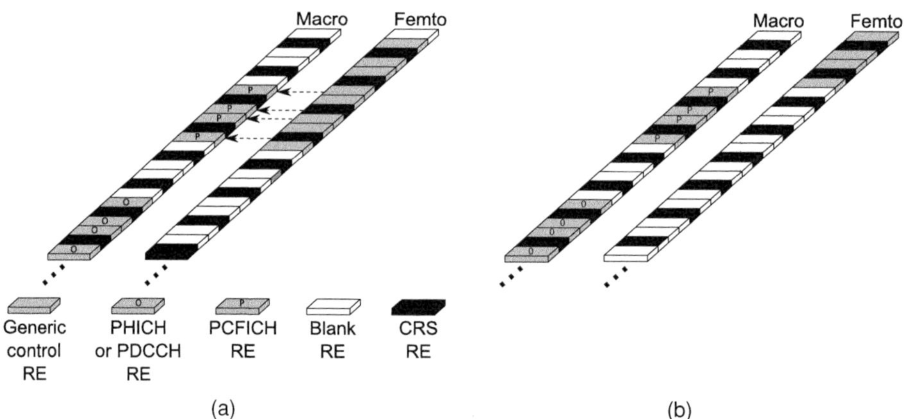

Figure 14.10 PCFICH interference mitigation via PCI manipulation. The PCFICH belonging to a trapped macro UE is interfered with by the CRS and control channels of a nearby HNB (a). Rearranging the occupied REs of the control region at the femto layer mitigates PCFICH interference for the vulnerable macro UE (b).

- It is assumed that the HNB is equipped with a DL receiver for power control purposes [26]. Since this receiver is also used to infer the PCI of the closest macro NB, no additional hardware is required at the HNB.
- This procedure poses no challenge for legacy (Rel-8/9) UEs since the HNB chooses the PCI and there is no additional burden to the UE.

The simplicity of this idea stems from the fact that regardless of how many macro UEs may be affected by one (or more) HNBs, each interfering HNB is capable of independently mitigating interference caused to the PCFICHs of the affected macro UEs without the need for any backhaul signal exchange. Thus the proposed scheme works for both sparse as well as densely deployed femtocell networks equally well.

By carefully selecting a suitable PCI, the HNB rearranges its control channels and CRSs such that interference to the PCFICH of the exposed macro UEs is significantly reduced. This is illustrated in Figure 14.10. This figure shows an example of parts of the first OFDM symbols of the macro and femto layers. If the uncoordinated configuration shown in Figure 14.10(a) is assumed, any macro UE that is in the vicinity of an active HNB will experience heavy interference on the REs on which it receives the PCFICH. Such a situation is responsible for the poor PCFICH performance observed in Figure 14.8. To counter this situation, the aggressor HNB, simply by choosing an appropriate PCI value, rearranges the contents of the control region so as to minimize the experienced PCFICH interference. This is shown in Figure 14.10(b).

In order to assess the performance of this technique, the same system setup detailed in Section 14.6 is used. It is assumed that every HNB chooses the most appropriate PCI so that collisions with the PCFICH of the most dominant neighboring NB (based on received reference signal power) are minimized. The algorithm used to choose the PCI value is out of the scope of this chapter. In the simulation, the PCI is chosen in a brute-force fashion.

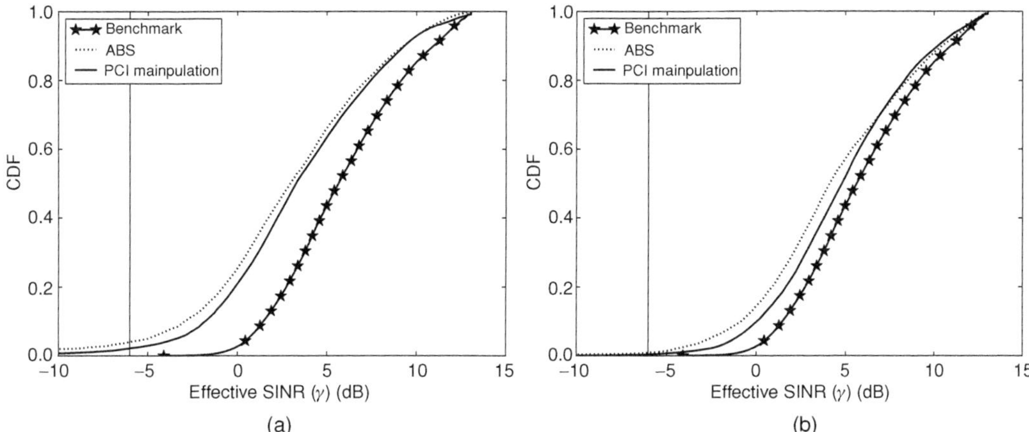

Figure 14.11 Performance of PCI manipulation versus the ABS configuration and the benchmark in the system (a) without power control; and (b) with power control.

Figure 14.11 renders the CDF of the effective SINR of all macro UEs in the system. The performance of the system employing the proposed PCI manipulation technique is compared against the system where ABSs are configured for the femto layer and the benchmark system where HNBs do not transmit at all. The system with the ABS configuration is chosen because it is seen from Figure 14.8 that the system configured with ABS approaches the benchmark system most closely. It is clearly seen that the benchmark significantly outperforms the other two schemes, due to the absence of femtocell interference. We see that employing ABS at the femto layer improves the effective SINR even further. Due to the lack of control channels with the ABS configuration, the only DL interference seen by the PCFICH of the trapped macro UEs originates from the CRSs belonging to the HNBs. It must be pointed out, however, that in the ABS configuration case, the HNBs are assigned PCIs without any coordination whatsoever. Therefore, even though there are only 16 REs assigned to the PCFICH, there is a probability that femto CRS to macro UE PCFICH collision occurs, which leads to a degraded PCFICH performance for the trapped macro UEs. The proposed PCI manipulation scheme carefully chooses the PCI of the HNBs such that it minimizes the collisions to the PCFICH of the trapped macro UEs. The selection of the PCI takes into account all three control channels *and* CRS and the latter point is responsible for the improved performance over ABS.

However, the achieved gains seen by the PCFICH of the trapped macro UEs via PCI manipulation (in comparison to the ABS scheme) come at the cost of increased interference to the CRS of the trapped macro UEs, which may lead to channel estimation errors. There are two approaches to alleviate this problem:

- CRS interference cancellation could be employed to cancel the CRS interference originating from the nearby HNB. However, this is not possible with legacy UEs and is therefore only applicable to a subset of macro UEs.

- The other option is to allow CRSs to collide with the PCFICH of trapped macro UEs. While this will degrade the performance, it will only degrade it to the level of the ABS performance – which is still acceptable.

The results presented in Figure 14.11 demonstrate that the proposed PCI manipulation procedure combines the advantages of two other state-of-the-art techniques: it enables the femto layer to continue transmitting data (not possible with the ABS configuration) and, at the same time, it shows a significant performance improvement over the sparse control region approach (see Figure 14.8). It must also be stressed that while the entire data portion of an ABS is lost, this is not the case with the proposed scheme. In an LTE frame, a maximum of six subframes can be configured as ABSs. Therefore, even though the improvement via PCI manipulation over the ABS configuration is not significant, the potential data loss resulting from using the ABS configuration can be as high as 60%. This, of course, is not the case with the proposed PCI manipulation scheme and is the main benefit of using this scheme.

14.9   Conclusion and discussion

This chapter has presented accurate Rel-8/9 LTE-compliant system-level simulations with co-channel femtocell deployment. This is important in the context of current 3GPP activity where legacy support of heterogeneous networks for future releases (Rel-10 and beyond) is a key consideration. Future LTE releases will be characterized by new control channels and methods designed specifically to mitigate femto-to-macro interference. However, legacy UEs will not be capable of exploiting these improvements. Therefore, it is important that UE-transparent interference management techniques are proposed and analyzed. With this in mind, several state-of-the-art techniques are introduced and investigated in this chapter. One of these techniques proposes that the control channel on the femto layer be made sparse so as to reduce the probability of collision on the various control channels of the macro layer. Another technique is DL power control on the femto layer.

Results show that a sparse control channel closely approaches the performance of a control channel limited to one OFDM symbol with the other symbols blank. However, it is also seen that without power control, the performance of the PCFICH is unsatisfactory. Using power control helps boost the PCFICH SINR, but this is still not enough. In conclusion, if the subframe on the femto layer contains no data, power control alone is enough to assure the control channel performance of trapped macro UEs. However, if there is data to transmit on the femto layer, then satisfactory trapped macro UE PDCCH and PHICH performance can be assured by using a sparse control channel on the femto layer in conjunction with power control.

This is, however, not the case with PCFICH. It is clear that since the success with decoding the entire subframe depends on the decoding success of the PCFICH, this control channel merits further attention. With this in mind, the latter part of this chapter presents a novel, yet backward-compatible, legacy-compliant, and freely scalable technique that effectively mitigates PCFICH interference for trapped macro UEs without introducing any additional signaling burden in the network. Results show that employing this technique concurrently with power control improves PCFICH performance to within 2 dB of the ideal case where the system does not contain femtocells.

References

[1] Z. Bharucha, H. Haas, A. Saul, and G. Auer, "Throughput enhancement through femto-cell deployment," *European Trans. Telecommun.*, vol. 21, no. 4, pp. 469–77, Mar. 2010.

[2] V. Chandrasekhar, J. G. Andrews, and A. Gatherer, "Femtocell networks: a survey," *IEEE Commun. Mag.*, vol. 46, no. 9, pp. 59–67, Sep. 2008.

[3] H. Mahmoud, I. Güandvenç, and F. Watanabe, "Performance of open access femtocell networks with different cell-selection methods," in *Proc. IEEE Vehicular Tech. Conf. (VTC)*, May 2010, pp. 1–5.

[4] J. Weitzen and T. Grosch, "Comparing coverage quality for femtocell and macrocell broadband data services," *IEEE Commun. Mag.*, vol. 48, no. 1, pp. 40–4, Jan. 2010.

[5] J. D. Hobby and H. Claussen, "Deployment options for femtocells and their impact on existing macrocellular networks," *Bell Labs Technical J.*, vol. 13, no. 4, pp. 145–60, Feb. 2009.

[6] R.-T. Juang, P. Ting, H.-P. Lin, and D.-B. Lin, "Interference management of femtocell in macro-cellular networks," in *Proc. Wireless Telecommun. Symp. (WTS)*, Tampa, USA, Apr. 2010, pp. 1–4.

[7] D. López-Pérez, A. Valcarce, G. de la Roche, and J. Zhang, "OFDMA femtocells: a roadmap on interference avoidance," *IEEE Commun. Mag.*, vol. 47, no. 9, pp. 41–8, June 2009.

[8] J. Espino and J. Markendahl, "Analysis of macro–femtocell interference and implications for spectrum allocation," in *Proc. IEEE Int. Symp. Personal, Indoor, Mobile Radio Commun. (PIMRC)*, Tokyo, Japan, Sep. 2009, pp. 2208–12.

[9] Z. Bharucha, A. Saul, G. Auer, and H. Haas, "Dynamic resource partitioning for downlink femto-to-macro-cell interference avoidance," *EURASIP J. Wireless Commun. Network. (Special issue on Femtocell Networks)*, vol. 2010, no. Article ID 143413, pp. 1–12, May 2010.

[10] 3GPP, "New work item proposal: enhanced ICIC for non-CA based deployments of heterogeneous networks for LTE," Mar. 2010.

[11] S. Sesia, M. Baker, and I. Toufik, *LTE: The UMTS Long Term Evolution: From Theory to Practice*. Wiley-Blackwell, July 2011.

[12] A. Gjendemsjø, D. Gesbert, G. E. Oien, and S. G. Kiani, "Binary power control for sum rate maximization over multiple interfering links," *IEEE Trans. Wireless Commun.*, vol. 7, no. 8, pp. 3164–73, Aug. 2008.

[13] R. Chang, Z. Tao, J. Zhang, and C.-C. Kuo, "A graph approach to dynamic fractional frequency reuse (FFR) in multi-cell OFDMA networks," in *Proc. IEEE Int. Conf. on Commun. (ICC)*, June 2009, pp. 1–6.

[14] M. Iwamura, K. Etemad, M. Fong, R. Nory, and R. Love, "Carrier aggregation framework in 3GPP LTE-Advanced," *IEEE Commun. Mag.*, vol. 48, no. 8, pp. 60–7, Aug. 2010.

[15] E. Dahlman, S. Parkvall, and J. Sköld, *4G LTE/LTE-Advanced for Mobile Broadband*, 1st edn., Academic Press, 2011.

[16] J. Huang, P. Xiao, and X. Jing, "A downlink ICIC method based on region in the LTE-Advanced System," in *Proc. IEEE Int. Symp. Personal, Indoor, Mobile Radio Commun. (PIMRC)*, Sep. 2010, pp. 420–3.

[17] 3GPP, "Physical channels and modulation (Release 9)," 3GPP TS 36.211 V 9.1.0 (2010-03), Mar. 2010.

[18] ——, "Multiplexing and channel coding (Release 9)," 3GPP TS 36.212 V 9.2.0 (2010-06), June 2010.

[19] 3rd Generation Partnership Project (3GPP), Technical Specification Group Radio Access Network, *Evolved Universal Terrestrial Radio Access (E-UTRA); Physical Channels and Modulation (Release 9)*, 3GPP TS 36.211 V9.1.0, 3GPP Std., Mar. 2010.

[20] J.-F. Cheng, A. Nimbalker, Y. Blankenship, B. Classon, and T. Blankenship, "Analysis of circular buffer rate matching for LTE turbo code," in *Proc. IEEE Vehicular Tech. Conf. (VTC)*, Sep. 2008, pp. 1–5.

[21] 3GPP, "Evolved universal terrestrial radio access (E-UTRA); physical layer procedures (Release 9)," 3GPP TS 36.213 V 9.1.0 (2010-03), Mar. 2010.

[22] ——, "Evolved universal terrestrial radio access (E-UTRA); medium access control (MAC) protocol specification (Release 9)," 3GPP TS 36.321 V 9.2.0 (2010-03), Mar. 2010.

[23] Z. Bharucha, G. Auer, and T. Abe, "Downlink femto-to-macro control channel interference for LTE," in *Proc. IEEE Wireless Commun. Networking Conf. (WCNC)*, Cancun, Mexico, Mar. 2011.

[24] Kyocera, "Range expansion performance and interference management for control channels in outdoor hotzone scenario," 3GPP TSG RAN WG1 Meeting #60b R1-102363, Apr. 2010.

[25] CATT, "Analysis of time-partitioning solution for control channel," 3GPP TSG RAN WG1 Meeting #61b R1-103494, 28 June–2 July 2010.

[26] 3GPP, "Evolved universal terrestrial radio access (E-UTRA); FDD Home eNodeB (HeNB) radio frequency (RF) requirements analysis (Release 9)," 3GPP TR 36.921 V 9.0.0 (2010-03), Mar. 2010.

[27] ——, "Simulation assumptions and parameters for FDD HeNB RF Requirements," 3GPP TSG RAN WG4 R4-092042, May 2009.

[28] ——, "Further advancements for E-UTRA physical layer aspects (Release 9)," 3GPP TR 36.814 V0.4.1 (2009-02), Sep. 2009. Retrieved June 2, 2009 from www.3gpp.org/ftp/Specs/.

[29] ITU-R Working Party 5D (WP5D) - IMT Systems, "Report 124, Report of correspondence group for IMT.EVAL," May 2008, United Arab Emirates.

[30] T. S. Rappaport, *Wireless Communications: Principles and Practice*, 2nd edn. Prentice Hall, Dec. 2001.

[31] K. Brueninghaus, D. Astély, T. Sälzer, S. Visuri, A. Alexiou, S. Karger, and G.-A. Seraji, "Link performance models for system level simulations of broadband radio access systems," in *Proc. IEEE Int. Symp. Personal, Indoor, Mobile Radio Commun. (PIMRC)*, vol. 4, Berlin, Germany, Sep. 2005, pp. 2306–11.

[32] Huawei, "Control channel performance evaluations for co-channel deployment with MeNBs and outdoor Picos," 3GPP TSG RAN WG1 Meeting #60bis R1–101983, Apr. 2010.

[33] CATT, "Interference coordination for DL CCH considering legacy UE," 3GPP TSG RAN WG1 Meeting #60bis R1–101783, Apr. 2010.

[34] Z. Bharucha, I. Ćosović, H. Haas, and G. Auer, "Throughput enhancement through femto-cell deployment," in *Proc. IEEE Int. Workshop on Multi-Carrier Systems & Solutions (MC-SS)*, Herrsching, Germany, May 2009, pp. 311–19.

15 Spectrum assignment and fairness in femtocell networks

Mustafa Cenk Erturk, İsmail Güvenç, Sayandev Mukherjee, and Huseyin Arslan

15.1 Introduction and prior art

15.1.1 Frequency allocation for heterogeneous networks

In heterogeneous networks, frequency resources can be allocated to different tiers in a co-channel (shared-spectrum) or dedicated channel (split-spectrum)[1] manner, or through a hybrid technique, which is a combination of the two approaches. In the co-channel approach shown in Figure 15.1(a), while the spectrum resources are fully reused in different tiers, cross-tier interference may cause crucial setbacks to the system. For example, macrocell users in the vicinity of closed subscriber group (CSG) femtocells are not allowed to connect to the femtocells, even if their link quality with these femtocells is good. Therefore, such macrocell users receive strong downlink interference from CSG femtocells and may fall into outage.

The split spectrum approach shown in Figure 15.1(b), on the other hand, partitions the allocated spectrum between multiple tiers. Each tier can use its own segment of resource and therefore there is no cross-tier interference [1]. However, the amount of bandwidth available to each tier is reduced. Hybrid methods as shown in Figure 15.1(c) use a mixture of co-channel and dedicated channel methods, and aim to reuse the spectrum resources whenever feasible. For example, in [2], the macrocell users are dedicated a component carrier (CC), referred as the "escape carrier," which is not used by the femtocell network. Any macrocell mobile station (MMS) that is closeby to a femtocell is scheduled within this escape carrier, if the interference observed from the femtocell network is above a threshold. Hence, user outages are prevented by scheduling victim users in dedicated resources, while the spectrum is still reused in co-channel CCs. The resources within a certain CC may also be partitioned into smaller chunks for similar interference mitigation purposes [3].

The performance of dedicated channel and co-channel femtocell/macrocell networks has been investigated and compared through computer simulations in [4, 5]. Both

---

[1] Throughout the chapter, the terms shared-spectrum and split-spectrum will be used interchangeably with co-channel and dedicated channel, respectively.

Small Cell Networks: Deployment, PHY Techniques, and Resource Management, ed. Tony Q. S. Quek, Guillaume de la Roche, İsmail Güvenç, and Marios Kountouris. Published by Cambridge University Press.
© Cambridge University Press 2013.

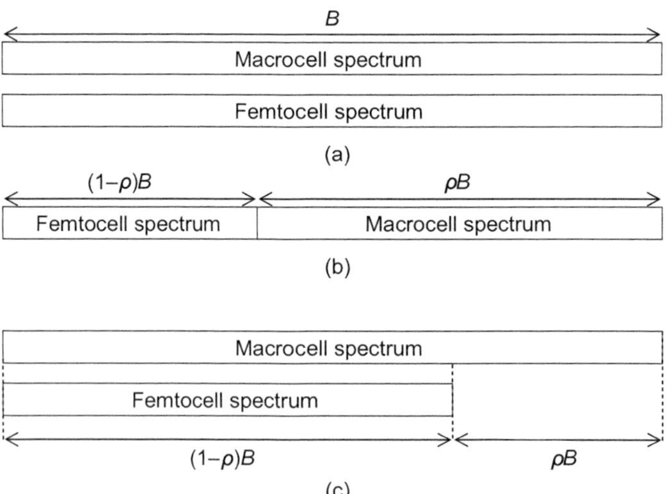

Figure 15.1 Different resource allocation approaches in heterogeneous networks: (a) co-channel approach; (b) dedicated channel approach; and (c) hybrid approach. Spectrum splitting ratio (SSR) is denoted by ρ.

papers show that co-channel deployment increases the total system throughput at the expense of some degradation in the throughput of macrocell users that are close to the femtocells. However, the impact of different spectrum splitting ratios (SSRs) on the overall network has not been studied in these works. Capacity cumulative distribution functions (CDFs) of indoor and outdoor users for different SSRs have been compared through computer simulations in [6], showing that for certain scenarios, performance close to the co-channel deployment can be obtained by appropriately setting the SSR value in a dedicated channel setting. Bharucha *et al.* [3] investigate the impact of dynamic resource partitioning for downlink femto-to-macrocell interference avoidance for co-channel femtocell deployments in [3]. The simulation results show that co-channel deployment with dynamic resource partitioning can benefit from the frequency reuse property to achieve high throughputs, and femtocells can switch to orthogonal resource utilization when a close-by macrocell user is detected. However, the so-called X2 interface between the macrocell base station (MBS) and the femtocell base station (FBS) is assumed to be available in order to exchange the interference coordination information.

15.1.2   Time-domain resource coordination

Similar to frequency-domain resource partitioning and sharing, time-domain resources may also be partitioned and shared among femtocells and macrocells. For example, as shown in Figure 15.2, 3GPP utilizes almost blank subframes (ABSs)[2] for mitigating

---

[2] The subframes are named as *almost* blank, since the reference signals are still transmitted in these subframes.

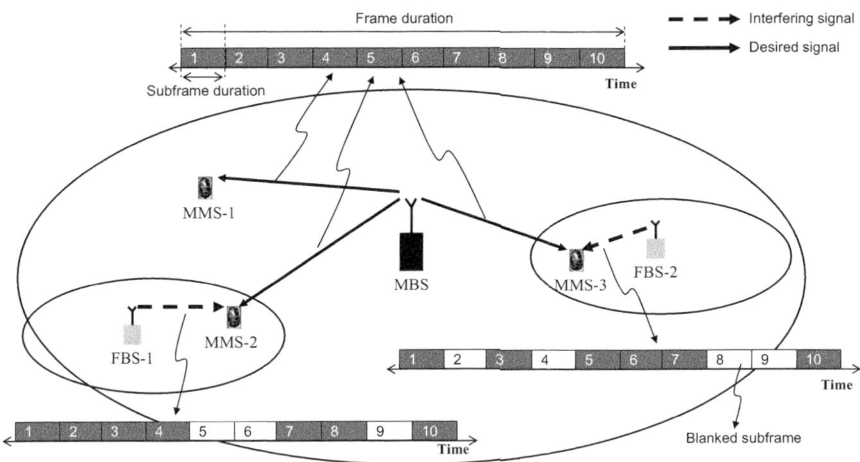

Figure 15.2 Use of blank subframes at femtocells for interference coordination in 3GPP.

interference to victim macrocell users [7–9], such as MMS-2 and MMS-3 in Figure 15.2. The femtocells FBS-1 and FBS-2 are configured not to schedule any transmission (other than the reference signals) in the ABSs for allowing protection of victim macrocell users. For example, in Figure 15.2, subframes 5, 6, and 9 are left blank in FBS-1, while subframes 2, 4, 8, and 9 are left blank in FBS-2. Then, victim macrocell users may be scheduled in macrocell resources overlapping with the blank subframes, i.e., MMS-2 may be scheduled in subframes 5, 6, and 9, while MMS-3 may be scheduled in subframes 2, 4, 8, and 9 of the macrocell, while there is no scheduling restriction for MMS-1.

Even though the ABS pattern may be dynamically changed (as fast as every 40 ms [10]) for picocells through the X2 interface in LTE, for femtocells, the solutions are less dynamic due to the absence of X2 interface between femtocells and macrocell. For example, through operation and management, different blank subframe patterns (known both to the femtocells and macrocell) may be used at different times of the day [10]. In [11], three alternative solutions are proposed in order for the MBS to know the blank subframe pattern in femtocells: (i) a single ABS pattern is configured for all femtocells in the macrocell coverage area (prevents adaptation of the femtocell to the traffic variations in its coverage); (ii) configured blank subframe pattern can be signaled in the system information of femtocells (not very desirable); and (iii) macrocell users may identify the blank subframe pattern through measurements, and report the identified pattern to the macrocell (increased complexity of the MMS). Nevertheless, regardless of whether a static or a dynamic approach is used, the number of blank subframes should be optimized in order to accommodate different numbers of users and their quality of service (QoS) requirements [8, 9], and the framework covered in this chapter can also be readily applied for time-domain resource configuration.

15.1.3   Fairness criteria for resource allocation

In order to maximize the capacity of a macrocell–femtocell network, adaptive access operation of femtocells [12], hybrid resource allocation [13], and adaptive transmit powers [14, 15] have formerly been proposed in the literature. In [16], a QoS parameter that captures the average capacities in femtocell and macrocell tiers has been defined, and spectrum allocation to different tiers is carried out based on this parameter. However, *fairness* of the overall system is not directly accounted for while allocating the spectrum among the tiers.

One of the key aspects of spectrum allocation in heterogeneous networks is to define a metric to measure and evaluate the degree of fairness and QoS in the overall system [17]. The fundamental work in the area was done by Jain [18], which analyzes all the properties of the fairness metric. Variance (σ^2), coefficient of variation (CoV), min-max ratio, and normalized distance are some of the other traditional fairness metrics that have been used for resource allocation problems in the literature [19]. Bandwidth assignment and scheduling related optimization problems using fairness criteria were investigated in [20]. Utility-based fairness indices [21] have been widely recognized due to their flexibility for various application types.

Fair resource allocation and associated criteria have also been investigated by standardization bodies. For example, normalized capacity CDFs of users are used as a fairness metric by 3rd Generation Partnership Project 2 (3GPP2). In this definition of fairness, a CDF of normalized user capacity (with respect to average capacity or maximum user capacity) is considered. Then, a fair scheduler's CDF plot of normalized throughput should lie to the right of a pre-prescribed line of reference. In order words, it ensures that the percentage of the users having very low data rate compared to average data rate does not go above a threshold value, and cell-edge users are not penalized [22].

15.1.4   Chapter outline

The goal of this chapter is to provide a fairness metric for heterogeneous network architectures and to optimize the SSR (ρ) in a dedicated channel approach as in Figure 15.1(b) and in the hybrid channel approach as in Figure 15.1(c), considering the fairness and QoS constraints.[3] First, sum capacities of different tiers in a heterogeneous network are expressed in closed form for all approaches, and capacity-maximizing spectrum splitting is investigated. To fairly allocate the resources to different tiers, a modified QoS-oriented fairness metric is introduced. This metric captures important characteristics of tiered network architectures such as the number of networks in each tier, the number of users in each network, and the QoS requirements of different tiers. Therefore, fairness in the tiered system is effectively captured when compared with the Jain's fairness index (JFI) [18]. Then, a spectrum splitting strategy that simultaneously considers capacity maximization, fairness constraints, and QoS constraints is proposed. For

---

[3] As mentioned in Section 15.1.2, discussions in the chapter also apply to time-domain resource allocation. However, the rest of the chapter will be written from the perspective of spectrum allocation.

Table 15.1 Description of parameters and notation [27].

Parameter	Description
ρ	Let B_M and B_F be the total bandwidth of a macrocell and femtocell network, respectively. Then, ρ is the portion of the accessed bandwidth for the macrocell network, i.e., $1-\rho$ is the portion for the femtocell network.
i, j, k	Indices for the tiers, networks in each tier, and users in each network, respectively.
$N_{N,i}$	Number of networks in i-th tier, ($i = 1, \ldots, T$), i.e., $N_{N,1} = 1$ for macrocell-tier, $N_{N,2}$ is the number of femtocell networks in femtocell-tier.
$N_{U,i,j}$	Number of users in i-th tier and j-th network, i.e., $N_{U,1,1}$ is the number of macrocell-tier users, $N_{U,2,j}$ is the number of users in the j-th femtocell.
N_{Tot}	Total number of users in the system. i.e., $N_{Tot} = \sum_{i=1}^{T} \sum_{j=1}^{N_{N,i}} N_{U,i,j}$.
$C_{i,j,k}$	Capacity of the k-th user in the i-th tier and j-th network. Note that, there is only one macrocell ($j = 1$, for $i = 1$) and several femtocells ($j = 1, \ldots, N_{N,2}$, for $i = 2$).
$B_{i,j,k}$	Bandwidth of the k-th user in the i-th tier and j-th network.
$SINR_{i,j,k}$	Signal to interference plus noise ratio (SINR) of the k-th user in the i-th tier and j-th network.
$P_{i,j,k}$	Received power of the k-th user in the i-th tier and j-th network.
$f_{QTFI}(\mathbf{C})$	QoS-oriented fairness metric. Note that $\mathbf{C}$ is the three-dimensional capacity matrix, which consists of $C_{i,j,k}, \forall i, j, k$.

different SSR values, sum capacities of macrocells and femtocells are obtained through analytical derivations and computer simulations, and compared for various scenarios. In the hybrid approach, resource management with max-min scheduling is investigated.

The remaining part of this chapter is organized as follows. In Section 15.2, the system model to provide total capacity of a macrocell–femtocell network is provided, and a QoS-oriented fairness metric for tiered network structures is proposed. In Section 15.3, capacities of co-channel, dedicated channel, and hybrid channel approaches are derived using a homogeneous Poisson point process (HPPP) and a max-min fair scheduler is introduced for the hybrid channel approach. Numerical results for various scenarios are presented in Section 15.4, followed by concluding remarks in the last section.

15.2   Fairness metric and system model for heterogeneous networks

Consider a two-tier macrocell–femtocell scenario (i.e., $T = 2$), where the macrocell-tier network is the tier-1 network and the femtocell-tier network is the tier-2 network. We follow the notation defined in Table 15.1 and evaluate the capacity within the coverage area of a given macrocell of interest (i.e., $j = 1$ for $i = 1$), surrounded by interfering macrocells. Moreover, $N_{N,2}$ femtocells ($j = 1, \ldots, N_{N,2}$ for $i = 2$) are assumed to be randomly distributed within the coverage area of the given macrocell. Our goal is to maximize

$$C_{Tot} = \sum_{i=1}^{T} \sum_{j=1}^{N_{N,i}} \sum_{k=1}^{N_{U,i,j}} C_{i,j,k}, \tag{15.1}$$

while considering the fairness metric and the QoS parameter.

15.2.1　Capacity of macrocell and femtocell

Using the notation given in Table 15.1, the total capacity for the femtocell-tier (tier-2) can be expressed as

$$C_{\text{Fem}} = C_2 = \sum_{j=1}^{N_{\text{N},2}} \sum_{k=1}^{N_{\text{U},2,j}} \underbrace{B_{2,j,k} \log_2 \left(1 + \frac{P_{2,j,k}}{I_{2,j,k} + B_{2,j,k} N_0} \right)}_{C_{2,j,k}}, \tag{15.2}$$

where $I_{2,j,k}$ denotes the interference power observed by the k-th user from the j-th femtocell, N_0 is the spectral density of noise, and $C_{2,j,k}$ is the capacity of femtocell user-k with the j-th femtocell.

Similarly, using the notation given in Table 15.1, the total capacity of the macrocell tier can be written as

$$C_{\text{Mac}} = C_1 = \sum_{j=1}^{N_{\text{N},1}} \sum_{k=1}^{N_{\text{U},1,j}} \underbrace{B_{1,j,k} \log_2 \left(1 + \frac{P_{1,j,k}}{I_{1,j,k} + B_{1,j,k} N_0} \right)}_{C_{1,j,k}}, \tag{15.3}$$

where $I_{1,j,k}$, and $C_{1,j,k}$ denote interference level and capacity for the k-th macrocell user in the j-th macrocell. For the sake of analytical tractability, we consider that the number of users in each femtocell is assumed to be fixed, i.e., $N_{\text{U},2,j} = N_{\text{U,F}}, \forall j$. Moreover, both macrocell users (MMS) and femtocell users (femtocell mobile station (FMS)) are assumed to be distributed uniformly within each circular macrocell and femtocell area.

15.2.2　QoS-orientation and fairness metric for tiered networks

In this section, we first define a fairness index and propose that a fair spectrum allocation can be achieved by considering the heterogeneous architecture of tiered networks. Then a QoS parameter is also added in the fairness metric to provide QoS orientation for the spectrum allocation.

Fairness index

Jain's fairness index [18] has been a widely used fairness criterion in the literature for resource allocation and can be written as

$$f(x) = \frac{\left(\sum_{i=1}^{N} x_i \right)^2}{N \sum_{i=1}^{N} x_i^2}, \tag{15.4}$$

where N denotes the total number of users and x_i denotes the received allocation for the i-th user. Some of the important properties of (15.4) are as follows: (i) population size independence; (ii) scale and metric independence; (iii) boundedness ($f(x) \in [1/N, 1], \forall x$); (iv) direct relationship; and (v) continuity (non-discrete).

Tiered network structures, such as those including femtocells, picocells, and relay networks overlaid with a macrocell network, introduce a multi-dimensional resource allocation problem. In tiered networks, where users are distributed among tiers and the

Table 15.2 Bounds for fairness indices [27].

FI	Lower bound	Upper bound
$f_{\mathrm{JFI}}(\mathbf{C})$	$1/N_{\mathrm{tot}}$	No explicit solution.[4]
$f_{\mathrm{TFI}}(\mathbf{C})$	$1/N_{\mathrm{tot}}$	1
$f_{\mathrm{WJFI}}(\mathbf{C})$	$\dfrac{\sum_{i=1}^{T}\sum_{j=1}^{N_{\mathrm{N},i}} 1/N_{\mathrm{U},i,j}}{\sum_{i=1}^{T} N_{\mathrm{N},i}}$	1
$f_{\mathrm{QTFI}}(\mathbf{C})$	$1/N_{\mathrm{tot}}$	1

networks within each tier, providing a global fairness index for the entire system requires a modified fairness criteria. Consider a T-tiered architecture where each tier has several networks, similar to the one defined in the previous subsection with the same notation defined in Table 15.1. We propose that a tiered fairness index (TFI) in such a system should be as follows

$$f_{\mathrm{TFI}}(\mathbf{C}) = \frac{\left(\sum_{i=1}^{T}\sum_{j=1}^{N_{\mathrm{N},i}}\sum_{k=1}^{N_{\mathrm{U},i,j}} N_{\mathrm{U},i,j}\, C_{i,j,k}\right)^2}{N_{\mathrm{Tot}}\sum_{i=1}^{T}\sum_{j=1}^{N_{\mathrm{N},i}}\sum_{k=1}^{N_{\mathrm{U},i,j}} N_{\mathrm{U},i,j}^2\, C_{i,j,k}^2}, \tag{15.5}$$

where $\mathbf{C}$ denotes the set of capacities of all the users in all tiers, and N_{Tot} is the total number of users in the entire system:

$$N_{\mathrm{Tot}} = \sum_{i=1}^{T}\sum_{j=1}^{N_{\mathrm{N},i}} N_{\mathrm{U},i,j}. \tag{15.6}$$

The difference of (15.5) from the JFI is that it is a global fairness index for a tiered network and it weighs the tiers and networks according to their numbers of users.[5]

Using the notation given in Table 15.1, (15.4) may be re-written by changing N with N_{Tot} as follows:

$$f_{\mathrm{JFI}}(\mathbf{C}) = \frac{\left(\sum_{i=1}^{T}\sum_{j=1}^{N_{\mathrm{N},i}}\sum_{k=1}^{N_{\mathrm{U},i,j}} C_{i,j,k}\right)^2}{N_{\mathrm{Tot}}\sum_{i=1}^{T}\sum_{j=1}^{N_{\mathrm{N},i}}\sum_{k=1}^{N_{\mathrm{U},i,j}} C_{i,j,k}^2}. \tag{15.7}$$

Since the JFI in (15.7) does not consider the number of users in each network for tiered scenarios, it does not satisfy the boundedness property (see Table 15.2). In other words, $f_{\mathrm{JFI}}(\mathbf{C})$ is no longer tightly bounded within $[1/N_{\mathrm{Tot}}, 1]$. While the number of users in each network varies, the upper bound of JFI changes. This property will be discussed in an example case study later in Section 15.4 (see Figure 15.3).

Finding the JFI for each network in each tier and obtaining a weighted summation of them could be another approach for a modified fairness index for tiered networks, which has an upper bound of 1 as opposed to JFI (see Table 15.2). The weighted sum

---

[4] Cannot be expressed independent of $\mathbf{C}$.
[5] Note that the number of users in each femtocell is assumed to be known.

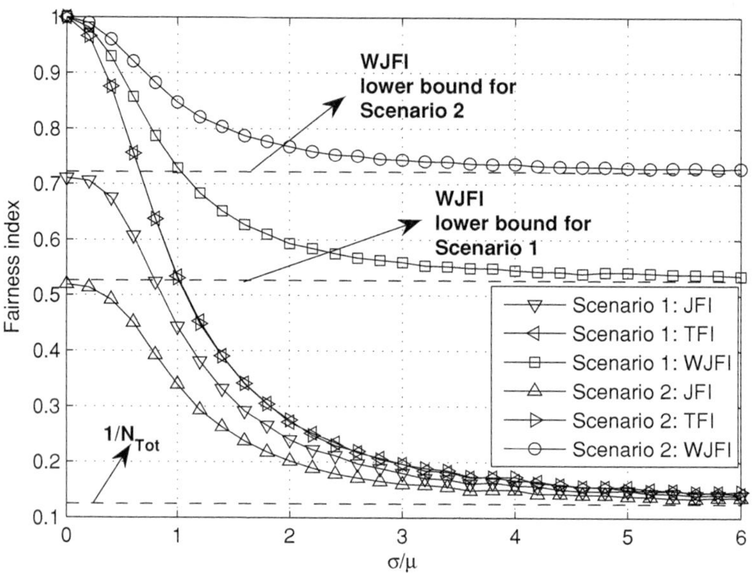

Figure 15.3 Fairness index vs. standard deviation (σ) /mean (μ) for normally distributed resource allocations for each user. ©2010 IEEE. Reprinted, with permission, from [27].

JFI (WJFI) could be written as

$$f_{\mathrm{WJFI}}(\mathbf{C}) = \frac{1}{\sum_{i=1}^{T} N_{\mathrm{N},i}} \times \sum_{i=1}^{T} \sum_{j=1}^{N_{\mathrm{N},i}} \frac{\left(\sum_{k=1}^{N_{\mathrm{U},i,j}} C_{i,j,k}\right)^2}{N_{\mathrm{U},i,j} \sum_{k=1}^{N_{\mathrm{U},i,j}} C_{i,j,k}^2}. \tag{15.8}$$

However, the WJFI does not consider the number of users in each network and weighs the fairness with the total number of networks in the system $(\sum_{i=1}^{T} N_{\mathrm{N},i})$. For instance, if one network (i.e., macrocell network or each one of the femtocell networks) has twice the number of users compared to another network, (15.8) does not consider this and provides equal weights for each network. While this metric has an upper bound of 1, it does not have a lower bound of $1/N_{\mathrm{Tot}}$. Table 15.2 shows that lower bound of the equation is increasing with the decrease in the number of users in each network, which is a very common case for a femtocell scenario.

QoS-oriented tiered fairness index (QTFI)

In tiered networks, it is typically expected that the tiers will have different QoS requirements. For example, in a two-tier network, femtocell users are expected to have significantly better throughput compared to macrocell users due to better link qualities and larger spectrum resources. Therefore, the QoS characteristics of each tier should also be considered within the fairness index in order to have a better representation of fairness within the whole system. Let β_i $(i = 2, \ldots, T)$ denote the QoS parameter defined as the ratio of the sum capacity in the first tier (e.g., macrocell tier) to the sum capacity in a different tier (e.g., femtocell tier)

$$
\beta_i = \frac{1/N_{\mathrm{N},1} \sum_{j=1}^{N_{\mathrm{N},1}} \sum_{k=1}^{N_{\mathrm{U},1,j}} C_{1,j,k}}{1/N_{\mathrm{N},i} \sum_{j=1}^{N_{\mathrm{N},i}} \sum_{k=1}^{N_{\mathrm{U},i,j}} C_{i,j,k}},
\tag{15.9}
$$

where $\beta_1 = 1$. Using this QoS parameter, a modified version of the proposed fairness index in (15.5) can be written as

$$
f_{\mathrm{QTFI}}(\mathbf{C}) = \frac{\left(\sum_{i=1}^{T} \sum_{j=1}^{N_{\mathrm{N},i}} \sum_{k=1}^{N_{\mathrm{U},i,j}} \beta_i N_{\mathrm{U},i,j} C_{i,j,k}\right)^2}{N_{\mathrm{Tot}} \sum_{i=1}^{T} \sum_{j=1}^{N_{\mathrm{N},i}} \sum_{k=1}^{N_{\mathrm{U},i,j}} \beta_i^2 N_{\mathrm{U},i,j}^2 C_{i,j,k}^2}.
\tag{15.10}
$$

Note that (15.10) converges to (15.5) while $\beta_i \to 1$ ($i = 2, \ldots, T$). Moreover, if the macrocell is the only tier in the system (i.e., $T = 1$, $N_{\mathrm{N},1} = 1$), then (15.5) converges to the JFI given in (15.7). This proves that the provided equations are the modified versions of the JFI in order to satisfy the boundedness property within $[1/N_{\mathrm{Tot}}, 1]$. Table 15.2 summarizes the lower and upper bounds of the above-mentioned fairness indices.

It is important to note that the proposed fairness indices $f_{\mathrm{TFI}}(\mathbf{C})$ and $f_{\mathrm{QTFI}}(\mathbf{C})$ are bounded and independent of the number of networks in the tiered-network structure. The JFI upper bound is not independent from the allocated resources, and therefore the upper bound cannot be expressed in closed form. Similarly, the lower bound of the WJFI depends on the number of networks and the number of users within the networks in each tier. As opposed to JFI and WJFI, proposed fairness indices TFI and QTFI provide a controlled metric for resource allocation problems in tiered networks.

15.3 Resource partitioning in macrocell–femtocell networks

In this section, we consider a two-tier macrocell–femtocell scenario (i.e., $T = 2$), where the macrocell network is the tier-1 network and the femtocell network is the tier-2 network. The goal is to split the total bandwidth B among tiers such that: (i) the capacity of the overall system is maximized; (ii) a level of global fairness is ensured between users in different tiers; and (iii) QoS requirements of users in different tiers in terms of relative data rates are satisfied. As shown in Figure 15.1(b) for the split-spectrum approach, the portion of the accessed bandwidth for the macrocell tier is $\rho = \frac{B_{\mathrm{M}}}{B}$, where $B_{\mathrm{M}} = \sum_{k=1}^{N_{\mathrm{U},1,1}} B_{1,1,k}$. Therefore, we have $B_{\mathrm{F}} = (1 - \rho)B = \sum_{k=1}^{N_{\mathrm{U},2,j}} B_{2,j,k}$, $\forall j$. For the hybrid approach as in Figure 15.1(c), the portion of the accessed bandwidth for the macrocell tier is the total bandwidth $B_{\mathrm{M}} = B$, and we have $B_{\mathrm{F}} = (1 - \rho)B$ where $B_{\mathrm{M}} = \sum_{k=1}^{N_{\mathrm{U},1,1}} B_{1,1,k}$ and $B_{\mathrm{F}} = \sum_{k=1}^{N_{\mathrm{U},2,j}} B_{2,j,k}$, $\forall j$. In both approaches, our goal when splitting the spectrum is to maximize

$$
C_{\mathrm{Tot}}(\rho) = \sum_{i=1}^{T} \sum_{j=1}^{N_{\mathrm{N},i}} \sum_{k=1}^{N_{\mathrm{U},i,j}} C_{i,j,k}(\rho),
\tag{15.11}
$$

while considering the fairness metric and QoS parameter.

15.3.1 Macrocell and femtocell deployment using homogeneous Poisson point process

In this section, we focus on a general analytical formulation of the macrocell–femtocell capacities by employing statistical models for MBS and FBS locations. We focus on an arbitrary mobile station (MS) (called a user equipment (UE) in 3GPP terminology) in this region and calculate the downlink capacity for macrocell and femtocell users for dedicated channel, co-channel, and hybrid channel scenarios. These results (which are functions only of the macro and femto relative densities, transmit powers, the parameters of the wireless channel, and the SSR) provide valuable insights for the architecture planning process for joint femto–macro deployments under different fairness and QoS criteria.

The MBS locations are assumed to be points of an HPPP on the plane with intensity λ, and have the following properties:

1. The number of MBSs $N(\mathbb{B})$ in any finite region $\mathbb{B}$ is Poisson($\lambda \times \text{area}(\mathbb{B})$), i.e.,

$$\mathbb{P}\{N(\mathbb{B}) = n\} = \mathrm{e}^{-\lambda \times \text{area}(\mathbb{B})} \frac{[\lambda \times \text{area}(\mathbb{B})]^n}{n!} \quad \textit{for } n = 0, 1, \ldots,$$

   with mean $\mathbb{E}N(\mathbb{B}) = \lambda \times \text{area}(\mathbb{B})$.
2. $\forall \mathbb{B}, \mathbb{B}': \mathbb{B} \cap \mathbb{B}' = \varnothing \Rightarrow N(\mathbb{B}), N(\mathbb{B}')$ are independent.
3. $\forall \mathbb{B}$, given $N(\mathbb{B}) = n$, these n MBS are i.i.d and uniformly distributed over $\mathbb{B}$.

Note that λ is in units of points per square meter. We model the locations of FBS by points of an independent HPPP with intensity λ' and all the FBSs are operating in CSG mode.

The wireless channel model we use in this study can be defined by the following assumptions:

1. Path-loss exponent is δ.
2. Fading in all macrocellular downlinks are i.i.d. Rayleigh with mean 1.[6]
3. All MBSs (FBSs) transmit with the same reference symbol power P_{RS} (P'_{RS}).
4. An MS at distance r from an MBS has reference symbol received power (reference signal received power (RSRP) in LTE terminology)

$$\text{RSRP}(r) = \frac{H}{r^{\delta}}, \quad H \sim \exp(P), \quad P \equiv K P_{RS} \tag{15.12}$$

   where the exponential distribution of H arises from the Rayleigh fading assumption, and K is a quantity that takes into account the relative heights of the transmitter and receiver on the link, etc., and is considered the same for all links from any MS location to any MBS. We are interested in MS locations whose distance from the nearest MBS exceeds some $r_{\min}$.
5. Similarly, an MS at distance r' from an FBS has an RSRP given by

$$\text{RSRP}'(r') = \frac{H'}{r'^{\delta}}, \quad H' \sim \exp(P'), \quad P' \equiv K' P'_{RS}. \tag{15.13}$$

---

[6] The analysis in this study can be applied to general models; however, the expressions are more complex, therefore we restrict ourselves for this case for brevity.

15.3.2 Results from the theory of Poisson point processes

In this subsection we will review the fundamental results from theory of HPPP that are applicable for the capacity analysis in this study.

Theorem 15.1 *Suppose that there are transmitters located at points of M independent HPPPs $1, \ldots, M$, with intensities $v_1, \ldots, v_M$, respectively, and that the MS must be a minimum distance of $d_{min,k}$ from the nearest transmitter of HPPP k, $k = 1, \ldots, M$. The fading coefficients on all transmitter–MS links are independent, and those on the links between the UE and the transmitters belonging to HPPP k are i.i.d. $Exp(\mu_k)$, $k = 1, \ldots, M$. Suppose the MS is at a distance of r from the nearest transmitter of HPPP1. Then the complementary cumulative distribution function (CCDF) of SIR(r), the signal to interference ratio (SIR) at the MS when served by the nearest transmitter of HPPP1, is given by*

$$\mathbb{P}\{\Gamma_1 > \gamma \mid R_1 = r_1\} = exp\left(-u\gamma^{\frac{2}{\delta}}\left[G\left(\frac{1}{\gamma^{\frac{2}{\delta}}}\right) + \sum_{k=2}^{M}\Theta_k G\left(\frac{m_k}{u\gamma^{\frac{2}{\delta}}\Theta_k}\right)\right]\right). \quad (15.14)$$

where $u = v_1\pi r_1^2$ and for $k = 1, \ldots, M$,

$$m_k = v_k\pi d_{min,k}^2, \qquad \Theta_k = \frac{v_k}{v_1}\left(\frac{\mu_k}{\mu_1}\right)^{2/\delta}, \quad (15.15)$$

$$G(y) = \int_y^\infty \frac{dx}{1 + x^{\frac{\delta}{2}}} = \left\{\begin{array}{ll} \pi/2 - tan^{-1}y & \delta = 4 \\ {}_2F_1(1, \frac{\delta}{2}; 1 + \frac{\delta}{2}; -x^{\frac{\delta}{2}})x\mid_y^\infty & \delta \neq 4 \end{array}\right\}, \quad (15.16)$$

and ${}_2F_1(a, b; c; z)$ is the hypergeometric function

$$_2F_1(a, b; c; z) = 1 + \sum_{n=1}^{\infty}\frac{z^n}{n!}\prod_{l=0}^{n-1}\frac{(a+l)(b+l)}{c+l}. \quad (15.17)$$

Proof. See [23] and Chapter 3 of this book.

To provide the mean capacity, we use the SINR (Γ) distribution (15.14)[7] in macrocell and femtocell tier capacities provided in (15.2), (15.3). Mean rate of a UE at distance $R_1 = r_1$ from nearest base station (BS) of HPPP1 can be written as

$$C_1(r_1) = \mathbb{E}_{\Gamma_1}\left[\log_2\left(1 + \Gamma_1(r_1)\right)\right] = \int_0^\infty \mathbb{P}\{\Gamma_1(r_1) > 2^x - 1 \mid R_1 = r_1\}dx. \quad (15.18)$$

Let $\lambda_{MS,1}$ be the density of the MSs for the HPPP1. Then, the aggregate rate over the region served by a single BS of HPPP1 can be written as

$$C_{1,BS} = \int_{d_{min,1}}^{d_{max,1}} 2\pi r_1\lambda_{MS,1}C_1(r_1)dr_1 \quad (15.19)$$

---

[7] Although (15.14) provides the SIR distribution, we can assume that the network is interference limited (SINR $\simeq$ SIR).

which, after a change of variable $t = r_1^2$, can be written as

$$C_{1,\mathrm{BS}} = \int_{d_{\min,1}^2}^{d_{\max,1}^2} \pi \lambda_{\mathrm{MS},1} \tilde{C}_1(t)\,\mathrm{d}t. \tag{15.20}$$

From (15.14), $\tilde{C}(t)$ in (15.20) is given by

$$\tilde{C}(t) = \int_0^\infty \exp\left(-\nu_1 \pi t (2^x - 1)^{\frac{2}{\delta}}\left[G\left(\frac{1}{(2^x - 1)^{\frac{2}{\delta}}}\right)\right.\right.$$
$$\left.\left. + \sum_{k=2}^{M} \Theta_k G\left(\frac{m_k}{\nu_k \pi t (2^x - 1)^{\frac{2}{\delta}} \Theta_k}\right)\right]\right) \mathrm{d}x. \tag{15.21}$$

Then, the capacity for a given tier can be written in a generalized form as follows

$$C_i = \frac{B N_{\mathrm{N},i}}{\pi \lambda_{\mathrm{MS},1}(d_{\max,1}^2 - d_{\min,1}^2)} \int_{d_{\min,1}^2}^{d_{\max,1}^2} \pi \lambda_{\mathrm{MS},1} \tilde{C}(t)\,\mathrm{d}t$$
$$\triangleq \mathbb{C}\left(B, N_{\mathrm{N},i}, M, \boldsymbol{\nu}, \boldsymbol{\mu}, \boldsymbol{m}, \boldsymbol{\Theta}, d_{\max,1}, d_{\min,1}\right) \tag{15.22}$$

where the total bandwidth is assumed to be distributed in a round-robin fashion in each MBS and FBS. Note that $\boldsymbol{\nu} = [\nu_1, \ldots, \nu_M]$, $\boldsymbol{\mu} = [\mu_1, \ldots, \mu_M]$, $\boldsymbol{m} = [m_1, \ldots, m_M]$, and $\boldsymbol{\Theta} = [\Theta_1, \ldots, \Theta_M]$ are vectors of $1 \times M$, with m_k and Θ_k as in (15.15). For a given area with radius R, the number of macrocells and femtocells can be calculated as $N_{\mathrm{N},1} = \lambda \pi R^2$ and $N_{\mathrm{N},2} = \lambda' \pi R^2$, respectively.

15.3.3   Co-channel macrocell/femtocell networks

Using (15.2) and the notation given in Table 15.1, the capacity of the femtocell tier can be written as

$$C_{\mathrm{Fem}} = \frac{B}{N_{\mathrm{U,F}}} \mathbb{E}\left[\sum_{j=1}^{N_{\mathrm{N},2}} \sum_{k=1}^{N_{\mathrm{U,F}}} \log_2\left(1 + \Gamma_{j,k}\right)\right]. \tag{15.23}$$

where $B_{2,j,k} = \frac{B_{\mathrm{F}}}{N_{\mathrm{U},2,j}} = \frac{B}{N_{\mathrm{U,F}}}$, $\forall j, k$. Then using (15.22) we can rewrite (15.23) as

$$C_{\mathrm{Fem}} = \mathbb{C}\left(B, N_{\mathrm{N},2}, M, \boldsymbol{\nu}, \boldsymbol{\mu}, \boldsymbol{m}, \boldsymbol{\Theta}, d_{\max,1}, d_{\min,1}\right), \tag{15.24}$$

where $M = 2$, HPPP1 is FBS locations, HPPP2 is MBS locations, $(\nu_1, \nu_2) = (\lambda', \lambda)$, $(\mu_1, \mu_2) = (P', P)$, $(m_1, m_2) = (\lambda' \pi r_{\min,f}^2, \lambda \pi r_{\min,m}^2)$, $\Theta_2 = \frac{\lambda}{\lambda'}(\frac{P}{P'})^{2/\delta}$, and $(d_{\max,1}, d_{\min,1}) = (r_{\max,f}, r_{\min,f})$. Note that $r_{\min,m}$ and $r_{\max,m}$ (or $r_{\min,f}$, $r_{\max,f}$) are the minimum and maximum possible distances between an MS and an MBS (or an FBS), respectively. Similarly, using (15.3), we have

$$C_{\mathrm{Mac}} = \frac{B}{N_{\mathrm{U,M}}} \mathbb{E}\left[\sum_{j=1}^{N_{\mathrm{N},1}} \sum_{k=1}^{N_{\mathrm{U,M}}} \log_2\left(1 + \Gamma_{j,k}\right)\right]. \tag{15.25}$$

where $B_{2,j,k} = \frac{B_F}{N_{U,2,j}} = \frac{B}{N_{U,F}}$, $\forall j, k$. Then using (15.22) we can rewrite (15.25) as

$$C_{\text{Mac}} = \mathbb{C}\left(B, N_{N,1}, M, \boldsymbol{v}, \boldsymbol{\mu}, \boldsymbol{m}, \boldsymbol{\Theta}, d_{\max,1}, d_{\min,1}\right) \tag{15.26}$$

where $M = 2$, HPPP1 is MBS locations, HPPP2 is FBS locations, $(v_1, v_2) = (\lambda, \lambda')$, $(\mu_1, \mu_2) = (P, P')$, $(m_1, m_2) = (\lambda \pi r^2_{\min,m}, \lambda' \pi r^2_{\min,f})$, $\Theta_2 = \frac{\lambda'}{\lambda}(\frac{P'}{P})^{2/\delta}$, and $(d_{\max,1}, d_{\min,1}) = (r_{\max,m}, r_{\min,m})$.

15.3.4   Dedicated channel macrocell/femtocell networks

Using (15.2) and the notation given in Table 15.1, the capacity of the femtocell tier can be written as

$$C_{\text{Fem}}(\rho) = \frac{B(1-\rho)}{N_{U,F}} \mathbb{E}\left[\sum_{j=1}^{N_{N,2}} \sum_{k=1}^{N_{U,F}} \log_2\left(1 + \Gamma_{j,k}\right)\right]. \tag{15.27}$$

Then using (15.22) we can rewrite (15.27) as

$$C_{\text{Fem}}(\rho) = \mathbb{C}\left(B(1-\rho), N_{N,2}, M, \boldsymbol{v}, \boldsymbol{\mu}, \boldsymbol{m}, 0, d_{\max,1}, d_{\min,1}\right) \tag{15.28}$$

where $M = 1$, HPPP1 is FBS locations, $v_1 = \lambda'$, $\mu_1 = P'$, $m_1 = \lambda' \pi r^2_{\min,f}$, and $(d_{\max,1}, d_{\min,1}) = (r_{\max,f}, r_{\min,f})$. Similarly, using (15.3)

$$C_{\text{Mac}}(\rho) = \frac{B\rho}{N_{U,M}} \mathbb{E}\left[\sum_{j=1}^{N_{N,1}} \sum_{k=1}^{N_{U,M}} \log_2\left(1 + \Gamma_{j,k}\right)\right]. \tag{15.29}$$

Then using (15.22) we can rewrite (15.29) as

$$C_{\text{Mac}}(\rho) = \mathbb{C}\left(B\rho, N_{N,1}, M, v, \mu, m, 0, d_{\max,1}, d_{\min,1}\right) \tag{15.30}$$

where $M = 1$, HPPP1 is MBS locations, $v_1 = \lambda$, $\mu_1 = P$, $m_1 = \lambda \pi r^2_{\min,m}$, and $(d_{\max,1}, d_{\min,1}) = (r_{\max,m}, r_{\min,m})$.

Therefore the total capacity for the dedicated channel scenario using (15.28) and (15.30) can be given as $C_{\text{Tot}}(\rho) = C_{\text{Mac}}(\rho) + C_{\text{Fem}}(\rho)$. Then, spectrum splitting ρ value that maximizes the $C_{\text{Tot}}(\rho)$ can be expressed as follows:

$$\rho_{\max} = \arg \max_{0 \le \rho \le 1} C_{\text{Tot}}(\rho). \tag{15.31}$$

Note that the objective function in (15.31) is a linear combination of (15.28) and (15.30). Since each femtocell reuses the spectrum more frequently, the capacity equation given in (15.28) includes a larger multiplying term $N_{N,2} > N_{N,1}$. Therefore, if the SINR levels of users in each tier are similar, objective function (15.31) will be maximized at $\rho = 0$. This issue is also investigated by calculating per-tier area spectral efficiencies (ASEs) in [24] and it is shown that capacity is maximized at extreme points without a fairness or QoS parameter. Such a partitioning is obviously unfair since it results in a greedy allocation to one of the tiers, which will be discussed in more detail in Section 15.4.

15.3.5   Hybrid approach for resource allocation

In this section we investigate the hybrid approach scenario as it is shown in Figure 15.1(c). Using (15.2) and the notation given in Table 15.1, the total capacity of the femtocell network can be calculated with a slight modification of (15.24), where the bandwidth of femtocell is ρB and femtocell users are always co-channel with the macrocell, that is,

$$C_{\mathrm{Fem}}(\rho) = \mathbb{C}\left(B(1-\rho), N_{\mathrm{N},2}, M, \boldsymbol{v}, \boldsymbol{\mu}, \boldsymbol{m}, \boldsymbol{\Theta}, d_{\max,1}, d_{\min,1}\right) \tag{15.32}$$

where $M = 2$, HPPP1 is FBS locations, HPPP2 is MBS locations, $(v_1, v_2) = (\lambda', \lambda)$, $(\mu_1, \mu_2) = (P', P)$, $(m_1, m_2) = (\lambda'\pi r^2_{\min,f}, \lambda\pi r^2_{\min,m})$, $\Theta_2 = \frac{\lambda}{\lambda'}(\frac{P}{P'})^{2/\delta}$, and $(d_{\max,1}, d_{\min,1}) = (r_{\max,f}, r_{\min,f})$. Note that $r_{\min,m}$ and $r_{\min,f}$ are the minimum distances between an MS–MBS and an MS–FBS, respectively, and $r_{\max,m}$ and $r_{\max,f}$ are the maximum distances between an MS–MBS, and an MS–FBS, respectively.

On the other hand, to calculate the macrocell capacity, the following steps should be followed:

- Consider a macrocell MS at distance r_1 from its nearest MBS. Let γ_c be the minimum rate for scheduling an MMS to dedicated channel portion of hybrid channel. Then the instantaneous rate of this MS from the MBS can be given as

$$C_{\mathrm{macro}}(r_1) = (1-\rho)C_{\mathrm{co}}(r_1)1\{C_{\mathrm{co}}(r_1) > \gamma_c\}$$
$$+ \rho C_{\mathrm{ded}}(r_1)1\{C_{\mathrm{co}}(r_1) \le \gamma_c\}. \tag{15.33}$$

Note that $C_{\mathrm{co}}(r_1)$, $C_{\mathrm{ded}}(r_1)$ are instantaneous (includes effects of fading) rates derived in (15.18) for co-channel and dedicated channel scenarios,[8] respectively.

- Therefore the mean rate for that MS can be written as

$$\bar{C}_{\mathrm{macro}}(r_1) = (1-\rho)\mathbb{E}\left[C_{\mathrm{co}}(r_1)1\{C_{\mathrm{co}}(r_1) > \gamma_c\}\right]$$
$$+ \rho\mathbb{E}\left[C_{\mathrm{ded}}(r_1)1\{C_{\mathrm{co}}(r_1) \le \gamma_c\}\right]. \tag{15.34}$$

- And finally, the aggregate rate for an MBS can be written as

$$C_{\mathrm{Mac}}(\rho) = \frac{B N_{\mathrm{N},1}}{\pi \lambda_{\mathrm{MS},1}(d^2_{\max,1} - d^2_{\min,1})} \int_{d_{\min,1}}^{d_{\max,1}} 2\pi \lambda_{\mathrm{MS},1} r_1 \bar{C}_{\mathrm{macro}}(r_1)\mathrm{d}r_1. \tag{15.35}$$

Therefore the macro-tier capacity can be provided as in (15.35), by solving for (15.34). Let $X = C_{\mathrm{co}}(r_1)1\{C_{\mathrm{co}}(r_1) > \gamma_c\}$. Then,

$$\mathbb{P}\{X > x\} = \begin{cases} \mathbb{P}\{\Gamma_{\mathrm{Co}}(r_1) > 2^{\gamma_c} - 1\}, & 0 < x \le \gamma_c \\ \mathbb{P}\{\Gamma_{\mathrm{Co}}(r_1) > 2^x - 1\}, & x > \gamma_c \end{cases}, \tag{15.36}$$

---

[8] As it is described in Sections 15.3.3 and 15.3.4, for the co-channel scenario, we have $M = 2$, HPPP1 is MBS locations, HPPP2 is FBS locations, $(v_1, v_2) = (\lambda, \lambda')$, $(\mu_1, \mu_2) = (P, P')$, $(m_1, m_2) = (\lambda\pi r^2_{\min,m}, \lambda'\pi r^2_{\min,f})$, $\Theta_2 = \frac{\lambda'}{\lambda}(\frac{P'}{P})^{2/\delta}$, and $(d_{\max,1}, d_{\min,1}) = (r_{\max,m}, r_{\min,m})$. On the other hand, for the dedicated channnel scenario, we have $M = 1$, HPPP1 is MBS locations, $v_1 = \lambda$, $\mu_1 = P$, $m_1 = \lambda\pi r^2_{\min,m}$, and $(d_{\max,1}, d_{\min,1}) = (r_{\max,m}, r_{\min,m})$.

from which we can write

$$\mathbb{E}[X] = \int_0^\infty \mathbb{P}\{X > x\}dx$$

$$= \gamma_c \underbrace{\mathbb{P}\{\Gamma_{\mathrm{Co}}(r_1) > 2^{\gamma_c} - 1\}}_{\text{substitute } \gamma = 2^{\gamma_c} - 1 \text{ in } (15.14)} + \underbrace{\int_{\gamma_c}^\infty \mathbb{P}\{\Gamma_{\mathrm{Co}}(r_1) > 2^x - 1\}dx}_{\text{integral in } (15.18) \text{ with } \gamma_c}\ . \qquad (15.37)$$

To calculate (15.34), we assume that interference from indoor femtocells to outdoor macrocell MSs is negligible due to the low power of femtocells and wall loss, therefore $C_{\mathrm{ded}}(r_1) \simeq C_{\mathrm{co}}(r_1)$. Then

$$\mathbb{E}\Big[C_{\mathrm{ded}}(r_1)\mathbb{1}\{C_{\mathrm{co}}(r_1) \le \gamma_c\}\Big] \simeq \mathbb{E}\Big[C_{\mathrm{co}}(r_1)\big(1 - \mathbb{1}\{C_{\mathrm{co}}(r_1) > \gamma_c\}\big)\Big]$$

$$= \mathbb{E}[C_{\mathrm{co}}(r_1)] - \underbrace{\mathbb{E}\Big[C_{\mathrm{co}}(r_1)\big(\mathbb{1}\{C_{\mathrm{co}}(r_1) > \gamma_c\}\big)\Big]}_{\mathbb{E}[X]}\ . \qquad (15.38)$$

Therefore (15.35) can be calculated using

$$\bar{C}_{\mathrm{macro}}(r_1) = (1-\rho)\mathbb{E}\Big[C_{\mathrm{co}}(r_1)\mathbb{1}\{C_{\mathrm{co}}(r_1) > \gamma_c\}\Big] + \rho\mathbb{E}\Big[C_{\mathrm{ded}}(r_1)\mathbb{1}\{C_{\mathrm{co}}(r_1) \le \gamma_c\}\Big]$$

$$= (1-2\rho)\mathbb{E}[X] + \rho\mathbb{E}\Big[C_{\mathrm{co}}(r_1)\Big], \qquad (15.39)$$

and the total capacity for the hybrid channel scenario using (15.32) and (15.35) can be given as $C_{\mathrm{Tot}}(\rho) = C_{\mathrm{Mac}}(\rho) + C_{\mathrm{Fem}}(\rho)$. Then, similar to (15.31), spectrum splitting ρ value that maximizes the $C_{\mathrm{Tot}}(\rho)$ can be expressed as

$$\rho_{\max} = \arg \max_{0 \le \rho \le 1} C_{\mathrm{Tot}}(\rho). \qquad (15.40)$$

In the hybrid channel scenario, jointly choosing γ_c and ρ becomes important since different sum capacities and fairness levels could be achieved with these two values. If the macrocell users that have lower SINRs can be scheduled to the dedicated channel portion, there will be no victim users. Therefore, the number of users assigned to the dedicated channel should be selected carefully since the amount of bandwidth corresponding to the dedicated channel should be kept low in order to efficiently reuse the resources in both tiers. In this chapter we consider a scheduler similar to the one discussed in [25]. First the macrocell MSs are sorted with respect to their maximum achievable co-channel capacities, and then $N_{\mathrm{U},M}^{(\rho)}$ mobile user equipments (MUEs) with worse capacities are scheduled in the dedicated channel portion (i.e., $N_{\mathrm{U},M}^{(1-\rho)}$ MUEs scheduled in co-channel portion). Note that, $N_{\mathrm{U},M}^{(\rho)}$ and γ_c has a direct relation, that is, γ_c determines the number of users that will be assigned to dedicated channel portion $N_{\mathrm{U},M}^{(\rho)}$. However, it is also important to note that the value of SSR (ρ) determines the bandwidth to be assigned for each user depending on co-channel or dedicated channel spectrum. We investigate the optimum solution of this problem in our computer simulations in detail. In our simulations, while selecting of $N_{\mathrm{U},M}^{(\rho)}$, we consider a max-min scheduler that maximizes the minimum capacity of macrocell users as follows:

$$N_{U,M}^{(\rho)} = \arg \max_{N_{U,M}^{(\rho)}} \left\{ \min_k \ C_{1,1,k} \right\}. \tag{15.41}$$

$$\rho = \arg \max_{0 \le \rho \le 1} \left\{ C_{\text{Tot}}(\rho, N_{U,M}^{(\rho)}) \right\}. \tag{15.42}$$

By using simple max-min capacity scheduling, the minimum capacity of MMSs are maximized by assigning them to a dedicated channel portion while also maximizing the overall capacity of the macrocell–femtocell network. The fairness and QoS orientation constraint in the network can also be introduced by using the fairness metric given in (15.10).

15.4   Numerical results

In this section, the numerical results for both analytical derivations and simulations are presented. First, we investigate the behavior of the discussed fairness indices for a particular scenario. Then, we present the optimum spectrum splitting strategy based on computer simulations and analytical derivations for co-channel, dedicated channel, and hybrid scenarios. Mathematical modeling is shown to be aligned with simulation results when the basic simulation parameters shown in Table 15.3 are used. Finally, to investigate different approaches in further detail, a 3GPP compatible simulator using the parameters given in [26] is used together with a max-min fair scheduler for the hybrid approach. Some of the critical parameters used for 3GPP aligned simulations are also summarized in Table 15.3. Both for simulation and theoretical results, round-robin scheduling is considered for dedicated and co-channel approaches. For the hybrid approach theoretical analysis, $\gamma_c = 10^4$ bps is a capacity threshold that determines the number of users to be scheduled for co-channel and dedicated channel portions in MBS, on the other hand for hybrid approach simulations, a max-min fair scheduler is considered for MBS and while a round-robin scheduler is considered for FBS.

15.4.1   Comparison of different fairness metrics

The effect of the number of networks and the number of users in each network with the bounds in Table 15.2 are investigated in a two-tier network case study ($T = 2$), where tier-1 has one network ($N_{N,1} = 1$) and tier-2 has two networks ($N_{N,2} = 2$). We consider two different scenarios to provide a better understanding for the metrics and their related bounds. In the first scenario, we assume that there are $N_{U,1,1} = 4$ users in tier-1, network-1, and $N_{U,2,1} = 3$, $N_{U,2,2} = 1$ users for tier-2, networks 1 and 2. Therefore, there are a total of eight users in the network for the first scenario. In the second scenario, we do not change the total number of users; however, we consider $N_{U,1,1} = 6$ users in tier-1, network-1 and $N_{U,2,1} = 1$, $N_{U,2,2} = 1$ users for tier-2, networks 1 and 2.

Table 15.3 Numerical parameters for analytical/simulation results.

	Description/value	
Parameter	Analytical results and basic simulator	3GPP compatible simulator
Macrocellular	HPPP for analytical and hexagonal layout with cell-center BSs for simulations.	Hexagonal layout with cell-center BSs.
Number of MBS	Infinite for analytical and 19 cell with wrap around for simulations.	19 cell with wrap around.
Inter-MBS distance	500 m in average. Therefore the density of macrocells (λ_{MBS}) is $\frac{1}{500\sqrt{3}/2} = 4.62 \times 10^{-6}$. Similarly density of femtocells $\lambda_{FBS} = 12 * \lambda_{MBS} = 5.54 \times 10^{-5}$.	500 m.
Number of FBS	12 per each macrocell.	12 per each macrocell.
FBS distribution	12 FBSs are randomly and uniformly distributed within each site.	4 FBSs are randomly and uniformly distributed within each sector.
Inter FBS distance	Varies FBS locations and path-loss model.	Varies FBS locations and path-loss model.
MBS–MMS minimum distance constraint	35 m [26].	35 m [26].
FBS–FMS minimum distance constraint	5 m [26].	5 m [26].
Bandwidth	10 MHz.	10 MHz.
DL transmit power mBS	60 dBm Tx power at MBS.	46 dBm Tx power at MBS, 14 dBi Antenna gain [26]. 3 sectors with 3-D antenna pattern. Antenna height 32 meters.
DL transmit power FBS	20 dBm.	20 dBm with antenna gain of 5 dBm.
Thermal noise density	-174 dBm/Hz.	-174 dBm/Hz.
Path-loss model (macrocell)	$128.1 + 40\log_{10}(R)$ (R in kilometers).	$128.1 + 37.6\log_{10}(R)$ (R in kilometers).
Path-loss model (femtocell)	$127 + 40\log_{10}(R)$ (R in kilometers).	$127 + 30\log_{10}(R)$ (R in kilometers).
Wall attenuation loss	20 dB.	20 dB.

The allocated resources for each user are assumed to be partitioned in a round-robin fashion within each network in all tiers and the capacity of each network is normally distributed with mean μ and variance σ^2 ($C_{i,j} \sim \mathcal{N}(\mu, \sigma^2)$). Figure 15.3 shows that proposed fairness index TFI is between $[1/N_{\text{Tot}}, 1]$ with controlled bounding, converging to WJFI at 1 for small standard deviation values. On the other hand TFI converges to JFI at 0.125 for increasing standard deviation. Moreover, the upper bound of JFI is decreased and the lower bound of WJFI is increased in Scenario 2 compared to Scenario 1. The

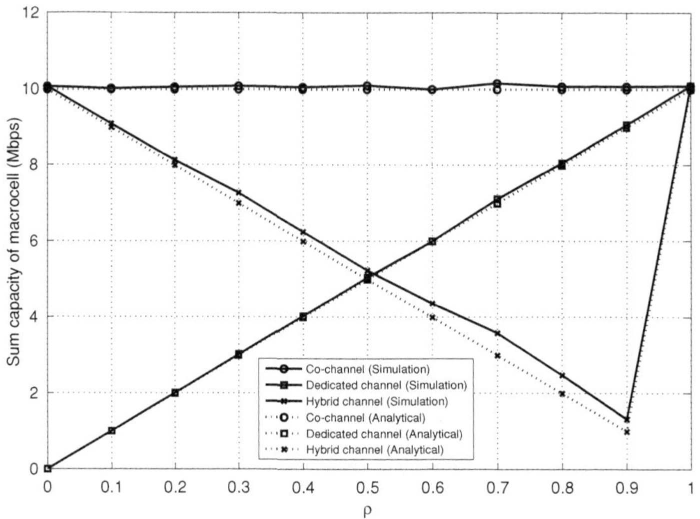

Figure 15.4 Sum capacity of macrocell users vs. SSR ρ.

non-even distribution of the users in networks increases the lower bound of the WJFI. For instance in Scenario 2, the networks 1 and 2 in tier 2 have only 1 user. Calculating the lower bound of WJFI according to Table 15.2 for Scenario 1 and Scenario 2 provides 0.527 and 0.722, respectively, which could also be tracked from Figure 15.3. Although an upper bound independent from allocated resource could not be achieved for JFI, Figure 15.3 shows that while the number of users in a network (for instance the number of users in tier 1 network 1 is very high compared to tier 2 networks) increases, the upper bound decreases.

15.4.2   Numerical results for analytical derivations

In this section we present the numerical results for equations derived through (15.23)–(15.39) and verify them through computer simulations. The simulation scenario includes analytical derivation assumptions and uses the parameters listed in Table 15.3. Note that the path-loss exponent (δ) for both macrocell and femtocell is selected as 4 in order to use the equations derived in previous sections. Three hundred users are dropped randomly and uniformly within each site, with two users for each CSG femtocell [26]. Hence, there are $300 - 12 \times 2 = 276$ macrocell users within each cell.

Figure 15.4 shows the sum capacity of a macrocell for different SSR values for co-channel, dedicated channel, and hybrid approaches. Results show that the co-channel scenario has better capacity when compared with dedicated and hybrid channel approaches. While ρ is increasing, the bandwidth assigned to macrocell users is increasing, and at $\rho = 1$, the capacity of the dedicated channel is almost same as with the co-channel approach. This result shows that for the given scenario, CSG indoor femtocell BSs are not causing severe interference to macro MSs, and therefore the co-channel scenario outperforms the dedicated channel scenario. On the other hand, for the hybrid channel

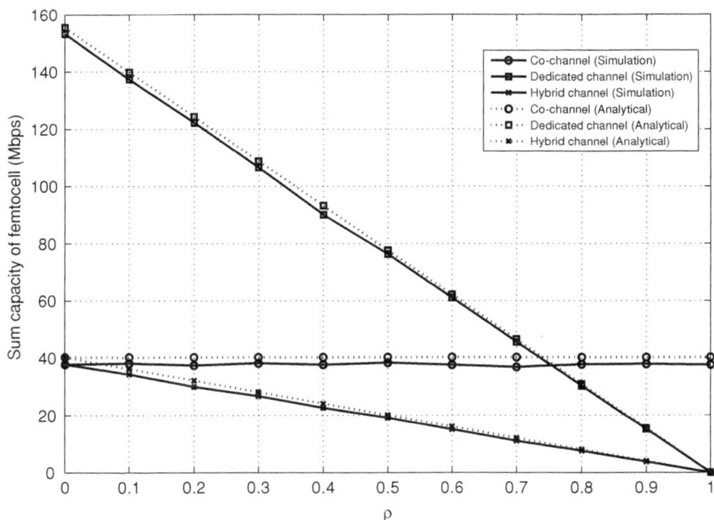

Figure 15.5 Sum capacity of femtocell users vs. SSR ρ.

scenario, for the given $\gamma_c = 10^4$ bps, increasing SSR decreases the macrocell capacity. For a fixed γ_c value, the number of users assigned to the dedicated channel portion is fixed and those users are the ones that have lowest capacity. Therefore, increasing the dedicated channel portion with increasing ρ decreases the capacity since the bandwidth assigned to a small number of users which have lower SINRs decreases the sum capacity in a macrocell. In the extreme cases $\rho = 0$, and $\rho = 1$, the hybrid approach converges to the co-channel and dedicated channel approaches. It is also important to note that simulation results and analytical results are aligned.

Figure 15.5 shows the sum capacity of a femtocell for different SSR values for co-channel, dedicated channel, and hybrid approaches. The co-channel capacity of a femtocell is greater than the dedicated channel for larger values of SSR, where femtocell bandwidth is less. Note that similar capacities can be achieved with co-channel and dedicated channel approaches for $\rho \simeq 0.75$. On the other hand, increasing SSR for the hybrid channel scenario also decreases the sum capacity of femtocells converging to co-channel at $\rho = 0$, and dedicated channel at $\rho = 1$. As a result, it can be concluded that for a fixed γ_c, and without fairness and QoS constraints, resource partitioning cannot be done effectively since extreme points are maximizing the capacity for both macrocells and femtocells.

15.4.3   Detailed computer simulations with max-min scheduling and fairness/QoS constraints

In the previous section, it was shown that although computer simulations and analytical results are aligned, the sum capacity is maximized at extreme points for both macrocell and femtocell networks. Therefore in this section we introduce the fairness criterion into

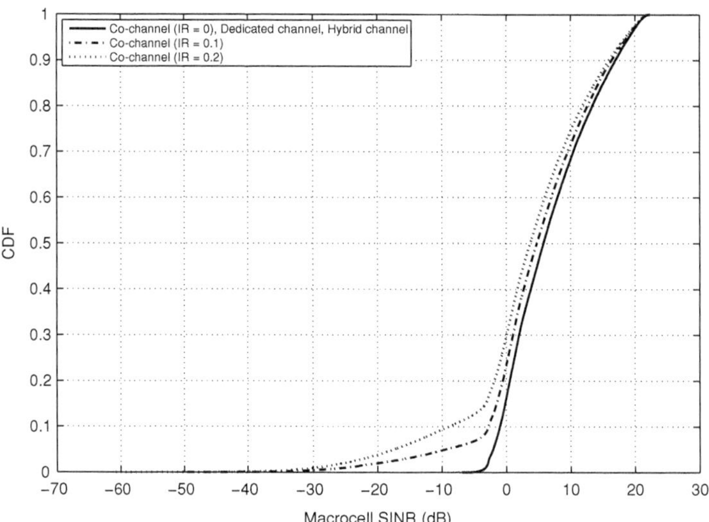

Figure 15.6 SINR of macrocell users vs. SSR ρ.

the optimization, while also considering a more applicable scenario where parameters are selected from [26]. This study also considers the case where a portion of macrocell MSs are inside the CSG femtocell area, which we called the indoor ratio. One hundred users are randomly and uniformly distributed within each sector, and there are two users associated with each closed-access femtocell [26]. This yields $100 - 4 \times 2 = 92$ macrocell users within the sector.

Figures 15.6 and 15.7 show the SINR distributions for macrocell and femtocell scenarios, respectively. In Figure 15.6, SINR CDFs for the co-channel scenario with indoor ratio (IR) $= 0$, dedicated channel scenario, and hybrid channel scenarios SINRs are aligned with the 3GPP benchmarks in [26]. In the co-channel scenario, for larger IR, SINRs of victim macrocell MSs get worse due to increasing interference. On the other hand, since the dedicated channel approach uses separate bandwidths for macrocell and femtocell such behavior is not observed. It is also important to note that if IR $= 0$, the co-channel and dedicated channel SINRs are the same, which validates the assumption in (15.38) for the CSG scenario with wall loss. Moreover, the hybrid approach with max-min fair scheduler also protects the victim macro MSs by assigning them to the dedicated portion of the hybrid approach. In Figure 15.7, the dedicated channel SINRs of femtocell MSs are compared to the co-channel approach. Since all femtocell users are co-channel with macrocell BSs, the SINRs of hybrid approach and co-channel approach are the same.

In Figures 15.8 and 15.9, the sum capacity of a macrocell and 5-percentile capacity of macrocell MSs are provided for various IR and SSR, respectively. Note that in both figures the hybrid approach converges to the co-channel at $\rho = 0$ and the dedicated channel at $\rho = 1$. The co-channel macrocell sum capacity decreases with increasing IR, and the dedicated channel capacity does not change with IR since femtocells do not

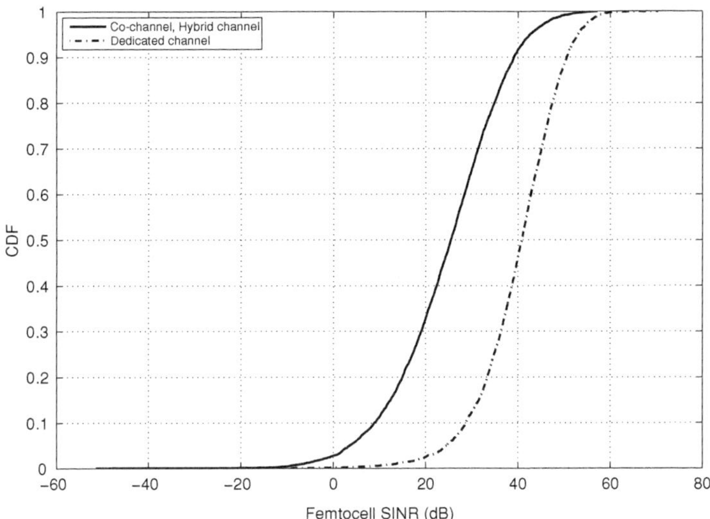

Figure 15.7 SINR of femtocell users vs. SSR ρ.

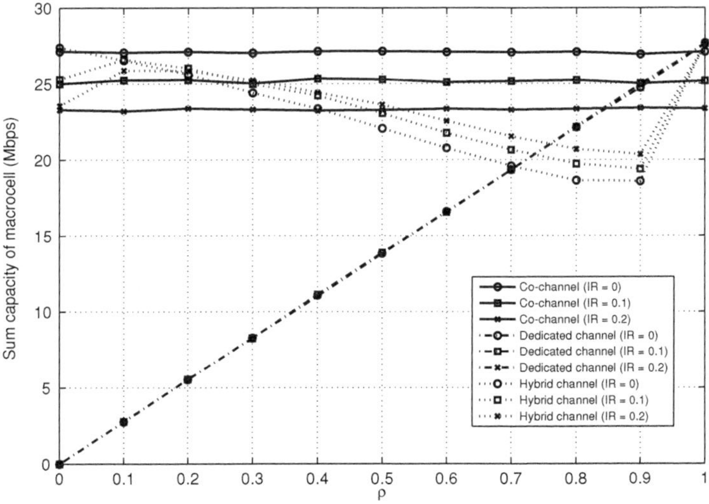

Figure 15.8 Sum capacity of macrocell vs. SSR ρ.

interfere with macrocells. Note that the hybrid approach does not let the sum capacity of the macrocell decrease below a level in Figure 15.8 and protects the victim users, which are under increased interference in Figure 15.9. Hybrid channel 5-percentile capacities are always better than both co-channel and dedicated channel. While protecting the victim users by the hybrid channel approach, the max-min fair scheduler also tries to maximize the sum capacity of macrocell.

In Figure 15.10, we present the femtocell sum capacities for various ρ values. Note that IR does not affect the femtocell capacities. The hybrid approach again converges

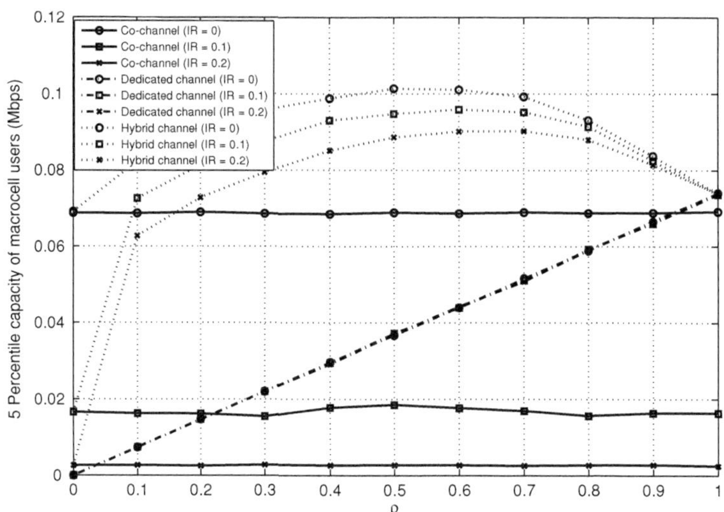

Figure 15.9 5-percentile capacity of macrocell vs. SSR ρ.

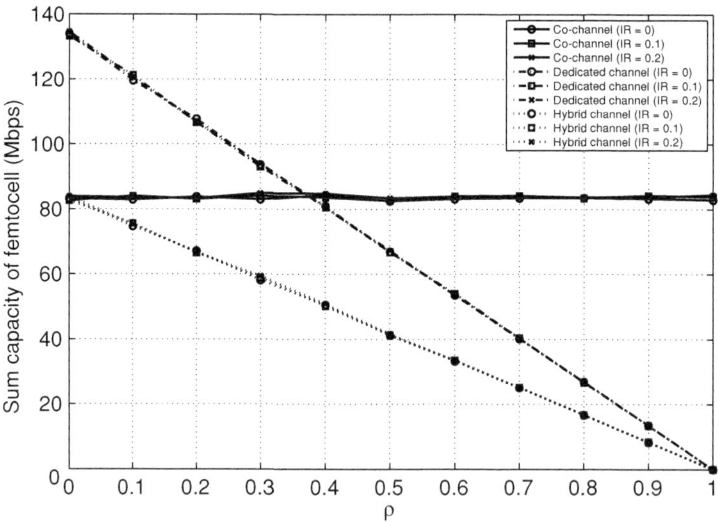

Figure 15.10 Sum capacity of femtocell vs. SSR ρ.

to the co-channel case at $\rho = 0$ and the dedicated channel at $\rho = 1$. Up to $\rho = 0.4$, the dedicated channel outperforms the co-channel scenario. Hybrid approach femtocell sum capacity is always lower than both the co-channel and dedicated channel scenario for femtocells. However since femtocells have better SINRs (Figure 15.7) compared to macrocells (Figure 15.6), the capacity of femtocells is still reasonable with this approach. One way to analyze this is to use the fairness metric defined in Section 15.2. Figure 15.11 presents the fairness level of a tiered network for co-channel, dedicated channel, and hybrid channel under various SSR and QoS parameters for IR = 0. Note that the hybrid

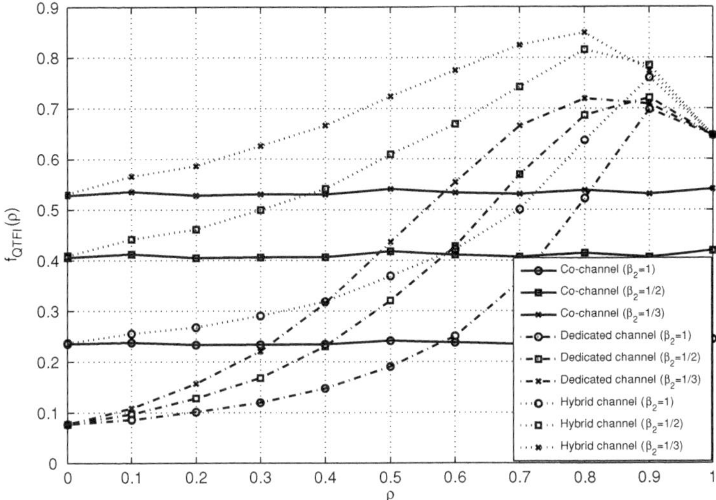

Figure 15.11 QoS-oriented fairness index ($f_{\mathrm{QTFI}}(\rho)$) vs. SSR ρ for various β_2.

approach $f_{\mathrm{QTFI}}(\rho)$ again converges to the co-channel case at $\rho = 0$ and the dedicated channel at $\rho = 1$.

In a co-channel scenario femtocell user capacities are always larger than macrocell users, and therefore their fairness metric may only be moderately improved by changing β in Figure 15.11. For example, for $\beta_2 = 1/3$, the expected femtocell user capacity is three times more than that of macrocell user; however, the $f_{\mathrm{QTFI}}(\rho)$ is still as low as 0.5. On the other hand, the hybrid channel approach's fairness metric is always above the co-channel and hybrid channel approaches. Note that for the hybrid channel approach the fairness maximization can be done for $\rho \approx 0.8$. This may also be traced by investigating Figures 15.8 and 15.10, since similar capacity values are achieved in both figures for $\rho \approx 0.8$. We also investigate the effect of IR on the fairness metric for $\beta_2 = 1/2$ in Figure 15.12. In the dedicated channel scenario, IR does not change the fairness of the system. On the other hand, while IR increases, the fairness of the system decreases for the hybrid and the co-channel scenario. Observe that the hybrid approach fairness is still greater than the dedicated channel fairness for all ρ and IR values.

15.5　Concluding remarks and discussion

In this chapter, using HPPPs, we study the sum capacities of co-channel, dedicated channel, and hybrid spectrum allocation methods for two-tier macrocell–femtocell networks. For dedicated channel and hybrid approaches, optimum partitioning of the available spectrum resources between the macrocell and femtocell networks is derived analytically and analyzed for various scenarios. The results show that without using fairness criteria, the capacity maximizing allocation is done by allocating the whole spectrum to femtocells due to their spectrum reuse capability. Since this approach leads to a very unfair

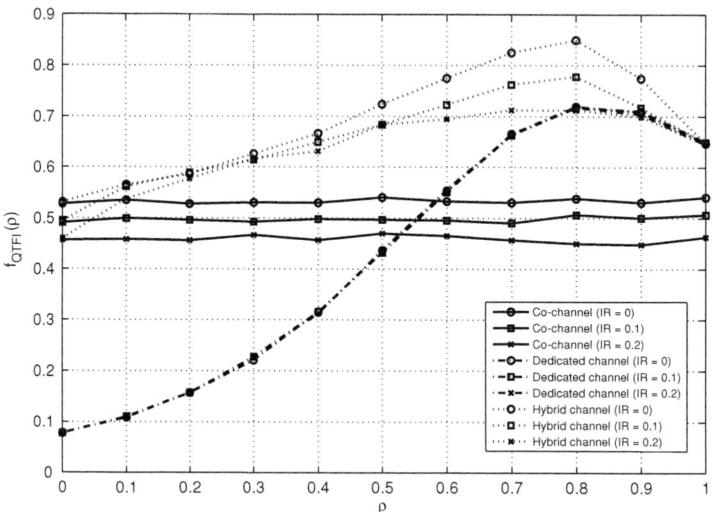

Figure 15.12 QoS-oriented fairness index ($f_{\mathrm{QTFI}}(\rho)$) vs. SSR ρ for various IR while $\beta_2 = 1/3$.

spectrum allocation, we propose a QoS-oriented fairness metric. By using this metric as a constraint for the spectrum allocation, we present a capacity maximizing spectrum allocation method, which guarantees a specific level of fairness and QoS. From a network provider point of view, partitioning of available resources with the hybrid approach yields the best tradeoff from capacity maximization, fairness, and QoS perspectives. The macrocell sum-capacity maximization is done at $\rho \approx 0.2$, macrocell 5-percentile capacity maximization is done at $\rho \approx 0.5$, and fairness maximization is done at $\rho \approx 0.8$. The findings in this chapter may also be easily extended to time-domain resource coordination among macrocells and femtocells as specified in 3GPP Release-10, where the duty cycle of ABSs may be optimized while jointly considering capacity maximization, fairness, and QoS constraints.

References

[1] V. Chandrasekhar, J. G. Andrews, and A. Gatherer, "Femtocell networks: a survey," *IEEE Commun. Mag.*, vol. 46, no. 9, pp. 59–67, Sep. 2008.

[2] Nokia Siemens Networks, "Macro+HeNB performance with escape carrier," 3GPP Standard Contribution (R1-101453), Feb. 2010.

[3] Z. Bharucha, A. Saul, G. Auer, and H. Haas, "Dynamic resource partitioning for downlink femto-to-macro-cell interference avoidance," *EURASIP J. Wirel. Commun. Netw.*, vol. 2010, no. 4, pp. 1–12, 2010.

[4] H. A. Mahmoud and I. Guvenc, "A comparative study of different deployment modes for femtocell networks," in *Proc. IEEE Indoor Outdoor Femtocells (IOFC) Workshop (co-located with PIMRC 2009)*, Tokyo, Japan, Sep. 2009, pp. 1–5.

[5] D. López-Pérez, A. Valcarce, A. Ladanyi, G. de la Roche, and J. Zhang, "Intracell handover for interference and handover mitigation in OFDMA two-tier macrocell-femtocell networks," *EURASIP J. Wirel. Commun. Netw.*, vol. 2010, no. 4, pp. 1–15, 2010.

[6] I. Demirdogen, I. Guvenc, and H. Arslan, "A simulation study of performance trade-offs in open access femtocell networks," in *Proc. IEEE Indoor Outdoor Femtocells (IOFC) Workshop (co-located with PIMRC 2010)*, Istanbul, Turkey, Sep. 2010, pp. 1–5.

[7] Samsung, "CSI measurement issue for macro–femto scenarios," 3GPP Standard Contribution (R1-106051), Jacksonville, FL, Nov. 2010.

[8] InterDigital Communications, LLC, "eICIC macro-femto: time-domain muting and ABS," 3GPP Standard Contribution (R1-105951), Jacksonville, FL, Nov. 2010.

[9] Ericsson, ST-Ericsson, "Details of almost blank subframes," 3GPP Standard Contribution (R1-105335), Xian, China, Oct. 2010.

[10] A. Damnjanovic, J. Montojo, Y. Wei, T. Ji, T. Luo, M. Vajapeyam, T. Yoo, O. Song, and D. Malladi, "A survey on 3GPP heterogeneous networks," *IEEE Trans. Wireless Commun.*, vol. 18, no. 3, pp. 10–21, June 2011.

[11] Samsung, "CSI measurement issue for macro-femto scenarios," 3GPP Standard Contribution (R1-106051), Nov. 2010.

[12] D. Choi, P. Monajemi, S. Kang, and J. Villasenor, "Dealing with loud neighbors: the benefits and tradeoffs of adaptive femtocell access," in *Proc. IEEE Global Telecommun. Conf. (GLOBECOM)*, Nov.–Dec. 2008.

[13] I. Guvenc, M. R. Jeong, F. Watanabe, and H. Inamura, "A hybrid frequency assignment for femtocells and coverage area analysis for co-channel operation," *IEEE Commun. Lett.*, vol. 12, no. 12, pp. 880–2, Dec. 2008.

[14] V. Chandrasekhar, J. G. Andrews, T. Muharemovic, Z. Shen, and A. Gatherer, "Power control in two-tier femtocell networks," *IEEE Trans. Wireless Commun.*, vol. 8, no. 8, pp. 4316–28, Aug. 2009.

[15] J. P. M. Torregoza, R. Enkhbat, and W.-J. Hwang, "Joint power control, base station assignment, and channel assignment in cognitive femtocell networks," *EURASIP J. Wirel. Commun. Netw.*, vol. 2010, no. 4, pp. 1–14, 2010.

[16] K. Huang, V. Lau, and Y. Chen, "Spectrum sharing between cellular and mobile ad hoc networks: transmission-capacity trade-off," *IEEE J. Sel. Areas Commun. (JSAC)*, vol. 27, no. 7, pp. 1256–67, Sep. 2009.

[17] M. Dianati, X. Shen, and S. Naik, "A new fairness index for radio resource allocation in wireless networks," in *Proc. IEEE Wirel. Commun. Netw. Conf. (WCNC)*, vol. 2, New Orleans, LA, Mar. 2005, pp. 712–17.

[18] D. C. R. Jain and W. Hawe, "A quantitative measure of fairness and discrimination for resource allocation in shared computer system," *DEC Technical Report 301*, 1984.

[19] D. Bertsekas and R. Gallager, *Data Networks*. Prentice-Hall, 1987.

[20] A. Kumar and J. Kleinberg, "Fairness measures for resource allocation," in *Proc. 41st Annual Symp. on Found. of Comp. Sci.*, Redondo Beach, CA, 2000, pp. 75–85.

[21] X. Gao, T. Nandagopal, and V. Bharghavan, "Achieving application level fairness through utility-based wireless fair scheduling," in *Proc. IEEE Global Telecommun. Conf. (GLOBECOM)*, vol. 6, San Antonio, TX, Nov. 2001, pp. 3257–61.

[22] S. Sesia, M. Baker, and I. Toufik, *LTE: The UMTS Long Term Evolution: From Theory to Practice*. Wiley-Blackwell, July 2011.

[23] S. Mukherjee, "UE coverage in LTE macro network with mixed CSG and open access femto overlay," in *Proc. IEEE Int. Conf. on Commun. (ICC)*, June 2011.

[24] V. Chandrasekhar and J. G. Andrews, "Spectrum allocation in two-tier networks," *IEEE Trans. Comm.*, vol. 57, no. 10, pp. 3059–68, Oct. 2009.

[25] I. Guvenc, "Capacity and fairness analysis of heterogeneous networks with range expansion, and interference coordination," *IEEE Commun. Lett.*, vol. 15, no. 10, pp. 1084–7, Oct. 2011.

[26] 3GPP TR 36.814 Release 9 2 V9.0.0 (2010-03), "Technical specification group radio access network; evolved universal terrestrial radio access (E-UTRA); further advancements for E-UTRA physical layer aspects (release 9)," Mar. 2010.

[27] M. Erturk, H. Aki, I. Guvenc, and H. Arslan, "Fair and QoS-oriented spectrum splitting in macrocell-femtocell networks," in *Proc. IEEE Global Telecommun. Conf. (GLOBECOM)*, Dec. 2010, pp. 1–6.

16 Self-organization and interference avoidance for LTE femtocells

David López-Pérez and Guillaume de la Roche

16.1 Introduction

Femtocell access points (FAPs) are foreseen to play a key role in the development and deployment of future orthogonal frequency division multiple access (OFDMA)-based cellular networks, e.g., Long Term Evolution (LTE) [1] and Wireless Interoperability for Microwave Access (WiMAX) [2], since they may deliver improved indoor coverage and network capacity [3]. Femtocell access points are low-cost, low-power, user-deployed small base stations (BSs), which provide wireless coverage of a cellular standard, and are connected to the network operator via a broadband connection, e.g., digital subscriber line (DSL), fiber optics, etc. Femtocells, as explained in the introductory chapter, offer a large number of advantages to future cellular networks. They may enhance indoor coverage, deliver both high data rates and new applications to users, and offload traffic from existing macrocell networks [4]. However, since FAPs are expected to be deployed in large numbers and because they may be installed by users in an uncoordinated manner, including self-organizing network (SON) capabilities in FAPs may be a key aspect for their successful operation of these devices [5].

A SON, defined as a network that requires minimal human involvement due to the automatic and/or autonomous nature of its functioning, integrates the processes of planning, configuration, optimization, and healing in a set of in-built automatic/autonomous functionalities. By using SON capabilities, operator intervention for network operation and maintenance can be reduced, thus minimizing deployment and operational costs of future cellular networks, which are major concerns of current mobile operators.[1] The rationale behind SON automation can be divided into two groups: (i) automating processes that are repetitive may save planning time and reduce execution effort, while (ii) automating processes that are too fast or complex for manual intervention may provide real-time network optimization. Thus, SON automation may lead to reduced expenses

---

[1] According to Motorola, 3G technologies have shown that operational costs represent around 30% of the total costs associated with a cellular network. This includes costs related to backhaul and network operation and maintenance [6].

Small Cell Networks: Deployment, PHY Techniques, and Resource Management, ed. Tony Q. S. Quek, Guillaume de la Roche, İsmail Güvenç, and Marios Kountouris. Published by Cambridge University Press. © Cambridge University Press 2013.

and enhanced performance. In this way, SON capabilities may help, e.g., to efficiently mitigate inter-cell interference in real time and in a distributed manner in massive and uncoordinated femtocell deployments [7].

In previous chapters, different techniques related to interference management have already been presented. However, this chapter focuses on the concept of SON, as it is seen from the point of view of current cellular standards. Today, the main standardization body is the 3rd Generation Partnership Project (3GPP) [8], which defines standards for LTE. The Small Cell Forum [9] also provides recommendations and common references about the functionalities that FAPs should implement. However, although these standardization bodies have performed large efforts in a wide range of topics, SON and interference management in femtocells networks is usually left up to vendor implementation (hence allowing competition and differentiation among them). As a consequence, the aim of this chapter is to provide an overview of the state of the art on SON capabilities for femtocells. In particular, we focus on LTE technology, since it is currently seen as the most suitable candidate for the fourth generation (4G) of cellular networks.

First of all, we present the current state of the art on SON capability types and standardization. Thereafter, we focus on interference management, which is probably the hardest challenge in large femtocell deployments. Finally, we present a novel and efficient distributed radio resource management technique, and illustrate its performance in an enterprise femtocell scenario.

16.2   Self-organizing femtocells

In order to mitigate inter-cell interference, the radio frequency (RF) parameters of femto-cells, a.k.a. Home enhanced NodeBs (HeNBs) in LTE, need to be optimized on a regular basis according to network, traffic, and channel variations. However, unlike traditional macrocells, a.k.a. enhanced NodeBs (eNBs) in LTE, which are installed and configured by engineers, HeNBs must be plug-and-play nodes capable of automatic/autonomous operation. Here is where the concept of SON kicks in, allowing HeNBs to independently and efficiently configure and optimize their own RF parameters, automating radio resource management procedures, and avoiding time-consuming human intervention.

When implementing SON capabilities in a network, there are three main approaches to do it:

1. *Centralized SON*, where a global manager is in charge of a given number of femtocells and is responsible for their optimization. In practice, a SON server is used, whose task is to control via backhaul the RF parameters of its subjugated femtocells.
2. *Distributed SON*, where there is no global manager, and the optimization of RF parameters is performed at the femtocell level by the HeNBs.
3. *Hybrid SON*, where both centralized and distributed approaches are combined.

In Figure 16.1 the principal concepts of centralized and distributed SON are illustrated. On the one hand, centralized techniques may benefit from reduced HeNB hardware

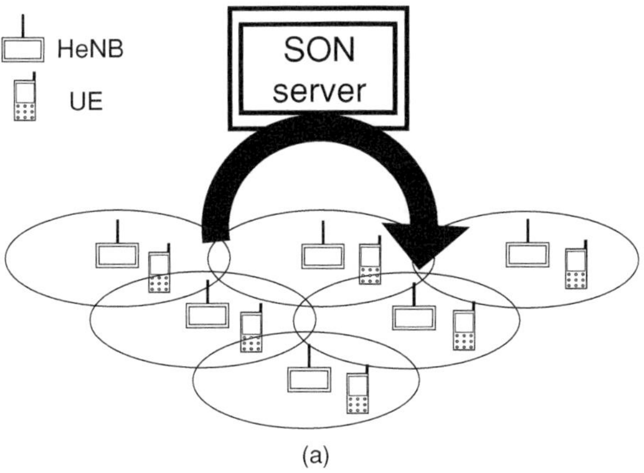

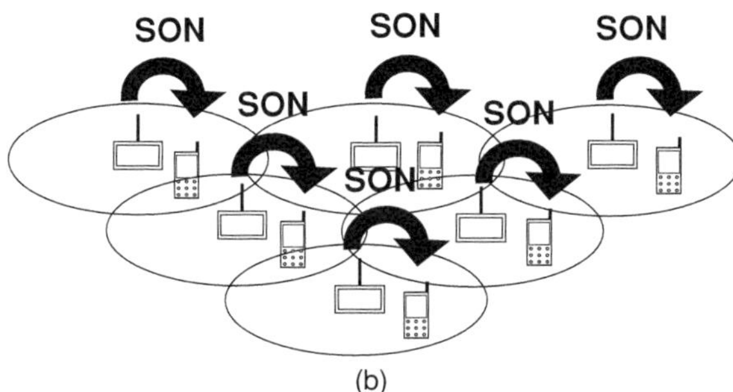

Figure 16.1 (a) Centralized SON where optimization is performed by a SON server, versus (b) distributed SON where optimization is performed at the HeNBs. Note that hybrid SON is a combination of both techniques.

complexity, because HeNBs do not need to incorporate specific optimizing capabilities (they are implemented at the SON server), and thus HeNBs can be manufactured at a lower cost. Moreover, centralized techniques may yield a better network performance, because the SON server has a global view of the interaction among cells and across the network. However, centralized optimization also has its own challenges. It imposes a single point of failure, and may increase deployment costs due to the need for a SON server. In addition, when the number of cells increases (thousands of HeNBs), signaling overhead and processing time at the SON server may grow beyond acceptable limits, thus becoming a bottleneck for network performance and introducing significant delays in the computation of optimal solutions. Since all solutions must be forwarded to the SON server, delay and burst issues may also affect network performance.

On the other hand, distributed techniques may reduce deployment costs and delays because no SON server is needed, and all backhaul signaling to/from the SON server is avoided. Moreover, distributed techniques may lead to an optimized performance in large femtocell deployments, because they allow for a real-time optimization due to reduced signaling overhead and processing time. However, distributed optimization also has its own challenges. For example, since HeNBs must be able to perform the optimization of their RF parameters in short times, they may be manufactured at a higher cost due to increased HeNB hardware complexity. In addition, obtaining optimum solutions at the network level while making decisions at the cell level is an intricate optimization problem, which often results in suboptimal solutions. Since standard and identical implementation of SON schemes in multi-vendor networks cannot be guaranteed, accurate monitoring of the environment is also necessary to guarantee a proper network operation.

Nowadays, most of macrocell SON deployments (e.g., high speed packet access (HSPA)) are based on centralized architectures [10]. During the last years, some companies, e.g., [11], [12], and [13], have also started to commercialize SON solutions dedicated to femtocells. These femtocell SON solutions mostly implement centralized SON techniques too, due to reduced HeNB hardware complexity and manufacturing costs. For instance, in [12], transmit powers are allocated through a SON server to femtocells based on the knowledge of their geographical positions, and adapted so that the coverage of neighboring femtocells does not overlap with each other. Moreover, in [14], a novel centrally controlled resource partitioning method is proposed based on graph coloring that assigns frequency subbands to neighboring femtocells, so that directly adjacent femtocells do not occupy the same subbands.

Since distributed and centralized techniques have both their own benefits and drawbacks, vendors and operators have also recently started to implement hybrid SON techniques based on a combination of these two approaches. In a hybrid approach, part of the SON schemes are executed at the SON server, while others are executed at the HeNB. For example, the values of initial RF parameters could be obtained from the SON server, while updates and refinements to these RF parameters in response to user equipment (UE) measurements could be carried out at the HeNBs [10]. Also in this line, complex optimization problems, such as the assignment of frequency bands, a.k.a. resource blocks (RBs) in LTE, may be performed by a global medium access control (MAC) scheduler controlled by the SON server, while power control may be executed independently on the HeNBs. An HeNB can at startup and periodically sense the received power from neighboring cells, and adjusts its maximum transmit power according to obtained measurements so that it provides the strongest signal in the desired coverage area (this power control has already been implemented in 3G femtocells as recommended by the Small Cell Forum [15]). As a result, hybrid SON usually requires the use of both a SON server and a SON agent implemented in the core network and HeNBs, respectively [12]. These hybrid approaches are more flexible and may allow vendors and operators to benefit from the advantage of both centralized and distributed techniques. The use of hybrid SON techniques is an initial step toward fully distributed SON techniques, which are likely to be preferred in the future due to their reduced deployment and operational costs, as well as their better operability.

16.2.1   Distributed self-organizing femtocells

In LTE, large numbers of HeNBs are expected to be deployed. This is why the concept of fully distributed SON HeNBs is highly regarded by current cellular operators, which aim at deploying plug-and-play HeNBs. As a result, the development of distributed SON schemes to manage both cross- and co-tier inter-cell interference in femtocell networks is a technical challenge that needs to be urgently addressed. Moreover, due to hardware limitations in HeNBs, which need to be manufactured at low cost to reach the market, it is important that the proposed distributed SON schemes meet the following requirements:

- *Operation under partial knowledge of network, traffic, and channel parameters.* Since it is not possible for a femtocell to accurately know all parameters from its neighboring cells and connected users, it must be able to operate under limited knowledge of such parameters.
- *Fast optimization.* The changes in the radio environment can be fast (switching on and off of neighboring cells, arrival and departure of users, shadowing and fading fluctuations, etc.), thus optimization must also be fast, e.g., run in less than 1 s.
- *Low complexity.* In order to build a low-cost HeNB, its hardware must be simple. Moreover, while an HeNB is optimizing its parameters, it must still be operational, thus optimization algorithms should not consume all available hardware resources.

Usually, the implementation of SON capabilities usually requires two phases:

- *Data collection.* During this phase, each HeNB collects data from its neighboring environment (i.e., related to connected UEs, neighboring eNBs and HeNBs, and the radio channel).
- *Optimization.* After data collection, the parameters of each HeNB are optimized by the SON server in the case of centralized SON, or the HeNBs in the case of distributed SON.

In the following sections, we present the state of the art with regard to SON techniques for femtocells as described by the Small Cell Forum and the 3GPP. Thereafter, we introduce our proposed solution for inter-cell interference coordination in an enterprise femtocell scenario and investigate its performance.

16.3   Self-organizing network femtocells in Femto Forum

In July 2007, the Small Cell Forum was founded to promote femtocell standardization. Today, most vendors and operators around the world are members of this organization. The Small Cell Forum initially focused on 3G femtocell standardization. However, due to the great interest of the mobile industry in 4G femtocells, the Small Cell Forum is also currently focusing on LTE femtocells, and provides recommendations for HeNBs. In the following, we summarize the work of the Small Cell Forum on data collection and interference management.

16.3.1   Data collection

Because the Small Cell Forum focuses on the standardization of FAPs, the solution investigated for collecting data is the network listening mode (NLM). Other possible solutions, involving other equipment rather than the FAP, such as UE measurement reports, are left up to 3GPP standardization, and are described in the following section.

In the Small Cell Forum, the NLM is referred to as the *network monitoring mode* [16]. According to the Small Cell Forum, an HeNB switched on in network monitoring mode should behave as a simplified UE, which should be able to scan the spectrum band and decode downlink (DL) signals from neighboring cells to thereafter perform the optimization of its RF parameters. The network monitoring mode should be executed at least at booting time and optionally during cell operation. For how long the network monitoring mode should be used and how often it should be executed are questions left up to vendor and operator implementation. A larger and more regular network monitoring mode operation would allow the HeNB to perform a more powerful RF parameter optimization as more and more reliable information would be available. However, this enhanced sensing may come at the expense of reduced performance, because FAPs running in network monitoring mode will not be able to achieve fully functional cell operation due to complexity limitations in the hardware.

The network monitoring mode operation can be classified in three different types, depending on the amount of information that the FAP is able to gather from the environment:

1. *Basic carrier detection.* In this case, an HeNB scans either the whole spectrum band or a given list of candidate frequencies. As a result, the receive strength signal indicator (RSSI) of each scanned carrier is measured. Such measurements allow the discerning of which carriers other cells are using.
2. *Long Term Evolution cell identification and power measurement.* In this case, given a list of candidate frequencies or physical cell identities (PCIs), the HeNB synchronizes to neighboring cells and identifies them. First, the primary synchronization signals (PSSs) are detected, and then for each candidate PSS, the secondary synchronization signal (SSS) is decoded. If both PSS and SSS are successfully derived, the cell is identified, and its PCI is obtained. Then, the reference signal received power (RSRP) of each identified cell is measured and a list of neighboring cells is built.
3. *Long Term Evolution cell information retrieving.* In the last form of network monitoring mode, information from the detected cells may also be decoded. In this case, HeNBs will decode the physical broadcast channel (PBCH) (which contains the master information block), and the physical downlink shared channel (PDSCH) (which contains the secondary information blocks) of neighboring cells, giving them access to key information of neighboring cells such as number of antennas, channel bandwidth, proprietary information, etc.

16.3.2   Optimization

The Small Cell Forum has also expended large efforts in interference characterization and management [15]. Within this context, the Small Cell Forum mainly focuses on evaluating possible interference scenarios and proposes techniques to mitigate inter-cell interference. the Small Cell Forum does not provide standard solutions, but gives very useful guidelines. For example, at the physical (PHY) layer, the proposed schemes are mainly based on power control, carrier selection, and multi-antenna techniques (the latter in case of considering multiple input multiple output (MIMO) femtocells). In case of power control, it is suggested that at booting time, HeNBs should adapt their maximum transmit power, so that their signal is stronger than that of the closest neighboring cell at a given targeted cell radius. At the MAC layer, the proposed schemes are mainly based on radio resource allocation, e.g., RB assignment, which can be performed in both the frequency- and time-domains due to the flexiblity of the LTE frame structure. Moreover, the Small Cell Forum has also investigated some solutions that rely on an X2 interference among macrocells and femtocells or between femtocells themselves. However, the 3GPP has not standardized yet the X2 interface for femtocells, as it is described in the following section.

16.4   Self-organizing network femtocells in 3GPP

In a similar way as the Small Cell Forum, the 3GPP does not propose standard solutions for implementing SON capabilities. Nevertheless, it clearly gives guidelines of possible solutions for data collection, and provides a complete list of use cases that may be optimized.

16.4.1   Data collection

In 3GPP, collecting data from neighboring cells and building/keeping an up-to-date list of neighboring HeNBs at a given cell is referred to as automatic neighbor relation (ANR). In [17] and [18] (respectively for frequency division duplex (FDD) and time division duplex (TDD) LTE), it is explained how an HeNB can be aware of the RF parameters of surrounding cells, using one of the following techniques or a combination of them:

1. *UE mode.* HeNBs can include basic UE capabilities, and switch on in a UE mode, during which they can "listen" to their neighboring cells. This is a similar solution to the network monitoring mode proposed by the Femto Forum and presented in the previous section.
2. *UE measurement reports.* In LTE, the concept of UE measurement report is also proposed, where each UE can report to its serving cell the list of surrounding cells and some related parameters. For example, UE measurement reports can be configured so that each user reports parameters (e.g., RSRP) related to its serving cell and/or to a list of selected neighboring cells.
3. *Message exchange between HeNBs.* A last approach, proposed by the 3GPP, but not standardized yet in LTE, is the exchange of messages between HeNBs. In this approach, each HeNB can report its list of neighboring cells and connected UEs to the neighboring HeNBs. This could be performed via a dedicated air interface or the user-provided backhaul. In dense femtocell deployment, this approach can be used

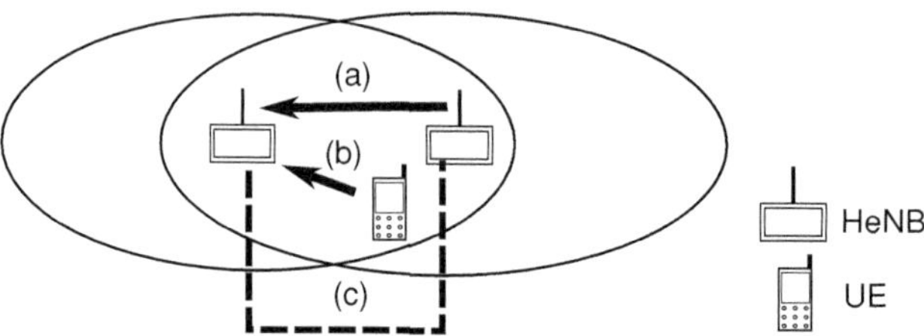

Figure 16.2 Techniques for HeNBs to collect data: (a) UE mode; (b) UE measurement reports; (c) message exchange with other cells via backhaul or wireless interface.

to efficiently manage, for instance, load balancing and inter-cell interference among cells. However, delay and overhead issues should be considered.

In Figure 16.2, the three techniques presented for data collection are illustrated. In practice, the UE mode is easy to run at booting time, but it is more complex to perform during cell operation due to hardware complexity reasons (while an HeNB is sensing, it must still be operational). The use of UE measurement reports may appear a better solution than the UE mode, since HeNBs only have to collect UE's fed back data, and because HeNBs benefit from information collected at the UE location. However, UE measurement reports also have their own drawbacks, since they will slightly decrease UE performance due to overhead, especially if SON has to be performed on a very regular basis. The last option, in which eHNBs directly exchange messages, may also lead to a better performance than the UE mode, but this technique is not standardized yet. In LTE standardization, there is a direct X2 connection between eNBs. However, this X2 interface has not been extended yet to HeNBs, since they may be interconnected through the public internet, thus creating security and latency issues. Due to these issues it may be that the best option for an inter-HeNB X2 is another wireless interface.

16.4.2   Optimization

When moving to more advanced LTE releases, SON becomes more and more important, and the 3GPP suggests some SON guidelines on parameter optimization. For example, according to the 3GPP, SON should work in three contexts [19]:

1. *Self-configuration.* In order to reduce human involvement in cell deployments, cells are automated with self-configuring capabilities. In this case, SON cells should automatically connect to the network and establish the necessary security contexts. Configuration files can then be downloaded from a server, thus providing an initial configuration for SON cells. Once the latest software release has been downloaded, SON cells should perform a self-test to check that everything works correctly. Finally, SON cells are taken into active service, where the configuration may still be improved using self-optimization.

2. *Self-optimization.* Since network load and user traffic may quickly vary depending on location and time (e.g., switch on and off of new cells, and/or arrival/departure, and user mobility), and because channel conditions can change rapidly due to shadowing and fading, SON cells must dynamically adapt their RF parameters to changing environments to optimize network performance. Because these changes can be detected during data collection (e.g., increase in traffic or interference, decrease in mobility events and/or dropped calls), SON cells can frequently tune their parameters (transmit power and/or RB assignment) online via self-optimizing approaches e.g., to mitigate interference and enhance capacity.

3. *Self-healing.* When some SON cells in the network become inoperative, self-healing mechanisms aim at reducing the impact from such failures, for instance, by adjusting parameters and algorithms in neighboring SON cells so that they can support the users that were supported by the damaged SON cell.

These three stages have been described from LTE Release 8 and beyond. The focus of LTE Release 8 was placed on procedures associated with initial equipment installation and integration, i.e., self-configuration, to support the commercial deployment of the first LTE networks. Examples of these procedures include: automatic inventory, automatic software download, automatic neighbor relation, and automatic physical cell ID assignment [10]. A key contribution of LTE Release 8 has been the ability to support SON features in multi-vendor network environments. From LTE Release 9, on the top of self-configuration, self-optimization was tackled and some RF parameters to be self-organized were identified. For example, some efforts were performed in the context of load balancing, where the target is to evenly distribute traffic load among network cells to improve capacity. Other LTE Release 9 use cases were: mobility robustness/handover optimization, random access channel (RACH) optimization, and inter-cell interference coordination [10]. In LTE Release 10, more RF parameters to be self-organized were included, providing a richer set of SON functions for heterogeneous networks (HetNets). The self-healing concept was defined in more detail too. However, most of the standardization work related to self-healing capabilities is expected to occur in LTE Release 11. Some LTE Release 10 use cases were: coverage and capacity optimization, enhanced inter-cell interference coordination, cell outage detection and compensation, self-healing functions, minimization of drive testing, and energy saving [10]. In LTE Release 11, more efforts are being performed in troubleshooting and optimization of multiple radio access technology (multi-RAT) and multi-tier HetNets. It should be noted that so far, most of the 3GPP work related to self-optimization and self-healing has been performed for macrocells, and that because there is no X2 interface defined yet for femtocells, as there is for macrocells, SON is harder to implement for femtocells. Indeed, due to delay in the femtocell backhaul, which makes synchronization among cells a challenging task, it is difficult to develop SON capabilities that are reactive enough to the fluctuating network, traffic, and channel conditions. This is a challenge that should be urgently addressed.

The list of parameters/use cases to be self-organized that was suggested by the 3GPP (either during self-configuration, self-organization, or self-healing) is as follows [19]:

- *Coverage and capacity optimization.* A typical SON task is to optimize the network in order to maximize coverage and capacity. The main idea here is to optimize the size of the cell so that the overlap between neighboring cells is reduced, while the cell capacity is increased. In macrocell scenarios, this task was performed using centralized planning tools, but in the femtocell case, it has to be done in an automatic and decentralized manner. Currently, most of the existing solutions are related to transmit power control optimization based on the network monitoring mode, where the femtocell will adapt its transmit power depending on the transmit power received from neighboring cells, so that its connected UEs can always receive a useful signal within a targeted radius.

- *Energy saving.* This task is related to the idea of green networking, where one of the targets is to avoid cells emitting when there is no UE to serve. This is ensured by using different kinds of idle modes at the HeNB. In this way, when no UE is detected for a long time period, the HeNB can partially switch off some parts of its hardware, so that only minimum detection tasks could be performed.

- *Interference reduction.* Interference reduction can be performed, for example, in two ways: through transmit power control or RB allocation. For instance, a solution could be that at booting time, each HeNB scans the whole spectrum and chooses for transmission the unused RBs. Operators can also choose to deploy the HeNBs in orthogonal bands to the macrocells to avoid overlaps, making the problem easier to solve, but leading to lower spatial reuse.

- *Automated configuration of PCIs.* In LTE, it is suggested that each cell at booting time should be able to automatically choose its own cell identity. This task is challenging due to limitations on the maximum number of cell identities (i.e., 504), which may lead to confusion/collision. Therefore, gathering data from neighboring cells to make a decision is important, mostly in scenarios with large femtocell densities.

- *Mobility robustness optimization.* In LTE, it is advised that HeNBs must also be able to optimize their handover (HO) parameters. This is performed with the help of HO failure and ping-pong data.

- *Mobility load balancing optimization.* In LTE, it is suggested that HeNBs must be able to optimize cell re-selection and HO parameters to cope with the unequal traffic load, and minimize the number of HOs and redirections needed to achieve load balancing.

- *ANR function.* As explained before, the update of the neighboring lists has critical influence on the performance of SON, hence HeNBs should sense the network as previously described.

- *Inter-Cell interference coordination (ICIC).* This technique is related to the ability of neighboring cells to have some kind of collaboration so that they allocate their transmit power and RBs in a way that inter-cell interference will be minimized. Performing ICIC in a distributed and fast manner is the key to successful femtocell performance.

- *Random access channel optimization.* The RACH collision probability is significantly affected by the RACH settings, making this a critical factor for call setup delays, data resuming delays from the uplink (UL) unsynchronized state, and HO delays. Thus, it is important that interference among RACHs is optimized.

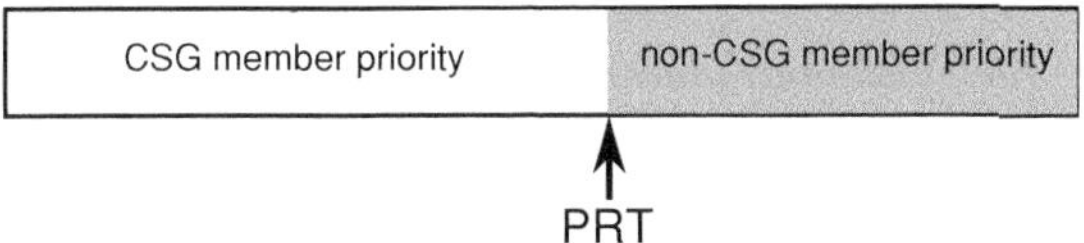

Figure 16.3 The priority region threshold (PRT) is utilized to regulate the quantity of resources opened to non-CSG UEs. In this case, the resources to share could be, for example, subframes in the time-domain or RBs in the frequency-domain.

16.5 Hybrid femtocells

In LTE Release 9, the concept of a hybrid access femtocell, whose principal objective is to mitigate interference, was introduced. Femtocells were initially standardized with two possible access modes [20]:

1. Non-closed subscriber group (CSG) mode, a.k.a open access, in which all UEs can access to all HeNBs.
2. Closed subscriber group (CSG) mode, a.k.a. closed access, in which only registered UEs, identified by the HeNB owners, can access to an HeNB.

It has been shown that CSG femtocells are often preferred by customers, leaving non-CSG femtocells as a less appealing option for operators. Nevertheless, CSG femtocells are challenging, since they generate complex interference scenarios. For instance, a non-registered UE entering in a household hosting a CSG femtocell will create a large UL interference to the nearby CSG femtocell, because it will use a large transmit power to connect to the closest non-CSG cell. This UE will also suffer from a large DL interference from the nearby CSG femtocell, since it is not allowed to connect to it.

As a result, the hybrid femtocell concept was introduced in LTE Release 9 [21, 22]. Hybrid HeNBs may provide different service levels to UEs, depending on whether they are registered. For example, considering that paging is the lowest service level, a hybrid HeNB may allow access to non-registered UEs just to receive paging messages. Moreover, hybrid HeNBs may provide the same service level to all UEs, but restrict the amount of resources allocated to different UE types, keeping part of their resources in CSG mode, and the remaining ones in non-CSG mode. In this case, a priority region threshold (PRT) (see Figure 16.3) is defined to select the amount of resources that should be shared between CSG and non-CSG UEs. This can be done in a dynamic manner, where femtocells adapt their PRT depending on collected UE statistics and channel conditions. Hybrid access femtocells are a good solution for operators to mitigate interference in the above-presented CSG femtocell scenarios (users will be allowed to connect to the nearby femtocell) and reduce load of traffic from macrocells.

16.6 Self-organizing femtocell networks in academic literature

In order to address the femtocell interference problem in a distributed manner, several approaches have been proposed in the academic papers. In [23] and [24], the authors propose the use of orthogonal and partially shared spectrum to remove cross-tier interference. Moreover, in [25] the authors implement a dynamic fractional frequency reuse approach that assigns users with bad geometry in neighboring femtocells to orthogonal bands in specified subframes. However, even though these approaches efficiently mitigate cross-tier interference, co-channel macrocell–femtocell deployments are preferred to make a better reuse of the scarce and expensive spectrum. In [26] and [27], the authors analyze the use of sector and multiple antenna elements to mitigate cross- and co-tier interference in co-channel macrocell–femtocell deployments. Nevertheless, these approaches need a more complex hardware due to antenna issues, thus increasing the femtocell cost. Thus RB allocation and power control have been heralded as the most promising ways of coping with inter-cell interference in co-channel macrocell–femtocell deployments [5]. Based on a uniform transmit power distribution among subcarriers, in [28] a network monitoring mode is employed at the femtocell to sense interference conditions, and assign RBs with low interference to femtocell users. In [29], UE measurement reports are used to better estimate interference at user locations and efficiently allocate RBs to femtocell users. Moreover, in order to allow a better spatial spectrum reuse, in [30] the authors propose a power control where the total transmit power of the femtocell is tuned in order to provide a constant femtocell radius according to the received signal strength from the closest neighboring macrocell as suggested in Section 16.2. Similarly, in [31] femtocells adjust their total transmit powers through a distributed Q-learning approach with docitive features to minimize interference toward neighboring macrocell users. However, all these schemes do not explore a joint RB and transmit power allocation.

In the following sections of this chapter, we will present an efficient distributed radio resource management scheme to deal with inter-cell interference based on a joint RB and transmit power allocation, and focus on the following deployment case:

- *Uncoordinated deployment*, where no cell planning was performed and HeNBs are randomly deployed.
- *Enterprise scenarios*, where many cells are deployed in the same building, and thus where interference is large due to the excessive overlap between cells created because of the uncoordinated deployment.
- *Co-channel deployment (spectrum sharing)*, where all enterprise femtocells reuse the same frequency bands.
- *Fully distributed optimization*, where no SON server is deployed and where neighboring cells do not communicate.

16.7   SON femtocells in enterprise scenarios

In the OFDMA literature, dynamic subchannel assignments with equal transmit power per subcarrier are usually preferred to complex joint dynamic subchannel and transmit power assignments due to mathematical tractability and easier implementation.

Moreover, these simpler dynamic subchannel assignments have also been sheltered by previous analyses [32, 33], which showed that improvements in system performance due to the allocation of different transmit powers to different subcarriers is low in scenarios with a wide range of users that demand diverse quality of service (QoS). Practical examples are current LTE deployments, where transmit power control is not considered in the DL, i.e., transmit power is uniformly distributed among all subcarriers. The same modus operandi applies to the femtocell literature, where total HeNB transmit power may be tuned (adaptive coverage), but afterwards it is uniformly distributed among subcarriers.

This chapter's contribution is to show that in distributed systems, allocating different transmit powers to different subchannels has remarkable SON properties. Using these SON properties, we propose allocation models, in which each HeNBs aims at minimizing its transmit power, allocating different transmit powers to different subchannels according to its user channel conditions and QoS requirements.

The joint allocation of subchannels and transmit power is an intricate optimization problem. This problem becomes even more complex due to the existence of link adaptation and several modulation and coding schemes (MCSs) in LTE and WiMAX standards e.g., LTE. Moreover, a user assigned to more than one subchannel must be allocated with the same MCS in all of them [1][2] (although channel conditions may differ from one subchannel to another). In the proposed allocation scheme, we consider these constraints and provide an optimization framework able to obtain optimal solutions at the cell level. For performance evaluation, we concentrate in an OFDMA femtocell enterprise network due to its challenging interference conditions.

16.7.1   System model and notation

Let us define an OFDMA femtocell enterprise network as a set of:

- Femtocells: $\mathcal{F} = \{F_1, \ldots, F_m, \ldots, F_n, \ldots, F_F\}$,
- Users of femtocell F_m: $\mathcal{U}^m = \{U_1^m, \ldots, U_u^m, \ldots, U_{U_m}^m\}$,
- Subchannels (SCs): $\mathcal{K} = \{1, \ldots, k, \ldots, K\}$,
- Modulation and coding schemes: $\mathcal{R} = \{1, \ldots, r, \ldots, R\}$

Signal quality modeling

Assuming that all subcarriers within a subchannel experience the same channel conditions, the signal to interference plus noise ratio (SINR) $\gamma_{m,u,k}$ of user u in subchannel k can be modeled as:

$$\gamma_{m,u,k} = \frac{P_{u,k}^m \cdot \Gamma_{m,u}}{w_{u,k} + \sigma^2} = \frac{P_{u,k}^m \cdot \Gamma_{m,u}}{\sum_{m'=1, m'\neq m}^{M} P_{u',k}^{m'} \cdot \Gamma_{m',u} + \sigma^2} \tag{16.1}$$

where $P_{u,k}^m$ is the power applied by femtocell F_m in each one of the subcarriers of subchannel k, allocated to user u, $\Gamma_{m,u}$ is the channel gain between femtocell F_m and

---

[2] This constraint simplifies the decoding process by minimizing signaling and enhances the performance of turbo codes due to the larger length of blocks.

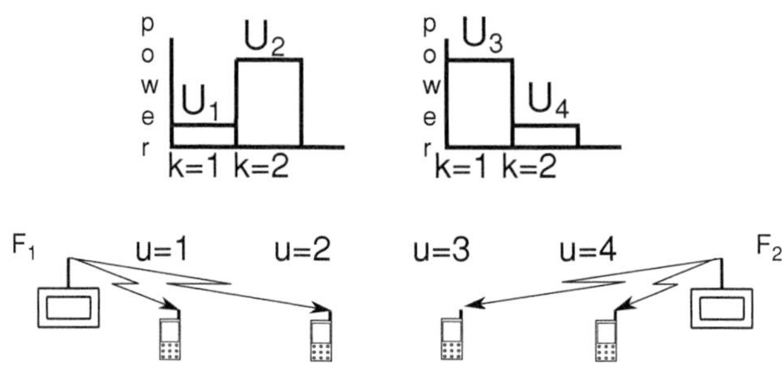

Figure 16.4 Users co-existing in the same subchannel, while meeting their SINR demands.

user u, $w_{u,k}$ is the interference suffered by user u in subchannel k, and σ^2 is the noise power.

User capacity modeling
Following previous assumptions, both the bit rate $BR_{u,r,k}$ and the throughput $TP_{u,r,k}$ of user u in subchannel k when using MCS r can be modeled as:

$$BR_{u,r,k} = \Theta \cdot \mathit{eff}_r = \frac{SR_{OFDM} \cdot SY_{OFDM}}{T_{\text{subframe}}} \cdot \mathit{eff}_r \tag{16.2}$$

$$TP_{u,r,k} = BR_{u,r,k} \cdot (1 - BLER(r, \gamma_{m,u,k})) \tag{16.3}$$

where Θ is a fixed parameter that depends on network configuration, being SR_{OFDM} and SY_{OFDM} the number of data subcarriers (frequency) and orthogonal frequency division multiplexing (OFDM) symbols (time) per subchannel, respectively, T_{subframe} is the frame duration in time units, eff_r is the efficiency in bits per symbol of the selected MCS r, and $BLER(r, \gamma_{m,u,k})$ is the block error rate (BLER) suffered by user u in subchannel k, which is a function of both MCS r and SINR $\gamma_{m,u,k}$.

Measurement report
We assume that user u sends at regular time intervals ($T_{u,mr}$) a measurement report (MR) MR_u to its serving femtocell F_m to assess user channel conditions. Each user's MRs indicates the power $w_{u,k}$ of the interference suffered by user u in all subchannels $\mathcal{K}$.

16.7.2 Basis of the proposed distributed resource allocation

In the proposed distributed resource allocation, the objective of each femtocell is to allocate radio resources to users, while meeting its users throughput demands and minimizing its overall transmit power. The reasons for minimizing transmit power are as follows:

1. Minimizing transmit power reduces interference. This is obvious from the Shannon theorem [34]:

$$C = B \cdot \log_2 \left(1 + \frac{P_{u,k}^m \cdot \Gamma_{m,u}}{w_{u,k} + \sigma^2} \right) \rightarrow w_{u,k} = \frac{P_{u,k}^m \cdot \Gamma_{m,u}}{2^{\frac{C}{B}} - 1} - \sigma^2 \tag{16.4}$$

where C is capacity in bps, and B is bandwidth in Hz. From this formula, it is easy to see that decreasing $P_{u,k}^m$ results in a decreased $w_{u,k}$.

2. A cell that targets at minimizing its transmit power allocates less transmit power to those users that are closer to the BS or have smaller throughput requirements. Thus and according to (16.4), interference is mitigated.

3. A cell that targets at minimizing its transmit power tends to use those subchannels that are not being used by its neighboring cells, because less transmit power is necessary in a less interfered or faded subchannel to achieve a targeted SINR.

Putting together the concepts behind statements (2) and (3), SON behaviors can be achieved in distributed systems, when each cell independently aims at minimizing its transmit power. In this line, in [35], it is shown that minimizing transmit power independently in each cell leads the network to self-organize into an efficient and stable frequency reuse pattern. Let us show a simple example to demonstrate this. Consider a network comprised of two subchannels k_1 and k_2, and two femtocells F_1 and F_2, giving service to two distinct users each. In each femtocell, one of the two users is close to the cell center, while the other one is at the cell edge (Figure 16.4). Then,

- *Case A (cells aiming at minimizing their transmit power).* According to statement (2), those users close to the cell center will be allocated less transmit power than those at the cell edge. According to statement (3), when a femtocell, e.g., F_1, decides how to assign k_1 and k_2 to its users, and observes stronger interference in one subchannel, e.g., k_1, than in the other one, i.e., k_2, due to the transmit power difference applied by femtocell F_2 when using k_1 and k_2. Then, this femtocell, F_1, will allocate its cell-edge user to subchannel k_2 and its cell-center user to subchannel k_1, since less transmit power will be needed to meet its user throughput requirements. Following this procedure, users at the cell edge are not much interfered due to the low transmit power allocated by the neighboring femtocell to the corresponding subchannel.
- *Case B (power uniformly distributed among subcarriers).* In this case, both cells will assign the same transmit power to their users independently of their positions and channel conditions. As a result, users located at the cell edge will then suffer from larger interference than in the previous Case A.

16.7.3   Resource allocation problem

In the following section, we define our model for the MCS, subchannel, and transmit power assignment in each femtocell, an optimization problem that will be referred from now on to as the resource allocation problem (RAP). The RAP does not imply communication between neighboring femtocells, and each femtocell tunes its resource allocation independently.

First of all, let us indicate that the transmit power $P_{u,k,r}^m$ that femtocell F_m has to allocate in each of the subcarriers of subchannel k to achieve the SINR threshold γ_r of MCS r can be modeled as:

$$P_{u,k,r}^m = \gamma_r \cdot \frac{w_{u,k} + \sigma^2}{\Gamma_{m,u}} \tag{16.5}$$

where the derivation of transmit power $P_{u,k,r}^m$ is straightforward from (16.1) since femto-cell F_m "knows" $\Gamma_{m,u}$ and $w_{u,k}$ due to user MRs (see Section 16.7.1). In order to mitigate the effects of channel fading, user feedback is averaged over time, thus avoiding fast variations of $P_{u,k,r}^m$ and the rapid fluctuations of cell assignments. In this case, average channel quality in the form of instantaneous $w_{u,k}$ averaged over tens of user MRs is used to derive $P_{u,k,r}^m$ (similar to [35]).

With that said, the RAP in femtocell F_m can be formulated as the following integer linear problem:

$$\min_{\chi_{u,k,r}} \sum_{u=1}^{U} \sum_{k=1}^{K} \sum_{r=1}^{R} P_{u,k,r}^m \cdot \chi_{u,k,r} \tag{16.6a}$$

subject to:

$$\sum_{u=1}^{U} \sum_{r=1}^{R} \chi_{u,k,r} \leq 1 \qquad \forall k \tag{16.6b}$$

$$\sum_{r=1}^{R} \rho_{u,r} \leq 1 \qquad \forall u \tag{16.6c}$$

$$\chi_{u,k,r} \leq \rho_{u,r} \qquad \forall u, k, r \tag{16.6d}$$

$$\sum_{k=1}^{K} \sum_{r=1}^{R} \Theta \cdot \textit{eff}_r \cdot \chi_{u,k,r} \geq T P_u^{req} \qquad \forall u \tag{16.6e}$$

$$\rho_{u,r} \in \{0, 1\} \qquad \forall u, r \tag{16.6f}$$

$$\chi_{u,k,r} \in \{0, 1\} \qquad \forall u, k, r \tag{16.6g}$$

where $\chi_{u,k,r}$ (16.6g) is a decision binary variable that is equal to 1 if user u uses MCS r in subchannel k or 0 otherwise; $\rho_{u,r}$ (16.6f) is a decision binary variable that is equal to 1 if user u makes use of MCS r or 0 otherwise; constraint (16.6b) makes sure that subchannel k is only assigned to at most one user u; constraints (16.6c) and (16.6d) together guarantee that each user is allocated to at most one MCS; and constraint (16.6e) makes sure that each user u achieves its throughput demand TP_u^{req}, which may differ for each user.

16.7.4   Coordinated resource allocation problem

Cell-edge users are more prone to inter-cell interference. Thus in order to minimize their inter-cell interference and enhance overall system performance, we also propose that neighboring cells coordinate their allocations in terms of subchannels and transmit power to their cell-edge users via a message passing approach, which may operate the femtocell gateway. In this way, the proposed cell independent transmit power optimization may converge quicker toward stable solutions. This is an upgrade to the previously presented model, which can only be used when femtocells benefit from efficient backhauls in terms of bandwidth and delay. In Section 16.7.8, we will benchmark the improvement in performance due to this coordination.

When cell-edge user u connects to femtocell F_m, it will be assigned to subchannel k with MCS r and transmit power $P_{u,k,r}^m$ based on the distributed resource allocation described in the previous section. According to this allocation, femtocell F_m can compute the maximum interference $w_{u,k}^{\max}$ that user u can suffer in subchannel k to achieve the SINR threshold γ_{u_r} of its targeted MCS r_u as:

$$w_{u,k}^{\max} = \frac{P_{u,k,r}^m \cdot \Gamma_{m,u}}{\gamma_{r_u}} - \sigma^2. \tag{16.7}$$

In order to avoid unfair transmit power decrease requests among neighboring femtocells, this maximum interference $w_{u,k}^{\max}$ may be equally shared/divided by the number N_m^{intf} of potential interfering cells of user u, i.e.,

$$w_{u,k}^{maxN} = \frac{w_{u,k}^{\max}}{N_m^{\text{intf}}}, \tag{16.8}$$

where N_m^{intrf} is the cardinality of $\Psi_m \in \mathcal{F}$, and $\Psi_m \in \mathcal{F}$ is the set of interfering cells of user u, it may may be comprised, e.g., of the adjacent cells sensed/detected by serving femtocell F_m, using a UE mode (see Section 16.4).

Femtocell F_m can compute the maximum transmit power $P_k^{m',\max}$ that interferer F_m' can apply to subchannel k to prevent the outage of cell-edge user u as:

$$P_k^{m',\max} = w_{u,k}^{maxN} \cdot \Gamma_{m',u} \tag{16.9}$$

where $\Gamma_{m',u}$ is the channel gain between femtocell F_m' and user u, and forward this information to each potential interferer $F_{m'} \in \Psi_m$.

In order to indicate to each potential interferer $F_{m'} \in \Psi_m$, that they should not use a transmit power $P_{u,k,r}^{m'}$ greater than $P_k^{m',\max}$ in subchannel k, the use of DL high interference indicator (HII) is proposed [36]. In this case, each DL HII will contain two values, constraint $P_k^{m',\max}$ and index k, and will be addressed to a specific femtocell $F_{m'} \in \Psi_m$. When receiving several DL HIIs for the same subchannel k, femtocell $F_{m'} \in \Psi_m$ will follow the lowest power constraint. To avoid sending numerous DL-HII messages in response to fluctuating channel conditions, transmit power assignments to cell-edge users will only be updated if $P_{u,k,r}^m$ significantly varies with respect to its current value, e.g., 1 dB. Femtocell F_m will indicate to all interfering cells $F_{m'} \in \Psi_m$ when the transmit power constraint expires using an DL HII where $P_k^{m',\max}$ is null.

This inter-femtocell coordination can be realized in RAP (16.6) by incorporating the following constraint to the problem:

$$P_{u,k,r}^m \leq P_k^{m,\max} \qquad\qquad \forall u, k, r \tag{16.10}$$

where constraint (16.10) guarantees that power constraints imposed by neighboring cells via DL-HII messages are respected. In order to distinguish the autonomous and coordinated version of RAP, let us call the former autonomous RAP (auRAP) and the latter cooperative RAP (coRAP).

16.7.5 Subchannel and power allocation subproblem

This section discusses an important subproblem of RAP referred to as the subchannel and power allocation subproblem (SPAP) that occurs when the MCS of each user is known a priori as part of the input of the problem.

An efficient solution to the SPAP has two important applications:

1. It can be used as a subroutine to solve the RAP problem, as will be presented in Section 16.7.6.
2. It can be used as a low latency subchannel and power allocation scheme, as will be presented in Section 16.7.7.

Assuming that an MCS r_u has been already given to user u, i.e., $\rho_{u,r}\forall u\forall r$ is known and fixed a priori as part of the input, the RAP transforms to an easier form.

Clearly, the used MCS r_u determines the number D_u of subchannels needed for satisfying the throughput requirement TP_u^{req} of user u. Namely,

$$D_u := \left\lceil \frac{\mathrm{TP}_u^{\mathrm{req}}}{\sum_{r=1}^{R} \Theta \cdot \mathit{eff}_r \cdot \rho_{u,r}} \right\rceil = \left\lceil \frac{\mathrm{TP}_u^{req}}{\Theta \cdot \mathit{eff}_{r_u}} \right\rceil. \tag{16.11}$$

In addition, let us introduce the binary decision variable $\phi_{u,k}$, which indicates whether user u makes use of subchannel k, i.e.,

$$\phi_{u,k} := \sum_{r=1}^{R} \chi_{u,k,r}. \tag{16.12}$$

Substituting them into (16.6a)–(16.6g), we obtain the following subchannel and transmit power allocation subproblem, i.e., SPAP:

$$C_S = \min_{\phi_{u,k}} \sum_{u=1}^{U} \sum_{k=1}^{K} P_{u,k,r_u}^m \cdot \phi_{u,k} \tag{16.6a*}$$

subject to:

$$\sum_{u=1}^{U} \phi_{u,k} \leq 1 \qquad\qquad \forall k \tag{16.6b*}$$

$$\sum_{k=1}^{K} \phi_{u,k} = D_u \qquad\qquad \forall u \tag{16.6e*}$$

$$P_{u,k,r}^m \leq P_k^{m,\max} \qquad\qquad \forall u, k, r \tag{16.10}$$

$$\phi_{u,k} \in \{0, 1\} \qquad\qquad \forall u, k \tag{16.6g*}$$

where constraint (16.10) is neglected if we are solving autonomous SPAP (auSPAP) or considered if we are solving cooperative SPAP (coSPAP).

Solving SPAP optimally

The following observation makes it possible to solve auSPAP optimally and efficiently.

Claim 16.1 *Let us define the following network flow problem [37] with vertex set*

$$V := \mathcal{U} \cup \mathcal{K} \cup \{s, t\}, \tag{16.13a}$$

being $s, t \in V$ the source and the sink of V, respectively.

edge set

$$E := \{(su) : u \in \mathcal{U}^m\} \cup \{(uk) : u \in \mathcal{U}^m, k \in \mathcal{K}\} \cup$$
$$\{(kt) : k \in \mathcal{K}\}, \tag{16.13b}$$

capacity function

$$cap(ab) := \begin{cases} D_b, & \textit{if } a = s, b \in \mathcal{U}^m \\ 1 & \textit{otherwise,} \end{cases} \tag{16.13c}$$

and cost function

$$cost(uk) := \begin{cases} P^m_{u,k,r_u}, & \textit{if } u \in \mathcal{U}^m, k \in \mathcal{K} \\ 0 & \textit{otherwise.} \end{cases} \tag{16.13d}$$

Then, the minimal cost network flow of V will provide an optimal solution to SPAP.

Following this claim, problem (16.6) can be formulated and efficiently solved as a minimum cost network flow (assignment problem), where users are assigned subchannels so as to minimize transmit power.

In order to solve this problem, the *network simplex* algorithm [38] implemented in the LEMON library [39] has been used for our simulations.

In a similar way, we can also efficiently solve coSPAP by replacing edge set (16.13b) by $E := \{(su) : u \in \mathcal{U}^m\} \cup \{(uk) : u \in \mathcal{U}^m, k \in \mathcal{K} \mid P^m_{u,k,h} \leq P^{m,\max}_k\} \cup \{(kt) : k \in \mathcal{K}\}$, where the edge between user u and subchannel k is broken if $P^m_{u,k,h} > P^{m,\max}_k$, and thus user u cannot use subchannel k.

16.7.6 Solving RAP and coordinated RAP

In the following section, a smart exhaustive search-based approach is proposed for solving RAP based on SPAP. The key idea behind this approach is that a smart search can be performed over the MCS assignment solution space, where for each MCS solution, the optimal subchannel and transmit power allocation can be obtained by solving SPAP (see Section 16.7.5).

In order to provide a better description of this approach, we define $\mathcal{S}_m$ as the vector that indicates the MCSs assigned to all users $\mathcal{U}^m$ of femtocell F_m. For an arbitrary $\mathcal{S}_m$, solving SPAP, the subchannel and transmit power assignment associated to $\mathcal{S}_m$ can be derived. Then, the quality of this MCS allocation $\mathcal{S}_m$ can be evaluated according to C_S (16.6a*), i.e., the cost of the subchannel and power assignment found by SPAP. Using

C_S, a smart search can be performed to find the MCS assignment that yields the lowest transmit power, where some assignments can be safely excluded:

- If an MCS $r_{\max}$ can be found, which is suitable to satisfy the throughput requirement TP_u^{req} of user u by using 1 subchannel, no higher order MCSs are allocated thereafter to user u, since allocating a higher order MCS would unnecessarily increase the required transmit power.
- If an MCS allocation S_m can be found, which requires more subchannels than are available to satisfy the throughput requirement of users, no other MCS allocation S' that can be derived from S_m by lowering the selected MCS of a single user is used. Moreover, allocating S'_m would also require more subchannels than are available.

In femtocell scenarios, this smart search may solve RAP optimally and sufficiently fast, thanks to the limited number of connected users per femtocell, and the speed of network simplex schemes for solving SPAP.

16.7.7   Resource management architecture

As it will be shown in the following section, RAP (i.e., smart search plus SPAP) can be solved in around half a second, while SPAP can be solved faster, in less than one millisecond. Therefore, we propose the following architecture:

- By solving RAP, the MCSs of users can be updated on a second-by-second basis to cope with per-cell time fluctuations of traffic load and user mobility.
- By solving SPAP, subchannel and transmit power allocations to users can be updated independently and on a much faster basis than the assignment of MCSs to cope with fast variations of the channel due to interference and fading.

In order to distinguish the autonomous and coordinated radio resource allocation architectures (RRAAs), let us name the former as autonomous RRAA (auRRAA) and the later as cooperative RRAA (coRRAA).

16.7.8   Simulations and results

The scenario used for our system-level simulations is an enterprise of $69\,\text{m} \times 32\,\text{m}$, hosting nine OFDMA femtocells. It is assumed that macrocells and femtocells are allocated to orthogonal spectrum resources or that there is no macrocell so that we can focus on the femtocell tier.

Figure 16.5 illustrates the femtocell locations within the scenario, while Table 16.1 provides details of simulation parameters. Path losses and shadowing were modeled according to the finite-difference time-domain (FDTD)-based model in [40], while BLER was modeled using look up tables (LUTs) from [41]. Simulations with 4, 6, or 8 static uniformly distributed users per femtocell were performed, where this represents a 50%, 75%, and 100% cell traffic load since the simulated network had 2.5 MHz bandwidth and eight subchannels. Users held their connections for a given time dictated by an exponential distribution of mean $\mu_p = 45\,\text{s}$, and then disconnected. When a user

Table 16.1 Simulation parameters.

Parameter	Value	Parameter	Value
Femtocells	9	UE ant. height	1.5 m
Simulation time	600 s	UE noise figure	9 dB
Scenario size	69 m × 32 m	UE body loss	0 dB
Carrier frequency	2.0 GHz	Path-loss model	FDTD-based model
Channel bandwidth	2.5 MHz	Shadowing	Predicted by FDTD
Frame duration	5 ms	Users per sector (U)	4, 6, 8
Data subcarriers	192	User distribution	Uniform
Subchannels (K)	8	Min. dist. UE to FAP	1 m
DL OFDM data symbols	39	Mean holding time (μ_p)	45 s
FAP Tx power(P_m^{tot})	20 dBm	Type of service	Full buffer
FAP ant. base gain	0 dBi	Min service TP	250 kbps
FAP ant. pattern	Omni	T_{outage}	4 s
FAP ant. height	1.5 m	User MR freq.	10 ms
FAP ant. tilt	–	NLM updating freq.	100 ms
FAP noise figure	5 dB	IM updating freq.	100 ms
FAP body loss	0 dB	Stolyar's updating freq.	100 ms
Thermal noise density	−174.0 dBm/Hz	T_{rap}	1 s
UE ant. gain	0 dBi	T_{spap}	100 ms
UE ant. pattern	Omni	γ^e	3 dB

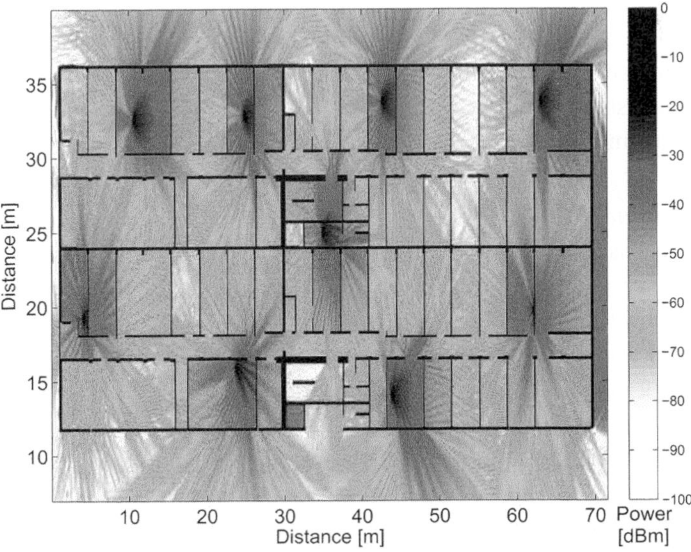

Figure 16.5 Office scenario with nine femtocells: best server.

disconnected, a new one appeared in a new random position. A full buffer model was
used to simulate the traffic of users, i.e., there was always data available to transmit
for a user, and their throughput demand was set to $\text{TP}_u^{\text{req}} = 250$ kbps. A user suffered
from outage if it could not transmit at a throughput larger than its demand for a time

$T_{\text{outage}} = 4\,$s. A user is considered as a cell-edge user if its sideband SINR is lower than $\gamma^e = 3\,$dB. In order to mitigate ping-pong effects, only those assignments that were at least 10 % better than the current ones in terms of cost function (16.6a*) were applied.

Approaches used for comparison

Four different radio resource management schemes were used for performance comparison. Note that in the first three schemes transmit power is uniformly distributed among subcarriers.

Random assignment

Subchannels are assigned randomly to users without taking into account any kind of information. Therefore, inter-cell interference coordination does not exist.

Network listening mode (NLM)

Each femtocell periodically measures the received strength of the interference in each subchannel, and subsequently allocates the subchannels suffering from the lowest interference to their users. Let us note that the information used for the scheduling is collected at the femtocell location, not at the user positions.

Interference minimization (IM)

Each femtocell periodically performs an optimization process, whose target is to allocate subchannels to users while minimizing the sum interference suffered by the femtocell. This scheme is assisted by user MRs. Therefore the information utilized for the radio resource management is taken at the user locations.

Stolyar's approach

This scheme is based on a completely distributed architecture and a dynamic allocation of transmit power to users, in which users are allocated to frequency subbands according to a transmit power minimization problem similar to SPAP [35]. Subbands in Stolyar's approach correspond to subchannels in our implementation. Within each subband frequency hopping is used. In order to free the largest number of subcarriers for a proper hopping, the least number of subcarriers are allocated per user to meet its throughput demand, thus leading to the usage of high-order MCSs and large transmit power per subcarrier.

A more detailed description of these approaches can be found in [42]. In the following, we will always refer to auRRAA instead of coRRAA, unless it is otherwise stated.

Running time and solution quality

One possible approach for solving RAP is to apply standard integer linear programming (ILP) techniques [43]. These techniques are able to solve RAP up to optimality, but the running times incurred by ILP solvers are unpredictable (exponential in the worst case), which renders them inappropriate for their use in femtocells.

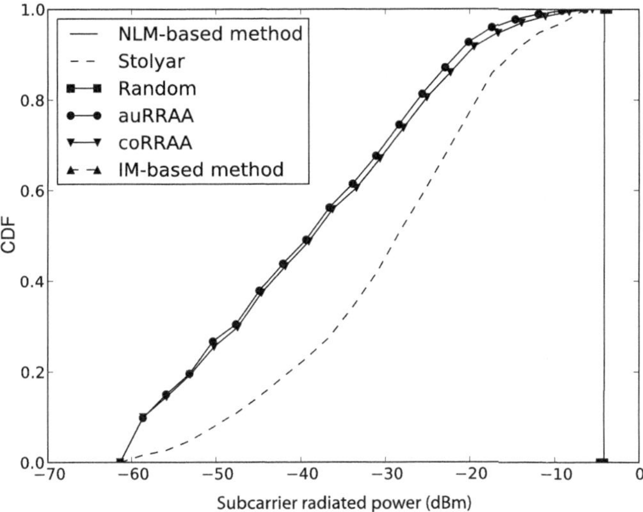

Figure 16.6 Cumulative distribution function of the transmit power per subcarrier.

In order to compare the performance of our two-level decomposition approach with that of an ILP, we extracted 100 instances from the simulations of the described scenario above. The computer used for this simulation contained an AMD Opteron 275 dual-core processor running at 2.2 GHz with 16 GB of RAM and the ILP solver was IBM ILOG CPLEX (version 9.130).

With regard to running times, the average running time of the ILP solver when solving auRAP was 23.4 s, but this running time varied significantly between instances, the maximum being 91.4 s. On the contrary, the average running time of network simplex when solving auSPAP was around 0.20 ms, whereas that of the exhaustive search when solving auRAP was around 0.39 s. These results show that our two-level decomposition approach provides a large running-time improvement over ILP solvers. Let us also remember that our two-level decomposition approach was able to find the optimal solution in all problem instances.

According to obtained results and the architecture proposed in Section 16.7.7, when running the proposed self-organizing approach in our simulations, each femtocell will independently solve auSPAP after a time uniformly distributed in [2, 100] ms after its last auSPAP update, and auRAP after a time uniformly distributed in [0.5, 1.0] s after its last auRAP update. Note that we use the same parameters for coSPAP and coRAP.

Transmit power and resulting interference

Figure 16.6 shows the cumulative distribution function (CDF) of the transmit power per subcarrier for the simulation setup with eight users per femtocell. The transmit power applied when using uniform power distributions (note that the CDFs of random, NLM, and IM are superimposed) is larger than when using the proposed SON approach, and does not vary depending on traffic and channel conditions, Stolyar's approach also uses more transmit power than the proposed SON approach, because its objective is to

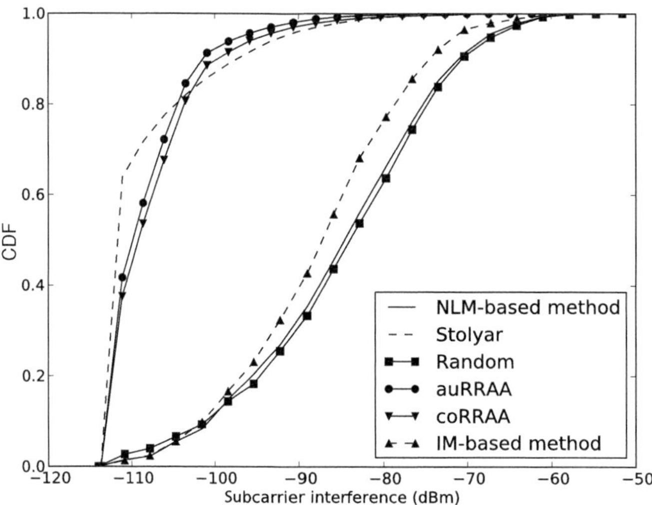

Figure 16.7 Cumulative distribution function of the received interference per subcarrier.

allocate as few subcarriers as possible within a subchannel to allow better hopping (thus allocating higher order MCSs and larger transmit power per subcarrier).

As a result of using lower transmit powers, the proposed SON approach reduces interference toward neighboring femtocells. This is corroborated by Figure 16.7, which illustrates the CDF of the interference suffered per subcarrier during the simulation. Additionally, this figure also illustrates that random, NLM, and IM perform similarly in terms of inter-cell interference mitigation. This is because when cells are fully loaded and the power is uniformly distributed, these schemes measure approximately the same interference in all subchannels during their data collection. Hence, no scheduling opportunity exists, and the performance of NLM and IM is similar to that of random assignments.

Since coRRAA results in fewer user outages/larger traffic load due to coordination, inter-cell interference and power consumption are slightly larger in coRRAA than in auRRAA. Details on network performance in terms of user outages will be presented later.

Stability of the proposed resource allocation

In order to show the stability of the proposed auRRAA, Figure 16.8 illustrates the sum of the transmit powers used by all femtocells over time. Note that in this experiment, we froze user generation after a certain time, i.e., 70 s. It can be seen that after this time femtocells' transmit power allocation reaches stability, meaning that the femtocell network converges to a stable state. There are no ping-pong effects in which a femtocell changes its resource allocation continuously in response to the changes made by neighboring femtocells. This is in line with the proof of convergence proposed in [35].

Figure 16.9 illustrates the quantity of power allocated by three neighboring cells in each one of the eight existing subchannels. This figure depicts how interference mitigation is not only achieved due to transmit power minimization, but also because

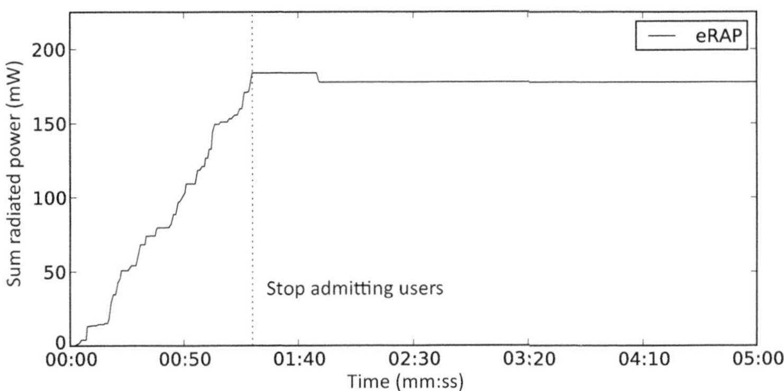

Figure 16.8 Sum of the radiated power.

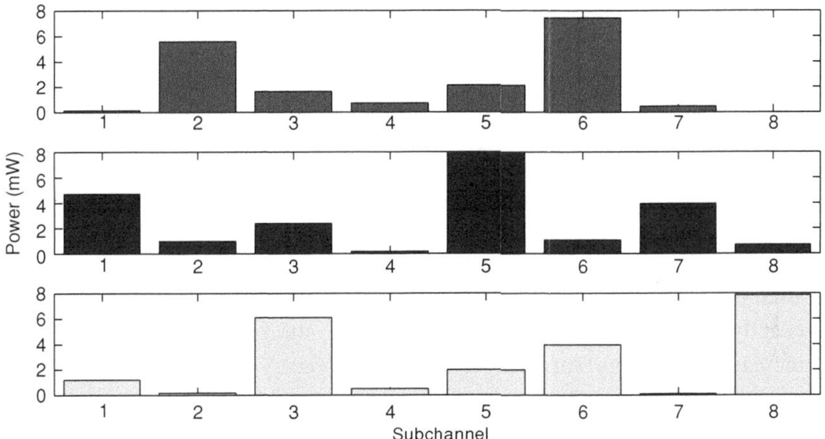

Figure 16.9 Power allocation of three neighboring enterprise femtocells.

the network settles into an efficient subchannel reuse pattern when using RAP. In this figure, it can be observed, as explained in Section 16.7.2 that cells tend to allocate high transmit power in those subchannels in which the neighboring cells assign low power and vice versa. In other words, there is an implicit coordination between cells in the allocation of resources to their cell-edge and cell-center users.

Signaling overhead

Let us recall that every DL HII contains two items, transmit power constraint $P_k^{m,\max}$ and subchannel index k. Assuming that 10 bits and 3 bits are used to encode $P_k^{m,\max}$ (1024 levels) and k (8 subchannels), respectively, the number of bits needed per exchanged DL HII is $(10+3)\cdot D_u$ bits, where D_u is the number of subchannels allocated to cell-edge user u. Let us note that according to our simulations, an average of 9.32, 14.07, and 18.91 DL HIIs per second were sent per femtocell for the 4, 6, and 8 users per cell cases, respectively. Therefore, the backhaul bit rate required for these three scenarios was an

Table 16.2 System-level simulation results.

Cell load	Scheme	Random	MNL	IM	Stolyar	auRRAA	coRRAA
4 users/cell	Outage	15.09 %	8.62 %	2.59 %	5.39 %	0.86 %	0.22 %
50% load	Users	31.13	33.11	35.20	34.44	35.75	35.87
	Mbps	7.65	8.24	8.77	8.56	8.83	8.91
6 users/cell	Outage	20.06 %	16.22 %	7.82 %	6.69 %	1.99 %	2.13 %
75% load	Users	43.65	45.15	49.64	49.92	52.71	52.79
	Mbps	10.73	11.18	12.35	12.33	12.87	13.02
8 users/cell	Outage	19.87 %	19.77 %	15.52 %	8.18 %	3.19 %	1.70 %
100% load	Users	58.36	58.59	60.47	65.93	69.63	71.00
	Mbps	14.40	14.48	14.98	16.23	16.78	17.45

average of 0.28, 0.42, and 0.57 kbps, respectively, which is well below current backhaul capabilities. Nonetheless, since femtocell-to-femtocell interfaces may be handled via user-provided backhauls, points of failure and delay issues may arise during DL-HII exchange that could compromise coRRAA performance.

System-level capacity performance

Table 16.2 depicts the network performance in terms of outage, average number of network connected users, and average network throughput when running all presented schemes. Three different scenarios were simulated where in each one 4, 6, or 8 users attempted to connect per femtocell. We can observe that these different scenarios follow exactly the same trend. However, when the number of users increased, outages also increased because of larger inter-cell interference.

coRRAA and auRRAA provided a significant improvement in network performance over all methods used for comparison. Specifically, coRRAA resulted in an average performance improvement over the third best method, i.e., Stolyar's, of 61 user outages, 5.07 connected users, and 1.22 Mbps (7.51%) and over the fourth best method, i.e., IM, of 130 user outages, 10.53 connected users, and 2.47 Mbps (16.48%).

Within the proposed models, i.e., coRRAA and auRRAA, coRRAA resulted in the best performance since it allows inter-cell communication, and thus a better inter-cell interference mitigation through DL HIIs. However, we can see that for the most challenging scenario, i.e., the scenario with eight users per cell, auRRAA performance is close to that of coRRAA. It incurred 15 outages more in 10 minutes, and it was 1.37 and 3.99 % worse in terms of average connected users and network throughput, respectively. Hence, because auRRAA does not require any signaling between femtocells, this scheme may still be more appealing for femtocell rollouts where backhaul QoS may not be guaranteed.

coRRAA, auRRAA, and Stolyar's outperformed all other schemes because they permit a better spatial spectrum reuse: minimizing transmit power at every cell and hence applying less transmit power to those subchannels allocated to users with good geometry or lower throughput demands allows for reduced inter-cell interference and creates new scheduling opportunities, e.g., neighboring cells will "observe" low interference in such subchannels, and they will allocate their worst geometry users to these subchannels.

This kind of spatial reuse is not possible by making use of uniform power distributions when cells are heavily loaded.

coRRAA and auRRAA provided better performance than Stolyar's approach due to their better interference mitigation. coRRAA and auRRAA performed better because:

- The cooperation provided by coRRAA introduces stability in the network. When new users with bad geometry appear in the network, they are allocated to subchannels that suffer from low interference due to cell cooperation. On the contrary, when using Stolyar's approach, ping-pong issues may appear, and thus it may take some time to reach a new stable solution through decentralized optimization.
- Assigning the least number of subcarriers within a subchannel, but with higher order MCSs and more transmit power, to improve the efficiency of frequency hopping increases interference compared to solutions in which all subcarriers within a subchannel are modulated with lower order MCS and the least transmit power. Moreover, because the least number of subcarriers is used, if one of them fails to provide its contribution to achieve the user throughput demand, users are likely to suffer from service disruption. However, when all subcarriers of a subchannel are used, robust coding (redundancy) can be used.

16.8   Conclusion

There is no doubt that future heterogeneous networks will require the use of SON femtocells. In this chapter, after introducing the concept of SON femtocells, we presented the state of the art about SON standardization from both Femto Forum and 3GPP perspectives. As far as standardization concerns, SON femtocells are still at a very early stage, and there is still a large number of challenges to be addressed. In the context of interference coordination, we presented a distributed SON approach applied to HeNBs. In this approach, each cell independently optimizes both its MCS, subchannel, and transmit power allocations, while achieving a good degree of coordination at the network level. It is strongly believed that similar solutions, requiring a low complexity and leading to stable solutions, will have to be used for future femtocell deployments.

References

[1] E. Dahlman, S. Parkvall, J. Sköld, and P. Beming, *3G Evolution: HSPA and LTE for Mobile Broadband*, 2nd edn. Academic Press, Oct. 2008.

[2] "IEEE standard for local and metropolitan area networks part 16: Air interface for fixed and mobile broadband wireless access systems amendment 2: Physical and medium access control layers for combined fixed and mobile operation in licensed bands and corrigendum 1," 2006.

[3] J. Zhang and G. de la Roche, *Femtocells: Technologies and Deployment*. John Wiley and Sons, Ltd, Jan. 2010.

[4] V. Chandrasekhar, J. G. Andrews, and A. Gatherer, "Femtocell networks: a survey," *IEEE Commun. Mag.*, vol. 46, no. 9, pp. 59–67, Sep. 2008.

[5] D. López-Pérez, A. Valcarce, G. de la Roche, and J. Zhang, "OFDMA femtocells: a roadmap on interference avoidance," *IEEE Commun. Mag.*, vol. 47, no. 9, pp. 41–8, Sep. 2009.

[6] Motorola, "LTE operations and maintenance strategy: using self-organizing networks to reduce OPEX," 2009.

[7] Self Optimisation and self-ConfiguRATion in wirelEss networkS (SOCRATES), http://www.fp7-socrates.eu.

[8] 3rd Generation Partnership Project (3GPP), http://www.3gpp.org.

[9] Femto Forum, http://www.femtoforum.org.

[10] 4G Americas, "Self-optimizing networks: the benefits of SON in LTE," 2011.

[11] Ubiquisys, http://http://ubiquisys.com.

[12] BeeSON Networks, http://www.beesonetworks.com.

[13] Tata Elxsi, http://www.tataelxsi.com.

[14] S. Uygungelen, G. Auer, and Z. Bharucha, "Graph-based dynamic frequency reuse in femtocell networks," in *Proc. IEEE Vehicular Tech. Conf. (VTC)*, Yokohama Japan, May 2011.

[15] Femto Forum, "Interference management in OFDMA femtocells," Whitepaper available at www.femtoforum.org, Mar. 2010.

[16] Femto Forum, "LTE Network Monitor Mode Specification," 2010.

[17] 3GPP, "TR36.921, evolved universal terrestrial radio access (E-UTRA); FDD Home eNode B (HeNB) radio frequency (RF) requirements analysis," 2010.

[18] ——, "TR36.922, evolved universal terrestrial radio access (E-UTRA); TDD Home eNode B (HeNB) radio frequency (RF) requirements analysis," 2010.

[19] ——, "TR 36.902, evolved universal terrestrial radio access network (E-UTRAN); self-configuring and self-optimizing network (SON) use cases and solutions," 2010.

[20] G. de la Roche, A. Valcarce, D. López-Pérez, and J. Zhang, "Access control mechanisms for femtocells," *IEEE Commun. Mag.*, vol. 48, no. 1, pp. 33–9, Jan. 2010.

[21] 3GPP, "RP-100222, Home NodeB and Home eNB enhancements," 3GPP TSG RAN meeting 47, 2009.

[22] ——, "R2-097408 stage3 CR for LTE hybrid cell idle mode mobility. Vodafona, Motorola, Deutsche Telekom, NEC, ST Ericsson, Qualcomm Europe, Huawei, Telecom Italia CR 36.304."

[23] V. Chandrasekhar and J. G. Andrews, "Spectrum allocation in two-tier networks," *IEEE Trans. Commun.*, vol. 57, no. 10, pp. 3059–68, Oct. 2009.

[24] J. D. Hobby and H. Claussen, "Deployment options for femtocells and their impact on existing macrocellular networks," *Bell Labs Technical J.*, vol. 13, no. 4, pp. 145–60, Feb. 2009.

[25] M. Y. Arslan, J. Yoon, K. Sundaresan, S. V. Krishnamurthy, and S. Banerjee, "Fermi: a femtocell resource management system for interference mitigation in OFDMA networks," in *Proc. ACM Int. Conf. Mobile Computing and Networking (MobiCom)*, Las Vegas USA, Sep. 2011.

[26] V. Chandrasekhar and J. G. Andrews, "Uplink capacity and interference avoidance for two-tier femtocell networks," *IEEE Trans. Wireless Commun.*, vol. 8, no. 7, pp. 3498–509, July 2009.

[27] H. Claussen and F. Pivit, "Femtocell coverage optimization using switched multi-element antennas," in *Proc. IEEE Int. Conf. on Commun. (ICC)*, Dresden, Germany, June 2009, pp. 1–6.

[28] J. Ling, D. Chizhik, and R. Valenzuela, "On resource allocation in dense femto-deployments," in *Proc. IEEE Int. Conf. Microwaves, Communications, Antennas and Electronics Systems (COMCAS)*, Tel Aviv Israel, Nov. 2009.

[29] D. López-Pérez, A. Jüttner, and J. Zhang, "Dynamic system level simulation of dynamic frequency planing for real time services in WiMAX networks," in *Int. Conf. on Wireless Communications and Mobile Computing (IWCMC)*, Leipzig, Germany, 2009, pp. 1397–403.

[30] H. Claussen, "Co-channel operation of macro- and femtocells in a hierarchical cell structure," *Int. J. Wirel. Inform. Netw.*, vol. 15, no. 3, pp. 137–47, Dec. 2008.

[31] A. Galindo-Serrano, L. Giupponi, and M. Dohler, "Cognition and docition in OFDMA-based femtocell networks," in *Proc. IEEE Global Telecommun. Conf. (GLOBECOM)*, Miami USA, Dec. 2010.

[32] G. Song and Y. Li, "Cross-layer optimization for OFDM wireless networks – part I: theoretical framework," *IEEE Trans. Wireless Commun.*, vol. 4, no. 2, pp. 614–24, Mar. 2005.

[33] J. Jang and K. B. Lee, "Transmit power adaptation for multiuser OFDM systems," *IEEE J. Sel. Areas Commun. (JSAC)*, vol. 21, no. 2, pp. 171–8, Feb. 2003.

[34] C. E. Shannon, "A mathematical theory of communication," *Bell System Technical J.*, vol. 27, pp. 379–423, July 1948.

[35] A. Stolyar and H. Viswanathan, "Self-organizing dynamic fractional frequency reuse in OFDMA systems," in *Proc. IEEE Int. Conf. on Computer Commun. (INFOCOM)*, Phoenix USA, Apr. 2008, pp. 691–9.

[36] S. Sesia, M. Baker, and I. Toufik, *LTE: The UMTS Long Term Evolution: From Theory to Practice*. Wiley-Blackwell, July 2011.

[37] R. K. Ahuja and T. L. M. J. B. Orlin, *Network Flows: Theory, Algorithms, and Applications*. Prentice Hall, 1993.

[38] G. B. Dantzig, *Linear Programming and Extensions*. Princeton University Press, 1963.

[39] "Library for efficient modeling and optimization in networks (LEMON)," http://lemon.cs.elte.hu.

[40] A. Valcarce, G. de La Roche, A. Jüttner, D. López-Pérez, and J. Zhang, "Applying FDTD to the coverage prediction of WiMAX femtocells," *EURASIP J. Wirel. Commun. Netw. Special issue on Advances in Propagation Modeling for Wireless Systems*, Feb. 2009.

[41] Forsk Atoll – Global RF Planning Solution, http://www.forsk.com.

[42] D. López-Pérez, "Practical models and optimization methods for inter-cell interference coordination in self-organizing cellular networks," Ph.D. dissertation, University of Bedfordshire, Luton UK, Apr. 2011.

[43] R. Rardin, *Optimization in Operations Research*. Prentice Hall, 1998.

Index